Second Edition

WATER POLLUTION and FISH PHYSIOLOGY

Alan G. Heath

Department of Biology
Virginia Polytechnic Institute and State University
Blacksburg, Virginia

CRC Press
Taylor & Francis Group
Boca Raton London New York

CRC Press is an imprint of the
Taylor & Francis Group, an **informa** business

First published 1995 by Lewis Publishers

Published 2019 by CRC Press
Taylor & Francis Group
6000 Broken Sound Parkway NW, Suite 300
Boca Raton, FL 33487-2742

© 1995 by Taylor & Francis Group, LLC
CRC Press is an imprint of Taylor & Francis Group, an Informa business

First issued in paperback 2019

No claim to original U.S. Government works

ISBN 13: 978-0-367-44892-9 (pbk)
ISBN 13: 978-0-87371-632-1 (hbk)

Visit the Taylor & Francis Web site at
http://www.taylorandfrancis.com

and the CRC Press Web site at
http://www.crcpress.com

Library of Congress Cataloging-in-Publication Data

Heath, Alan G.
  Water pollution and fish physiology / Alan G. Heath. — 2nd ed.
     p.   cm.
  Includes bibliographical references and index.
  ISBN 0-87371-632-9
     1. Fishes—Effect of water pollution on.   2. Fishes—Physiology.
  I. Title.
  SH174.H43   1995
  597′.024—dc20   95-16292
                                                                      CIP

Library of Congress Card Number 95-16292

Preface

The genesis of this book came from my having taught a postgraduate course to biology and fisheries students on the physiological action of water pollutants in fish for many years. The literature is scattered in numerous journals and books, and while there have been several excellent symposia volumes published, most of the papers therein have been presentations of primary research. As with many healthy disciplines, this one has experienced a tremendous growth over the past 20 years; especially during the 1980s and early 1990s. Recently, there have been a number of good reviews dealing with fairly specific aspects of the subject (e.g., effects of pollutants on osmoregulation); however, there are no books which attempt to look at the whole field. Therefore, the objective of this book is to provide a reasonably concise synthesis of what is known about how pollutants affect physiological processes in fish.

As in the first edition, this revised and updated second edition begins with a discussion of some concepts that are important in understanding pollution biology and fish physiology. These concepts are often implied though rarely mentioned explicitly. Following this brief chapter, an analysis of the physiological responses to environmental hypoxia is provided. This is discussed early in the book because polluted waters are often lacking in dissolved oxygen and many toxic chemicals at acute concentrations induce an hypoxic condition in the fish. Each of the subsequent chapters is generally devoted to a specific physiological process (e.g., energetics, uptake and accumulation of contaminants, reproduction). Each of these begins with a review of some basic physiology applied to fish followed by a more detailed discussion of how various pollutants affect these functions. This second edition has had two additional chapters added: immunology and acid toxicity, which reflect the large amount of past and present research conducted in these fields. Throughout, the emphasis is on the mechanisms of sublethal effects, rather than lethal ones, for those are what the fish will usually confront in contaminated waterways. The book closes with a critical look at some physiological and biochemical measurements that are, or could be, utilized in work on water pollution control.

The literature coverage is through 1993, but I have not attempted to be all-inclusive, for one could easily get bogged down in details and therefore lose sight of generalizations. Indeed, some of the individual chapters could easily be expanded into a whole book unto themselves. Where the literature is extensive, I have tried to cite examples of trends. The most recent studies and reviews have been emphasized so that those who wish to pursue a topic in further depth can quickly get into the literature.

In preparing an interpretive treatise of this sort, there is always the danger of "over" interpreting the results of others. At times, I have been rather free with speculations. I have tried to make it obvious where I am speculating and apologize in advance to those whose ideas may have been inadvertently used without adequate attribution.

Several people critically read specific chapters. For this I am grateful to Drs. Douglas Anderson (National Fisheries Research Laboratory, Leetown, West Virginia), Gary Atchison (Iowa State University), Joe Cech (University of California, Davis), Brian Eddy (University of Dundee, Scotland), Mark Greeley, Jr., (Oak Ridge National Laboratory, Tennessee), Steven Koenig (West Virginia University), James McKim (USEPA, Duluth), and Chris Wood (McMaster University). Their perceptive suggestions and corrections have been extremely helpful as my level of expertise in some of these areas is clearly limited. Of course, all blame for errors of omission and commission must rest with me.

Alan G. Heath
Blacksburg, Virginia

The Author

Alan G. Heath, Ph.D., is a professor in the Department of Biology at Virginia Polytechnic Institute and State University, Blacksburg, Virginia.

Dr. Heath received his B.S. degree from San Jose State University in California in 1958 and M.S. and Ph.D. degrees from Oregon State University in 1961 and 1963, respectively. After a year as an National Institutes of Health postdoctoral fellow at the University of Miami Marine Laboratory, he moved to Virginia Tech, where he has been ever since. Dr. Heath served as a Senior U.S. Environmental Protection Agency Postdoctoral Fellow in Bristol, England for a year and recently spent a year as a Visiting Scholar in the Department of Wildlife and Fisheries Biology at the University of California, Davis.

Dr. Heath's research, often in collaboration with his graduate students, has been primarily devoted to laboratory investigations on the sublethal physiological responses of fish to environmental hypoxia or to the presence of waterborne chemical toxicants often present in polluted waters. His professional affiliations include Society of Environmental Toxicology and Chemistry, American Fisheries Society, American Society of Zoologists, American Institute of Biological Sciences, AAAS, and Sigma Xi.

Table of Contents

Chapter 1
Some Introductory Concepts
  I. Types of Water Pollution 1
    A. Putrecible Organic Materials 1
    B. Excessive Nutrition 1
    C. Suspended Solids 2
    D. Toxic Chemicals 2
        1. Metals 2
        2. Chlorine 2
        3. Cyanides 3
        4. Ammonia 3
        5. Detergents 3
        6. Acids 3
        7. Pesticides 3
        8. Polychlorinated Biphenyls 4
        9. Petroleum Hydrocarbons 4
       10. Pulpmill Effluents 4
       11. Miscellaneous 5
    E. Thermal Pollution 5
  II. The Relationship Between Aquatic Toxicology and Fish
    Physiology 5
  III. Levels of Biological Organization 6
  IV. Importance of Dose and Duration of Exposure 7
  V. Stress 9
  VI. Toxic Mode of Action 10
  References 11

Chapter 2
Environmental Hypoxia
  I. Introduction 15
  II. Minimum Levels of Oxygen Required for Fish Life 16
  III. Interaction of Hypoxia and Toxicity of Pollutant
    Chemicals 17

  IV. Gill vs. Cutaneous Respiration 18
   V. Adjustments in Ventilation 20
  VI. Adjustments by the Gills to Hypoxia 24
 VII. Transport of Oxygen by the Blood 24
VIII. Cardiovascular Changes During Hypoxia 26
  IX. Respiratory Regulation and Conformity 27
   X. Anaerobic Metabolism 29
  XI. Swimming Speed 31
 XII. Behavior 33
XIII. Blood and Urine 34
XIV. Histopathology 36
 XV. Acclimation to Hypoxia 36
References 37

Chapter 3
Respiratory and Cardiovascular Responses
    I. Introduction 47
   II. Overview of Normal Respiratory Physiology 47
  III. Histopathology of Gill Lamellae Exposed to Pollutants 49
  IV. Ventilation Changes in Response to Pollutants 52
   V. Physiological Mechanisms of Changes in Ventilation 55
  VI. Circulatory Physiology 58
 VII. Cardiac Responses to Pollutants 59
References 62

Chapter 4
Hematology
    I. Introduction 67
   II. Fish Blood Cells and their Measurement 67
      A. Hematocrit 68
      B. Blood Cell Count 68
      C. Hemoglobin Concentration 68
      D. Blood Cell Differential Count 68
  III. Chemicals that Cause Anemia 69
      A. Cadmium 69
      B. Lead 70
      C. Mercury 71
      D. Chloramine 71
      E. Nitrite 72
      F. Pulpmill Effluent 72
      G. Miscellaneous 72
      H. Biological Significance of Anemia 73
  IV. Chemicals Causing an Increase in Hematological Variables 73
      A. Acid 73
      B. Pesticides 74
      C. Copper 74
   V. Concluding Remarks 74
References 75

Chapter 5
Uptake, Accumulation, Biotransformation, and Excretion of
Xenobiotics
  I. Introduction... 79
 II. Uptake from the Environment 80
     A. Metals 81
     B. Organics 87
III. Transport Within the Fish of Metals and Organics 91
 IV. Accumulation of Metals in Different Organs............................ 91
  V. Regulation of Metal Concentration...................................... 96
 VI. Glutathione and Metal Detoxification................................... 97
VII. Involvement of Metallothionein in Metal Accumulation and
     Acclimation to Metals 99
VIII. Bioconcentration of Organic Pollutants 101
 IX. Biotransformation of Organic Contaminants 106
  X. Excretion of Organic Contaminants 110
References 113

Chapter 6
Liver
  I. Introduction.. 125
 II. Structure of the Liver.. 125
III. Alterations in Liver/Somatic Index 126
 IV. Histopathological Effects of Pollutants 127
  V. Major Functions of the Liver.. 128
     A. Interconversion of Foodstuffs 128
     B. Storage of Glycogen 128
     C. Removal and Metabolism of Foreign Chemicals in
        the Blood 129
     D. Formation of Bile.. 129
     E. Synthesis of Many Plasma Proteins 129
     F. Synthesis of Cholesterol for Use in Steroid Hormones
        and Cell Membranes 129
     G. Exocrine Pancreatic Secretion 129
     H. Metabolism of Hormones 129
 VI. Effects of Pollutants on Liver Function............................... 129
     A. Plasma Clearance by the Liver of Specific Dyes.............. 130
     B. Plasma Enzymes of Liver Origin 130
     C. Changes in Plasma Concentration of Chemicals
        Produced or Excreted by Liver 132
VII. Ascorbic Acid and Pollutant Exposure 134
References 136

Chapter 7
Osmotic and Ionic Regulation
  I. Overview of Some General Principles 141
 II. Effects of Pollutants on Osmotic and Ionic Regulation........... 144
     A. Copper... 145

    B. Zinc 150
    C. Cadmium 151
    D. Mercury 153
    E. Chromium 155
    F. Lead 155
    G. Aluminum 156
    H. Tin 157
    I. Manganese and Iron 157
    J. Detergents 157
    K. Phenol 158
    L. DDT 158
    M. Fenvalerate 158
    N. Petroleum Hydrocarbons 159
    O. Ammonia 160
    P. Nitrite 160
    Q. Chlorine and Ozone 161
  III. Mucus 161
  IV. Chloride Cell Proliferation 162
   V. Some Summary Comments Regarding Osmoregulatory
        and Electrolyte Alterations 162
References 164

Chapter 8
Physiological Energetics
    I. Introduction.. 171
   II. General Concepts 171
  III. Methods of Measuring Energy Expenditure in Fish 174
  IV. Effects of Metals on Metabolic Rate 176
    A. Copper... 176
    B. Miscellaneous Metals 180
   V. Gill Tissue Metabolism: Effects of Metals and Possible
       Relation of Gill Metabolism to Whole-Body Metabolic Rate 182
  VI. Effects of Pesticides on Whole-Body and Individual
       Tissue Respiration 184
 VII. Effects of Petroleum Hydrocarbons on Energy Metabolism 189
VIII. Methods Applicable to Measurement of Energy Expenditure
       in the Field 189
  IX. Effects of Pollutants on Larval and Juvenile Growth 191
   X. Swimming Performance 195
  XI. Changes in Carbohydrate, Lipid and Protein Energy Stores ... 200
    A. Introduction... 200
    B. Effect of Pulpmill Effluent... 202
    C. Metals 203
    D. Petroleum Hydrocarbons and Surfactants 205
    E. Pesticides.. 205
    F. Miscellaneous Pollutants 207
    G. Concluding Comment 207
References 208

Chapter 9
Alterations in Cellular Enzyme Activity, Antioxidants, Adenylates,
and Stress Proteins
  I. Introduction 217
 II. Some Comments About Enzyme Methodology 217
III. Alterations in Cellular Enzyme Activity Resulting from
     Metal Exposure 219
 IV. Enzyme Effects from Organic Chemicals 225
  V. Concluding Comments on Enzyme Effects 230
 VI. Antioxidants 231
VII. Adenylates 232
VIII. Stress Proteins 233
References 234

Chapter 10
Acid Pollution
  I. Introduction 239
 II. Spawning 240
III. Embryonic Development and Hatching 242
 IV. Larvae from Hatching Through Swim-Up 242
  V. Juvenile and Adult: Blood Acid-Base Balance and
     Electrolyte Changes from Acute Exposures 244
 VI. Juvenile and Adult: Blood Changes from Chronic
     Exposure 247
VII. Hormonal Responses 249
VIII. Ventilation and Blood Gases 250
 IX. Oxygen Consumption, Swimming Performance, and
     Swim Bladder Inflation 251
  X. Behavior 252
 XI. Concluding Comment 253
References 253

Chapter 11
The Immune System
  I. Overview of Fish Immunology 259
 II. Effects of Pollutants on Immune Function 260
    A. Metals 261
       1. Cadmium 261
       2. Copper 262
       3. Miscellaneous Metals 262
    B. Organic Pollutants 263
    C. Miscellaneous Pollutants 264
    D. Immune Effects in Fish from Contaminated
       Natural Waters 265
III. Hormonal Modulation of Immune Response 266
References 266

Chapter 12
Behavior and Nervous System Function
    I. Introduction 271
   II. Locomotor Activity 272
       A. Methods for Measuring Locomotor Activity 272
       B. Metals 273
       C. Miscellaneous Chemicals 274
  III. Avoidance of or Attractance to Waterborne Chemicals 276
  IV. Sensory Receptors 279
   V. Feeding and Predator–Prey Behavior 282
  VI. Aggression 284
 VII. Learning 285
VIII. Optomotor Response 287
  IX. Acetylcholinesterase 288
   X. Concluding Comments 289
References 290

Chapter 13
Reproduction
    I. Introduction 299
   II. Brief Overview of Fish Reproductive Physiology 299
       A. Reproductive Strategies 300
       B. Hormonal Control of Gametogenesis 300
       C. Spawning 301
       D. Development 302
  III. Action of Pollutants on Reproductive Processes 303
       A. Hormonal Controls 303
       B. Egg Production 308
       C. Transfer of Contaminants into Eggs from Adults 309
       D. Effects on Sperm and Fertilization 310
       E. Egg Activation 310
       F. Oocyte and Embryo Energy Metabolism 311
       G. Changes In Pollutant Sensitivity During Embryonic
          Development 312
       H. Hatching Time and Viable Hatch 313
       I. Larvae 314
       J. Concluding Comment 317
References 317

Chapter 14
Use of Physiological and Biochemical Measures in Pollution Biology
    I. Introduction 325
   II. Water Quality Criteria 326
  III. Biomonitoring of Fish in the Field and Mesocosms 327
       A. Cellular and Tissue Chemistry 329
       B. Histopathology 331

   C. Hematology, Immunology, and Blood Chemistry 331
   D. Challenge Tests 334
   E. Behavior Tests 335
   F. Conclusions 335
 IV. Early Warning Systems 336
References 337

Index...339

Chapter

1

Some Introductory Concepts

I. TYPES OF WATER POLLUTION

For all practical purposes, water pollution is the addition by humans of something to the water that alters its chemical composition, temperature, or microbial composition to such an extent that harm occurs to resident organisms or to humans (Lloyd, 1992). While chemical pollution has implications for human health, both directly from toxic chemicals in drinking water, and indirectly from the accumulation of toxic compounds by organisms that are then eaten by people, this book will not deal with these types of pollution, nor with the introduction of pathogenic microbes and carcinogenic chemicals into waterways. Instead, an attempt will be made to look at pollution "through the eye of the fish" from a functional standpoint.

The brief survey of pollutants presented here is not meant to be comprehensive, but instead is meant as an introduction to the sorts of pollution that may have physiological effects on fish.

A. PUTRECIBLE ORGANIC MATERIALS

Putrecible organic materials are characteristic of untreated or inadequately treated domestic and industrial waste. Oxygen is required for the microbial decomposition of this organic matter and the quantitative measure of this oxygen requirement is referred to as the biochemical oxygen demand (BOD). As the BOD gets larger due to a greater organic load, unless there is considerable mixing of the water, a condition of abnormally low dissolved oxygen (hypoxia) occurs. It is this environmental hypoxia that is of primary interest from the standpoint of fish physiology.

B. EXCESSIVE NUTRITION

In locations where there is agricultural runoff or non-biodegradable detergents being added to the water, the growth of phytoplankton is stimulated due to excess amounts of plant nutrients. This eutrophication process results in large daily changes in dissolved oxygen from photosynthesis during the daylight hours and respiration at night. The utilization of oxygen by phytoplanktonic respiration at night can produce conditions of very low dissolved oxygen in the hours just before daybreak.

C. SUSPENDED SOLIDS

Silt suspended in the water column is probably the most prevelant of the suspended solids. It generally results from runoff where land has been disturbed by plowing or excavation. Ground up wood fibers can also be a significant form of suspended solid pollution.

D. TOXIC CHEMICALS

The conditions producing a low dissolved oxygen concentration and toxic chemicals are the most important types of water pollution that affect fish. There are some 65,000 industrial chemicals in use and 3–5 new ones enter the marketplace each day (World Commission on Environment and Development, 1987). Fortunately, a very small percentage of those chemicals enter waterways, but the possibilities are immense. The major classes of toxic chemicals of concern for fish are metals, chlorine, cyanides, ammonia, detergents, acids, pesticides, polychlorinated biphenyls, petroleum hydrocarbons, pulpmill effluents, and other miscellaneous chemicals.

1. Metals

There has been a tendency in the literature on water pollution to speak of nearly all metals as "heavy metals", although some have tried to avoid this designation (Nieboer and Richardson, 1980). Here we will not attempt to separate heavy metals from any others. (Metaloids such as selenium and arsenic are included with the metals.)

When speaking of metals the expression "trace metals" is often used (e.g., Leland and Kuwabara, 1985). This reflects the important fact that many metals are required for normal physiological function in animals but only at trace concentrations, and these concentrations vary considerably between different species. Important trace metals include copper, iron, zinc, iodine, manganese, cobalt, selenium, tin, and chromium. Altered physiological function results when one or more of these reach sufficiently high concentrations in cells.

Metals enter waterways from a wide variety of industrial effluents and old mines. Acid precipitation also causes leaching of metals from surrounding soils (Norton, 1982; Spry and Wiener, 1991). The metals of most concern for studies of the effect of pollution on fish physiology include copper, zinc, tin (primarily the methylated form), cadmium, mercury (both the methylated and non-methylated forms), chromium, lead, nickel, arsenic, and aluminum. An important problem when working with metals in water is that they tend to complex with organic and inorganic chemicals and this may reduce their bioavailability to resident organisms (Leland and Kuwabara, 1985). Thus a simple analysis for total metal could actually overestimate the bioavailability. A comprehensive review of the chemistry and biology of metals in natural waters has been published by Moore and Ramamoorthy (1984).

In the past there has been some tendency to lump all the metals together when talking about their "physiological mode of action". We now know this is not valid, as will become evident in later chapters, although there is considerable overlap in physiological effects for many of them.

2. Chlorine

The concern here is not for the chloride ion, but rather with the chemicals formed when chlorine gas is introduced into water either for antifouling in industrial cooling systems or for disinfection of sewage effluents. The free gas does not exist in water for any significant period of time, but quickly forms HOCl or OCl–, which are commonly called "free chlorine". In the presence of ammonia, some or all of the free chlorine is converted into monochloramine (NH_2Cl) which is known as "combined chlorine". Both free and combined chlorine are oxidants with the former being the strongest. Total residual chlorine is the sum of the free and combined concentrations. The relative stability and toxicity of these forms of chlorine differs considerably. Free chlorine is more toxic but the combined form is more stable and thus stays around longer (Hall et al., 1981).

3. Cyanides
The cyanide radical occurs in many industrial wastes, particularly those involved with the manufacturing of synthetic fabrics and plastics and the processing of metals. "Free cyanide" (CN ion and HCN) occurs mostly as molecular hydrogen cyanide, unless the pH is above about 9. The toxicity of cyanide to fish and other organisms has been reviewed by Leduc (1984) and Eisler (1991).

4. Ammonia
This compound not only occurs in many effluents, but also results from the natural decomposition of organic matter. Ammonia gas forms ammonium hydroxide in water which in turn can dissociate into ammonium and hydroxyl ions. It is the non-ionized form of ammonia that is toxic to fish, but the toxicity is complicated by the fact that the degree of dissociation depends on the pH and temperature of the water. Increasing the pH or temperature increases the toxicity because more of the ammonia will be in the non-ionized form. The pH and temperature in natural waters often change rapidly during the day so ammonia toxicity becomes difficult to predict. In addition, the gill surface of fish will have a much higher carbon dioxide concentration than the surrounding water and the carbonic anhydrase in mucus will catalyze the formation of carbonic acid from that carbon dioxide. The resulting pH decrease at the surface will then affect the ammonia toxicity (Lloyd, 1992).

Oxidation of ammonia by bacteria produces nitrite which is further oxidized into nitrate. Nitrite is far more toxic than nitrate (which has almost no toxicity), but it is present usually in only trace amounts in natural waters. In aquaculture facilities which use nitrification to convert ammonia to nitrate, however, the process may become impeded and nitrite may accumulate (Russo, 1985).

5. Detergents
In 1965, there was a shift by the detergent industry from the alkylbenzene sulfonates (ABS) to the more biodegradable linear alklylate sulfonates (LAS). This commendable attempt at reducing their environmental impact is not unequivocally a good thing. The LAS is four times as toxic to fish as is ABS (Pickering, 1966), but fortunately, the toxicity is lost upon biodegradation.

6. Acids
The main effects on aquatic biota of acids are due to a simple change in the pH of the water. In addition, however, there may also be an indirect effect due to altered toxicity of certain pollutants (e.g., metals) (Baker, 1982). Acid pollution from mine drainage and acid rain is an increasing problem in many parts of both developed and developing countries. A tremendous body of information on the effects of acid on all forms of aquatic life is rapidly accumulating (e.g., Morris et al., 1989). Because of that, we devote a whole chapter in this book to the subject.

7. Pesticides

The pesticides of interest here are primarily the insecticides, herbicides, and wood preservatives. Tributyltin and chlorine are also used as antifouling agents, a type of pesticidal activity, but are taken up elsewhere. Briggs (1992) provides a good general guide to all types of pesticides including both technical and common names.

Insecticides fall into four general types: organochlorine, organophosphate, carbamate, and botanicals. The organochlorine insecticides include DDT, aldrin, chlordane, dieldrin, endrin, heptachlor, lindane, methoxychlor, and toxaphene. Because of their environmental persistence and high toxicity, most are no longer legally used in the U.S. but are still extensively used in some other countries.

The organophosphates include diazinon, malathion, parathion, methyl parathion, dichlorvos, Dursban®, etc. This is a steadily expanding list. Important carbamate insecticides are Sevin® and carbofuran.

Botanical insecticides include rotenone, pyrethrum, and allethrin. The term "botanical" refers to the fact they are derived from plants, although there has been considerable development of synthetic forms.

The herbicides and fungicides include among others amitrol, diquat, endothall, molinate, silvex, and paraquat.

Wood preservatives include pentachlorophenol and 2-(thiocyanomethylthio) benzothiazole (TCMTB).

The acute toxicity of many of these pesticides to fish and other aquatic life has been reviewed by Livingston (1977) and Murty (1986). As a group, the acute toxicity of the organochlorine insecticides tends to be considerably greater than for the organophosphates. The 96-h LC50s for organochlorine compounds are generally in the low microgram per liter range whereas the LC50s for the organophosphate compounds are in the range of low milligram per liter. Pyrethroids are similar to organochlorine insecticides in toxicity to fish, and the herbicides have, with a few exceptions, relatively low toxicities for fish, but when used for weed control, they can cause depletion of oxygen (Murty, 1986). As we shall see in later chapters, nearly all pesticides can have some subtle and not so subtle physiological effects under conditions of chronic exposure.

8. Polychlorinated Biphenyls

These compounds, commonly called PCBs, have generated considerable interest primarily due to their toxicity to humans. They are also quite toxic to fish and other aquatic life. The term "Aroclor" with a four-digit number after it refers to a specific PCB formulation.

9. Petroleum Hydrocarbons

The composition of crude oil is complex and varies from region to region. The major components are aliphatic hydrocarbons, cyclic paraffin hydrocarbons, aromatic hydrocarbons, naphtheno-aromatic hydrocarbons, resins, asphaltenes, heteroatomic compounds, and metallic compounds. The aromatic and naphtheno-aromatic hydrocarbons are considered to be the most toxic components in oil (Anderson, 1979), and they increase in percent of the total content of the oil during the refining process. Extensive discussions of the sources, fates, and biological effects of petroleum hydrocarbons are found in Neff and Anderson (1981).

10. Pulpmill Effluents

In the U.S., pulp and paper mills are the largest dischargers of conventional pollutants subject to national effluent standards (General Accounting Office, 1987). This waste is often called kraft mill effluent (KME) and it results from the digestion of wood in an alkaline mixture which may be followed by bleaching. The resulting effluent possesses a complex mixture of organic and inorganic salts which has a considerable BOD. It also has several toxic substances such as resin acids and chlorinated phenolics (Lindstrom-Seppa and Oikari, 1990).

11. Miscellaneous
With the huge number of chemicals that are introduced into the industrial stream every year, there are those that do not fall into any group listed above. Fortunately, most of these currently do not appear in waterways to a great extent, but potentially they could.

E. THERMAL POLLUTION
Elevated temperatures occur from clearing of cover over streams and heated effluents from steam power generating plants. The available literature on the effects of temperature on fish is huge. Raney et al. (1972) compiled a bibliography on this subject that included over 4000 references, and the number of papers has expanded exponentially since then. A good, concise summary of this extremely large topic is that by Houston (1982). Because he is a fish physiologist, his treatise has a physiological "flavor" to it. Other good reviews of the effects of temperature on various physiological functions in fish include Crawshaw and Hazel (1984) and Hazel (1993). Space limitations here will permit only a consideration of temperature in relation to its interaction with hypoxia and with toxic chemicals on the fish.

II. THE RELATIONSHIP BETWEEN AQUATIC TOXICOLOGY AND FISH PHYSIOLOGY

Toxicology is the study of poisons, their identification, chemistry, degree of toxicity, and physiological actions. The major aim of the aquatic toxicologist, as with other toxicologists, is to protect the organisms that potentially may be the recepients of some harmful chemical in the environment. Mammalian (classical) toxicology has a long and distinguished history while aquatic toxicology is a much younger discipline. Its history has been briefly reviewed by Macek (1980) who traces the early development in the 1930s through the stimulus that occurred from water quality legislation in the 1960s and finally in the 1970s when several parallel branches evolved. One of these involves the use of aquatic organisms as animal models for human toxicological problems, which is a merging of biomedical research and aquatic toxicology.

Aquatic toxicology has adopted many of the techniques of its classical predecessor. One of the major techniques is the acute bioassay whereby the concentration lethal to 50% of a population of fish or invertebrate in a given exposure time is determined (LC50). The procedures for carrying out aquatic bioassays have become rather well standardized (APHA, 1992), but the excellent short paper by Sprague (1973) is still useful.

It has been said many times, but needs emphasis here: in a bioassay, the organisms are actually acting as a sort of chemical measuring device which also integrates other conditions into the measure such as temperature, disease, dissolved oxygen, etc. The bioassay then is useful for comparing the gross effects of various chemicals on different species and populations. These acute tests, along with chronic bioassays in which a population is exposed for weeks, months, or even through one or more generations to different concentrations of a toxicant, have been used as tools in establishing water quality criteria (Alabaster and Lloyd, 1980; EPA, 1986). Such criteria are then used by government regulators to formulate legal water quality standards. In a nutshell then, much of aquatic toxicology has primarily been aimed at the important function of determining the maximum amount of some toxicant that can be permitted in the environment without causing significant harm to the resident biota.

From approximately the mid-1970s aquatic toxicology has increasingly used the tools of the physiologists. This is partly to understand why a fish or invertebrate is debilitated, but it is also because of a realization that there are many sublethal effects that may occur without necessarily resulting in death of the individual organism; or as Jan Prager was quoted (Sindermann, 1979, p. 438) as saying: "Death is too extreme a criterion for determining whether a substance is harmful to marine biota." While bioassays extending over several generations are useful, they are also extremely time consuming and expensive to carry out. Thus, there has been considerable interest in developing physiological and biochemical tests, or more commonly known as biomarkers, to assess the "health" of aquatic animals (Adams, 1990; McCarthy and Shugart, 1990; Huggett et al., 1992).

Fish physiologists traditionally have had little interest in aquatic toxicology. Their concern has been the understanding of the organ systems of various fish species and their physiological adaptations to environmental variables such as temperature and dissolved oxygen. In the multivolume work entitled *Fish Physiology* (Hoar and Randall, 1969–1992) the effects that pollution has on these organisms rarely is mentioned. The chapter by von Westernhagen (1988) is a notable exception. Such seeming neglect may be due to at least three factors: (1) the authors are, with a few exceptions, not professionally involved with work on pollution, (2) the database has, until recently, been severely limited and the quality of some of the work was not especially high, and (3) it is customary in physiology books (whether on humans or other animals) to not spend much if any time on the effects of toxicants.

Fish physiology is now becoming an integral part of aquatic toxicology. From a purely physiological standpoint, the presence of pollutants in the environment at sublethal concentrations can be considered as another extremely interesting environmental variable to which a fish will physiologically respond.

III. LEVELS OF BIOLOGICAL ORGANIZATION

When investigating the effects of pollution on fish (or other organisims) it is useful to keep in mind the spectrum shown in Figure 1. (The expressions "levels of complexity" or "levels of integration" are often used to designate this topic.) Starting from the left, foreign chemicals, or other environmental stressors, such as elevated temperatures, exert their primary effects at the enzyme level, or they may alter some other cell function such as permeability of membranes. These changes affect cell integrity, ultramicroscopic structure, and grosser functions such as energy expenditure or secretion rate of a hormone. If these changes are severe enough, many cells may die resulting in histological lesions which are visible using light-microscopic techniques. Because organs are composed of many types of cells, effects on one or more of these types will be reflected in changes in organ function. For example, many pollutants cause a thickening and even necrosis (cell death) of the gill epithelium. This in turn produces a reduction in permeability of the gill to oxygen which thereby affects respiratory function for the whole animal. A failure of homeostasis may then be seen. Some organs show compensatory changes (e.g., increased breathing rate) when homeostasis is altered as an attempt to bring the internal condition back toward normal. In this example, the initial gill damage is a pathological effect which then causes one or more physiological responses.

Gene Function Enzyme Activity Membrane Perm.	Cell Integrity and Metabolism	Histological Lesions	Organ Function	Homeostasis	Growth and Reproduction	Ecology and Behavior

Figure 1 Levels of biological complexity in the study of the effect of some environmental factor, including pollution. The extent of complexity increases as one progresses from left to right.

Moving further across the levels of organization spectrum (Figure 1), chronic exposure to a pollutant may depress growth. Reproduction is one of the processes of fish that is most sensitive to pollution, particularly the larval stages. Anything that effects the nervous system will alter behavior, and many substances directly cause alterations in the functions of the nervous system. They may affect behavior indirectly as well by affecting other organ functions such as osmoregulation and metabolism of sex hormones. Finally, changes in the function of a group of organisms in an ecosystem cause effects on other organisms, whether they be predators or prey.

In the levels of organization spectrum it is important to realize that no level is more important than another. As Bartholomew (1964, p. 8) said so well: "...each level offers unique problems and insights; each level finds its explanation of mechanism in the levels below, and its significance in the levels above." (Also see Jorgensen, 1983.) As a rule, the higher the level, the more generalized the response. So if one wishes to assess the general "health" of an organism, higher levels are appropriate; however, if one is interested in studying more specific actions of various things and wishes to understand mechanisms, lower levels are investigated.

IV. IMPORTANCE OF DOSE AND DURATION OF EXPOSURE

Figure 2 illustrates the general effect of environmental concentration of a chemical or altered physical parameter such as dissolved oxygen on some measurable response in the organism. Concentrations below the sublethal response threshold are best called a "no effect" level rather than a "safe" level as has unfortunately been done in some studies. The sublethal threshold will vary with the response that is being measured, and due to the fact that only small changes are being measured, random (sometimes called stochastic) processes will make it difficult to specify with precision (Dinman, 1972). Within the sublethal range a wide variety of reversible and irreversible processes take place. This is the area of most interest to physiologists and many pollution biologists. Prolonged exposure within the upper end of the sublethal zone may cause death through a general weakening of the animal so it becomes more susceptible to disease and/or predation.

The lethal concentration (LC50) is defined as that concentration which causes death to half the test animals within a specified period of time (frequently 96 h). The higher one goes into the lethal range, the more rapidly death occurs. This resistance time can be quantified as the median lethal time (LT50) which refers to the time required for 50% of the test population to succumb to the experimental condition. Exposure to concentrations that produce death in 96 h or less are usually called acute exposures, whereas the concentrations in the sublethal zone are referred to as sublethal, or chronic if the time of exposure exceeds 96 h.

In physiological studies, one measures either rate functions (e.g., breathing rate, swimming speed, oxygen consumption) or the concentrations of something (e.g.,

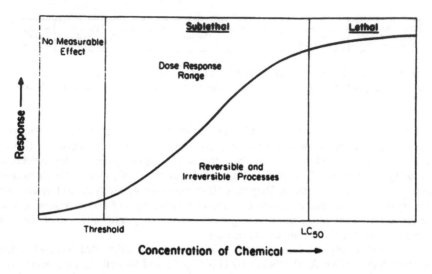

Figure 2 Idealized diagram of the effect of dose on the response as measured by some physiological change (including percent dead). (Modified from Waldichuck, M., *The Assessment of Sublethal Effects of Pollutants in the Sea*, Cole, H. A., Ed., *Philos. Trans. R. Soc. London B*, 286, 397, 1979.)

serum electrolytes, serum glucose, liver glutathione). A physiological response then is a change in one or more of these measures caused by the altered environmental condition. The response may be initiated by the fish as a means to maintain homeostasis, or the response may reflect a breakdown of some physiolgical function. In that case, it may be better to designate it as an effect, rather than a response. Thus, a physiological effect of a pollutant (e.g., increased loss of blood electrolytes) may initiate a physiological response (e.g., increased cortisol) to correct the altered internal state of the animal.

The duration of exposure to an experimental condition may have a considerable impact on both the qualitative character of a physiological change and its quantitative aspects. Figure 3 shows a generalized view of the major sorts of changes that may occur in a physiological measure (e.g., blood glucose or breathing rate) during the period of experimental exposure. If the concentration of pollutant is sufficiently high, death may ensue and be reflected in the physiological variable going rapidly one way or the other. This does not necessarily mean this was the physiological mode of death, as some workers have claimed, for other even more important things may have taken place but were not measured.

If the exposure is to sublethal levels, then either an increase or decrease in the variable may occur, usually over a period of hours or days. This may be followed by a return toward normal which we can call recovery, even though the exposure continues. Another variation shown that is not uncommon, is the phenomenon of initial acceleration followed by inhibition. This is generally observed where there is an initial physical irritation to sensitive tissues, or where there is a "psychological stress" (Schreck, 1981) followed by toxicological effects.

In general, the effect of increasing exposure concentration is to move the curves shown in Figure 3 to the left and to increase the amplitude of the change. Clearly then, attention

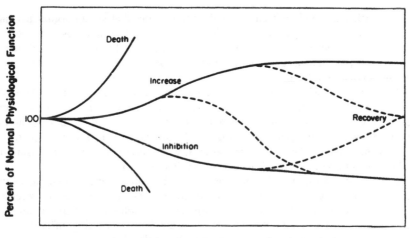

Figure 3 Generalized illustration of the types of changes that can occur in some physiological variable as a result of pollution exposure. See text for explanation.

must be given to when one measures a particular physiological variable during the period of exposure as it will have a very profound effect on the data obtained both quantitatively and qualitatively. What this argues for is measurements taken at several time intervals during a given exposure.

V. STRESS

The concept of stress in physiological systems has an extensive history. The "father" of stress studies is generally considered to be Hans Selye, although others such as Cannon (1929) set the stage. Selye (1950) developed the idea that a mammal when subjected to almost any kind of stress exhibits a generalized group of physiological responses. There is a rapid elevation in adrenalin and noradrenalin which mobilizes muscle glycogen into blood sugar, causes blood pressure to rise, and in general causes the body to undergo the "fight or flight" response. If the stressful condition continues, the adrenal cortex is stimulated to release increased amounts of cortisol which sustain the changes caused by the adrenalin and also cause a mobilization of some of the body protein into plasma amino acids and an assortment of other physiological changes. The hormonal changes are referred to as primary effects of the stress and the other physiological alterations produced by them are called secondary effects. Collectively, they are called the general adaption syndrome (GAS), because it is a group of stereotyped responses which do not differ with the original cause. Similar changes have been observed in fish (Donaldson, 1981; Mazeaud and Mazeaud, 1981; Schreck, 1990) and are further discussed in Chapter 8. A bibliographic database for personal computers on the subject is available from Davis and Schreck (1994).

The terminology used in the physiology of stress has been confusing because some refer to stress as the cause of the responses (e.g., thermal stress). Other workers call the responses themselves stress (i.e., the animal is showing physiological stress) and the causes become known as stressors. Pickering (1981) gives a good short review of this problem, particularly with reference to fish, pointing out that strong arguments can be made for each of the various terminologies. In more recent years most who deal with stress in fish consider the animal is under stress when it exhibits diminished function, which implies potential death of the individual or depressed reproduction leading to a declining population (Heath, 1990).

Determination of whether and how much an animal is under stress revolves partly around the problem of defining normality. There is considerable variation in what might be termed normal values for physiological measures. These variations may be caused by time of day when measured, psychological disturbances, age, season, and that all-encompassing blanket called biological variability. The problem is not at all insurmountable, but needs to be faced by all who work in this field.

Assuming one can define the "normal" range by using some animals as controls, which are treated the same as the experimentals except that they do not receive the stressor, then the problem becomes both statistical and biological. Statistical tests, such as the Student's t test, can tell if the experimentals are doing something different from the controls (assuming one is measuring a relevant variable). The statistical tests will not, however, tell whether the change is important for the animals. A 10% increase in the hemoglobin concentration, for example, may be statistically significant but have little functional relevance to the organism. There are a great deal of data in the literature which show statistically significant changes in some variable in the organisms subjected to a stressor, but an evaluation of the biological significance is usually lacking. Admittedly, such an evaluation may have a large subjective element to it but would give additional perspective to the data.

VI. TOXIC MODE OF ACTION

There is a popular misconception that a given chemical will have a single mode of toxic action or a single target organ in the organism. Such a conclusion is probably not true for any chemical. While it is true that harmful chemicals may tend to act mostly on certain organs or physiological functions, there is usually more than one being affected. This may in part be due to the fact that when one function is affected, this can immediately bring about a series of other changes that may reflect altered homeostasis or compensatory responses to that altered homeostasis. Even at the level of individual cellular enzymes, most foreign chemicals inhibit or stimulate several enzymes, although some are more sensitive than others to a given chemical (see Chapter 9).

The concept of target organ was probably first developed from work on drugs where the aim is to develop a drug to act on specific organ(s). Toxic chemicals in the environment tend to accumulate in particular organs but the organs most affected by that chemical are not necessarily the ones with the highest concentration. For example, DDT tends to accumulate in fat where it produces no known effect; DDT acts instead on nerves and on the process of electrolyte and amino acid uptake in the gut. Note that if an investigator had only looked for effects on the nervous system, the conclusion might have been that this is a neurotoxin only, whereas the latter effects may be more important at chronic levels of DDT where it might inhibit the uptake of necessary nutrients.

A foreign chemical will generally produce a group of physiological effects which is referred to as a syndrome. This concept has been useful in grouping chemicals into, "fish acute toxicity syndromes" (FATS), which can then be used for making predictions of toxic action based on chemical and physical characteristics of the xenobiotic (McKim et al., 1987). For example, at very high environmental concentrations, a respiratory irritant causes increased coughing and elevated arterial carbon dioxide (among other things), whereas a respiratory uncoupler causes increased oxygen consumption and ventilation volume (among other things).

One of the more interesting and important conclusions that is beginning to emerge from work on fish is that modes of action may differ markedly depending on the exposure concentration. For example, when fish are exposed to a high concentration of waterborne zinc (1.5 mg/L), a rapid drop in blood oxygen and pH occurs over a period of less than 12 h, and they die of hypoxia. At a lower level of zinc (0.8 mg/L) exposure for 3 days, they experience slight blood alkalosis and no change in blood oxygen, however, some mortality still occurs from unknown mechanisms (Spry and Wood, 1984). Therefore, the concentration and duration of exposure can greatly affect the syndrome observed in the fish.

REFERENCES

Adams, S. M., Ed., *Biological Indicators of Stress in Fish,* American Fisheries Symposium 8, Bethesda, MD, 1990.

Alabaster, J. S. and Lloyd, R., *Water Quality Criteria for Freshwater Fish,* Butterworths, London, 1980.

Anderson, J. W., An assessment of knowledge concerning the fate and effects of petroleum hydrocarbons in the marine environment, in, *Marine Pollution: Functional Responses,* Vernberg, W. B., Thurberg, F. P., Calabrese, A. and Vernberg, F. J., Eds., Academic Press, New York, 1979, 3.

APHA, *Standard Methods for the Examination of Water and Wastewater,* 18th Edition, American Public Health Association, Washington, D.C., 1992.

Baker, J. P., Effects on fish of metals associated with acidification, in, *Acid Rain/Fisheries,* Johnson, R. E., Ed., American Fisheries Society, Bethesda, MD, 1982, 165.

Bartholomew, G. A., The roles of physiology and behavior in the maintenance of homeostasis in the desert environment, in, *Homeostasis and Feedback Mechanisms,* Symposia Society Experimental Biology, No. 18, Cambridge University Press, New York, 1964, 7.

Briggs, S. A., *Basic Guide to Pesticides, Their Characteristics and Hazards,* Hemisphere Publishing, Washington, D.C., 1992.

Cannon, W. B., *Bodily Changes in Pain, Hunger, Fear and Rage: An Account of Recent Researches Into the Function of Emotional Excitement,* Appleton, New York, 1929.

Crawshaw, L. E. and Hazel, J. R., Temperature effects in fish, *Am. J. Physiol.,* 246, (Regulatory, Integrative, Comparative Physiology 15), R439, 1984.

Davis, L. E. and Schreck, C. B., Annotated bibliography on stress and transportation of fishes, and downstream passage of salmonids, Oregon Cooperative Fishery Research Unit, Oregon State University, Corvallis, OR, 97331, 1994.

Dinman, B. D., "Non-concept" of "no-threshold": chemicals in the environment, *Science,* 175, 495, 1972.

Donaldson, E. M., The pituitary-interrenal axis as an indicator of stress in fish, in, *Stress and Fish,* Pickering, A. D., Ed., Academic Press, New York, 1981, chap. 2.

Eisler, R., Cyanide hazards to fish, wildlife and invertebrates: A synoptic review, *U.S. Fish Wildl. Serv. Biol. Rep.,* 85 (1.23), 55, 1991.

EPA, *Quality Criteria for Water 1986,* United States Environmental Protection Agency, Office of Water Regulations and Standards, Wasnington, D.C., EPA 440/5-86-001.

General Accounting Office, *Water Pollution: Application of National Cleanup Standards to the Pulp and Paper Industry,* National Technical Information Service PB87-193231, U.S. Department of Commerce, Washington, D.C., 36, 1987.

Hall, L. W., Burton, D. T. and Liden, L. H., An interpretative literature analysis evaluating the effects of power plant chlorination on freshwater organisms, *CRC Crit. Rev. Toxicol.,* 9, 1, 1981.

Hazel, J. R., Thermal biology, in, *The Physiology of Fishes,* Evans, D. H., Ed., CRC Press, Boca Raton, FL, 1993, chap. 14.

Heath, A. G., Summary and perspectives, *American Fisheries Society Symposium,* 8, 183, 1990.

Hoar, W. S. and Randall, D. J., Eds., *Fish Physiology*, Vols. 1–10, 1969–1985.

Houston, A. H., *Thermal Effects Upon Fishes*, National Research Council Canada, 1982.

Huggett, R. J., Kimerle, R. A., Mehrle, P. M. and Bergman, H. L., Eds., *Biomarkers: Biochemical Physiological, and Histological Markers of Anthropogenic Stress*, Lewis Publishers, Boca Raton, FL, 1992.

Jorgensen, C. B., Ecological physiology, background and perspectives, *Comp. Biochem. Physiol.*, 75A, 5, 1983.

Leduc, G., Cyanides in water: toxicological significance, in, *Aquatic Toxicology*, Vol. 2, Weber, L. J., Ed., Raven Press, New York, 1984, 153.

Leland, H. V. and Kuwabara, J. S., Trace metals, in, *Fundamentals of Aquatic Toxicology, Methods and Applications*, Rand, G. M. and Petrocelli, S. R., Eds., Hemisphere Publishing, Washington, D.C., 1985, chap. 13.

Lindstom-Seppa, P. and Oikari, A., Biotransformation and other toxicological and physiological responses in rainbow trout (*Salmo gairdneri*) caged in a lake receiving effluents of pulp and paper industry, *Aquat. Toxicol.* 16, 187, 1990.

Livingston, R. J., Review of current literature concerning acute and chronic effects of pesticides on aquatic organisms, *CRC Crit. Rev. Environ. Control*, 7, 325, 1977.

Lloyd, R., *Pollution and Freshwater Fish*, Fishing News Books, Oxford, 1992.

Macek, K. J., Aquatic toxicology: fact or fiction?, *Environ. Health Perspect.*, 34, 159, 1980.

Mazeaud, M. M. and Mazeaud, F., Adrenergic responses to stress in fish, in *Stress and Fish*, Pickering, A. D., Ed., Academic Press, New York, 1981, chap. 3.

McCarthy, J. F. and Shugart, L. R., Eds., *Biomarkers of Environmental Contamination*, Lewis Publishers, Boca Raton, FL, 1990.

McKim, J. M., Bradbury, S. P. and Niemi, G. J., Fish acute toxicity syndromes and their use in the QSAR approach to hazard assessment, *Environ. Health Perspect.*, 71, 171, 1987.

Moore, J. W. and Ramamoorthy, S., *Heavy Metals in Natural Waters: Applied Monitoring and Impact Assessment*, Springer-Verlag, New York, 1984.

Morris, R., Taylor, E., Brown, D. and Brown, D., Eds., *Acid Toxicity and Aquatic Animals*, Cambridge University Press, New York, 1989.

Neff, J. M. and Anderson, J. W., *Responses of Marine Animals to Petroleum and Specific Petroleum Hydrocarbons*, Applied Science, London, 1981.

Murty, A. S., *Toxicity of Pesticides to Fish*, CRC Press, Boca Raton, FL, 1986.

Nieboer, E. and Richardson, D. H. S., The replacement of the nondescript term "heavy metals" by a biologically and chemically significant classification of metal ions, *Environ. Poll.*, 1, 3, 1980.

Norton, S. A., The effects of acidification on the chemistry of ground and surface waters, in, *Acid Rain/ Fisheries*, Johnson, R. E., Ed., American Fisheries Society, Bethesda, MD, 1982, 93.

Pickering, Q. H., Acute toxicity of alkylbenzene sulfonate and linear alkylate sulfonate to the eggs of the fathead minnow, *Pimephales promelas, Air Water Pollut.*, 10, 385, 1966.

Pickering, A. D., Introduction: the concept of biological stress, in, *Stress and Fish*, Pickering, A. D., Ed., Academic Press, New York, 1981, chap. 1.

Raney, E. C., Menzel, B. W. and Weller, E. C., *Heated Effluents and Effects on Aquatic Life with Emphasis on Fishes; a Bibliography*, U.S. Atomic Energy Commission, TID-3918, 1972.

Russo, R. C., Ammonia, nitrite and nitrate, in, *Fundamentals of Aquatic Toxicology, Methods and Applications*, Rand, G. M. and Petrocelli, S. R., Hemisphere Publishing, Washington, D.C., 1985, chap. 15.

Schreck, C. B., Stress and compensation in teleostean fishes: response to social and physical factors, in, *Stress and Fish*, Pickering, A. D., Ed., Academic Press, New York, 1981, chap. 13.

Schreck, C. B., Physiological, behavioral and performance indicators of stress, *Am. Fish. Soc. Symp.*, 8, 29, 1990.

Selye, H., *The Physiology and Pathology of Exposure to Stress*, a treatise based on The Concepts of The General-Adaptation-Syndrome and the Diseases of Adaptation, Acta, Montreal, 1950.

Sindermann, C. J., An opinion about research activities and needs concerning physiological effects of pollutants in the environment, in, *Marine Pollution: Functional Responses*, Vernberg, W. B., Thurberg, F. P., Calabrese, A. and Vernberg, F. J., Eds., Academic Press, New York, 1979, 437.

Sprague, J. B., The ABC's of pollutant bioassay using fish, *Biological Methods for the Assessment of Water Quality*, ASTM STP 528, American Society for Testing and Materials, 6, 1973.

Spry, D. J. and Wiener, J. G., Metal bioavailability and toxicity to fish in low-alkalinity lakes: a critical review, *Environ. Pollut.,* 71, 243, 1991.

Spry, D. J. and Wood, C. M., Acid-base, plasma ion and blood gas changes in rainbow trout during short term toxic zinc exposure, *J. Comp. Physiol. B,* 154, 149, 1984.

von Westernhagen, H., Sublethal effects of pollutants on fish eggs and larvae, in, *Fish Physiology,* Vol. XI, Part A, Hoar, W. S. and Randall, D. J., Eds., Academic Press, New York, 1988, chap. 4.

Waldichuk, M., Review of the problems, in, *The Assessment of Sublethal Effects of Pollutants in the Sea,* Cole, H. A., Ed., *Philos. Trans. R. Soc. London B,* 286, 397, 1979.

World Commission on Environment and Development, *Our Common Future,* Oxford University Press, Oxford, 1987.

Chapter 2

Environmental Hypoxia

I. INTRODUCTION

Hypoxia refers to any condition in which the amount of oxygen is measurably below air saturation levels. Anoxia means no oxygen and should be reserved for such conditions. Aquatic biologists generally think in terms of dissolved oxygen expressed as milligrams per liter (or the equivalent ppm) concentration of oxygen in the water. Physiologists usually measure oxygen in the environment or body fluids as partial pressure (Po_2) which is expressed in mmHg (= torr), or as pascals (1 mmHg = 133.32 pascals) (Bridges and Butler, 1989). While it is reasonably safe to assume a direct relationship between dissolved oxygen concentration and Po_2, this holds true only at a given temperature and salinity. As either of these increases, the solubility, and thus concentration, of oxygen decreases at any given Po_2.

Environmental hypoxia is taken up early in this book because (1) dissolved oxygen is often low in polluted waters and (2) many of the physiological responses of fish to chemical pollutants at acute concentrations, are similar to those produced in response to enviromental hypoxia. Therefore, a treatment of this topic is a good introduction to the types of actions pollutant chemicals may exert on the functions of various organs.

There are several potential causes of environmental hypoxia, some of which are not due to human activities (i.e., they are "natural") (Boutilier, 1990). For example, in thermally stratified eutrophic lakes, the Po_2 of the hypolimnion is almost always hypoxic (Barnes and Mann, 1991), and during the winter in lakes that are frozen over, respiration can cause depletion of oxygen in the water trapped below the ice (Pennak, 1968).

In lakes and stagnant streams where the concentration of nutrients are high, algal blooms may cause a considerable decrease in oxygen. The opposite of hypoxia, a supersaturation of oxygen (hyperoxia), may occur during midday in some of these ponds due to photosynthesis and warming of the water (Garey and Rahn, 1970). Also, wherever there is a large amount of putrescible organic matter in the water from industrial or domestic waste, microbial respiration utilizes a large percentage of the dissolved oxygen (i.e., the biochemical oxygen demand is elevated) (Poppe, 1990; Warren, 1971).

Hypoxia is not limited to freshwater habitats. Oxygen levels in the ocean vary with depth, temperature, salinity, and productivity (Bushnell et al., 1990). Nutrient enrichment and unique meteorological conditions as well as phytoplankton blooms can produce severely hypoxic conditions in marine habitats (Boesch, 1983; Swanson and Sinderman, 1979). Conditions of near zero dissolved oxygen have been observed due to pulpmill wastes confined in a partially enclosed saltwater bay (Swanson and Sinderman, 1979), and diurnal changes in oxygen can occur in intertidal areas with high nutrient loads (Truchot and Duhamel-Jowe, 1980).

II. MINIMUM LEVELS OF OXYGEN REQUIRED FOR FISH LIFE

Many of the earlier studies of fish and hypoxia were devoted to determining the minimum levels of oxygen required by fish. Doudoroff and Shumway (1970) reviewed much of this and provided extensive tabular data on lethal levels of oxygen for a wide variety of species. They also discussed some of the physiological effects of low oxygen. Davis (1975) provides a somewhat less extensive but still very useful review of this topic. His primary aim was the formulation of criteria for dissolved oxygen for Canadian fish and invertebrates. The approach used by Davis was to examine the literature looking for threshold levels of dissolved oxygen that caused changes in some physiological parameter such as reduced swimming stamina, increases or decreases in metabolic rate, reduced blood oxygen saturation, etc. The presumption is that at any level of oxygen below that threshold, the organisms will be expending excess energy to maintain homeostasis and thus experience some physiological stress.

Frequently, freshwater fish have been grouped into salmonids and non-salmonids with regard to their minimum oxygen requirements. The former are considered to be less tolerant of hypoxia than the latter. This broadly held assumption is borne out in the data summarized in the above-mentioned reviews, especially for the levels that are lethal. According to Davis (1975), the average physiological threshold for salmonids is a Po_2 of 120 mmHg and for non-salmonds it is 95 mmHg. Assuming temperatures of 15 and 25°C, respectively, those translate into dissolved oxygen concentrations of 7.8 and 5.2 ppm. The two groups do not often cohabitate but oxygen may not always be the reason. Temperature may in many areas be a more limiting factor for salmonid distribution than oxygen.

The minimum oxygen requirements of pelagic marine species has been little studied. There is little reason to presume they are especially well-adapted for low oxygen, although some certainly are quite tolerant of hypoxia (Wu and Woo, 1984). In the deeper oceanic regions there are large eutrophic oxygen-minimum zones where oxygen levels can even approach zero; fish from these areas are generally quite tolerant of low oxygen (Douglas et al., 1976; Yang et al., 1992).

Some species of freshwater fish are extremely resistant to low levels of oxygen, or even anoxia. The cyprinids are especially notable as they include the crucian carp (*Cyprinus carpio*), which can survive up to 6 months in cold water in the absence of oxygen (Blazka, 1958; Holopainen et al., 1986) and the common goldfish (*Carassius auratus*), which survives total anoxia for up to 22 h at 20°C (Van den Thillart et al., 1983). Another species that exhibits remarkable tolerance of anoxia is the toadfish (*Opsanus tau*), a marine species. These fish can survive an average of 20 h in oxygen-free water at 22°C (Ultsch et al., 1981).

In general, relatively low temperatures are required for high tolerance to anoxia, but there are exceptions. For example, Mathur (1967) reported that *Rasbora daniconius* (another cyprinid) can survive about 3 months of anoxia at 33°C.

Among freshwater fish, sensitivity of eggs and larvae to low oxygen varies with the particular stage of development. The early embryo is relatively resistant to low concentrations of oxygen, but as the embryo grows, its sensitivity to hypoxia increases to a maximum at hatching (Doudoroff and Shumway, 1970). The early fish embryo obtains its energy mostly by means of anaerobic glycolysis, and then as development proceeds, aerobic respiration becomes more important (Boulekbache, 1981). Thus, this ontogenetic change in sensitivity to hypoxia by the fish embryo appears to be related at least in part to the dominant mode of energy metabolism that it uses at a particular developmental stage. It should be noted here that although early fish embryos may tolerate a lack of oxygen rather well, it is often the most sensitive stage in the whole life cycle to the "insult" of a chemical pollutant (see Chapter 13).

Doudoroff and Shumway (1970) summarize a considerable amount of work on salmonids of several species and conclude that any reduction of oxygen below air saturation may produce delays in hatching and smaller than normal fry. These fry, however, are usually viable and not deformed unless the oxygen levels are below 2–3 mg/L. The measurement of oxygen levels in the water of a stream will not indicate the true oxygen availability to the salmonid embryos as these are buried in the streambed gravels where oxygen concentrations are often considerably less than that of the flowing water. Some warmwater species seemingly require higher oxygen concentrations for normal development than do salmonids (Doudoroff and Shumway, 1970).

The larvae of fish are far less able to tolerate hypoxia than the adults of the same species (Davis, 1975). As the larvae develop, marked changes in hypoxia tolerance can occur over just a few days. Spoor (1984) found that newly hatched smallmouth bass (*Micropterus dolomieui*) larvae are comparatively resistant to hypoxia (90% survived 6 h at 1 mg O_2/L at 20°). However, starting with the second day posthatch, they became increasingly sensitive to the lack of oxygen and this reached a maximum at the fourth day (none survived 3 h at 1 mg/L dissolved oxygen.). This high sensitivity continues until day 10; thereafter there occurs a rather sudden decrease in sensitivity to hypoxic conditions. The larvae do not start to breathe until the fourth day when they are least able to tolerate hypoxia (Spoor, 1984).

III. INTERACTION OF HYPOXIA AND TOXICITY OF POLLUTANT CHEMICALS

As a general rule, a given chemical becomes more toxic at lower levels of dissolved oxygen, but for most chemicals, the effect appears to be modest (Sprague, 1984; Rattner and Heath, 1994). Because the number of chemicals tested is quite small, however, and data on sublethal effects are rare, such a generalization should be treated as quite tentative. Because our interest here is primarily sublethal effects of chemicals, the following four studies are relevant.

In 32-day tests on the effects of 1,2,3-trichlorobenzene on larval fathead minnows, Carlson (1987) reported that when the dissolved oxygen (DO) was lowered to 4.5 ppm, the chemical caused a much greater effect on growth and survival of the larvae than when tested at near saturation for oxygen, but the threshold acute toxic concentration was unchanged. From an environmental standpoint, the sublethal effect is probably more important and it should be noted that a DO of 4.5 ppm is not very low.

Paraquat (an herbicide) causes accumulation of free radicals in cells as a result of its metabolism. This in turn induces formation of more of the enzyme superoxide dismutase as a mechanism for free radical removal. Severe hypoxia alone in carp induced superoxide dismutase in gill, liver, and brain tissue. Paraquat alone stimulated dismutase activity in only the gill tissue. Combining hypoxia and paraquat caused an additive effect in gill tissue but no further stimulation in the other two tissues (Vig and Nemcsok, 1989). Perhaps the most interesting finding was the induction of superoxide dismutase by hypoxia alone, a phenomenon that deserves further investigation.

Acetylcholinesterase in the nervous system of fish and other animals is greatly inhibited by organophosphorus pesticide poisoning (Mayer et al., 1992). There was also a report (Malyarevskaya, 1979) that hypoxia caused an inhibition of this enzyme in perch, however, we were unable to confirm this finding in my laboratory using trout (unpublished observations). Hoy et al. (1991) also found no effect of hypoxia on acetylcholinesterase activity in trout, but if the fish were exposed to the organophosphorus compound dichlorovos and hypoxia, a greater degree of inhibition occurred than if the exposures took place in water saturated with oxygen. This may have been due to a more rapid uptake of the poison by the fish because of hyperventilation in the hypoxic water, a mechanism that undoubtedly applies in many hypoxia-toxicity interactions.

Finally, we examine the complex relationship between acute toxicity of anthracene and dissolved oxygen. The toxicity of this polycyclic aromatic hydrocarbon is induced by UV light, a process that is also proportional to the amount of oxygen present. In an apparent contradiction of this physical phenomenon, maximum toxicity to fish occurs at an intermediate DO. McCloskey and Oris (1991) hypothesize that at intermediate DO levels, both hyperventilation due to mild hypoxia and photoinduced toxicity combine to cause the greatest effect on the fish. At a high DO, ventilation is reduced so less poison is brought to the gills and at low DO, the lack of oxygen in the water suppresses toxicity.

Exposure to a sublethal concentration of a toxic chemical can affect subsequent responses of a fish to hypoxia. Phenol causes histopathological changes in gill tissues and thereby reduces the ability of fish to tolerate hypoxia (Hlohowskyj and Chagnon, 1992). Within the gill, if transport of oxygen by the blood is compromised by some chemicals such as nitrate, there is a reduced ability to tolerate hypoxia (Watenpaugh and Beitinger, 1986). Finally, sublethal exposure to copper for a week results in an amplified stress response of bluegill (*Lepomis macrochirus*) to a rapid hypoxia exposure (Heath, 1991).

The remainder of this chapter will be devoted to the physiological responses of fish to hypoxia or anoxia. We will move from the processes of respiratory gas exchange and transport during hypoxia to the biochemical changes invoved in anaerobic metabolism during anoxia. Some emphasis will be placed on the adaptations that enable certain species to be better able than others to function under conditions of oxygen lack.

IV. GILL VS. CUTANEOUS RESPIRATION

The site of oxygen uptake in adult fish is primarily the gills. The skin, however, can be an important oxygen exchanger in some species. For example, in the buried plaice (*Pleuronectes platessa*), a significant amount of the oxygen uptake takes place through the skin under normoxic conditions. If the oxygen tension is lowered, the cutaneous oxygen uptake stays relatively constant while the gill, and thus total, oxygen uptake declines (Steffensen et al., 1981). Such an increased utilization of skin for oxygen uptake in the hypoxic plaice can be contrasted with the common carp (*C. carpio*) in which the opposite occurs. Cutaneous oxygen uptake is directly related to ambient oxygen over both hypoxic and hyperoxic conditions (Figure 1). Under normoxia Takeda (1989) found that this species obtains about 10% of its total oxygen uptake through the skin. Most if not all of that, however, is utilized directly by the cutaneous tissues (Nonnotte, 1981), which appears to also be true for a variety of teleosts including the plaice (Steffensen et al., 1985). Teleost skin has a very high rate of weight-specific metabolism, 1.7–1.9 times that of the intact fish (Steffensen et al., 1985). The energy released may be utilized in mucus production or osmoregulation, although Steffensen and Lomholt (1985) found that changing the salinity of the surrounding water had little effect on skin respiratory rate so it is not currently clear why the skin has such a high energy requirement.

Cutaneous oxygen uptake might be important for those species that occasionally go out on land because gill tissue typically collapses in air. Such amphibious species include some of the catfish (*Ictalurus*) (Nonnotte, 1984) and European eels, although in the latter, cutaneous tissue utilization of oxygen was equal to or greater than that taken from the air;

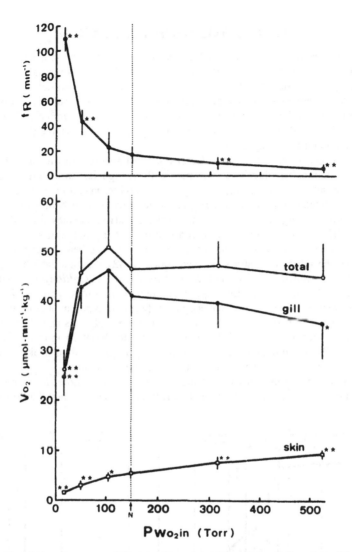

Figure 1 Respiratory frequency (fR), total O_2 uptake (Vo_2 total), gill O_2 uptake (Vo_2 gill), and cutaneous O_2 uptake (Vo_2 skin) as a function of ambient Po_2 (PwO_2 in) in the carp. Significance of difference from normoxic levels: ``$p < 0.01$; `$p < 0.05$. (From Takeda, T., *Comp. Biochem. Physiol.*, 94A, 205, 1989. With permission from Elsevier Publishers.)

the skin therefore cannot be considered as an oxygen exchanger under these conditions (Nonnotte, 1984).

Larval fish rely exclusively on cutaneous respiration immediately after hatching. The rate of development of gill respiration varies greatly between different species (Rombough, 1988); in the cyprinids, at least, this development is a gradual process wherein the cutaneous respiration may remain important well into the juvenile stages (El-Fiky and Weiser, 1988).

Rombough (1992), measured intravascular and skin-water interface Po_2 with micro-electrodes in these larvae and found the intravascular Po_2s of the larvae to be considerably lower than those of adults. He attributed this difference to the relatively large diffusional boundary layer in the larvae which is not ventilated as would occur with gills, so it becomes stagnant.

V. ADJUSTMENTS IN VENTILATION

As oxygen moves from the water which is passing over the gills to the site of utilization in the cells, it encounters a series of resistances (Figure 2). Environmental hypoxia lowers the starting point on the left of this curve, but by increasing the ventilation, the boundary layer next to the lamellae is more rapidly replaced with fresh water and the drop in oxygen tension due to interlamellar water convection is presumably reduced (i.e., the first resistance is lowered). Therefore, under mildly hypoxic conditions, the Po_2 of the arterial blood may remain relatively unchanged.

As the oxygen level in the water decreases, most species start to increase their gill ventilation volume. The threshold Po_2 varies with the species; limited data suggest that species better adapted for low oxygen (e.g., carp) tend to have a lower threshold, which is probably related to the oxygen-binding characteristics of their hemoglobin (Figure 3). The bottom part of Figure 3 illustrates the point that various species alter gill ventilation volume in different ways. Since ventilation volume is the product of stroke volume and breathing frequency (the latter sometimes called the opercular beat), changes in either one or both may be utilized. Trout (Smith and Jones, 1982), channel catfish (*I. punctatus*) (Burggren and Cameron, 1980), and the marine dragonet (*Callionymus*) (Hughes and Umezawa, 1968) alter ventilation largely by changes in stroke volume. The dragonet actually decreases its ventilation frequency in response to hypoxia, but the stroke volume increases to such an extent that ventilation volume rises (Hughes and Umezawa, 1968). Some species, such as the bluegill (*L. macrochirus*) (Heath, 1973) and carp (Lomholt and Johansen, 1979) rely more on changes in breathing frequency than stroke volume when responding to changes in level of oxygen or metabolic demand.

The sturgeon is a somewhat interesting case in that two studies have reported completely contradictory findings. Burggren and Randall (1978) claimed that gill ventilation

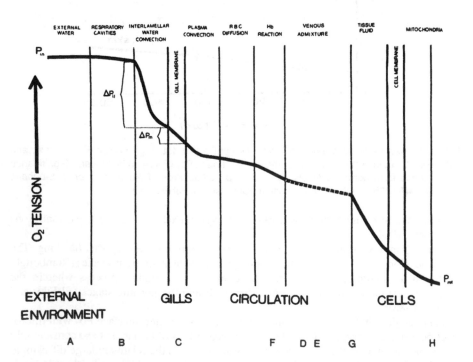

Figure 2 Diagram showing approximate changes in Po_2 from the environment to the mitochondria of the cells. (From Hughes, G. M., *Am. Zool.*, 13, 475, 1973. With permission.)

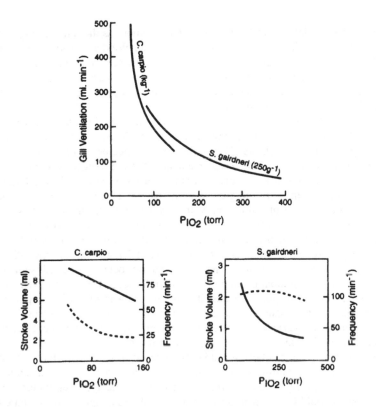

Figure 3 Relationship between gill ventilation and partial pressures of O_2 of inhaled water (PIO$_2$) in carp (*C. carpio*) and rainbow trout *(O. mykiss)*. The lower panels indicate the relative contribution of breathing rate (dotted lines) and stroke volume (solid lines) to change in gill ventilation. (Redrawn from Shelton, G., Jones, D. and Milsom, W., *Handbook of Physiology*, Sect. 3, Vol. II, Part 2, Fishman, A. et al., Eds., American Physiology Society, Bethesda, MD, 1986.)

volume in this species declined in hypoxia. They measured it utilizing a plastic membrane sutured around the mouth to separate the buccal water from opercular water. Recently, Nonnotte et al. (1993) performed essentially the same experiments wherein the fish were exposed to a gradual lowering of DO, but ventilation amplitude was estimated using pressure transducers rather than the membrane system of the Burggren and Randall (1978) study. Nonnotte et al. (1993) found that the ventilation increased both in amplitude and frequency in an almost linear relationship to ambient DO. They suggest that the presence of the membrane in the earlier study reduced the ability of the fish to respond to the hypoxia. Decreases in ventilation in response to hypoxia have been observed (Wu and Woo, 1984) but it would appear to be a rare phenomenon. We will return to the question of oxyconformity later in this chapter.

Control of ventilation volume in fish has been much investigated, but the mechanisms are only beginning to be understood. Randall concluded in his 1982 review (based mostly from work on trout) that the ventilation changes seen in response to environmental hypoxia are based on the oxygen content of the arterial blood, rather than its Po$_2$. The receptor(s) that detect the blood oxygen content were believed to lie in the post-gill arterial complex. In is now believed (Burleson et al., 1993) that there are two sets of oxygen-sensitive receptors involved in ventilatory reflexes in trout: one set measures the water oxygen and the other the arterial blood and both groups are located in the first gill arch.

In trout, the arterial Po_2 tracks that of the environment as the latter changes (Holeton and Randall, 1967), whereas the oxygen content of the blood at any given environmental Po_2 will depend on the oxygen dissociation characteristics of the blood (to be discussed below). Thus, the relationship between ventilation and environmental oxygen may depend largely on the affinity of the blood for oxygen in a given species if the ventilation is controlled by blood oxygen concentration.

Randall's (1982) conclusion that ventilation is controlled by arterial oxygen content, rather than Po_2, does not appear to apply to all species. More recent work on carp (Glass et al., 1990; Williams et al., 1992) has shown this species regulates its breathing in response to changes in Po_2 of the water independent of the oxygen content of the blood.

Whether a given species controls ventilation based on arterial Po_2, arterial oxygen content, or water Po_2 may depend greatly on the oxygen affinity of the blood. Shelton et al. (1986), in their review of respiratory control in fish, note that those species with blood that has rather low oxygen affinity (e.g., trout), maintain high arterial Po_2s and that this oxygen tension is rather dependent on that of the water. Those species with high oxygen affinities (e.g., carp and *Silurus*) maintain low arterial Po_2 and this is little influenced by the water Po_2 (Figure 4). A point that seemingly has not been considered is that a species such as the carp, which monitors water Po_2, might increase ventilation even when there was no need for such a response because its hemoglobin was still >80% saturated (Figure 5), and thereby wastes considerable amounts of energy. Intuitively, it seems like it would make more homeostatic sense to monitor the variable that needs to be kept as constant as possible, namely the oxygen content of the arterial blood.

In recent years there has been considerable interest in the involvement of catecholamines in respiratory control. Aota et al. (1990) reported that plasma adrenalin and noradrenalin rose during hypoxia in trout and that part of the hyperventilation in response to that hypoxia could be eliminated by adrenergic blocking agents. In seeming contrast to this finding, Perry and Kinkead (1990) reported hyperventilation in trout under mild hypoxia with little change in plasma catecholamines. In a follow-up study (Kinkead and Perry, 1991) they administered boluses of catecholamines into the blood of trout while the fish were experiencing hypoxia and observed an actual inhibition of ventilation, rather than a stimulation. Perry and Kinkead (1990; Kinkead et al., 1991) also found that Atlantic cod (*Gadus morhua*) respond to severe hypoxia with a strong catecholamine rise but this seemed to have no influence on ventilation because adrenergic blocking agents had no effect on the hypoxic hyperventilation. Randall and Perry, in their 1992 review (p. 280) note that "...there is evidence that catecholamine infusion has an effect on breathing in fish, causing either an increase or a decrease in rate depending on the species,

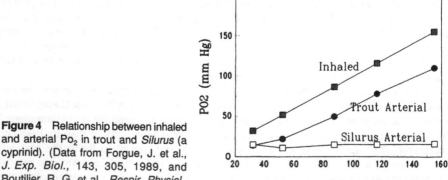

Figure 4 Relationship between inhaled and arterial Po_2 in trout and *Silurus* (a cyprinid). (Data from Forgue, J. et al., *J. Exp. Biol.*, 143, 305, 1989, and Boutilier, R. G. et al., *Respir. Physiol.*, 71, 69, 1988.)

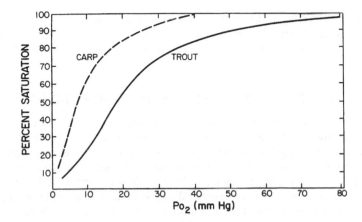

Figure 5 Comparison of oxygen dissociation curves for the hemoglobin from carp (*C. carpio*) and rainbow trout (*O. mykiss*). Carp data at pH 7.9 and 20°C, from Weber, R. E. and Lykkeboe, G., *J. Comp. Physiol.*, 128, 127, 1978. Trout at 3 mmHg CO_2 and 15°C, from Cameron, J. N., *Comp. Biochem. Physiol.*, 38A, 699, 1971.

the time of year, and the physiological state of the animal." Thus, the physiological relevance of catecholamines for ventilatory control remains uncertain in spite of considerable work. Indeed, Randall and Taylor (1991) argue that it does have relevance while Perry et al. (1992) argue against such a conclusion. So, while the involvement of catecholamines in the control of ventilation during hypoxia may or may not be important, it will be seen below that catecholamines have a considerable influence on respiratory gas exchange both at the gill and in the process of oxygen transport by the blood.

It has generally been assumed that arterial PCO_2 or pH has little influence on fish respiration. When elevations in respiration have been seen associated with elevated carbon dioxide, this was attributed to reduced oxygen loading due to the Bohr or Root effects. Perry and Wood (1989, p. 2962), however, argue that "...there now exists sufficient evidence to indicate that CO_2 and (or) pH also can stimulate Vw through mechanisms independent of O_2." Obviously, with the extreme diversity of fish species adapted to quite different sorts of habitats, it perhaps is not surprising that they would exhibit a diversity of approaches to respiratory control.

Some fish species, such as members of the Scombridae family, and some salmon exhibit a transition from active branchial ventilation to ram gill ventilation during swimming at high speeds. As they reach a threshold swimming velocity, ventilatory movements cease and the mouth is held open so the water is forced over the gills by the forward movement of the fish (Roberts, 1975). This shift in ventilatory mode provides a significant energetic savings while swimming (Steffensen, 1985). Roberts (1975) reported that hypoxia had little effect on the transition velocity for Scombrids and concluded that it was controlled by mechanoreceptors. More recently, Steffensen (1985) explored this relationship between transition velocity and ambient Po_2 in rainbow trout and sharksuckers (*Echeneis naucrates*). Some populations of rainbow trout exhibit ram ventilation while others do not, so there is probably a genetic component to the behavior in this species. He used a group that exhibited the ventilation mode and found that both it and the sharksucker shifted to ram ventilation at a higher velocity as the ambient Po_2 was lowered. Clearly, there is thus a chemoreceptor, perhaps along with a mechanoreceptor, controlling this mode of ventilation.

In work on two species of tuna which exhibit obligate ram ventilation, Bushnell and Brill (1991) found that swimming speed and mouth gape increased with hypoxia. While the increased swimming speed might appear to be a mechanism to increase ventilation volume, the authors maintain, based on their indirect measurements of ventilation volume, that it is instead, an escape behavior aimed at removing the fish from the hypoxic water.

VI. ADJUSTMENTS BY THE GILLS TO HYPOXIA

The respiratory surface area of the gills of fishes appears to be related to the normal activity of the species. Those that are active and therefore require high levels of oxygen uptake have the largest gill surface areas (Hughes, 1966). As was discussed above, some species of fish are far more tolerant of low oxygen than others. It might be expected that they would have larger gill surface areas, but this does not appear to be the case. For example, the trout, which is an active oxyphilic species, has a larger gill surface area than the more hypoxia-resistant catfish. It also has a thinner epithelial diffusion distance between the water and blood (Laurent and Perry, 1991).

While there may not be a morphological difference in gill structure that aids hypoxia-resistant species, fish have been shown to enhance both lamellar surface area and reduce the diffusion distance from water to blood in response to both short- and long-term hypoxia. The lamellae of resting normoxic fish are partially buried in the filament thereby reducing the respiratory gas exchange surface. Hypoxia causes them to become more erect and project further out of the filament. This may be due to a combination of hemodynamic and intrafilamental smooth muscle contractions (Laurent and Perry, 1991).

An environmental factor that can alter the gill surface area is the salt content of the water. Trout in freshwater have less lamellar surface area than those in water with more NaCl. This is due to a greater number of chloride cells in the freshwater-adapted trout. Thomas et al. (1988) found that those in natural waters having NaCl at only 0.1 mM/L were less tolerant of hypoxia (presumably due to the greater number of chloride cells) in that their arterial Po$_2$ was more sensitive to ambient Po$_2$ than those in water with higher salt concentrations (1 mM/L).

The control of blood circulation through the gill of fish is complex involving local effects of the oxygen content of the water as well as hormonal and neural mechanisms. In normoxic rainbow trout, only 60% of the gill lamellae (the structures where gas exchange occurs) are perfused with blood (Booth, 1978), so any mechanism to increase this during hypoxia would appear to be helpful, as it would increase the surface area for gas exchange. Environmental hypoxia appears to produce a vasoconstrictor effect in specific lamellae, and since the point of vasoconstriction is efferent to the lamellae, it can cause a dilation of previously unperfused lamellae and thereby increase the overall number of these structures receiving blood (Butler and Metcalf, 1983). The catecholamines, adrenaline and noradrenaline, reduce the resistance to blood flow through the gills and also cause a recruitment of unperfused lamellae. Adrenaline, in addition to its effects on gill circulation, may enhance the permeability of the gill epithelium to oxygen (Butler and Metcalf, 1983).

Another possible mechanism for microcirculatory adjustments in the gills involves prostaglandins. Mustafa and Jensen (1992) recorded elevated levels of the unstable prostaglandin thromboxane A2 in blood downstream from the gills of trout following hypoxia. They note that these hormones are involved in mammalian microcirculatory adjustments to stress.

VII. TRANSPORT OF OXYGEN BY THE BLOOD

An important adaptation for taking up oxygen that all fish, except Antarctic ice fish (family Chaenichthyidae), show is the presence of hemoglobin in their erythrocytes. The amount of oxygen carried by the hemoglobin in relation to the Po_2 is not linear (Figure 5). This means that as the arterial Po_2 decreases, as during hypoxia, the oxygen content of the blood remains virtually unchanged until it reaches a Po_2 where the curve starts downward. Thus, under mild hypoxia, assuming no change in circulation, the number of molecules of oxygen carried to the tissues should remain nearly unchanged.

As is illustrated in Figure 5, the oxygen dissociation curves are different for different species of fish. The hemoglobin of carp shows a higher affinity for oxygen than does that of trout, because it achieves a higher level of saturation (and thereby content of oxygen) at any given Po_2. This is clearly a key adaptation enabling the carp to function better than the trout in conditions of hypoxia. It does mean, however, that the tissues must function at a lower Po_2 in order for the hemoglobin to give up oxygen to them. Garey and Rahn (1970) recorded a tissue Po_2 for carp of approximately 12 mmHg and for trout between 30 and 40 mmHg. These measurements are probably on the high side of the true values as they estimated the tissue Po_2 by measuring the gas tensions in a bubble of air injected into the peritoneal cavity and allowed it to equilibrate with the surrounding tissues. The Po_2 of mixed venous blood taken from indwelling catheters was 3 mmHg for the carp (Garey, 1967) and 19 mmHg for trout (Stevens and Randall, 1967). Notice in Figure 2 that if the tissues were at these oxygen tensions, that would allow a somewhat better than 50% unloading of the oxygen from the hemoglobin, which probably comes close to approximating the true situation in the resting fish.

Several factors, both environmental and internal, cause the oxygen dissociation curves to shift to the right or left (lowering or raising the oxygen affinity, respectively). As the pH is lowered, or the CO_2 is raised, the curve shifts to the right and sometimes downward due to the well-known Bohr and Root effects, respectively. Increasing the ambient temperature lowers the oxygen affinity of the blood in most fish; so warm, acidic, hypoxic conditions are especially difficult for fish to cope with.

It is common to see blood acidosis in fish exposed to severe hypoxia (Thomas and Hughes, 1982; Tetens and Lykkeboe, 1985), although the opposite can also occur (Perry and Thomas, 1991). The source of plasma acidity may be muscle lactic acid and protons released from erythrocytes in exchange for plasma sodium. The sodium/proton exchange in erythrocytes is apparently stimulated by catecholamines, especially noradrenaline (Fievet et al., 1988).

The effect of extrusion of protons from the erythrocytes into the plasma is to raise the intracellular pH above that of the extracellular pH within minutes and this causes an increase in oxygen affinity of the hemoglobin (Fievet et al., 1988; Tetens and Lykkeboe, 1985). The increase in catecholamines also causes erythrocyte swelling and a decreased concentration of intracellular organic phosphates which further improves oxygen affinity of the hemoglobin (Perry and Thomas, 1991). The cellular swelling is mediated by the second messenger cAMP and the effect, at least in carp, is potentiated by lower arterial oxygen tensions (Salama and Nikinmaa, 1990; Thomas and Perry, 1992). Thus, these "stress hormones" not only enhance the flow of blood through the gills and the permeability of gill tissue to oxygen, they also aid the blood in removing oxygen from water where it is in low concentration.

The percentage of the available dissolved oxygen that is removed from the water during its passage over the gills is referred to as the percentage utilization (%U) or extraction efficiency. In normoxic water, this figure ranges from 23% in the lamprey to over 80% in the carp (see Campagna and Cech, 1981, for review). Increasing the ventilation volume has generally been reported to decrease the %U, however, Campagna and Cech (1981) have noted that the extent of this decrease is less in those species well adapted for living in hypoxic conditions, such as carp and Sacramento blackfish (*Orthodon microlepidotis*). Using a quantitative measure of this decrease, plus the extent of hyperventilation during hypoxia, and the blood affinity for oxygen, they have calculated a "respiratory efficiency index" that can be used to compare species as to their tolerance for environmental hypoxia. On this basis, ranging from most tolerant to least are the white sturgeon, carp, Pacific lamprey, striped mullet, channel catfish, rainbow trout, and starry flounder. A comparative evaluation of the importance of catecholamines in this adaptation would be interesting.

VIII. CARDIOVASCULAR CHANGES DURING HYPOXIA

The classical response of the fish heart to environmental hypoxia is a distinct slowing of the heart rate (bradycardia) (reviewed in Randall, 1982; Fritsch, 1990). There appears from recent work, however, to be a few exceptions to this generalization. Glass et al. (1991), using a fairly slow rate of imposition of the hypoxia, found a tachycardia in carp at Po_2s down to 50 mmHg. Below 30 mmHg, bradycardia occurred but the authors present evidence that this may have involved some myocardial dysfunction from hypoxemia. Fritsch (1990), using an abrupt lowering of water Po_2, observed bradycardia in two species of marine teleosts, but not in a third one. An important factor that ought to be considered is the rate at which the oxygen level is lowered in the test chambers. A rapid decrease in oxygen tension, as has been used by many workers, is likely to initiate excitement and cause release of catecholamines. The importance of this is well exemplified in the study by Borch et al. (1993). Contrary to what most others have observed, they recorded a tachycardia in rainbow trout during hypoxia (60 mmHg) which was imposed very slowly. At lower levels of oxygen, bradycardia eventually occurred, possibly due to hypoxemia acting directly on the heart muscle.

When bradycardia is observed it is initially caused by the cardiac branch of the vagus nerve (Randall and Smith, 1967; Wood and Shelton, 1980), so it is not a direct effect of low oxygen on the heart itself except at extremely low levels of oxygen. During bradycardia, the cardiac output (amount of blood pumped per minute) may go down (Farrell, 1982), however, due to an increase in stroke volume, the cardiac output can remain unchanged (Wood and Shelton, 1980) or it may even increase (Itazawa and Takeda, 1978), depending on the species and level of hypoxia.

In teleosts, the receptors for oxygen tension that initiate the hypoxic slowing of the heart are located on the first gill arch in the region of the efferent vessel and are distinct from the receptors that initiate ventilatory responses to hypoxia (Daxboek and Holeton, 1978; Burleson and Milsom, 1993). Burleson and Milsom (1993) claim the receptors are externally oriented so they measure water Po_2 rather than that of blood.

Because bradycardia in response to hypoxia is such a widespread phenomenon in fishes, the question of its physiological value has repeatedly come up, especially when it is observed that cardiac output sometimes remains unchanged or increases. Vagotomy in the dogfish (*Scyliorhynus canicula*) eliminated the response but did not affect oxygen transfer under hypoxic conditions (Short et al., 1979), so the bradycardia was no direct benefit in this species. With increases in mean systemic blood pressures and enhanced stroke volume due to Starling's Law of the heart, the pulse pressure will increase. Butler and Metcalf (1983) suggest this change in pulse pressure may aid lamellar recruitment in the gills of some species. Finally, many fish species do not have a separate coronary circulation, therefore, the heart muscle must obtain oxygen from the venous blood within the heart chambers (Davie and Farrell, 1991). Slowing the heartbeat during hypoxia would permit more time for this diffusion of oxygen to take place; so the bradycardia response might be a mechanism to maintain heart function rather than to facilitate oxygen uptake from the water.

During hypoxia, a rise in blood pressure is often seen even with a slowing of the heart. This is caused by peripheral vasoconstriction mediated possibly by a mixture of neural stimulation and circulating catecholamines (Kinkead et al., 1991; Satchell, 1991). As is discussed below, this peripheral vasoconstriction can influence kidney function in the hypoxic fish.

IX. RESPIRATORY REGULATION AND CONFORMITY

The main reason for adjusting the various processes involved in taking up oxygen from the water is to increase respiratory independence (or regulation). When the resting oxygen consumption of a fish species that is an oxyregulator is measured at several different environmental Po_2s, there is usually observed a range of oxygen levels where the consumption of oxygen is unaffected, or it may go up somewhat due to the increased energy demand of hyperventilation and overall restlessness (e.g., Petersen and Petersen, 1990). At Po_2s below some value referred to as the critical oxygen tension (PC or TC), the animal is unable to maintain normal aerobic respiration and the oxygen consumption begins to decline. The position of the PC is a function of ventilatory and circulatory changes and on the oxygen affinity of the hemoglobin. It will be emphasized below that it is also very much a function of the amount of physical activity the fish engages in during hypoxic exposure and the amount of time given for adjusting to the lowered oxygen content.

It has been taken for granted for some time that the PC is lower for fish species that are better adapted for hypoxia (Beamish, 1964; Jones et al., 1970). Thus, Beamish (1964) reported a PC of 90 mmHg for trout and 60 mmHg for carp. However, Ott et al. (1980) reexamined the question by measuring oxygen consumption at a given Po_2 only after the fish had become "adjusted" to the new Po_2. Under these conditions, the trout and carp did not differ and both exhibited a seemingly low PC of 20 mmHg. This suggests that when fish are exposed to hypoxia rather slowly, it may allow enough time for changes in the efficiency of oxygen extraction from the water to take place.

Within a species, the PC depends on whether one is measuring the resting or active consumption of oxygen. When fish are forced to swim at a maximum cruising speed, oxygen consumption may increase 8- to 10-fold or more (Brett and Groves, 1979), and the PC moves to higher environmental levels of oxygen. This may actually exceed the point of saturation of oxygen in the water so, in essence, the fish becomes a conformer (Beamish, 1964; Kaufmann and Wieser, 1992).

Several workers (reviewed by Ott et al., 1980) have reported a rise in the PC with an increase in temperature. Carefully conducted studies are not in agreement as to the effect of temperature on PC. Ott et al. (1980) found little effect on the rainbow trout and carp, whereas Cech et al. (1979a,b) report a raising of the PC with increased temperature in largemouth bass and Sacramento blackfish. Ott et al. (1980) attribute their seemingly contradictory findings to the possibility that others were actually measuring a level of metabolism well above the "standard" rate, perhaps more of a "routine" rate of metabolism. Assuming that explanation to be valid, the argument can be made that "routine" rates of metabolism probably come closer to representing that actually seen in nature. A further factor that may be important is that some species exhibit a lowering of the PC in the summer.

Rombough (1988b) determined the PC for steelhead (*Salmo gairdneri*) embryos and larvae and found that it changed dramatically with stage of development (Figure 6). The early embryo had a very low PC but this rose to a maximum at hatching apparently due to increasing metabolic demand of the growing embryo. The resistance to oxygen flow then drops abruptly at hatching as the animal escapes the confines of the egg capsule. The PC continued to decline during subsequent development of the larvae even though growth continued. This improvement in ability to remove dissolved oxygen is related to the expansion of gill area (Rombough, 1988a). Note also in Figure 6 that the PC at a given stage of development increased with temperature. Other than Rombough's study, there

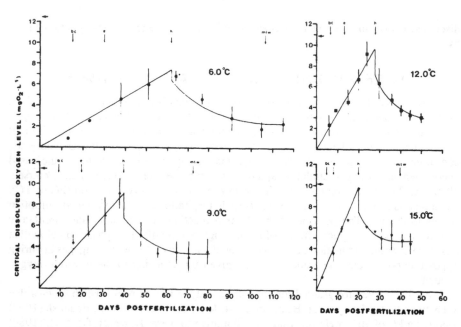

Figure 6 Critical dissolved oxygen levels (P_c) for steelhead embryos and alevins incubated at four temperatures; h = hatch; mtw = maximum tissue weight. Error bars give 95% confidence limits for P_c. Horizontal arrows indicate oxygen concentration at 100% saturated. (From Rombough, P. J., *Can. J. Zool.*, 66, 651, 1988b. With permission.)

has been very little work on the ontogenetic changes in respiratory capability in fish. Given that there are so many species with quite diverse developmental patterns (Balon, 1984) and ecologies, this appears to be a fruitful area for further investigation.

The opposite of oxyregulation is conformity. A metabolic conformer would be an animal in which resting oxygen consumption was directly proportional to Po_2. There have been occasional examples of this observed, starting with an early (and often cited) report (Hall, 1929) that the toadfish (*O. tau*) is a conformer. However, Ultsch et al. (1981) reexamined this question and found that the toadfish is actually a rather good regulator (i.e., has a wide range of respiratory independence) providing it is given time to adjust to the lowered oxygen levels. The channel catfish (*I. punctatus*) has been reported to be a regulator (Burggren and Cameron, 1980) and a conformer (Gerald and Cech, 1970). Marvin and Heath (1968) observed respiratory conformity in a different species of catfish (*I. nebulosis*) that tolerates hypoxia much better than does the channel catfish, and Hughes et al. (1983) found conformity in the carp at 10°C, but it showed fairly good respiratory regulation at 20°C.

Several marine and estuarine species have been claimed to exhibit respiratory conformity (Courtois, 1976; Subrahmanyam, 1980; Wu and Woou, 1984). However, in all of these studies, the Po_2 was lowered rapidly over a few minutes, or hours at the most. In a series of experiments where it was lowered more slowly on the lingcod (*Ophiodon elongatus*), Farrell and Daxboeck (1981) demonstrated that this species can exhibit a wide range of respiratory independence (i.e., it is a regulator), and Heath (1964) found "good" regulation in the marine grouper (*Epinephelus*).

When all is said and done, the extent of conformity or regulation seen with a given species undoubtedly depends on a combination of factors including the species, temperature, duration of hypoxic exposure, physical activity of the fish, and on how confined the fish are in the respirometer chambers when the measurements are made. Ham (1993) has demonstrated how stepwise discriminant analysis can be used to sort out the influence of things like feeding level, body mass, temperature, and pH on the relationship between oxygen consumption and oxygen availability. While it might seem logical to carry out a comparative study using several species in exactly the same experimental design, "appropriate" temperatures and respirometer design would have to be different for each species, which would make comparisons challenging. Perhaps more attention should be paid to the actual ecology of the animals so laboratory simulations can have more environmental realism.

The foregoing account of the debate on respiratory regulation and conformity may sound like a trivial academic exercise. It actually has important physiological and ecological implications for fish in water with low dissolved oxygen. Fish are, of course, basically aerobic organisms and the essence of much of this discussion is that they do all they can to avoid hypoxic conditions in their tissues. Some do this better than others but all may, on occasion, encounter conditions where the transport of oxygen to the tissues does not meet their requirements. Assuming the PC is based on a truly resting fish, at any environmental Po_2 below that, energy demand must be partially met by anaerobic metabolism, or the energy demand must be reduced.

Metabolic depression would be a good strategy if the animal has the opportunity to become quiescent, as during low-temperature dormancy (Crawshaw, 1984) and is "practiced" even at higher temperatures by at least some of those species that are well adapted for living in conditions of low or zero oxygen, such as the goldfish and crucian carp (Van Waversveld et al., 1989; Nilsson et al.,1993). Recent findings on crucian carp (Nilsson, 1990) suggest elevations during anoxia in the brain of inhibitory neurotransmitters along with a concomitant decline in excitatory amino acids. These changes provide a mechanism for reducing muscular activity.

X. ANAEROBIC METABOLISM

Adenosine triphosphate (ATP) is the energy currency in cells and this can be produced by either aerobic or anaerobic metabolism, or a combination of the two. Aerobic metabolism, which is measured as oxygen consumption, is about 18 times as efficient as anaerobic metabolism in generating ATPs. This means that fish subjected to a period of severe hypoxia often show a rapid depletion of glycogen (Heath and Pritchard, 1965; Jogensen and Mustafa, 1980) a major, but not the only, (van Waversveld et al., 1989) anaerobic fuel. Concurrent with the reduction in glycogen there occurs a rise in tissue lactic acid, a major endproduct of this type of metabolism.

Measures of the concentration of lactic acid in tissues or blood gives a rough indication of the extent of anaerobiosis that is occurring. When fish are exposed to a gradually lowered oxygen tension (over a period of hours) a threshold is encountered where a rise in lactic acid begins to take place (Burton and Heath, 1980; Boutilier et al., 1988) This threshold oxygen tension correlates reasonably well with the relative oxygen affinity of the blood for each of the species in question. The threshold may or may not correspond to the PC, because as was discussed in the preceding section, different authors have reported markedly different PC values for the same species. As was found with attempts to define a PC for various fish species, the anaerobic threshold is greatly influenced by the rapidity with which the hypoxic condition is imposed (Heath et al., 1980). By giving the fish a longer time to adjust to a given oxygen level, the threshold is effectively moved to a lower Po_2. Most of the more recent studies have used exposure periods at a given Po_2 of 8–24 h which is much longer than the duration of exposures used by earlier investigators.

The environmental threshold for anaerobic metabolism is clearly temperature dependent (Burton and Heath, 1980) in that it moves to a lower Po_2 at colder temperatures of acclimation. This is probably a reflection of a lesser energy demand as well as an effect of temperature on the oxygen affinity of the blood. Although the anaerobic threshold is shifted downward by a lower temperature, Van den Thillart et al. (1983) have shown that the ability to generate ATP anaerobically in goldfish muscles is much less at 5°C than at either 10 or 20°C. This change in anaerobic capacity with temperature was estimated, in part, from the median survival times of the fish in anoxia which were 45, 65, and 22 h, respectively. Thus, 10°C would appear to be a sort of optimum temperature for survival of this species in anoxic conditions.

The energy status of a tissue can be assessed by measuring the concentration of the adenylates (ATP, ADP, and AMP) and then using these values to calculate a dimensionless number called the energy charge, which ranges from zero to one (Atkinson, 1977). Skeletal muscle (Boutilier et al., 1988) and heart muscle (Koke and Anderson, 1986) from resting fish show little change in energy charge when the animal is exposed to hypoxia, but liver from the same animal exhibits large decreases, almost entirely due to a precipitous drop in concentration of ATP (Van Waarde et al., 1983; Vetter and Hodson, 1982). This difference between muscle and liver is probably due to a greater capacity for anaerobic metabolism in muscle, which usually "uses" this ability to obtain energy during bursts of swimming, rather than hypoxia. Teleost heart muscle evidently also has a high anaerobic capacity (Koke and Anderson, 1986).

Brain tissue of vertebrate animals is traditionally considered to be very sensitive to a lack of oxygen. In mammals, the enzymes for anaerobic glycolysis are present, but the levels of glycogen are quite low and are quickly depleted during oxygen deprivation (Dunn and Bondy, 1974). However, the bullhead catfish, which is quite resistant to anoxia, has four times the glycogen levels of the rainbow trout, which is sensitive to anoxia. Thus, the catfish is endowed with considerably more fuel in the brain to support anaerobiosis there (DiAngelo and Heath, 1987). During anoxia, the brain glycogen becomes depleted in both species of fish but only the catfish exhibits much of an increase of lactic acid in its brain. This suggests that the catfish brain tissue has a greater anaerobic capacity than does the trout. Surprisingly, the trout showed almost no decrease in brain ATP during anoxia, whereas the catfish exhibited a significant loss of this adenylate even though it survives anoxia five times longer than the trout (DiAngelo and Heath, 1987). Evidently, the trout dies even while its brain energy status is near normal, so the cause of anoxic death (i.e., they stop breathing) in this species remains unknown.

In a follow-up study (Heath, 1988), in vitro energy metabolism of brain and liver tissue from the same two fish species was investigated to examine the hypothesis that the Pasteur effect would be less in the species more tolerant of anoxia (Hochachka and Guppy, 1987). The Pasteur effect refers to an enhanced rate of glycolysis when oxygen is in inadequate supply either due to exercise or lack of environmental oxygen. A reduced Pasteur effect would imply metabolic depression in that tissue and is presumably adaptive for anoxic conditions (Hochachka and Guppy, 1987). However, anoxic catfish brain tissue exhibited a slightly greater rate of anaerobiosis than did that from trout although the Pasteur effect was eliminated at cold temperatures in both species. It was also absent in liver tissue from both species at all temperatures. Based on enzyme activity measurements in goldfish brain, this species may also exhibit a large Pasteur effect (Storey, 1987). It therefore appears that whole-animal metabolic depression can occur during hypoxia or anoxia, but brain tissue of fish may not necessarily experience such a depression except in cold temperatures.

Typical anaerobic metabolism (glycolysis) which yields lactic acid as an endproduct is obviously too inefficient for long-term use. Yet there are invertebrates that may live for many days buried in anoxic mud. Biochemical studies have shown they use modified metabolic pathways in their cells which produce succinate, alanine, propionate, and some other endproducts with little or no accumulation of lactic acid (Hochachka, 1980). More importantly, these alternate metabolic pathways produce two to four times as many ATPs per gram of fuel over the traditional glycolysis.

The finding of more efficient metabolic pathways for the generation of ATP in some invertebrates stimulated a similar search for them in fish. The results were exciting but not in quite the expected way. While some species (e.g., flounder and crucian carp) accumulate succinate and/or alanine in selected tissues under severe hypoxia (Jorgensen and Mustafa, 1980; Johnston, 1975), for the most part, fish appear to do things differently than the invertebrates. The goldfish (*Carassius auratus*) and crucian carp (*C. carassius*) have received by far the most attention, primarily because they are very good at tolerating anoxia. Goldfish exposed to anoxia initially accumulate lactic acid and experience a severe reduction in muscle glycogen, a classic Pasteur effect, but this does not persist under continued anoxia. Instead, a very large amount of metabolic CO_2 is produced. Indeed, CO_2 production is greater in anoxic goldfish than in controls breathing well-aerated water (Van den Thiellart and Van Waarde, 1985).

In animal cells in which metabolic pathways are well known, carbon dioxide is a byproduct of the Kreb's cycle which requires oxygen as an ultimate hydrogen acceptor. Therefore, in these fish there must be another hydrogen acceptor which can be used in the absence of oxygen. Mourik et al. (1982) present evidence that this acceptor is acetaldehyde which the anaerobic mitochondria produces from pyruvate. Part of the CO_2 production is coupled with the formation of ethanol in the same manner as done by yeast (Shoubridge and Hochachka, 1980). However, much more ethanol is produced than can be accounted for by breakdown of glycogen alone, and a variety of pieces of biochemical evidence indicate that amino acids are serving as substrates for ethanol and CO_2 production during anoxia in this species (Van Waarde, 1988). Ammonia is also produced in the anoxic goldfish muscle by the deamination of adenosine monophosphate. A further peculiar aspect of anaerobic metabolism in goldfish is that found by Shoubridge (1980) in which the heart and brain produce lactic acid in the process of obtaining energy while the muscle actually uses lactic acid for energy. In summary, the goldfish tissues (particularly muscle) have the ability to utilize under anoxic conditions both traditional glycolysis and Kreb's cycle but with the production of ethanol and CO_2 as endproducts, instead of lactate or in addition to ethanol.

The ethanol metabolic pathway has also been confirmed in crucian carp, but was not found in common carp (*C. carpio*) or a wide variety of Northern European and tropical teleosts (Wissing and Zebe, 1988). Wissing and Zebe did, however, discover the ethanol pathway in the bitterling (*Rhodeus amarus*), a European cyprinid. The bitterling, however, survives anoxia no better than does the common carp. Thus, the presence of the ethanol pathway does not automatically permit long anoxia survival.

The ethanol pathway does not actually yield a more efficient energy metabolism. Rather, based on *in vivo* ^{31}P-NMR measurements in crucian compared with common carp, its main advantage appears to be that it retards acid buildup in both muscle cells and blood during anoxia (Van den Thillart and Van Waarde, 1991).

XI. SWIMMING SPEED

It was pointed out earlier in this chapter that hypoxia (if severe enough) reduces the active oxygen uptake rate of fish, so an effect of low oxygen on sustained swimming would also be predicted. In a relatively early study it was found that reducing the concentration of oxygen to about 3 ppm did not prevent salmon and largemouth bass from swimming for several hours at submaximal speeds (Dahlberg et al., 1968). However, maximum sustained swimming speed is another matter. This is measured by forcing the fish to swim in a water tunnel where the speed of the current can be increased incrementally. The fish is allowed to swim at a particular speed for a fixed interval (frequently 10 or 30 min) then the speed is increased to the next step. The maximum sustained swimming speed is determined as that speed where the fish fails to maintain position in the current.

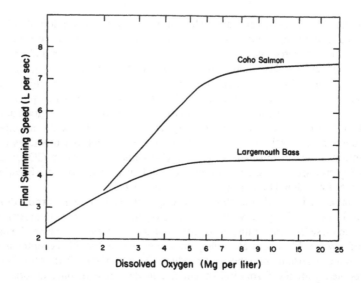

Figure 7 The effect of dissolved oxygen concentration on final (maximum) swimming speed of coho salmon (*O. kisutch*) at 20°C and largemouth bass (*Micropterus salmoides*) at 25°C. Velocity increments were at 10-min intervals. (Redrawn from Dahlberg, M. L. et al., *J. Fish. Res. Bd. Can.*, 25, 49, 1968.)

Figure 7 shows data from coho salmon and largemouth bass, the latter a species more resistant to hypoxia than the salmon. The shape of the curves are similar to those obtained from oxygen consumption studies of resting fish in that there is a threshold concentration of oxygen below which the maximum swimming speed becomes depressed. However, the point of inflection in the swimming speed curve seems low compared to that obtained from active oxygen consumption data which tends to approach or exceed saturation for oxygen (Beamish, 1964). In other words, the maximum swimming speed may be less sensitive than active oxygen consumption to the effects of hypoxia. If so, the utilization of some anaerobic metabolism to supplement the aerobic metabolism may be a possible mechanism here.

The contribution of anaerobic metabolism toward the energy requirement for sustained swimming in oxygen-saturated water is somewhat unclear and undoubtedly varies between species (Jones and Randall, 1978). When the oxygen in the water is limiting (i.e., at levels below the PC for active oxygen uptake), then the fish may compensate for a reduction in aerobic metabolism by increased anaerobiosis. This could tend to push the inflection point for maximum swimming speed to a lower concentration of oxygen than would be reflected in the curves for active oxygen uptake. Still because of the low efficiency of anaerobiosis, such a finding would only be evident where the time intervals for swimming were relatively short, otherwise, the fuel reserves would be rapidly depleted.

The actual reason for a reduced sustained maximum swimming speed under low oxygen tensions is unknown. Simple fatigue, whereby muscle glycogen becomes depleted and alterations in cellular pH occur, generally results in a cessation of swimming (Kutty, 1968), whereas Smit (1971) and Fry (1971) have hypothesized a depressing effect of low oxygen on the nervous system as a reason for the slowed swimming. Because nervous tissue is classically considered to be the tissue most sensitive to oxygen lack, such a hypothesis may have merit.

Recent studies on cyprinid larvae (Kaufmann and Wieser, 1992) reveal an interesting phenomenon; as expected, critical swimming speed was reduced under hypoxia but at the same time the net metabolic cost of swimming was about 30% lower. Evidently, when the ambient oxygen is low, maintenance functions not related to swimming are greatly reduced which gives the effect of lowering the cost of swimming. The logical follow-up question is: To what extent does this apply to adult fish?

Because reductions in dissolved oxygen can reduce maximum swimming ability, supersaturation of oxygen might be expected to enhance it. However, several studies on adult fish (reviewed in Jones and Randall, 1978) and cyprinid larvae (Kaufmann and Wieser, 1992) have not supported this hypothesis, nor does supersaturation affect growth or feed conversion in rainbow or cutthroat trout (Edsall and Smith, 1990).

XII. BEHAVIOR

Measurements on the effects of environmental hypoxia on behavior of fish fall into five groups: (1) changes in spontaneous muscular activity, (2) increased use of aquatic surface respiration, (3) increased use of air breathing (in those capable of doing it), (4) avoidance of areas of low dissolved oxygen, and (5) selection of cooler temperatures. Overall, there has not been a great deal of systematic study of behavioral effects.

It was mentioned above that some fish will decrease muscular activity when in water low in oxygen as a way to decrease the oxygen demand, although the increased demand from ventilation may cause the oxygen consumption to actually go up slightly. Fish almost always feed less when under hypoxia (Kramer, 1987) and this may partially account for a decreased level of activity as well as a decline in growth rate.

Some fish may show increased activity in hypoxia which could be a non-specific "fear" response to a changed environment (Schreck, 1981). They may also be attempting to find water with higher levels of oxygen (Kramer, 1987; Petersen and Petersen, 1990). If guppies are given access to the surface, they increase activity under hypoxia, but activity is decreased if access to the surface is prevented (Weber and Kramer, 1983). A rather specific reason for increased activity occurs with three-spined sticklebacks during reproduction. The males guard the nest and ventilate the eggs by fanning water over them. This fanning behavior increases during periods of low oxygen (Reebs et al., 1984), which indicates a well-developed sense of the level of dissolved oxygen.

Aquatic surface respiration takes advantage of the fact that the upper few millimeters of water often have much higher levels of oxygen than the deeper layers due to the fact that oxygen diffuses through water very slowly. It can permit survival in waters that otherwise would be lethal (Kramer, 1987). It is, however, largely limited to rather small species and thus makes them more vulnerable to predation both from birds above and larger fish and reptiles from below.

Air breathing is found in what are referred to as bimodal fish (i.e., they use both gills and some air-breathing organ). Most of these are in tropical freshwater environments where hypoxia or anoxia is common. The morphological arrangements which have been evolved to permit air breathing in fish are quite diverse. They include modifications of a variety of organs including the pharynx, swim bladder, intestine, and skin (Johansen, 1970). Most bimodal fish use water respiration if the levels of dissolved oxygen are high, presumably because it is less energetically expensive for them than is air breathing, but as dissolved oxygen gets lower, the increased costs of water ventilation makes air breathing more efficient.

Because dissolved oxygen is so variable in aquatic habitats and spatial heterogeneity is considerable, it might be expected that fish would be able to detect and avoid areas of low dissolved oxygen. Earlier studies (e.g., Hoglund, 1961) suggested that the Po_2 had to be so low as to cause respiratory distress before this occurred, however, later studies showed that many fish can clearly detect and avoid low oxygen (reviewed in Kramer, 1987). This ability has been shown both in field studies (Suthers and Gee, 1986; Phil et al., 1991) and the laboratory (Kramer, 1987; Spoor, 1990). Spoor (1990) observed that brook trout (*Salvelinus*) tested in a gradient avoided oxygen concentrations below 4 mg/L and preferred 5 mg/L or higher. Interestingly, Hallock et al. (1970) present circumstantial evidence that migrating salmon would not move through an area until the dissolved oxygen rose above 4.5–5.0 mg/L. This apparent threshold of around 4–5 mg/L for avoidance of hypoxic waters by salmonids deserves further study.

The avoidance of areas of low Po_2 may be beneficial under some circumstances, but Phil et al. (1991) point out that dimersal fish leave the deeper areas on a seasonal basis which correlates with levels of dissolved oxygen. As increased eutrophication occurs due to organic pollution, this may cause the duration of hypoxia in these areas to be extended. Because these same areas are important nursery grounds for some species, it could have a considerable negative impact on the fisheries.

One final behavioral aspect that has only recently received any attention is the effect of hypoxia on selected temperature. Nearly all fish have preferred temperatures that vary with the species and thermal acclimation history (Reynolds, 1977). A wide variety of animals, both invertebrate and vertebrate (even protozoa!) behaviorally select cooler temperatures when subjected to low oxygen availability (Wood, 1991). This includes fish although only a very few species have been investigated. For example, the goldfish begins to choose a cooler temperature at a Po_2 threshold of approximately 35 torr and as the Po_2 is lowered still further, the selected temperature can drop as much as 10°C (Wood, 1991). This should be beneficial as it would reduce the oxygen demand due to the well-known Q10 effect on metabolism. Rausch and Crawshaw (1990) found that goldfish exposed to anoxia initially chose a cooler temperature, but then moved to a temperature of around 19°C. This is probably related to the effect of temperature on the anaerobic capacity of this species (Van den Thillart et al., 1983). This capacity drops rapidly at temperatures below 10°C.

It should be recalled that the goldfish is rather unique in that it produces ethanol during anaerobic respiration. It is therefore interesting to note that this species responds to the presence of ethanol in the water with a reduction in selected temperature (O'Connor et al., 1988). The authors are apparently unaware of the unique metabolic capability of this species, but it is tempting to hypothesize that the goldfish is responding to ethanol as if it were hypoxia.

XIII. BLOOD AND URINE

Increases in tissue metabolites such as lactic acid will be reflected in increased levels of these in the blood. Severe hypoxia in trout can result in increases of 18-fold or more in the level of blood lactic acid (Heath and Pritchard, 1965; Dunn and Hochachka, 1986). While this is only half as high as that reached in trout exercised to exhaustion (Black et al., 1962), it still raises the possibility of acid-base alterations. In studies utilizing an extracorporeal circulation which permitted continuous measurements of blood gases and pH in trout, Thomas and associates (1986, 1988) found that exposure to severe hypoxia (Po_2 = 40 torr) produces an initial blood alkalosis presumably due to hyperventilation. Within 5 min this is followed by an acidosis which initially precedes the rise in blood lactic acid. Further blood acidosis is clearly caused by diffusion of lactic acid from white muscle and adrenergic stimulation of H^+/Na^+ exchange in erythrocytes (Fievet et al., 1987). The latter mechanism serves to remove hydrogen ions from erythrocytes and thereby maintain the intracellular pH in these cells. This helps prevent any compromise in oxygen transport by the hemoglobin, as was discussed earlier. There is also a marked rise in plasma sodium from the tissues which serves to balance the lactate ions. Most of the hydrogen ions produced by the dissociation of lactic acid are buffered intracellularly (Thomas et al., 1986).

It has been known for a long time that fish require several hours for blood lactic acid levels to return to normal following exercise (Black et al., 1962) or hypoxia (Heath and Pritchard, 1965). However, the acid-base status of the blood following hypoxia returns to control levels much more rapidly. This is seemingly achieved by redistribution of ions between extracellular and intracellular compartments rather than by branchial or renal mechanisms (Thomas et al., 1986). It should also be mentioned here that trout that are exposed to deep hypoxia for even longer times (i.e., several hours) experience a compensation that restores plasma pH within 2 h after the initial metabolic acidosis (Thomas and Hughes, 1982), even while the animal is continuing to experience the hypoxia.

The lactic acid that appears in the blood is metabolized by some tissues for energy, but this capacity varies with different fish species and tissue types (e.g., Shoubridge, 1980). Apparently, very little of the lactate is converted to liver glycogen (Dando, 1969) as is common in mammals. Some lactic acid is excreted via the urine but this seems to not account for much of the loss from the blood (Hunn, 1969; Thomas et al., 1986).

Blood glucose may or may not be elevated during hypoxia (Heath and Pritchard, 1965; Swift, 1981; Dunn and Hochachka, 1986). To a large extent, elevations in blood glucose may reflect a generalized stress response to a variety of environmental conditions and is brought about by increased levels of adrenalin and cortisol (Thomas, 1990). Thus, the rate of change in environmental oxygen could have a considerable effect on any changes in blood glucose; a rapid hypoxic stress would be more likely to induce hyperglycemia because it is "perceived" by the fish as a sudden change in its environment (Schreck, 1981).

In trout, the urine flow rate increases immediately upon exposure to hypoxia. This initial increase seems to be a non-specific, stress-induced diuresis which may cause a hemoconcentration of the blood (Swift and Lloyd, 1974). If the hypoxia is prolonged (e.g., 24 h) there appears to be an increase in fish permeability to water. Because hypoxia can cause physical damage to the gill tissue (see Section XIV), the increased permeability may be a result of that damage. The increased urine flow is accompanied by an elevation of electrolytes in the urine (Hunn, 1969), but this appears to have little effect on the regulation of blood electrolytes (Thomas et al., 1986).

Urine flow rate in carp changes in a very different manner than in trout. Kakuta and Murachi (1992) exposed carp to a gradually induced hypoxia over 6 h and noted a decline in glomerular filtrate and urine volume. The urine decline paralleled the DO level in the water very closely. During the hypoxia exposure, concentrations of various ions rose in the urine; even urinary lactic acid rose threefold. During hypoxia in the carp, plasma catecholamines rose several orders of magnitude. The authors point out that these hormonal increases probably caused peripheral vasoconstriction and reduced renal blood flow, and thus a reduced glomerular filtration rate.

The final aspects of blood chemistry to be considered are the hemoglobin concentration, hematocrit, and blood cell count. While these are obviously related to each other, it is possible to see larger changes in one than the other (e.g., the animal may produce a large number of cells lacking in hemoglobin). Increases in one or more of these factors have been reported many times in hypoxic fish, although not all species exhibit this (Weber and Jensen, 1988). The cause of the increase during acute hypoxic exposure is water loss from the blood and/or release of cells from storage points in the liver and spleen (Swift, 1981; Wells and Weber, 1990). Splenic contraction occurs in trout in response to acute hypoxia but not chronic exposure (Wells and Weber, 1990). The effect of these responses is to increase the oxygen-carrying capacity of the blood, but it does not change the affinity for oxygen. Under acclimation to hypoxia, changes in both capacity and affinity occur (see below).

XIV. HISTOPATHOLOGY

Histopathological lesions in various tissues have frequently been reported following a chemical exposure, but fish that have experienced hypoxia alone have not generally been examined for these. When channel catfish were exposed to 1.5 ppm dissolved oxygen for up to 72 h necrosis, hemorrhage, hyperplasia, hypertrophy, and hyperanemia were observed in gills, liver, kidney and spleen (Scott and Rogers, 1980). Somewhat less severe damage was reported by Drewett and Abel (1983) in brown trout sampled after death in acutely hypoxic water (0.3–1.5 ppm). Among other things, they noted numerous small breaks in the gill epithelium which were evident only under electron microscopic observation.

The significance of the histological effects produced by hypoxia remains to be elucidated but does raise the question as to what degree are the changes in various physiological factors (e.g., urine flow rate) due to these lesions. Also, hypoxia may produce effects on gill histopathology which are synergistic with chemical exposure (see Chapter 3).

XV. ACCLIMATION TO HYPOXIA

It is well known that acclimation to high altitudes enhances the ability of terrestrial animals, including humans, to function under those conditions of low oxygen (West and Lahiri, 1984). Acclimation to hypoxia would seem to be a very important ability for those fish that experience chronic conditions of low oxygen.

Shepard (1955) elegantly showed that acclimation of young speckled trout to low dissolved oxygen made them able to tolerate a lower level of oxygen in their environment. This was indirectly shown to be due to an enhanced ability to extract oxygen from the water. The acclimation process required several days, or if it was done gradually, 20–33 h were required to acclimate to a change of 1.0 ppm oxygen. He further found that acclimated trout were able to resist a lethal level of hypoxia longer than controls, which suggests the anaerobic capacity of the acclimated fish was also increased. This latter compensation may depend on the level of hypoxia to which they were acclimated as more severe levels appear to weaken trout (Smith and Heath, 1980).

Studies of the physiological mechanisms of acclimation to hypoxia at the "whole animal" level have largely been devoted to the eel, carp, goldfish, killifish, and flounders (Greaney et al, 1980; Lomholt and Johansen, 1979; Kerstens et al., 1979; Wood and Johansen, 1973; Wood et al., 1975; Jensen and Weber, 1985). A common observation is that the acclimated fish have a lower resting oxygen uptake rate and the critical oxygen tension (PC) is shifted to a lower Po_2. The lower oxygen uptake of the resting fish after acclimation seems to reflect less spontaneous activity in these animals.

Extraction efficiency (percent of oxygen removed from water as it passes over the gills) is generally enhanced with acclimation to low oxygen. This is achieved through an increase in the effective gill surface and an increased affinity and capacity of the blood for oxygen. The changes in the blood are brought about through increases in hematocrit (presumably from more red blood cell production) and lowered erythrocyte nucleoside triphosphate (NTP) levels (Weber and Lykkeboe, 1978; Wood et al., 1975; Smit and Hattingh, 1981). The latter causes the increased affinity for oxygen and requires a variable period of time (hours to days, depending on species) to come into play. The NTP in trout erythrocytes is ATP whereas the main modulator in fish species more tolerant of hypoxia is GTP, which has a greater effect on affinity than does ATP (Weber and Jensen, 1988). The increased ability of the blood to carry oxygen results in a large decrease in cardiac output in acclimated eels (Wood and Johansen, 1973).

Although acclimation to hypoxia has been shown to enhance the ability of fish to extract oxygen from hypoxic water, two studies indicate this provides no improvement in swimming performance of either goldfish or rainbow trout in hypoxic waters (Kutty, 1968; Bushnell et al., 1984). Johnston and Bernard (1982) found that acclimation of tench to hypoxia caused a reduction in the number of capillaries per fiber for both slow and fast muscles, thus the lack of an effect on aerobic swimming could rest with a reduced diffusion surface area in the muscle. Johnston and Bernard associated this loss of capillaries with the decreased resting oxygen consumption rate of fish acclimated to hypoxia, which raises the question of cause and effect. Does the decreased oxygen availability cause the loss of capillaries which then reduces oxygen uptake rate, or does the lowered oxygen uptake from reduced spontaneous activity cause the reduction in capillaries?

It has long been known that fish growth is quite dependent on adequate levels of oxygen (Weatherley and Gill, 1987). In part, this is due to a shift in metabolism away from anabolism which results in declines in free amino acids and protein in plasma and free histidine in muscle (Medale et al., 1987).

Two relevant studies show there are changes in certain cellular enzyme activities during acclimation to hypoxia. Greaney et al. (1980) found in *Fundulus* that acclimation to hypoxia enhanced the glycolytic and biosynthetic capacity of the liver (but there were no changes in white muscle). Most of these changes returned to control levels sometime between 28 and 35 days of continued low oxygen exposure. Smith and Heath (1980) acclimated trout and carp to hypoxia for 7–10 days and found increased succinate anaerobiosis in carp red muscle but not in the trout. This alternate form of anaerobic metabolism which is found among facultative ancrobic invertebrates (see above) is more efficient than the usual glycolysis so it is clearly adaptive.

The two studies above suggest that acclimation to low oxygen improves the anaerobic metabolism abilities of some species of fish. It will be interesting to learn how widespread among fish species this is and how important the dynamics are (see Greaney et al., 1980). For example, a chronic exposure to a mild hypoxia might be expected to affect aerobic processes such as respiratory gas exchange mechanisms, whereas repeated brief acute exposures to severe hypoxia might induce a greater anaerobic capacity.

REFERENCES

Aota, S., Holmgren, K. D., Gallaugher, P. and Randall, D. J., A possible role for catecholamines in the ventilatory responses associated with internal acidosis or external hypoxia in rainbow trout, *Oncorhynchus mykiss, J. Exp. Biol.,* 151, 57, 1990.

Atkinson, D. E., *Cellular Energy Metabolism and its Regulation,* Academic Press, New York, 1977.

Baker, J. N., Role of hemoglobin affinity and concentration in determining hypoxia tolerance of mammals during infancy, hypoxia, hyperoxia, and irradiation, *Am. J. Physiol.,* 189, 281, 1957.

Balon, E. K., Reflections on some decisive events in the early life of fishes, *Trans. Am. Fish. Soc.,* 113, 178, 1984.

Barnes, R. S. K. and Mann, K. H., Eds., *Fundamentals of Aquatic Ecology,* Blackwell Scientific, London, 1991.

Beamish, F. W. H., Respiration of fishes with special emphasis on standard oxygen consumption. III. Influence of oxygen, *Can. J. Zool.,* 42, 355, 1964.

Black, E. C., Connor, A. R., Lam, K. and Chiu, W., Changes in glycogen, pyruvate and lactate in rainbow trout (*Salmo gairdneri*) during and following muscular activity, *J. Fish. Res. Bd. Can.,* 19, 409, 1962.

Blazka, P., The anaerobic metabolism of fish, *Physiol. Zool.,* 31, 117, 1958.

Boesch, D. F., Implications of oxygen depletion on the continental shelf of the northern Gulf of Mexico, *Coastal Ocean Pollut. Assess. News,* 2, 25, Marine Science Research Center, SUNY, Stony Brook, New York, 1983.

Booth, J. H., The distribution of blood flow in gills of fish: application of a new technique to rainbow trout (*Salmo gairdneri*), *J. Exp. Biol.,* 73, 119, 1978.

38

Borch, K., Jensen, F. B. and Andersen, B., Cardiac, ventilation rate and acid-base regulation in rainbow trout exposed to hypoxia and combined hypoxia and hypercapnia, *Fish Physiol. Biochem.*, 12, 101, 1993.

Boulekbache, H., Energy metabolism in fish development, *Am. Zool.*, 21, 377, 1981.

Boutilier, R. G., Respiratory gas tensions in the environment, in, *Advances in Comparative and Environmental Physiology*, Boutilier, R. G., Ed., Springer-Verlag, Berlin, 1990, chap. 1.

Boutilier, R. G., Dobson, G., Hoeger, U. and Randall, D. J., Acute exposure to graded levels of hypoxia in rainbow trout (*Salmo gairdneri*): metabolic and respiratory adaptations, *Respir. Physiol.*, 71, 69, 1988.

Brett, J. R. and Groves, T. D. D., Physiological energetics, in, *Fish Physiology*, Vol. VIII, Hoar, W. S. and Randall, D. J., Eds., Academic Press, New York, 1979, chap. 6.

Bridges, C. R. and Butler, P. J., Eds., *Techniques in Comparative Respiratory Physiology*, Cambridge University Press, New York, 1989.

Burggren, W. W. and Randall, D. J., Oxygen uptake and transport during hypoxic exposure in the sturgeon (*Acipenser transmontanus*), *Respir. Physiol.*, 34, 171, 1978.

Burggren, W. W. and Cameron, J. N., Anaerobic metabolism, gas exchange, and acid-base balance during hypoxic exposure in the channel catfish, *Ictalurus punctatus, J. Exp. Zool.*, 213, 405, 1980.

Burleson, M. L. and Milsom, W., Sensory receptors in the first gill arch of rainbow trout, *Respir. Physiol.*, 93, 97, 1993.

Burton, D. T. and Heath, A. G., Ambient oxygen tension (Po_2) and transition to anaerobic metabolism in three species of freshwater fish, *Can. J. Fish. Aquat. Sci.*, 37, 1216, 1980.

Bushnell, P. G., Steffensen, J. F. and Johansen, K., Oxygen consumption and swimming performance in hypoxia-acclimated rainbow trout (*Salmo gairdneri*), *J. Exp. Biol.*, 113, 225, 1984.

Bushnell, P. G., Brill, R. W. and Bourke, R. E., Cardiorespiratory responses of skipjack tuna (*Katsuwonus pelamis*), yellowfin tuna (*Thunnus albacares*), and bigeye tuna (*Thunnuus obesus*) to acute reductions of ambient oxygen, *Can. J. Zool.*, 68, 1857, 1990.

Bushnell, P. G. and Brill, R., Responses of swimming skipjack (Katsuwonus pelamis) and yellowfin (*Thunnus albacares*) tunas to acute hypoxia, and a model of their cardiorespiratory function, *Physiol. Zool.*, 64, 787, 1991.

Butler, P. J. and Metcalf, J. F., Control of respiration and circulation, in, *Control Processes in Fish Physiology*, Rankin, J. C., Pitcher, T. J. and Duggan, R. T., Eds., Wiley and Sons, New York, 1983, chap. 3.

Cameron, J. N., Oxygen dissociation characteristics of the blood of the rainbow trout (*Salmo gairdneri*), *Comp. Biochem. Physiol.*, 38A, 699, 1971.

Campagna, C. G. and Cech, J. J., Gill ventilation and respiratory efficiency of Sacramento blackfish, *Orthodon microlepidotus* (Ayres), in hypoxic environments, *J. Fish Biol.*, 19, 581, 1981.

Carlson, A. R., Effects of lowered dissolved oxygen concentration on the toxicity of 1,2,3-trichlorobenzene to fathead minnows, *Bull. Environ. Toxicol. Contam. Toxicol.*, 38, 667, 1987.

Cech, J. J., Campagna, C. and Mitchell, S. J., Respiratory responses of largemouth bass (*Micropterus salmoides*) to environmental changes in temperature and dissolved oxygen, *Trans. Am. Fish. Soc.*, 108, 166, 1979a.

Cech, J. J., Mitchell, S. J. and Massingill, M. J., Respiratory adaptations of Sacramento blackfish, *Orthodon microlepidotus* (Ayres), for hypoxia, *Comp. Biochem. Physiol.*, 63A, 411, 1979b.

Courtois, L. A., Respiratory responses of *Gillichthys mirabilis* to changes in temperature, dissolved oxygen and salinity, *Comp. Biochem. Physiol.*, 53A, 7, 1976.

Crawshaw, L. I., Low-temperature dormancy in fish, *Am. J. Physiol.*, 246, R479, 1984.

Dahlberg, M. L., Shumway, D. L. and Doudoroff, P., Influence of dissolved oxygen and carbon dioxide on swimming performance of largemouth bass and coho salmon, *J. Fish. Res. Bd. Can.*, 25, 49, 1968.

Dando, P. R., Lactate metabolism in fish, *J. Mar. Biol. Assoc. U.K.*, 49, 209, 1969.

Davie, P. S. and Farrell, A. P., The coronary and luminal circulations of the myocardium of fishes, *Can. J. Zool.*, 69, 1993, 1991.

Davis, J. C., Minimal dissolved oxygen requirements of aquatic life with emphasis on Canadian species: a review, *J. Fish. Res. Bd. Can.*, 32, 2295, 1975.

Daxboeck, C. and Holeton, G., Oxygen receptors in the rainbow trout (*Salmo gairdneri*), *Can. J. Zool.*, 56, 1254, 1978.

DiAngelo, C. R. and Heath, A. G., Comparison of *in vivo* energy metabolism in the brain of rainbow trout, *Salmo gairdneri*, and bullhead catfish, *Ictalurus nebulosus*, during anoxia, *Comp. Biochem. Physiol.* 88B, 297, 1987.

Doudoroff, P. and Shumway, D. L., Dissolved oxygen requirements of freshwater fishes, *FAO Fish. Tech. Paper*, 86, 291 p, 1970.

Douglas, E. L., Friedl, W. A. and Pickwell, G. V., Fishes in oxygen minimum zones: blood oxygenation characteristics, *Science*, 191, 957, 1976.

Drewett, N. and Abel, P. D., Pathology of lindane poisoning and of hypoxia in the brown trout, *Salmo trutta L.*, *J. Fish Biol.*, 23, 373, 1983.

Dunn, A. and Bondy, S. C., *Functional Chemistry of the Brain*, Spectrum Publications, New York, 1974.

Dunn, J. F. and Hochachka, P., Metabolic responses of trout (*Salmo gairdneri*) to acute environmental hypoxia, *J. Exp. Biol.*, 123, 229, 1986.

Edsall, D. and Smith, C., Performance of rainbow trout and Snake River cutthroat trout reared in oxygen-supersaturated water, *Aquaculture*, 90, 251, 1990.

El-Fiky, N. and Weiser, W., Life styles and patterns of development of gills and muscles in larval cyprinids (Cyprinidae; Teleostei), *J. Fish Biol.*, 33, 135, 1988.

Farrell, A. P. and Daxboeck, C., Oxygen uptake in the lingcod, *Ophiodon elongatus*, during progressive hypoxia, *Can. J. Zool.*, 59, 1272, 1981.

Farrell, A. P., Cardiovascular changes in the unanesthesized lingcod (*Ophidon elongatus*) during short-term, progressive hypoxia and spontaneous activity, *Can. J. Zool.*, 60, 933, 1982.

Fievet, B., Motais, R. and Thomas, S., Role of adrenergic-dependent H$^+$ release from red cells during acidosis induced by hypoxia in trout, *Am. J. Physiol.*, 252, R269, 1988.

Forgue, J., Burtin, B. and Massabau, J., Maintenance of oxygen consumption in resting *Silurus glanis* at different levels of ambient oxygenation, *J. Exp. Biol.*, 143, 305, 1989.

Fritsch, R., Effects of hypoxia on blood pressure and heart rate in three marine teleosts, *Fish Physiol. Biochem.*, 8, 85, 1990.

Fry, F. E. J., The effect of environmental factors on the physiology of fish, in, *Fish Physiology*, Hoar, W. S. and Randall, D. J., Eds., Vol. VI., Academic Press, New York, 1971, chap. 1.

Garey, W. F., Gas exchange, cardiac output and blood pressure in free swimming carp (*Cyprinus carpio*), Ph.D. Dissertation, State University of New York at Buffalo, New York, 1967.

Garey, W. F. and Rahn, H., Gas tensions in tissues of trout and carp exposed to diurnal changes in oxygen tension of the water, *J. Exp. Biol.*, 52, 575, 1970.

Gerald, J. W. and Cech, J. J., Respiratory responses of juvenile catfish (*Ictalurus punctatus*) to hypoxic conditions, *Physiol. Zool.*, 43, 47, 1970.

Glass, M. L., Andersen, N. A., Kruhoffer, M., Williams, E. M. and Heisler, N., Combined effects of environmental Po$_2$ and temperature on ventilation and blood gases in the carp *Cyprinus carpio*, *J. Exp. Biol.*, 148, 1, 1990.

Glass, M. L., Rantin, F. T., Verzola, M. M., Fernandes, M. N. and Kalinin, A. L., Cardiorespiratory syunchronization and myocardial function in hypoxic carp, *Cyprinus carpio L.*, *J. Fish Biol.*, 39, 143, 1991.

Greaney, G. S., Place, A. R., Cashon, R. E., Smith, G. and Powers, D. A., Time-course of changes in enzyme activities and blood respiratory properties of killifish during long-term acclimation to hypoxia, *Physiol. Zool.*, 53, 136, 1980.

Hall, F. G., The influence of varying oxygen tensions upon the rate of oxygen consumption in marine fishes, *Am. J. Physiol.*, 88, 212, 1929.

Hallock, R. J., Elwell, R. and Fry, D. H., Migrations of adult king salmon *Oncorhynchus tshawytscha* in the San Joaquin delta as demonstrated by the use of sonic tags, *Calif. Dept. Fish Game Bull.*, 151, 1, 1970.

Ham, K. D., The effect of temperature, feeding and pH on respiratory responses of bluegill to gradual hypoxic challenge. Abstract of paper at 1993 Soc. Environ. Toxicol. Chem. meeting.

Heath, A. G., Heart rate, ventilation, and oxygen uptake in a marine teleost in various oxygen tensions, *Am. Zool.*, 7, 85, 1964.

Heath, A. G., Ventilatory responses of teleost fish to exercise and thermal stress, *Am. Zool.*, 13, 491, 1973.

Heath, A. G., Anaerobic and aerobic energy metabolism in brain and liver tissue from rainbow trout (*Salmo gairdneri*) and bullhead catfish (*Ictalurus nebulosus*), *J. Exp. Zool.*, 248, 140, 1988.

40

Heath, A. G., Effect of water-borne copper on physiological responses of bluegill (*Lepomis macrochirus*) to acute hypoxic stress and subsequent recovery, *Comp. Biochem. Physiol.,* 100C, 559, 1991.

Heath, A. G. and Pritchard, A. W., Effects of severe hypoxia on carbohydrate energy stores and metabolism in two species of fresh-water fish, *Physiol. Zool.,* 38, 325, 1965.

Heath, A. G., Burton, D. T. and Smith, M. J., Anaerobic metabolism in fishes: environmental thresholds and time dependence, *Rev. Can. Biol.,* 39, 123, 1980.

Hlohowskyj, I. and Chagnon, N., Reduction in tolerance to progressive hypoxia in the central stoneroller minnow following sublethal exposure to phenol, *Water, Air, Soil Pollut.,* 60, 189, 1991.

Hochachka, P. W., *Living without Oxygen,* Harvard University Press, Cambridge, MA, 1980.

Hochachka, P. W. and Guppy, M., *Metabolic Arrest and the Control of Biological Time,* Harvard University Press, Cambridge, MA, 1987.

Hoglund, L. B., The reactions of fish in concentration gradients, Institute of Freshwater Research, Drottningholm Report No. 43, 1961.

Holeton, G. and Randall, D. J., The effect of hypoxia upon the partial pressure of gases in the blood and water afferent and efferent to the gills of rainbow trout, *J. Exp. Biol.,* 46, 317, 1967.

Holeton, G. F. and Randall, D. J., Changes in blood pressure in the rainbow trout during hypoxia, *J. Exp. Biol.,* 46, 297, 1967.

Holopainen, I. J., Hyvarinen, H. and Piironen, J., Anaerobic wintering of crucian carp (*Carrassius carassius* L.) II. Metabolic products, *Comp. Biochem. Physiol.,* 83A, 239, 1986.

Hoy, T., Horsberg, T. E. and Wichstrom, R., Inhibition of acetylcholinesterase in rainbow trout following dichlorvos treatment at different water oxygen levels, *Aquaculture,* 95, 33, 1991.

Hughes, G. M., The dimensions of fish gills in relation to their function, *J. Exp. Biol.,* 45, 177, 1966.

Hughes, G. M., Respiratory responses to hypoxia in fish, *Am. Zool.,* 13, 475, 1973.

Hughes, G. M. and Umezawa, S., On respiration in the dragonet *Callionymus lyra, J. Exp. Biol.,* 49, 565, 1968.

Hughes, G. M., Albers, C. and Gotz, K. H., Respiration of the carp, *Cyprinus carpio* L., at 10 and 20 degrees and the effects of hypoxia, *J. Fish Biol.,* 22, 613, 1983.

Hunn, J., Chemical composition of rainbow trout urine following acute hypoxia, *Trans. Am. Fish. Soc.,* 98, 20, 1969.

Itazawa, Y. and Takeda, T., Gas exchange in carp gills in normoxic and hypoxic conditions, *Respir. Physiol.,* 35, 263, 1978.

Jensen, F. B. and Weber, R. E., Kinetics of the acclimational response of tench to combined hypoxia and hypercapnia, *J. Comp. Physiol.,* 156B, 197, 1985.

Johansen, K., Air breathing in fishes, in, *Fish Physiology,* Hoar, W. and Randall, D., Eds., Vol. IV, Academic Press, New York, 1970, chap. 9.

Johnston, I. A. and Bernard, L. M., Routine oxygen consumption and characteristics of the myotomal muscle in tench: effects of long-term acclimation to hypoxia, *Cell Tissue Res.,* 227, 161, 1982.

Johnston, I. A., Anaerobic metabolism in the carp (*Carassius carassius* L.), *Comp. Biochem. Physiol.,* 51B, 235, 1975.

Jones, D. R., Randall, D. J. and Jarman, G. M., A graphical analysis of oxygen transfer in fish, *Respir. Physiol.,* 10, 285, 1970.

Jones, D. R. and Randall, D. J., The respiratory and circulatory systems during exercise, in, *Fish Physiology,* Vol. VII, Hoar, W. S. and Randall, D. J., Eds., Academic Press, New York, 1978, chap. 7.

Jorgensen, J. B. and Mustafa, T., The effect of hypoxia on carbohydrate metabolism in flounder (*Platichthys flesus* L.). I. Utilization of glycogen and accumulation of glycolytic end products in various tissues, *Comp. Biochem. Physiol.,* 67B, 243, 1980.

Kakuta, I. and Murachi, S., Renal responses to hypoxia in carp, *Cyprinus carpio:* Changes in glomerular filtration rate, urine and blood properties and plasma catecholamines of carp exposed to hypoxic conditions, *Comp Biochem. Physiol.,* 103A, 259, 1992.

Kaufman, L., Catastrophic change in species-rich freshwater ecosystems, the lessons of Lake Victoria, *Bioscience,* 42, 846, 1992.

Kaufmann, R. and Wieser, W., Influence of temperature and ambient oxygen on the swimming energetics of cyprinid larvae and juveniles, *Environ. Biol. Fish.,* 33, 87, 1992.

Kerstens, A., Lomholt, J. P. and Johansen, K., The ventilation, extraction, and uptake of oxygen in undisturbed flounders, *Platichthys flesus:* responses to hypoxia acclimation, *J. Exp. Biol.,* 83, 169, 1979.

Kinkead, R. and Perry, S., The effects of catecholamines on ventilation in rainbow trout during hypoxia or hypercapnia, *Respir. Physiol.*, 84, 77, 1991.

Kinkead, R., Fritsche, R., Perry, S. and Nillson, S., The role of circulating catecholamines in the ventilatory and hypertensive responses to hypoxia in the Atlantic cod, *Gadus morhua, Physiol. Zool.*, 64, 1087, 1991.

Koke, J. and Anderson, D., Changes in metabolite levels and morphology of teleost ventricular myocytes due to hypoxia, ischaemia, and metabolic inhibitors, *Cytobios*, 45, 97, 1986.

Kramer, D. L., Dissolved oxygen and fish behavior, *Environ. Biol. Fishes*, 18, 81, 1987.

Kutty, M. N., Influence of ambient oxygen on the swimming performance of goldfish and rainbow trout, *Can. J. Zool.*, 46, 647, 1968.

Laurent, P. and Perry, S., Environmental effects on fish gill morphology, *Physiol. Zool.*, 64, 4, 1991.

Lomholt, J. P. and Johansen, K., Hypoxia acclimation in carp—how it affects oxygen uptake, ventilation, and oxygen extraction from water, *Physiol. Zool.*, 52, 38, 1979.

Malyarevskaya, A. Y., Specific and nonspecific changes induced in fish by various toxic agents, *Hydrobiol. J.*, 15, 52, 1979.

Marvin, E. E. and Heath, A. G., Cardiac and respiratory responses to gradual hypoxia in three ecologically distinct species of freshwater fish, *Comp. Biochem. Physiol.*, 27, 349, 1968.

Mathur, G. B., Anaerobic respiration in a cyprinoid fish, *Rasbora daniconius, Nature*, 214, 318, 1967.

Mayer, F. L., Versteeg, D. J., McKee, M. J., Folmar, L. C., Graney, R. L., McCume, D. C. and Rattner, B. A., Physiological and nonspecific biomarkers, in, *Biomarkers, Biochemical Physiological, and Histological Markers of Anthropogenic Stress*, Huggett, R. J., Kimerle, R. A., Mehrle, P. M. and Bergman, H. L., Eds., Lewis Publishers, Boca Raton, FL, 1992, chap. 1.

McCloskey, J. T. and Oris, J. T., Effect of water temperature and dissolved oxygen concentration on the photo-induced toxicity of anthracene to juvenile bluegill sunfish (*Lepomis macrochirus*), *Aquat. Toxicol.*, 21, 145, 1991.

McKim, J., Early life stage toxicity tests, in *Fundamentals of Aquatic Toxicology*, Rand, G. M. and Petrocelli, S. R., Eds., Hemisphere Publishing, Washington, D.C., 1985, chap. 3.

Medale, F., Parent, J. and Vellas, F., Responses to prolonged hypoxia by rainbow trout (*Salmo gairdneri*). I. Free amino acids and proteins in plasma, liver and white muscle, *Fish Physiol. Biochem.*, 3, 183, 1987.

Mourik, J., Raeven, P., Steur, K. and Addink, A. D. F., Anaerobic metabolism of red skeletal muscle of goldfish, *Carassius auratus* L.): mitochondrial produced acetaldehyde as anaerobic electron acceptor, *FEBS Lett.*, 137, 11, 1982.

Mustafa, T. and Jensen, F., Effect of hypoxia on *in vivo* thromboxane and prostacyclin levels in arterial blood of rainbow trout, *Oncorhynchus mykiss, J. Fish Biol.*, 40, 303, 1992.

Nilsson, G., Long-term anoxia in crucian carp: changes in the levels of amino acid and monamine neurotransmitter in the brain, catecholamines in chromaffin tissue, and liver glycogen, *J. Exp. Biol.*, 150, 295, 1990.

Nilsson, G., Rosen, P. and Johansen, D., Anoxic depression of spontaneous locomotor activity in crucian carp quantified by a computerized imaging technique, *J. Exp. Biol.*, 180, 153, 1993.

Nonnotte, G., Cutaneous respiration in six freshwater teleosts, *Comp. Biochem. Physiol.*, 70A, 541, 1981.

Nonnotte, G., Cutaneous respiration in the catfish, *Ictalurus melas, Comp. Biochem. Physiol.*, 78A, 515, 1984.

Nonnotte, G., Maxime, V., Truchot, J., Williot, P. and Peyraud, C., Respiratory responses to progressive ambient hypoxia in the sturgeon, *Acipenser baeri, Respir. Physiol.*, 91, 71, 1993.

O'Connor, C. S., Crawshaw, L., Bedichek, R. and Crabbe, J., The effect of ethanol on temperature selection in the goldfish, *Carassius auratus, Pharmacol. Biochem. Behav.*, 29, 243, 1988.

Ott, M. E., Heisler, N. and Ultsch, G. R., A re-evaluation of the relationship between temperature and the critical oxygen tension in freshwater fishes, *Comp. Biochem. Physiol.*, 67A, 337, 1980.

Pennak, R. W., Field and experimental winter limnology of three Colorado mountain lakes, *Ecology*, 49, 505, 1968.

Perry, S. F. and Kinkead, R., The role of circulating catecholamines versus blood/water respiratory status in the control of breathing in teleosts, *Physiologist*, 33, A29, 1990.

Perry, S. F. and Wood, C. M., Control and coordination of gas transfer in fishes, *Can. J. Zool.*, 67, 2961, 1989.

Perry, S. F. and Thomas, S., The effects of endogenous or exogenous catecholamines on blood respiratory status during acute hypoxia in rainbow trout (*Oncorhynchus mykiss*), *J. Comp. Physiol.*, 161B, 489, 1991.

Perry, S. F., Kinkead, R. and Fritsche, R., Are circulating catecholamines involved in the control of breathing in fishes?, *Rev. Fish Biol. Fish.*, 2, 65, 1992.

Petersen, J. K. and Petersen, G. I., Tolerance, behavior and oxygen consumption in the sand goby, *Pomatoxchistus minutus* (Pallas), exposed to hypoxia, *J. Fish Biol.*, 37, 921, 1990.

Phil, L., Baden, S. and Diaz, R., Effects of periodic hypoxia on distribution of dismersal fish and crustaceans, *Marine Biol.*, 108, 349, 1991.

Poppe, W. L., Dissolved oxygen in streams and reservoirs, Research Jr., WPCF 62, 555, 1990.

Prosser, C. L., Barr, L. M., Pinc, R. D. and Lauer, C. Y., Acclimation of goldfish to low concentrations of oxygen, *Physiol. Zool.*, 30, 137, 1957.

Randall, D. J., The control of respiration and circulation in fish during exercise and hypoxia, *J. Exp. Biol.*, 100, 275, 1982.

Randall, D. J. and Perry, S. F., Catecholamines, in, *Fish Physiology*, Vol. 12B, Hoar, W. S. and Randall, D. J., Eds., Academic Press, New York, 1992, chap. 4.

Randall, D. J. and Smith, J. C., The regulation of cardiac activity in fish in a hypoxic environment, *Physiol. Zool.*, 40, 104, 1967.

Randall, D. J. and Taylor, E. W., Evidence of a role for catecholamines in the control of breathing in fish, *Rev. Fish Biol. Fish.*, 1, 139, 1991.

Rattner, B. A. and Heath, A. G., Factors affecting contaminant toxicity in aquatic and terrestrial vertebrates, in, *Handbook of Ecotoxicology*, Hoffman, D. J., Rattner, B. A., Burton, A. G. and Cairns, J., Eds., Lewis Publishers, Boca Raton, FL, 1994.

Rausch, R. and Crawshaw, L. I., Anoxia and hypoxia lower the selected temperature of goldfish, *Physiologist*, 33, A-62, 1990.

Reebs, S. G., Whoriskey, F. G. and Fitzgerald, G. J., Diel patterns of fanning activity, egg respiration, and the nocturnal behavior of male three-spined sticklebacks, *Gasterosteus aculeatus* L., *Can. J. Zool.*, 62, 329, 1984.

Reynolds, W. W., Temperature as a proximate factor in orientation behavior, *J. Fish. Res. Bd. Can.*, 34, 734, 1977.

Roberts, J., Active branchial and ram gill ventilation in fishes, *Biol. Bull.*, 148, 85, 1975.

Rombough, P. J., Respiratory gas exchange, aerobic metabolism, and effects of hypoxia during early life, in, *Fish Physiology*, Vol. 11A, Hoar, W. S. and Randall, D. J., Eds., Academic Press, New York, 1988a, chap. 2.

Rombough, P. J., Growth, aerobic metabolism, and dissolved oxygen requirements of embryos and alevins of steelhead, *Salmo gairdneri, Can. J. Zool.*, 66, 651, 1988b.

Rombough, P. J., Intravascular oxygen tensions in cutaneously respiring rainbow trout, *Oncorhynchus mykiss*, larvae, *Comp. Biochem. Physiol.*, 101A, 23, 1992.

Salama, A. and Nikinmaa, M., Effect of oxygen tension on catecholamine-induced formation of cAMP and on swelling of carp red blood cells, *Am. J. Physiol.*, 259, (*Cell Physiol.* 28, C726), 1990.

Satchell, G., *Physiology and Form of Fish Circulation*, Cambridge University Press, Cambridge, 1991.

Schreck, C. B., Stress and compensation in teleostean fishes: response to social and physical factors, in, *Stress and Fish*, Pickering, A. D., Ed., Academic Press, New York, 1981, chap. 13.

Scott, A. L. and Rogers, W. A., Histological effects of prolonged sublethal hypoxia on channel catfish *Ictalurus punctatus, J. Fish Dis.*, 3, 305, 1980.

Shelton, G., Jones, D. and Milsom, W., Control of breathing in ectothermic vertebrates, in, *Handbook of Physiology*, Sect. 3, Vol II, Part 2, Fishman, A., Cherniack, N., Widdicombe, J. and Geiger, S., Eds., American Physiology Society, Bethesda, MD, 1986, chap. 28.

Shepard, M. P., Resistance and tolerance of young speckled trout (*Salvelinus fontinalis*) to oxygen lack, with special reference to low oxygen acclimation, *J. Fish Res. Bd. Can.*, 12, 387, 1955.

Short, S., Taylor, E. W. and Butler, P. J., The effectiveness of oxygen transfer during normoxia and hypoxia in the dogfish (*Scyliorhinus canicula* L.) before and after cardiac vagotomy, *J. Comp. Physiol.*, 132, 289, 1979.

Shoubridge, E. A., The metabolic strategy of the anoxic goldfish, Ph.D. Thesis, University of British Columbia, Vancouver, 1980.

Shoubridge, E. A. and Hochachka, P. W., Ethanol: novel endproduct of vertebrate anaerobic metabolism, *Science*, 209, 308, 1980.

Skidmore, J. F., Respiration and osmoregulation in rainbow trout with gills damaged by zinc sulphate, *J. Exp. Biol.*, 52, 481, 1970.

Smit, G. L. and Hattingh, J., The effect of hypoxia on haemoglobins and ATP levels in three freshwater fish species, *Comp. Biochem. Physiol.,* 68A, 519, 1981.

Smit, H., Amelink-Koutstaal, J. M., Vijverberg, J. and von Vaupel-Klein, J. C., Oxygen consumption and efficiency of swimming goldfish, *Comp. Biochem. Physiol.,* 39A, 1, 1971.

Smith, M. J. and Heath, A. G., Responses to acute anoxia and prolonged hypoxia by rainbow trout *(Salmo gairdneri)* and mirror carp *(Cyprinus carpio)* red and white muscle: use of conventional and modified metabolic pathways, *Comp. Biochem. Physiol.,* 66B, 267, 1980.

Smith, F. M. and Jones, D. R., The effects of changes in blood oxygen carrying capacity on ventilation volume in the rainbow trout *(Salmo gairdneri), J. Exp. Biol.,* 97, 325, 1982.

Spoor, W. A., Oxygen requirements of larvae of smallmouth bass, *Micropterus dolomieui* Lacepede, *J. Fish. Biol.,* 25, 587, 1984.

Spoor, W. A., Distribution of fingerling brook trout, *Salvelinus fontinalis,* in dissolved oxygen concentration gradients, *J. Fish Biol.,* 36, 363, 1990.

Sprague, J. B., Factors that modify toxicity, in *Fundamentals of Aquatic Toxicology,* Rand, G. M. and Petrocelli, S. R., Eds., Hemisphere Publishing, Washington, D.C., 1984, chap. 6.

Steffensen, J. F., Lomholt, J. P. and Johansen, K., The relative importance of skin oxygen uptake in the naturally buried plaice, *Pleuronectes platessa,* exposed to graded hypoxia, *Respir. Physiol.,* 44, 269, 1981.

Steffensen, J. F., The transition between branchial pumping and ram ventilation in fishes: energetic consequences and dependence on water oxygen tension, *J. Exp. Biol.,* 114, 141, 1985.

Steffensen, J. F. and Lomholt, J. P., Cutaneous oxygen uptake and its relation to skin blood perfusion and ambient salinity in the plaice, *Pleuronectes platessa, Comp. Biochem. Physiol.,* 81A, 373, 1985.

Stevens, E. D. and Randall, D. J., Changes in the gas concentrations in blood and water during moderate swimming activity in rainbow trout, *J. Exp. Biol.,* 46, 329, 1967.

Storey, K. B., Tissue-specific controls on carbohydrate catabolism during anoxia in goldfish, *Physiol. Zool.,* 60, 601, 1987.

Subrahmanyam, C. B., Oxygen consumption of estuarine fish in relation to external oxygen tension, *Comp. Biochem. Physiol.,* 67A, 129, 1980.

Suthers, I. M. and Gee, J., Role of hypoxia in limiting diel spring and summer distribution of juvenile yellow perch *(Perca flavescens)* in a prairie marsh, *Can. J. Fish. Aquat. Sci.,* 43, 1562, 1986.

Swanson, R. L. and Sinderman, C. J., Eds., Oxygen Depletion and Associated Benthic Mortalities in New York Bight, 1976, *NOAA Professional Paper* 11, U.S. Department of Commerce, Rockville, MD, 1979.

Swanson, R. L. and Parker, C. A., Physical environmental factors contributing to recurring hypoxia in the New York bight, *Trans. Am. Fish. Soc.,* 117, 37, 1988.

Swift, D. J., Changes in selected blood component concentrations of rainbow trout, *(Salmo gairdneri)* Richardson, exposed to hypoxia or sublethal concentrations of phenol or ammonia, *J. Fish Biol.,* 19, 45–61, 1981.

Swift, D. J. and Lloyd, R., Changes in urine flow rate and haematocrit value of rainbow trout *(Salmo gairdneri)* Richardson, exposed to hypoxia, *J. Fish. Biol.,* 6, 379, 1974.

Takeda, T., Cutaneous and gill O_2 uptake in the carp, *Cyprinus carpio,* as a function of ambient Po_2, *Comp. Biochem. Physiol.,* 94A, 205, 1989.

Tetens, V. and Lykkeboe, G., Acute exposure of rainbow trout to mild and deep hypoxia: oxygen affinity and oxygen capacity of arterial blood, *Respir. Physiol.,* 61, 221, 1985.

Thomas, S. and Hughes, G. M., A study of the effects of hypoxia on acid-base status of rainbow trout blood using an extracorporeal blood circulation, *Respir. Physiol.,* 49, 371, 1982.

Thomas, S., Fievet, B., Claireaux, G. and Motais, R., Adaptive respiratory responses of trout to acute hypoxia. I. Effects of water ionic composition on blood acid-base status response and gill morphology, *Respir. Physiol.* 74, 77, 1988.

Thomas, S. and Perry, S. F., Control and consequences of adrenergic activation of red blood cell Na^+/H^+ exchange on blood oxygen and carbon dioxide transport in fish, *J. Exp. Zool.,* 263, 160, 1992.

Thomas, S., Fivet, B. and Motais, R., Effect of deep hypoxia on acid-base balance in trout: role of ion transfer processes, *Am. J. Physiol.,* 250, R319, 1986.

Thomas, P., Molecular and biochemical responses of fish to stressors and their potential use in environmental monitoring, *Am. Fish. Soc. Symp.,* 8, 9, 1990.

Truchot, J. P. and Duhamel-Jowe, A., Oxygen and carbon dioxide in the marine intertidal environment: diurnal and tidal changes in rockpools, *Respir. Physiol.*, 39, 241, 1980.

Ultsch, G., Boschung, H. and Ross, M., Metabolism, critical oxygen tension and habitat selection in darters (*Etheostoma*), *Ecology*, 59, 99, 1978.

Ultsch, G. R., Jackson, D. C. and Moalli, R., Metabolic oxygen conformity among lower vertebrates: the toadfish revisited, *J. Comp. Physiol.*, 142, 439, 1981.

Umezawa, S. and Hughes, G. M., Effects of water flow and hypoxia on respiration of the frogfishes, *Histrio histrio* and *Phrynelox tridens*, *Jpn. J. Ichthyol.*, 29, 421, 1983.

Van den Thillart, G., Adaptations of fish energy metabolism to hypoxia and anoxia, *Mol. Physiol.*, 2, 49, 1982.

Van den Thillart, G. and Van Waarde, A., Teleosts in hypoxia: aspects of anaerobic metabolism, *Mol. Physiol.*, 8, 393, 1985.

Van den Thillart, G. and Van Waarde, A., pH changes in fish during environmental anoxia and recovery: the advantages of the ethanol pathway, in *Physiological Strategies for Gas Exchange and Metabolism*, Woakes, A. J., Grieshaber, M. K. and Bridges, C. R., Eds., Society for Experimental Biology Seminar Series 41, Cambridge University Press, Cambridge, 1991, 173.

Van den Thillart, G., Kesbeke, F. and Van Waarde, A., Influence of anoxia on the energy metabolism of goldfish *Carassius auratus*, *Comp. Biochem. Physiol.*, 55A, 329, 1976.

Van den Thillart, G., Van Berge-Henegouwen, M. and Kesbeke, F., Anaerobic metabolism of goldfish, *Carassius auratus* (L.): ethanol and CO_2 excretion rates and anoxia tolerance at 20, 10 and 5°C, *Comp. Biochem. Physiol.*, 76A, 295, 1983.

Van Waarde, A., Biochemistry of non-protein nitrogenous compounds in fish including the use of amino acids for anaerobic energy production, *Comp. Biochem. Physiol.*, 91B, 207, 1988.

Van Waarde, A., Van den Thillart, G. and Kesbeke, F., Anaerobic energy metabolism of the European eel, *Anguilla anguilla* L., *J. Comp. Physiol.*, 149, 469, 1983.

Van Waversveld, J., Addink, A. and Van den Thillart, G., Simultaneous direct and indirect calorimetry on normoxic and anoxic goldfish, *J. Exp. Biol.*, 142, 325, 1989.

Vetter, R. D. and Hodson, R. E., Use of adenylate concentrations and energy charge as indicators of hypoxic stress in estuarine fish, *Can. J. Fish. Aquat. Sci.*, 39, 535, 1982.

Vig, E. and Nemcsok, J., The effects of hypoxia and paraquat on the superoxide dismutase activity in different organs of carp, *Cyprinus carpio*, *J. Fish Biol.*, 35, 23, 1989.

Warren, C., *Biology and Water Pollution Control*, W. B. Saunders, Philadelphia, 1971.

Watenpaugh, D. E. and Beitinger, T. L., Resistance of nitrate-exposed channel catfish, *Ictalurus punctatus*, to hypoxia, *Bull. Environ. Contam. Toxicol.*, 37, 802, 1986.

Weatherley, A. H. and Gill, H. S., *The Biology of Fish Growth*, Academic Press, New York, 1987.

Weber, R. E. and Lykkeboe, G., Respiratory adaptations in carp blood. Influences of hypoxia, red cell organic phosphates, divalent cations and CO_2 on hemoglobin-oxygen affinity, *J. Comp. Physiol.*, 128, 127, 1978.

Weber, R. E. and Kramer, D. L., Effects of hypoxia and surface access on growth, mortality, and behavior of juvenile guppies *Poecilia reticulata*, *Can. J. Fish. Aquat. Sci.*, 40, 1583, 1983.

Weber, R. E. and Jensen, F. B., Functional adaptations in hemoglobins from ectothermic vertebrates, *Annu. Rev. Physiol.*, 50, 161, 1988.

Wells, R. and Weber, R., The spleen in hypoxic and exercised rainbow trout, *J. Exp. Biol.*, 150, 461, 1990.

West, J. B. and Lahiri, S., Eds., *High Altitude and Man*, American Physiology Society, Washington, D.C., 1984.

Williams, E., Glass, M. and Heisler, N., Blood oxygen tension and content in carp, *Cyprinus carpio*, during hypoxia and methaemoglobinaemia, *Aquacult. Fish. Mangt.*, 23, 679, 1992.

Wissing, J. and Zebe, E., The anaerobic metabolism of the bitterling *Rhodeus amarus* (Cyprinidae, Teleostei), *Comp. Biochem. Physiol.*, 89B, 299, 1988.

Wood, C. M. and Shelton, G., The reflex control of heart rate and cardiac output in the rainbow trout: interactive influences of hypoxia, haemorrhage, and systemic vasomotor tone, *J. Exp. Biol.*, 87, 271, 1980.

Wood, S. C., Interactions between hypoxia and hypothermia, *Annu. Rev. Physiol.*, 53, 71, 1991.

Wood, S. C. and Johansen, K., Blood oxygen transport and acid-base balance in eels during hypoxia, *Am. J. Physiol.*, 225, 849, 1973.

Wood, S. C., Johansen, K. and Weber, R. E., Effects of ambient P_{O_2} on hemoglobin-oxygen affinity and red cell ATP concentrations in a benthic fish, *Pleuronectes platessa, Respir. Physiol.,* 25, 259, 1975.

Wu, R. S. S. and Woo, N. Y. S., Respiratory responses and tolerance to hypoxia in two marine teleosts, *Epinephelus akaara* (Temminck and Schlegel) and *Mylio macrocephalus* (Basilewsky), *Hydrobiologia,* 119, 209, 1984.

Yang, T. H., Lai, N. C., Grahm, J. B. and Somero, G. N., Respiratory, blood and heart enzymatic adaptations of *Sebastolobus alascanus* (Scorpaeinidae; Teleostei) to the oxygen minimum zone: a comparative study, *Biol. Bull.,* 183, 490, 1992.

Zinkl, J. G., Lockhart, W. L., Kenny, S. A. and Ward, F. J., Effects of cholinesterase inhibiting insecticides on fish, in, *Cholinesterase-Inhibiting Insecticides—Impact on Wildlife and the Environment,* Mineau, P., Ed., Elsevier Science Publishers, Amsterdam, 1991, 233.

Chapter **3**

Respiratory and Cardiovascular Responses

I. INTRODUCTION

The respiratory system provides the most extensive interface of a fish with the aquatic environment. Because of this and its delicate epithelium, it is frequently the first system to be affected by pollutants. When death occurs as a result of acute exposure, it is often due to failure of respiratory homeostasis. The circulatory system provides the means of transport for essentially everything in an animal's body. Because control of the rate of blood flow and its distribution in a fish is based primarily on homeostasis of respiratory gases, the cardiovascular responses to pollution are included in this chapter with the respiratory system.

II. OVERVIEW OF NORMAL RESPIRATORY PHYSIOLOGY

The typical teleost fish respiratory system consists of the buccal and paired opercular cavities and the gills suspended between them (Figure 1). Ventilation is provided by the coordinated expansion and contraction of the buccal and opercular cavities which, by their movements, work as a "double pump" providing a nearly constant flow of water over the gills throughout a single respiratory cycle. Ventilation volume is adjusted primarily in response to changes in arterial oxygen (discussed in Chapter 2).

The gills consist of horizontal flat filaments which are supported in the water stream by the bony gill arches. On the filaments are found the secondary lamellae which range in frequency along a given filament from 10–60/mm; the higher numbers are found in the more active species (Hughes, 1982). Because of the structure of the gills, they provide a large surface area for the movement of oxygen, carbon dioxide, electrolytes, water, ammonia, and hydrogen ions between the blood and water. Thus, the gills are really a multipurpose organ directly involved in a variety of functions including respiratory gas exchange, osmoregulation, acid-base balance, and nitrogenous waste excretion. The majority of these fluxes of chemicals take place across the epithelial surface of the lamellae, which because of its multiple function, has been referred to as the "renaissance epithelium" (Bentley, 1980).

The cross-sectional structure of a "typical" lamellum is shown in Figure 2a. The blood is separated from the water by two outer layers of epithelial cells, a basement membrane,

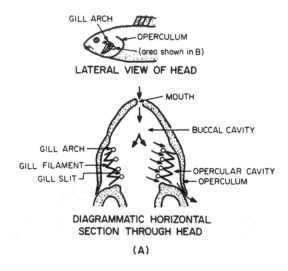

LATERAL VIEW OF HEAD

DIAGRAMMATIC HORIZONTAL
SECTION THROUGH HEAD

(A)

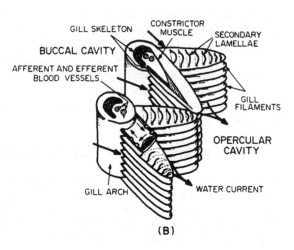

(B)

Figure 1 Major features of the respiratory system of a teleost fish. (A) The general arrangement of the buccal and opercular cavities with the associated gill structures. Arrows indicate the direction of water flow. (B) Enlarged view of segments of two gill arches showing the two rows of filaments on each arch and the secondary lamellae projecting dorsally and ventrally from each filament. (Redrawn from Hughes, G. M., *New Sci.*, 11, 346, 1961.)

and the flanges of the pillar cells. The total thickness of these layers ranges from 1–10 μm depending on the species of fish (Hughes, 1980). The outer epithelial cells have microridges and microvillae when viewed with a scanning electron microscope; these may help to anchor mucus and increase the surface area for exchange of gases (Hughes, 1979). Blood flows in the lamellum in a direction opposite to that of the water through a plexus formed by the pillar cells. This countercurrent flow aids in the achievement of a high utilization efficiency of oxygen (i.e., percent of the dissolved oxygen removed from the water in one passage over the gills).

Recent studies have revealed that the fluxes across the gill membranes alter the chemical characteristics of the water in the branchial boundary layer. For example, under most conditions, due to the excretion of protons, the water next to the lamellar epithelium is more acid than that midstream between the lamellae and this can affect the flux of other ions such as ammonia (Randall et al., 1991). This localized microenvironment also has implications for modeling of toxicant uptake from the water via the gills (Chapter 5).

Several comprehensive reviews of the normal physiology and histology of the fish respiratory system have been published (Cameron, 1989; Perry and Wood, 1989; Randall, 1990; Perry and McDonald, 1993). Additionally, the proceedings of a symposium on

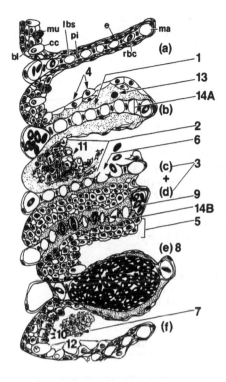

Figure 2 Composite diagram of the common gill lesions induced by pollutants. Six lamellae are shown (a–f), the top one of which is characteristic of a normal rainbow trout. The lesions are numbered as follows: (1) epithelial lifting; (2) necrosis; (3) lamellar fusion; (4) hypertrophy; (5) hyperplasia; (6) epithelial rupture; (7) mucus secretion; (8) lamellar aneurism; (9) vascular congestion; (10) mucus cell proliferation; (11) chloride cells damaged early; (12) chloride cell proliferation; (13) leukocyte infiltration of epithelium; (14A) lamellar blood sinus dilates; and (14B) lamellar sinus constricts. Abbreviations: bl = basal lamina; cc = chloride cell; e = typical lamellar epithelial cells; lbs = lamellar blood sinus; pi = pillar cell; rbc = erythrocyte. (From Mallatt, J., *Can. J. Fish. Aquat. Sci.*, 42, 630, 1985. With permission.)

environmental effects (excluding for the most part, pollutants) on gills have been published in *Physiological Zoology*, Vol. 64 (1), 1991.

A wide variety of pollutants cause impaired regulation of plasma electrolytes and this is due mostly to effects on the gills, a topic considered at length in Chapters 7 and 10. Some aspects of the effects of pollutants on fish respiratory physiology are reviewed by Rankin et al. (1982) and Satchell (1984).

III. HISTOPATHOLOGY OF GILL LAMELLAE EXPOSED TO POLLUTANTS

Because respiratory gases must pass through the lamellar epithelium by diffusion alone, this surface is quite delicate compared with the rest of the surface of the fish. Moreover, there is a large flow of water over the lamellae even in a resting fish [e.g., approximately 70 ml/min in a 0.5-kg trout (McKim and Goeden, 1982)]. There is thus ample opportunity for dissolved or suspended materials to come in contact with the delicate lamellae. In the 1920s and 1930s there were reports that fish exposed to heavy metals (e.g., zinc) in the water suffocated due to damage to the gills and/or clogging of the lamellae with mucus [see Jones (1964) for review of this earlier work]. A variety of histological studies since then have in part confirmed that view and revealed that many pollutants, both inorganic and organic, affect gill tissue (Table 1). This list is not meant to be all inclusive, nor does it include all the references in which damage has been reported for a given chemical. It is included here to give an indication of the wide variety of substances that can cause gill histopathology. (See Mallatt, 1985, for a more comprehensive listing.) It is important to realize that there are marked quantitative differences between the chemicals. For example, chromium is not nearly as harmful to gill tissue as is nickel, when fish are immersed in solutions of comparable toxicity (Hughes et al., 1979).

Table 1   Some Substances Reported to
Cause Gill Histopathology in Fish

Substance	Ref.
Acid	Daye and Garside (1977)
Ammonia	Kirk and Lewis (1993)
Arsenic	Sorenson et al. (1979)
Beryllium	Jagoe et al. (1993)
Cadmium	Stromberg et al. (1983)
Chlorine	Bass et al. (1977)
Chlorothalonil	Davies (1987)
Copper	Wilson and Taylor (1993)
Detergent	Abel and Skidmore (1975)
Formalin	Smith and Piper (1972)
Iron	Cruz (1969)
Lead	Sippel et al. (1983)
Malathion	Richmonds and Dutta (1987)
Mercury	Olson et al. (1973)
Nickel	Hughes et al. (1979)
Ozone	Wedemeyer et al. (1979)
Paraquat	Meyers and Hendricks (1985)
Permethrin	Kumaraguru et al. (1982)
Petroleum	Prasad (1991)
Phenol	Kirk and Lewis (1993)
Zinc	Tuurala and Soivio (1982)

Nearly all studies of gill histopathology resulting from exposure to some pollutant involved the fish receiving the test chemical in the water. However, gill damage can also be caused by the fish eating contaminated food. The insecticide permethrin caused essentially equivalent histopathology in trout which had either eaten contaminated food or been immersed in solutions of the chemical (Kumaraguru et al., 1982). Thus, this toxicant at least, can reach the gills by way of the circulation in sufficient concentration to cause damage. Perhaps others have a similar potential but this seemingly has not been investigated. Obviously, from the list in Table 1, a broad spectrum of substances causes alterations in gill structure. The lesions include necrosis, hyperplasia, inflammation, epithelial lifting, cell swelling, and hypersecretion of mucus (Figure 2). A typical chronology of damage from acute exposure to the test chemical is first a lifting of the outer layer of the lamellar epithelium (desquamation), usually starting in the area of the chloride cells. Edematous spaces are formed between the layers of epithelium (Figure 2) and these may become infiltrated with leukocytes. Eventually the whole epithelium sloughs off and the lamellum loses rigidity. On the blood side of the lamellum the central spaces collapse, but the marginal channel often remains normal until the rest of the lamellum is essentially destroyed. Acute exposure to some chemicals can cause rapid destruction of the gill lamellae within a few hours. Death then may follow as a probable result of blood hypoxia (see below). However, exposure with some substances such as ammonia causes relatively little gill damage unless concentrations are quite high, but death still takes place, obviously from other causes (Smart, 1978). Thus, gill histopathology may or may not indicate "cause of death", depending on the extent of the lesions encountered.

Some authors have reported clubbing of the ends of the lamellae and a tendency of adjacent lamellae to stick together. Increased discharge of mucus has been observed by some but not by others. Examination with scanning electron microscopy of lamellae from trout exposed to acid revealed loss of microridge patterns on the epithelium (Jagoe and Haines, 1983). All of these changes will reduce diffusion of oxygen (and possibly carbon dioxide) so blood gases will be affected (discussed below).

It would be nice to find there are certain histological changes diagnostic for particular toxicants, and this has been attempted (Abel and Skidmore, 1975; Wobeser, 1975). Mallatt (1985) has done an extensive quantitative and qualitative review of 133 published studies of fish gill pathologies involving over two dozen chemicals. He concludes that the structural alterations are a stereotyped physiological reaction to environmental stressors. Indeed, it may be that nearly any pollutant can cause gill damage if in sufficient concentration in the water. Furthermore, there seem to be far more similarities than differences in the gill pathologies from acute exposures, and even non-chemical stressors, such as hypoxia and temperature change, can produce some of the same tissue changes. More recently, Kirk and Lewis (1993) have used scanning electron microscopy of gills from fish exposed to three different pollutants and claim some distinct differences. They tested copper, phenol and ammonia at two different concentrations and time intervals of either 2 or 24 h. All three chemicals caused disorganization of lamellae and proliferation of mucus cells. High concentrations of copper (1.0 mg/L) caused lamellar fusion and swelling in the distal tips of filaments. Lamellar epithelium also exhibited hypertrophy. Phenol at 10 mg/L stripped the epithelium away from lamellae and ammonia caused the formation of distinctive circular depression in the epithelium. All these changes were after rather high doses; effects of chronic exposures may be quite different.

The histological lesions induced by chronic exposures to chemicals are much less known than those from acute exposures and may be more pollutant specific, especially if they are studied at the level of transmission or scanning electron microscopy; and at the level of light microscopy, there is a need for more quantification of the changes seen. Sublethal exposure to various chemicals can cause swelling of the epithelium which increases the diffusion distance for oxygen transfer. Hughes and Perry (1976) proposed a light-microscopic method for quantifying this and Hughes et al. (1979) subsequently applied it with limited success to a study of rainbow trout exposed to nickel, chromium, and cadmium. Unfortunately, the method is tedious and quite labor intensive. Newer computerized image analysis equipment greatly improves this procedure, however, it has rarely been utilized.

Nikl and Farrell (1993) demonstrated well how gill morphometric data can be used to understand the relationship between histopathological changes and physiological capacity. Juvenile salmonids were exposed to a wood preservative at several concentrations for 96 h after which their maximum aerobic swimming speed was measured in a water tunnel. Figure 3 summarizes the data obtained showing a fairly good association between interlamellar distance or the blood-water diffusion distance and the percentage decrease in swim speed.

It should be pointed out that there are at least two potential problems that may be encountered in studies of histopathology in gill tissue of freshwater fish following chronic exposure to a pollutant. One is the susceptibility of the fish to bacterial gill disease. This disease is frequently brought on by environmental stress and is characterized by proliferation of the gill epithelium and clubbing and fusing of the lamellae (Snieszko, 1980). These very same things are frequently observed in work on toxicants and could be incorrectly interpreted as a direct effect of some test chemical. A second source of error has been noted by Mitchell and Cech (1983) who found that exposure to ammonia alone did not cause histopathology of the gills of channel catfish, but when this was combined with a low level of monochloramine (a form of residual chlorine) in the water, acute tissue hyperplasia occurred. The levels of monochloramine used (which alone did not cause gill damage) were the same as those often present in charcoal-filtered domestic tap water, a technique widely used in laboratories for removal of chlorine in flowing water systems for experimental aquaria. This same synergistic effect might take place in tests with other pollutants in laboratories where charcoal filtering of chlorinated tap water is practiced.

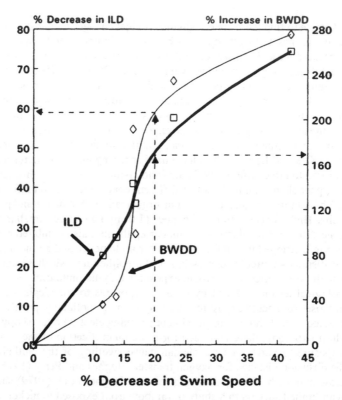

Figure 3 Comparison of percentage reductions in interlamellar distance (ILD) and percentage increases in water blood-water diffusion distance (BWDD) associated with a given reduction in chinook salmon swimming speed after exposure to TCMTB, a wood preservative. Measured histological changes relating to a 20% reduction in critical swimming speed are indicated by the dashed line. (From Nikl, D. and Farrell, A., *Aquat. Toxicol.*, 27, 245, 1993. With permission.).

IV. VENTILATION CHANGES IN RESPONSE TO POLLUTANTS

Ventilation refers to the movement of water over the gills. It is synonymous with breathing; the use of the term "respiration" for this process should be avoided since it is a general term that means different things to different people (e.g., many equate it with rate of oxygen consumption).

As discussed earlier, the fish respiratory system moves water over the gills by the coordinated expansion and contraction of the buccal and opercular cavities. In order to measure this, one would ideally determine the rate of water flow through the operculae, but this is technically difficult or impossible to do with most species; so other more indirect means are usually used which measure some aspect of the ventilatory movements (Heath, 1972). The simplest method is to merely count visually the opercular beat, providing one can see the fish in the experimental apparatus. However, the "subject" often will react negatively to the presence of an observer and visual counting is quite labor intensive. Because movement of the water is accomplished by pressure changes in the opercular and buccal cavities, the process of ventilation can be studied by detecting these pressure changes with suitable transducers and displaying them on a strip-chart recorder. This requires cannulation of at least one of the cavities with a thin catheter tube and subsequent partial restraint of the fish so it will not get the tube(s) tangled (Sellers et al., 1975). The advantage of this method is that it gives an indication of relative changes in stroke volume (depth of breathing) as well as breathing frequency. The actual ventilation volume is the product of these two measures and some fish species, such as the rainbow trout, may change stroke volume more than frequency when subjected to some stress (Shelton et al., 1986). Because the fish must be partially restrained, however, it does impose an additional factor that may concern some workers, and for long-term experiments (i.e., more than a week), erosion of the tissue around the catheter can be troublesome.

In studies on the effect of pollutants on ventilation, a method that has been used extensively, especially for biomonitoring of water quality, is one referred to as "dual external electrodes" (Heath, 1972). For this purpose electrodes are placed at both ends of a small aquarium containing a single fish. The breathing movements of the fish cause a cyclic potential change between the electrodes which is amplified and displayed on a suitable oscillographic recorder. Carlson (1982) has analyzed this method in considerable detail and coined the term "electrobranchiogram" (EBG) to refer to the oscillographic record obtained (Figure 4). Both the frequency of ventilation and that of coughing can be detected in this manner, and occasionally, the electrocardiogram is superimposed on the record. Even the breathing of fry can be detected with this technique (Thomas and Rice, 1979). A further advantage of this method is that it lends itself to automation whereby a computer can be used to count the ventilation frequency (but generally not the coughs). Such an application has considerable potential for early warning systems in biological monitoring (Cairns and van der Schalie, 1980; Gruber et al., 1991).

Figure 4 shows the record of a cough, a common occurrence in fish exposed to a wide variety of pollutants (Table 2). Hughes (1975) has studied coughing in rainbow trout in some detail including electromyographic analyses of the muscles involved. A cough occurs at a random position during a respiratory cycle and is characterized as a rapid expansion and contraction of the buccal and opercular cavities. These movements produce a two- to fivefold increase in the reversal phase of the differential pressure across

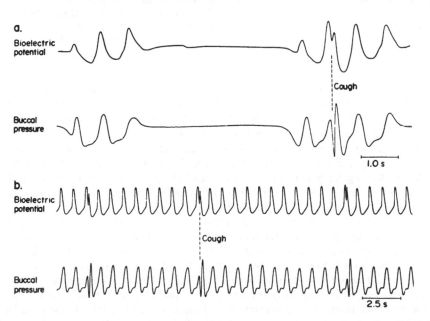

Figure 4 Simultaneous recordings of bioelectric potentials and buccal water pressure changes associated with ventilatory movements in the bluegill. (a) Fish in non-contaminated water and (b) fish after 1 h exposure to 10 mg/L waterborne zinc. Note frequency of breathing and coughing was increased in the exposed fish. (From Carlson, R. W., *Environ. Pollut. Ser. A*, 29, 35, 1982. With permission.)

Table 2 Chemicals Reported to Cause Changes in Ventilation and/or Coughing

Chemical	Cough or ventilation	Species	Ref.
Acrolein	C	Bluegill	Carlson (1990)
Aluminum	C,V	Rainbow trout	Malte and Weber (1988)
Ammonia	V	Rainbow trout	Smart (1978)
Anthracene	C	Bluegill	McCloskey and Oris (1991)
Benzaldehyde	C,V	Bluegill	Carlson (1990)
Cadmium	C,V	Bluegill	Bishop and McIntosh (1981)
Carbaryl	V	Catfish	Arunachlam et al. (1980)
Clay	V	Sunfish	Horkel and Pearson (1976)
Chlorine	C,V	Rainbow trout	Bass and Heath (1977)
Chlorothalonil	V	Rainbow trout	Davies (1987)
Coal dust	C	Rainbow trout	Hughes (1975)
Copper	C	Brook trout	Drummond et al. (1973)
Chromium	C,V	Rainbow trout	van der Putte et al. (1982)
Crude oil	C,V	Atlantic salmon	Barnett and Toews (1978)
Cyanide	V	Rainbow trout	Sawyer and Heath (1988)
Cypermethrin	C	Rainbow trout	Bradbury et al. (1991)
DDT	C	Rainbow trout	Lunn et al. (1976)
Endosulfan	C	Rainbow trout	Bradbury et al. (1991)
Fenitrothion	C,V	Rainbow trout	Klaverkamp (1982)
Fenvalerate	C	Rainbow trout	Bradbury et al. (1991)
Oil + dispersant	C,V	Atlantic salmon	Barnett and Toews (1978)
Mercury	C	Brook trout	Drummond et al. (1974)
Methyl parathion	V	Tilapia	Bashamohideen et al. (1987)
Naphthalene	V	Pink salmon fry	Thomas and Rice (1979)
Paraquat	V	Catfish fry	Tortorelli et al. (1990)
Pulpmill effluent	C,V	Sockeye salmon	Davis (1973)
Toluene	V	Pink salmon	Thomas and Rice (1979)
Surfactants	C,V	Bluegill	Maki (1979)
Wood pulp	C	Rainbow trout	Hughes (1975)
Zinc	C,V	Rainbow trout	Hugh and Tort (1985)

the gills. This phase normally occupies only about 10% of the normal cycle and is of low amplitude. Thus, the cough produces a quick reversal of water flow that presumably aids in freeing the gills of suspended matter. Backward coughs can also be observed whereby the water exits via the opercular openings (Hughes and Adney, 1977a). Some fish show a low frequency of coughing, even in presumably clean water. The interval between individual coughs can be quite regular and when the frequency increases from chemical irritation, this regularity is maintained (Bass and Heath, 1977). In rainbow trout, there appears to be a maximum cough frequency of around 15–20 per minute, no matter what the extent of stimulus (Bass and Heath, 1977).

Table 2 lists several of the substances that have been observed to affect ventilation frequency, coughing, or both. The table does not include all of the reports for a given chemical; in most cases the most recent study was cited. In addition, while acid effects on ventilation have been examined by a number of workers, the results are very dependent on the level of carbon dioxide in the water (see Chapter 10). As with histological damage to the sensitive gill tissues, nearly any pollutant can probably cause changes in ventilation if the test concentration is high enough. It is of more interest to deal with incipient lethal and sublethal concentrations as these will probably have more relevance to what may be encountered in nature. In most cases, those studies cited in Table 2 included a dose that was sublethal.

There are several aspects of the typical ventilatory response to a pollutant: (1) the percent change in coughing frequency, if present, is usually greater than the percent change in ventilation frequency, although some chemicals such as cyanide or ammonia do not cause coughing, even at lethal levels; (2) at low concentrations, an initial irritation response may be followed by a sort of acclimation to the toxicant; (3) conversely, if the chemical affects the nervous system, but only after accumulation, the frequency changes may not take place until some level of accumulation has taken place; and (4) the change in frequency of ventilation or coughing is usually an increase, rather than a decrease from control rates, although when concentrations are sublethal, there have been several examples of decreases in ventilation seen (e.g., Bashamohideen et al., 1987; Tortorelli et al., 1990). Also, the character of the response may depend to some extent on the fish species. While most people have observed hyperventilation in the presence of cadmium, Moffitt (1990) reported a decreasing ventilation frequency in brown bullhead catfish (*Ictalurus nebulosus*) experiencing cadmium stress. Because this species may show a slowing of ventilation under environmental hypoxia (Marvin and Heath, 1968), the response to cadmium may not be unexpected.

The ventilatory change may be proportional to the dose even at very low levels. Bishop and McIntosh (1981) report a good correlation between the threshold for an increase in cough frequency in water containing cadmium and the maximum acceptable toxicant concentration (MATC) obtained by growth and reproduction studies. A similar type of correlation with the MATC was seen with coughing of rainbow trout in water containing aluminum (Ogilvie and Stechey, 1983) and ventilation frequency of bluegill immersed in surfactant solutions (Maki, 1979). Thus, these responses could be a useful method for estimating water quality criteria, providing it is recognized by the investigators that not all substances cause changes in ventilation. For example, in tests on DDT, dieldrin, and methyl carbamate with rainbow trout, only DDT gave useful results with the ventilatory response. The other two pesticides produced no response until clearly lethal concentrations were used (Lunn et al., 1976). More recently, Carlson (1990) found that malathion caused no ventilatory response in bluegill whereas carbaryl, another acetylcholinesterase inhibitor like malathion, caused hyperventilation but only at fairly high concentrations. It should also be realized that the initial response to some chemicals may be lost after about a day of continuous exposure so measurements taken at only one time interval of exposure could be misleading.

V. PHYSIOLOGICAL MECHANISMS OF CHANGES IN VENTILATION

There is more than one potential physiological cause for the ventilatory responses observed in fish emersed in various chemicals:

1. Simple irritation of the sensitive epithelial tissues of the gills and buccal cavity can cause coughing and hyperventilation. The latter could be explained as a sort of psychological alarm response (Schreck, 1981). Under more prolonged exposure, the alarm component, and maybe even the irritation effects may be lost, so the frequency of the two ventilatory factors may return toward the control level.

2. Changes in demand for metabolic energy by the fish will be reflected in increased or decreased ventilation. If this is the sole cause of an observed response, then coughing will not necessarily occur. Of course, irritation from the chemical can cause more spontaneous activity (see Chapter 8) which would elevate the demand for energy, but, in general, changes in ventilation should not be interpreted as being due to altered metabolic demand without supporting data on oxygen consumption. This is because reason number three (below) may be a more important cause of the observed changes.

3. Impaired gas exchange in the lamellae will cause hyperventilation because this is interpreted by the arterial oxygen receptors in the first gill arch (Burleson and Milsom, 1993) as a hypoxic condition. The histopathology seen in gills that have been exposed to a pollutant strongly suggests this is a common mechanism by which hyperventilation is induced because most of these histological changes increase the diffusion distance for oxygen from the water to the blood. A reduction in the flux of oxygen from the water into the blood from fish with damaged gills can be confirmed by taking serial samples for Po_2 measurement of arterial blood from trout with catheters in their dorsal aortas. Skidmore (1970) was the first to do this using an extremely high concentration of zinc (40 ppm) and found the oxygen tension of the blood dropped to less than 20 mmHg at death. Using lower concentrations of zinc, which approximated that of the 48-h LC50, Sellers et al. (1975) and Spry and Wood (1984) also observed a very low arterial blood Po_2 following 12–24 h of exposure. Under these acute exposures, accumulation of zinc in the blood is negligible (Spry and Wood, 1984) but lactic acid levels rise, which indicates anaerobic glycolysis. Anaerobic metabolism is to be expected if there is an impairment of oxygen flux from the water into the blood, whether this is produced by environmental hypoxia (Chapter 2) or a diffusion barrier at the gill surface.

In fish with gills damaged by zinc, accumulation of lactic acid and carbon dioxide in the blood produced a combined metabolic and respiratory acidosis (Spry and Wood, 1984). Even though blood pH decreased in the trout exposed to zinc, the extent of this change was no greater than that observed in trout that swam to exhaustion (Turner et al., 1983), a condition from which they readily recover. Thus, acidosis is not the cause of death. All these findings are consistent with the hypothesis originally formulated long ago by Carpenter (1927) that hypoxemia (low blood oxygen) is the cause of death from acute exposure to zinc. A similar conclusion is appropriate for high doses of free chlorine (Bass and Heath, 1977), but not for monochloramine (combined chlorine) at the same concentration (Travis and Heath, 1981). Based on numerous findings of gill histopathology (discussed previously), this hypoxemia is undoubtedly a major contributing factor to mortality from acute concentrations of a large number of substances, but other physiological "failures", such as in blood electrolyte homeostasis or in nervous system function, may be of equal or greater importance in the etiology of death, especially at lower, but still lethal environmental levels of some pollutant (e.g., Wilson and Taylor, 1993).

Some idea of how fast the oxygen tension in the blood can drop during exposure to a pollutant, and then recover following it, was determined by Bass and Heath (1977). Rainbow trout were subjected to "pulses" of chlorine which rose to a peak concentration in 30 min and then declined to 0 in about 45 min after the peak. These were repeated at 8-h intervals. (This simulates the mode of effluent release from many steam-powered electric generating plants.) At the peak of the chlorine pulse, arterial Po_2 had dropped to 40 mmHg from a control level of 102 mmHg, but then just before the second pulse, Po_2 had nearly recovered to that of the controls. The extent of recovery of blood oxygen between chlorine pulses, however, became less and less with each subsequent pulse until death occurred at an arterial Po_2 below 20 mmHg.

Recovery of gill function following a chemical "insult" has been studied little. It can require days (Skidmore and Tovell, 1972) or weeks (Hughes et al., 1979; Scheier and Cairns, 1966) for histological morphology to completely return to normal when the fish is removed from contaminated water. The findings mentioned above on the effects of chlorine suggest that if the exposure is of short enough duration, at least a partial restoration of capacity to diffuse oxygen can be achieved within a matter of hours. In order to determine if the same thing would hold true for a very different type of chemical, in a preliminary study in my laboratory, rainbow trout with catheters in their dorsal aortas were exposed to 2 ppm of zinc for 11 h at which time the arterial Po_2 averaged 40 mmHg. The changes in arterial Po_2 during recovery were then followed. These fish exhibited almost complete recovery of arterial oxygen within 10 h after this rather acute exposure to a toxicant. This implies that there was no actual sloughing off of the respiratory epithelia in the gills with this short-term exposure, but rather a swelling of the epithelial layer and/or an accumulation and subsequent loss of mucus (Ultsch and Gros, 1979), which increased the diffusion distance for oxygen. Using a 10-fold higher zinc concentration (Hughes and Tort, 1985) reported that following exposures of 120 min to the dissolved metal it required several days for ventilation and coughing to return to the control level. Such findings as these may have relevance to industrial spills where the situation is corrected quickly.

Lower concentrations of zinc (equivalent to the 96-h LC50) produce only a slight reduction in arterial Po_2 after 3 days of exposure but a significant accumulation of zinc in the blood. Spry and Wood (1984) attribute death at this concentration to some internal actions of zinc, independent of the damage to the respiratory system. They also noted a slight blood alkalosis at the lower level of zinc exposure (recall that acute exposure caused acidosis). It might be assumed that hyperventilation produced this alkalosis by "blowing off" blood CO_2. However, fish are not like terrestrial vertebrates in that they are unable to regulate blood pH and carbon dioxide to much of an extent by respiratory means (Cameron, 1979). Spry and Wood (1984) propose several alternative reasons for the alkalosis including accumulation of ammonia, alterations in the chloride-bicarbonate, and sodium-hydrogen electroneutral exchanges at the gill surface and/or an inhibition of carbonic anhydrase activity in gill tissue. The work of Spry and Wood (1984) on fish exposed to two different lethal concentrations of a toxicant points up what may be an important generalization; the mechanisms of toxic death can be very different depending on the concentration of the chemical. Very high doses of zinc, acid (Milligan and Wood, 1982), and probably many other chemicals, cause death by hypoxemia due to impaired oxygen uptake by the gills, whereas somewhat lower doses produce mortality more slowly by the chemical getting inside and causing alterations in other physiological machinery directly.

Of course, a combination of hypoxemia and other dysfunctions, such as ionoregulatory, may cause death as has been recently demonstrated for acute copper exposure. Wilson and Taylor (1993) exposed trout to a high dose of copper and found declines in plasma electrolytes and arterial Po_2. These results are similar to what occurs with acute acid exposure where death results from a secondary hemoconcentration, rapid elevation in arterial blood pressure, and probable heart failure (see Chapter 10).

One of the significant questions in work on pollution is to what extent does a long-term exposure to some chemical affect the ability of the organism to handle a subsequent environmental stressor of a different sort, such as low dissolved oxygen. Majewski and Giles (1981) investigated this in trout which had been exposed to 6.4 mg/L cadmium for 30 days. After the exposure, the fish were fitted with dorsal aortic catheters and allowed to recover in non-contaminated water for 3 days. Then, they were subjected to a gradually lowered environmental oxygen tension while the arterial Po_2 was monitored. The fish that had received the cadmium treatment exhibited a slightly lower arterial oxygen tension (6–10 mmHg) than controls at most of the environmental levels of oxygen. This difference is slight, but statistically significant, indicating that even an exposure at 40% of the incipient lethal level produced a measurable reduction in oxygen diffusion capacity across the gills. It seems doubtful that this amount of respiratory impairment would affect the fish's ability to function in the environment at rest, but it certainly could have an inhibitory effect on a fish swimming actively in hypoxic waters.

VI. CIRCULATORY PHYSIOLOGY

The function of the circulatory system is intimately related to the process of respiratory gas exchange and the control mechanisms for this system are largely based on the detection of changes in blood oxygen. Thus, any environmental factor that alters the process of oxygen uptake can be expected to affect circulation. Of course, this is not to exclude the possibility of some chemical acting directly on the heart and/or blood vessels. Further, adrenergic excitation, as a stereotyped response to stress (Mazeaud and Mazeaud, 1981), will have marked cardiovascular results.

The teleost heart has four chambers arranged in series. The sequence by which blood passes through the heart is sinus venosus, atrium, ventricle, and bulbus arteriosus. The ventricle provides the arterial blood pressure while the bulbus dampens the rise and fall of blood pressure caused by the individual heartbeats and thereby probably protects the delicate gill capillaries from damage. It also helps maintain a constant blood flow throughout a large portion of the cardiac cycle which probably aids in the uptake of oxygen by the gills. The pacemaker tissue in the typical teleost heart is diffuse rather than being nodal as in mammals. Also, because most fish are poikilothermic, the intrinsic beat is temperature dependent. The intrinsic beat is altered to meet varying demands by changes in autonomic neural activity. All fish hearts, except hagfishes, have extensive vagal (parasympathetic cholinergic) innervation and most teleost hearts apparently have sympathetic (adrenergic) innervation. Many species of fish appear to have hearts that are under the influence of considerable parasympathetic tone which slows the intrinsic beat. Thus, alterations in rate of heartbeat could, theoretically, be achieved largely by changes in the level of this activity alone.

The ventral aorta, which is an extension of the bulbus, has approximately twice the blood pressure of the dorsal aorta which is on the efferent side of the gills. This pressure drop across the gills is variable depending on autonomic nervous activity and the action of hormones such as catecholamines and acetylcholine. Working out the pathways for blood flow through the gill filaments and lamellae has proven to be difficult in part because the gills possess a highly variable network of vessels (Laurent, 1984). A major problem is in understanding how the number of lamellae perfused with blood is increased or decreased in response to the rise and fall of internal demand for oxygen or changes in availability of environmental oxygen.

Blood flows from the ventral aorta to the afferent filament arteries, through the lamellae to the efferent arteries, and then to the dorsal aorta. Blood can, however, effectively by-pass a lamellum by going through its basal channel rather than through the lamellum proper. At rest, only 60% of the lamellae are perfused and these are the ones located more proximally on the filaments (Booth, 1978). Catecholamines (adrenaline and noradrenaline) increase the number of lamellae perfused and cause a reduction in vascular resistance to blood flow through the gills (Booth, 1979). Because these changes increase the permeability of the gills to water, it is considered likely that they also produce an enhancement in oxygen permeability. Acetylcholine, as would be released by parasympathetic activity, has the opposite effect of the catecholamines. The respiratory surface area can also be changed by non-neurohumoral mechanisms as well, because an increase in blood pressure, as occurs during exercise, causes recruitment of more lamellae (Randall, 1982). Further details on the fish circulatory system can be found in reviews by Klaverkamp (1982), Butler and Metcalf (1983), Satchell (1991), Hoar et al. (1992), and Farrell (1993).

This brief overview of the circulatory physiology of the fish emphasizes the close relationship between the processes of circulation and respiration. It also shows how there are points where pollutants could have a considerable effect on gill function via alterations in circulation through the gill but this is an area of research that has received little attention, perhaps in part because the techniques are difficult to master.

An example of the sort of study that can be revealing is that of Bolis and Rankin (1980) in which they found that the detergent linear alkylate sulfonate (LAS) caused a vasodilation of gill vasculature that was apparently mediated through the beta adrenergic receptors. The detergent also, however, inhibited the normal vasodilatory action of catecholamines in the gills of eels (*Anguilla anguilla*) and brown trout (*Salmo trutta*). It is possible that such an effect could take place with other chemicals as well and would reduce the ability of the gill vasculature to respond to changes in oxygen demand or supply.

VII. CARDIAC RESPONSES TO POLLUTANTS

There has been some interest in the effect of pollutants on heart rate in fish. By implanting thin electrode wires in the ventral side of the body and allowing the fish to remain awake in a chamber which prevents extensive swimming movements, reasonably "normal" electrocardiographic records can be obtained. From these, the heart rate is counted. With suitable electrodes or pressure transducers, the ventilatory activity can be simultaneously recorded along with the ECG (Bass and Heath, 1977; Hughes and Adney, 1977b; Bradbury et al., 1991). When fish are exposed to a variety of different chemicals in acute concentrations, a slowing (bradycardia) of the heart usually occurs simultaneously with an increase in ventilation. These chemicals include residual chlorine (Bass and Heath, 1977), cyanide (Figure 5A), DDT and dieldrin (Lunn et al., 1976), organophosphorus insecticides (Klaverkamp, 1982), the herbicide paraquat (Tortorelli et al., 1990), oil spill dispersant (Kiceniuk et al., 1978), and zinc (Hughes and Adney, 1977a). Dose dependency was shown with paraquat and catfish fry heart rate (Tortorelli et al., 1990) and cyanide (Sawyer and Heath, 1988) (Figure 5A); it is not known if a similar dependency would be found with other chemicals. Most of the chemicals that induce a bradycardia also produce acute internal hypoxia. In a sense, cyanide is an exception in that blood oxygen probably remains unchanged, but the effect on the animal is similar to hypoxia and the receptors that bring about the hypoxic reflexes are affected similarly by cyanide (Burleson and Smatresk, 1990). Where internal hypoxia occurs as a result of the pollutant exposure, the bradycardia seen may be a result of vagus nerve inhibition of the heart, as occurs with exposure to environmental hypoxia (see Chapter 2). This neurogenic origin of the bradycardia in response to a chemical in the water was confirmed in the study on cyanide (Sawyer and Heath, 1988). There it was shown that the initial cardiac slowing could be blocked by injecting atropine (an acetylcholine antagonist) through a catheter into the pericardium during cyanide exposure. At higher doses of cyanide, reductions in heart rate occurred apparently due to a direct effect of cyanide on the heart. At least some pollutants can have a direct effect on heart contractility. This has been shown for both cadmium and zinc, metals that block calcium channels in excitable tissues such as heart muscle (Tort and Madsen, 1991).

The neurotoxin from the organism that causes red tide (*Chattonella marina*) causes fish death apparently by inducing a severe bradycardia which results in anoxia because of reduced gill blood circulation. Endo et al. (1992) further showed that the bradycardia is due to vagal stimulation because it was blocked by atropine injection. *In vitro* studies showed that fractions of the neurotoxin depolarized the vagus nerve directly rather than acting as an acetylcholine agonist.

Note that the data displayed in Figure 5B on brown bullhead catfish (*I. nebulosus*) is considerably different than that seen with rainbow trout shown in Figure 5A. Instead of a bradycardia, the catfish responded to cyanide with an increased heart rate. The tachycardia seen in the catfish may have been due to a gustatory stimulus. When an extract of fish food was added to the inflowing water it caused tachycardia in the catfish but not the trout.

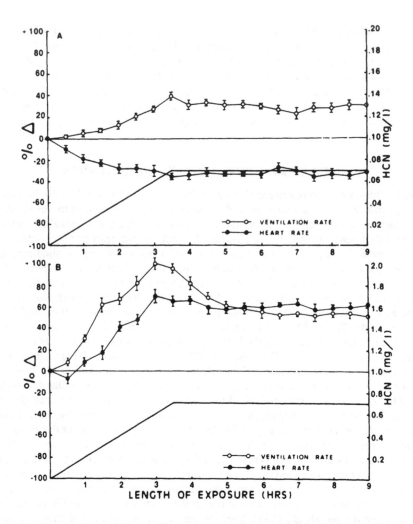

Figure 5 Changes in ventilation and heart rates of rainbow trout (A) and brown bullhead catfish (B) exposed to a linearly increasing cyanide concentration for 3.5 h and then held at a constant level throughout the remainder of the experiment. Points are the means (±1 SE) of four fish. Cyanide concentration is indicated by the solid line without points. Note that the concentration used with the catfish is an order of magnitude higher than that in the trout experiments. (From Sawyer, P. L. and Heath, A. G., *Fish Physiol. Biochem.*, 4, 203, 1988. With permission.)

Catfish have acute gustatory sensitivity (Atema, 1971) and because cyanide has a bitter taste and odor, it may have produced an adrenergic stress response in this species. These data also illustrate how different species may respond to the same environmental stimulus in very different ways.

Very high concentrations of pollutants can produce tachycardia in trout, at least initially. This has been shown for ammonia (Smart, 1978) where trout exposed to very high doses (death occurred between 1 and 3 h) experienced elevated heart rates even though blood oxygen was reduced to less than half that of controls. The fish struggled considerably and this caused an elevated rate of oxygen consumption, which was measured. Arillo et al. (1981) found elevated renin activity in plasma of trout which had been exposed to sublethal levels of ammonia. This enzyme catalyzes the formation of angiotensin, a powerful vasoconstrictor. They speculated that the elevated blood pressures that Smart observed upon ammonia exposure were due to this renin-angiotensin mechanism, although a larger part of this response was probably due to simple struggling of the fish. One might also speculate that angiotensin could cause vasoconstriction in the gills and thereby reduce respiratory gas exchange. This hypothesis gains some support from the observation that acute levels of ammonia apparently cause a severe lowering of dorsal aortic Po_2 in the absence of gill damage (Smart, 1978).

Increased heart rate and blood pressure can certainly result from stress-induced catecholamine release from autonomic nerves and/or interrenal cells. This was nicely shown by Milligan and Wood (1982) in trout exposed for several days to water of pH 4–4.5, but the cardiovascular effects caused by acid were also found to not be exclusively a result of catecholamines as adrenergic blocking agents only partially blocked them. During acid exposure, whole blood viscosity increased greatly due to a loss of plasma water and a consequent rise in plasma protein concentration and hematocrit. This cascade of effects was originally triggered by a loss of blood electrolytes through the gills (see Chapter 10). According to Milligan and Wood (1982), death of adult trout from acid water (pH 4–6) may be due to a combination of increasing blood pressure and decreasing plasma volume, which ultimately leads to circulatory failure. (The cause of death at a pH below 4 may be hypoxemia because of severe gill damage.)

Recently, Wilson and Taylor (1993) have reported that trout exposed to a dose of waterborne copper sufficient to cause death in 24 h experienced physiological disturbances quite similar to those seen with acute acid exposure. This included an immediate tachycardia (78% increase) that was seemingly due to loss of vagal tone rather than catecholamines, because there was no initial change in blood pressure. With continued exposure, the blood pressure rose due to hemoconcentration and probable catecholamine vasoconstriction. Ultimately, death resulted from circulatory failure brought on primarily from inoregulatory collapse.

In an investigation of a more chronic type of toxicant exposure on circulation, Majewski and Giles (1981) found that trout in water with a cadmium dose 40% of the incipient lethal level exhibited a 25% increase in heart rate after 3 days exposure. This persisted for several months of continuous cadmium exposure. Because ventilation was also elevated throughout this time, the rate of oxygen consumption was probably increased, although this was not measured. The increased heart rate would be explained by this mechanism.

The generality can probably be made that chemical-dose combinations that cause internal hypoxia (e.g., acute doses of zinc) will result in bradycardia, while those combinations that cause elevations in oxygen consumption (see Chapter 8) and/or catecholamines will result in tachycardia. Small transient changes in heart rate mean little because fish respond to mild disturbances such as noises usually with a short-term bradycardia followed by a tachycardia (personal observations). In the future, slow changes in concentration of a test pollutant may be more revealing of action than abrupt changes, and would probably have greater environmental realism.

For some time there has been interest in the benefit, if any, of the well-known bradycardia response of a fish to environmental hypoxia (Chapter 2). It would appear that such a response would be clearly beneficial to a fish experiencing internal hypoxia that had been produced by a reduction in the transfer of oxygen in the gills. Blood normally spends about 1 second in the lamellum in which time it must become oxygenated (Hughes et al., 1981). If a gill was suffering from a reduced capacity for oxygen transfer, the time needed for oxygenation might be prolonged (Figure 6), thus by slowing the blood flow rate through the gill, better oxygenation of the blood could occur.

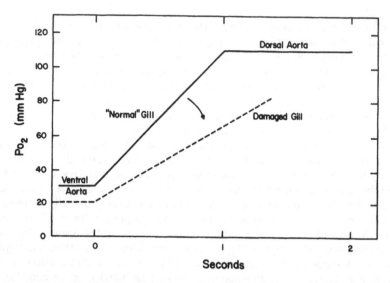

Figure 6 Hypothetical schematic of the effect pollutant damage to a fish gill has on the Po_2 of blood as it passes through a lamellum. In the normal gill, the 1-s interval in which the blood is in the lamellum permits the blood to essentially reach equilibrium with the water by the time it reaches the dorsal aorta. In the damaged gill, diffusion is slowed so equilibrium is not reached before the blood enters the dorsal aorta. (Values for the normal gill are from Holeton, G. F. and Randall, D. J., *J. Exp. Biol.*, 46, 317, 1967.) A decrease in venous Po_2 is assumed based on an increased rate of oxygen consumption in fish exposed to substances that cause gill damage (see Chapter 8).

REFERENCES

Abel, P. D. and Skidmore, F., Toxic effects of an anionic detergent on the gills of rainbow trout, *Water Res.*, 9, 759, 1975.

Arillo, A., Uva, B. and Vallarino, M., Renin activity in rainbow trout (*Salmo gairdneri*) and effects of environmental ammonia, *Comp. Biochem. Physiol.*, 68A, 307, 1981.

Arunachlam, S. et al., Toxic and sublethal effects of carbaryl on a freshwater fish, *Mystus vattatus, Arch. Environ. Contam. Toxicol.*, 9, 307, 1980.

Atema, J., Structures and functions of the sense of taste in catfish (*Ictalurus natalis*), *Brain Behav. Evol.*, 4, 273, 1971.

Barnett, J. B. and Toews, D., The effects of crude oil and the dispersant, Oilsperse 43, on respiration and coughing rates in Atlantic salmon (*Salmo salar*), *Can. J. Zool.*, 56, 307, 1978.

Bashamohideen, M., Obilesu, K. and Reddy, M., Behavioral changes induced by malathion and methyl parathion in the freshwater fish *Tilapia mossambica, Environ. Ecol.*, 5, 403, 1987.

Bass, M. L., Berry, C. and Heath, A. G., Histopathological effects of intermittent chlorine exposure on bluegill (*Lepomis macrochirus*) and rainbow trout (*Salmo gairdneri*), *Water Res.*, 11, 731, 1977.

Bass, M. L. and Heath, A. G., Cardiovascular and respiratory changes in rainbow trout, *Salmo gairdneri*, exposed intermittently to chlorine, *Water Res.*, 11, 497, 1977.

Bentley, P. J., The contributions of Jean Maetz to the biology of transport across epithelial membranes, in *Epithelial Transport in the Lower Vertebrates*, Lahlou, B., Ed., Cambridge University Press, New York, 1980, 7.

Bishop, W. E. and McIntosh, A. W., Acute lethality and effects of sublethal cadmium exposure on ventilation frequency and cough rate of bluegill (*Lepomis macrochirus*), *Arch. Environ. Contam. Toxicol.*, 10, 519, 1981.

Bolis, L. and Rankin, J. C., Interactions between vascular actions of detergent and catecholamines in perfused gills of European eel, *Anguilla anguilla* L. and brown trout, *Salmo trutta* L., *J. Fish. Biol.*, 16, 61, 1980.

Booth, J. H., The distribution of blood flow in gills of fish: application of a new technique to rainbow trout (*Salmo gairdneri*), *J. Exp. Biol.*, 73, 119, 1978.

Booth, J. H., The effects of oxygen supply, epinephrine, and acetylcholine on the distribution of blood flow in trout gills, *J. Exp. Biol.*, 83, 31, 1979.

Bradbury, S., Carlson, R., Miemi, G. and Henry, T., Use of respiratory-cardiovascular responses of rainbow trout (*Oncorhynchus mykiss*) in identifying acute toxicity syndromes in fish. Part 4. Central nervous system seizure agents, *Environ. Toxicol. Chem.*, 10, 115, 1991.

Burleson, M. and Smatresk, M., Effects of sectioning cranial nerves IX and X on cardiovascular and ventilatory reflex responses to hypoxia and NaCN in channel catfish, *J. Exp. Biol.*, 154, 407, 1990.

Burleson, M. and Milsom, W., Sensory receptors in the first gill arch of rainbow trout, *Respir. Physiol.*, 93, 97, 1993.

Butler, P. J. and Metcalf, J. F., Control of respiration and circulation, in *Control Processes in Fish Physiology*, Rankin, J. C., Pitcher, T. J. and Duggan, R. T., Eds., Wiley and Sons, New York, 1983, chap. 3.

Cairns, J. and van der Schalie, W. H., Biological monitoring. I. Early warning systems, *Water Res.*, 14, 1179, 1980.

Cameron, J. N., Excretion of CO in water-breathing animals, a short review, *Mar. Biol. Lett.*, 1, 3, 1979.

Cameron, J., *The Respiratory Physiology of Animals*, Oxford University Press, New York, 1989.

Carlson, R. W., Some characteristics of ventilation and coughing in the bluegill *Lepomis macrochirus* Rafinesque, *Environ. Pollut. Ser. A*, 29, 35, 1982.

Carlson, R. W., Ventilatory patterns of bluegill (*Lepomis macrochirus*) exposed to organic chemicals with different mechanisms of toxic action, *Comp. Biochem. Physiol.*, 95C, 181, 1990.

Carpenter, K. E., The lethal action of soluble metallic salts on fishes, *Br. J. Exp. Biol.*, 4, 378, 1927.

Cruz, J. A., About the possibility of iron adsorption by fishes and its pathological action, *Ann. Limnol.*, 5, 187, 1969.

Davies, P., Physiological, anatomic and behavioural changes in the respiratory system of *Salmo gairdneri* on acute and chronic exposure to chlorothalonil, *Comp. Biochem. Physiol.*, 88C., 113, 1987.

Davis, J. C., Sublethal effects of bleached kraft pulp mill effluent on respiration and circulation in sockeye salmon (*Oncorhynchus nerka*), *J. Fish. Res. Bd. Can.*, 30, 369, 1973.

Daye, P. G. and Garside, E. T., Histopathologic changes in superficial tissues of brook trout, *Salvelinus fontinalis* (Mitchell), exposed to acute and chronic levels of pH, *Can. J. Zool.*, 54, 2140, 1977.

Drummond, R. A., Spoor, W. A. and Olson, G. F., Some short term indicators of sublethal effects of copper on brook trout, *Salvelinus fontinalis*, *J. Fish. Res. Bd. Can.*, 30, 698, 1973.

Drummond, R. A., Olson, G. F. and Batterman, A. R., Cough response and uptake of mercury by brook trout, *Salvelinus fontinalis*, exposed to mercuric compounds at different hydrogen-ion concentrations, *Trans. Am. Fish. Soc.*, 103, 244, 1974.

Endo, M., Onoue, Y. and Kuroki, A., Neurotoxin-induced cardiac disorder and its role in the death of fish exposed to *Chattoneilla marina*, *Mar. Biol. (Berlin)* 112, 371, 1992.

Farrell, A. P., Cardiovascular system, in, *The Physiology of Fishes*, Evans, D. H., Ed., CRC Press, Boca Raton, FL. 1993, chap. 8.

Gruber, D., Diamond, J. and Parsons, M., Automated biomonitoring, *Environ. Auditor*, 2, 229, 1991.

Heath, A. G., A critical comparison of methods for measuring fish respiratory movements, *Water Res.*, 6, 1, 1972.

Hoar, W. S. and Randall, D. J., *Fish Physiology*, Vol. 10 (Part A), Academic Press, New York, 1984.

Hoar, W. S., Randall, D. J. and Farrell, A. P., Eds., *Fish Physiology: Cardiovascular Systems*, Vol. 12, Academic Press, New York, 1992.

Holeton, G. F. and Randall, D. J., The effects of hypoxia upon the partial pressure of gases in the blood and water afferent and efferent to the gills of rainbow trout, *J. Exp. Biol.*, 46, 317, 1967.

Horkel, J. D. and Pearson, W. D., Effects of turbidity on ventilation rates and oxygen consumption of green sunfish, *Lepomis cyanellus*, *Trans. Am. Fish. Soc.*, 105, 107, 1976.

Hughes, G. M., How a fish extracts oxygen from water, *New Sci.*, 11, 346, 1961.

Hughes, G. M., Coughing in the rainbow trout (*Salmo gairdneri*) and the influence of pollutants, *Rev. Suisse Zool.*, 82, 47, 1975.

Hughes, G. M., Scanning electron microscopy of the respiratory surfaces of trout gills, *J. Zool. (London)*, 188, 443, 1979.

Hughes, G. M., Functional morphology of fish gills, in *Epithelial Transport in the Lower Vertebrates*, Lahlou, B., Ed., Cambridge University Press, New York, 1980, 15.

Hughes, G. M., An introduction to the study of gills, in *Gills*, Houlihan, D. F., Rankin, J. C. and Shuttleworth, T. J., Eds., Cambridge University Press, New York, 1982, 1.

Hughes, G. M. and Adney, R. J., Variations in the pattern of coughing in rainbow trout, *J. Exp. Biol.*, 68, 109, 1977a.

Hughes, G. M. and Adney, R. J., The effects of zinc on the cardiac and ventilatory rhythms of rainbow trout (*Salmo gairdneri*, Richardson) and their responses to environmental hypoxia, *Water Res.*, 11, 1069, 1977b.

Hughes, G. M. and Perry, S. F., Morphometric study of trout gills: a light-microscopic method suitable for the evaluation of pollutant action, *J. Exp. Biol.*, 64, 447, 1976.

Hughes, G. M. and Tort, L., Cardio-respiratory responses of rainbow trout during recovery from zinc treatment, *Environ. Poll. Series* A, 37, 255, 1985.

Hughes, G. M., Perry, S. F. and Brown, V. M., A morphometric study of effects of nickel, chromium and cadmium on the secondary lamellae of rainbow trout gills, *Water Res.*, 13, 665, 1979.

Hughes, G. M., Horimoto, M., Kikuchi, Y., Kakiuchi, Y. and Koyamo, T., Blood-flow velocity in microvessels of the gill filaments of the goldfish (*Carassius auratus* L.), *J. Exp. Biol.*, 90, 327, 1981.

Jagoe, C. H. and Haines, T. A., Alterations in gill epithelial morphology of yearling sunapee trout exposed to acute acid stress, *Trans. Am. Fish. Soc.*, 112, 689, 1983.

Jagoe, C. H., Matey, V., Haines, T. and Komov, V., Effect of beryllium on fish in acid water is analogous to aluminum toxicity, *Aquat. Toxicol.*, 24, 241, 1993.

Jones, J. R. E., *Fish and River Pollution*, Butterworths, London, 1964.

Kiceniuk, J. W., Penrose, W. R. and Squires, W. R., Oil spill dispersants cause bradycardia in a marine fish, *Mar. Pollut. Bull.*, 9, 42, 1978.

Kirk, R. and Lewis, J., An evaluation of pollutant induced changes in the gills of rainbow trout using scanning electron microscopy, *Environ. Technol.*, 14, 577, 1993.

Klaverkamp, J. F., The physiological pharmacology and toxicology of fish cardiovascular systems, in, *Aquatic Toxicology*, Weber, L. J., Ed., Raven Press, New York, 1982, chap. 1.

Kumaraguru, A. K., Beamish, F. W. H. and Ferguson, H. W., Direct and circulatory paths of Permethrin (NRDC-143) causing histopathological changes in the gills of rainbow trout (*Salmo gairdneri* Richardson), *J. Fish Biol.*, 20, 87, 1982.

Lang, T., Peters, G., Hoffman, R. and Meyer, E., Experimental investigation on the toxicity of ammonia: effects on ventilation frequency, growth, epidermal mucous cells and gill structure of rainbow trout, *Salmo gairdneri*, *Dis. Aquat. Orgs.*, 3, 159, 1987.

Laurent, P., Gill internal morphology, in, *Fish Physiology*, Vol. 10A, Hoar, W. S. and Randall, D. J., Eds., Academic Press, New York, 1984, 73.

Lunn, C. R., Toews, D. P. and Pree, D. J., Effects of three pesticides on respiration, coughing and heart rates of rainbow trout (*Salmo gairdneri*), *Can. J. Zool.*, 54, 214, 1976.

Majewski, H. S. and Giles, M. A., Cardiovascular-respiratory responses of rainbow trout (*Salmo gairdneri*) during chronic exposure to sublethal concentrations of cadmium, *Water Res.*, 15, 1211, 1981.

Maki, A. W., Respiratory activity of fish as a predictor of chronic fish toxicity values for surfactants, in, *Aquatic Toxicology*, ASTM ATP 667, Marking, L. L. and Kimberle, R. A., Eds., American Society for Testing and Materials, Philadelphia, 1979, 77.

Mallatt, J., Fish gill structural changes induced by toxicants and other irritants: a statistical review, *Can. J. Fish. Aq. Sci.*, 42, 630, 1985.

Malte, H. and Weber, R., Respiratory stress in rainbow trout dying from aluminum exposure in soft, acid water, with or without added sodium chloride, *Fish Physiol. Biochem.*, 5, 249, 1988.

Marvin, D. and Heath, A., Cardiac and respiratory responses to gradual hypoxia in three ecologically distinct species of freshwater fish, *Comp. Biochem. Physiol.*, 27, 349, 1968.

Mazeaud, M. M. and Mazeaud, F., Adrenergic responses to stress in fish, in, *Stress and Fish*, Pickering, A. D., Ed., Academic Press, New York, 1981, chap. 3.

McCloskey, J. and Oris, J., Effect of water temperature and dissolved oxygen concentration of the photo-induced toxicity of anthracene to juvenile bluegill sunfish, *Lepomis macrochirus*, *Aquat. Toxicol.*, 21, 145, 1991.

McKim, J. M. and Goeden, H. M., A direct measure of the uptake efficiency of a xenobiotic chemical across the gills of brook trout (*Salvelinus fontinalis*) under normoxic and hypoxic conditions, *Comp. Biochem. Physiol.,* 72C, 65, 1982.

Meyers, T. and Hendricks, J., Histopathology, in, *Fundamentals of Aquatic Toxicology, Methods and Applications,* Rand, G. and Petrocelli, S., Eds., Hemisphere Publishing, Washington, D.C., 1985, 283.

Milligan, C. L. and Wood, C. M., Disturbances in haematology, fluid volume distribution and circulatory function associated with low environmental pH in the rainbow trout, (*Salmo gairdneri*), *J. Exp. Biol.,* 99, 397, 1982.

Mitchell, S. J. and Cech, J. J., Ammonia-caused gill damage in channel catfish (*Ictalurus punctatus*): confounding effects of residual chlorine, *Can. J. Fish. Aq. Sci.,* 40, 242, 1983.

Moffitt, B., Thermoregulatory and respiratory responses of brown bullhead catfish to cadmium ion stress, *Physiologist,* 33, A105, 1990.

Nikl, D. L. and Farrell, A. P., Reduced swimming performance and gill structural changes in juvenile salmonids exposed to 2-(thiocyanomethylthio)benxothiazole, *Aquat. Toxicol.,* 27, 245, 1993.

Ogilvie, D. M. and Stechey, D. M., Effects of aluminum on respiratory responses and spontaneous activity of rainbow trout, *Salmo gairdneri, Environ. Toxicol. Chem.,* 2, 43, 1983.

Olson, K. R., Fromm, P. O. and Franz, W. L., Ultrastructural changes of rainbow trout gills exposed to methyl mercury or mercuric chloride, *Fed. Proc.,* 32, 261, 1973.

Perry, S. F. and McDonald, G., Gas exchange, in, *The Physiology of Fishes,* Evans, D., Ed., CRC Press, Boca Raton, FL, 1993, chap. 9.

Perry, S. F. and Wood, C. M., Control and coordination of gas transfer in fishes, *Can. J. Zool.,* 67, 2961, 1989.

Prasad, M., SEM study on the effects of crude oil on the gills and air breathing organs of climbing perch, *Anabas testudineus, Bull. Environ. Contam. Toxicol.,* 47, 882, 1991.

Randall, D. J., The control of respiration and circulation in fish during excercise and hypoxia, *J. Exp. Biol.,* 100, 275, 1982.

Randall, D. J., Control and coordination of gas exchange in water breathers, in, *Advances in Comparative and Environmental Physiology,* Boutilier, R. G., Ed., Springer-Verlag, Berlin, 1990, 253.

Randall, D. J., Lin, H. and Wright, P., Gill water flow and the chemistry of the boundary layer, *Physiol. Zool.,* 64, 26, 1991.

Rankin, J. C., Stagg, R. M. and Bolis, L., Effects of pollutants on gills, in, *Gills,* Houlihan, D. F., Rankin, J. C. and Shuttleworth, T. J., Eds., Cambridge University Press, New York, 1982, 207.

Richmonds, C. and Dutta, H., Histopathological effects of malathion on the gills of the bluegill, *Lepomis macrochirus, Am. Zool.,* 27(4) 126A, 1987.

Satchell, G. H., Respiratory toxicology of fishes, in, *Aquatic Toxicology,* Weber, L. J., Ed., Raven Press, New York, 1984, 1.

Satchell, G. H., *Physiology and Form of Fish Circulation,* Cambridge University Press, New York, 1991.

Sawyer, P. and Heath, A., Cardiac, ventilatory and metabolic responses of two ecologically dissimilar species of fish to waterborne cyanide, *Fish Physiol. Biochem.,* 4, 203, 1988.

Scheier, A. and Cairns, J., Persistence of gill damage in *Lepomis gibbosus,* following a brief exposure to alkyl benzene sulfonate, *Notulae Nat.,* 391, 1, 1966.

Schreck, C. B., Stress and compensation in teleostean fishes: responses to social and physical factors, in, *Stress and Fish,* Pickering, A. D., Ed., Academic Press, New York, 1981, chap. 13.

Sellers, C. M., Heath, A. G. and Bass, M. L., The effect of sublethal concentrations of copper and zinc on ventilatory activity, blood oxygen and pH in rainbow trout (*Salmo gairdneri*), *Water Res.,* 9401, 1975.

Shelton, G., Jones, D. R. and Milsom, W. K., Control of breathing in ecothermic vertebrates, in, *Handbook of Physiology, Section 3. The Respiratory System,* Vol. II, Part 2, Fishman, A. P., Cherniack, N., Widdicombe, J. and Geiger, S., Eds., American Physiological Society, Bethesda, MD, 1986, 857.

Sippel, A., Geraci, J. and Hodson, P., Histopathological and physiological responses of rainbow trout (*Salmo gairdneri*) to sublethal levels of lead, *Water Res.,* 17, 1115, 1983.

Skidmore, J. F. and Tovell, P. W. A., Toxic effects of zinc sulphate on the gills of rainbow trout, *Water Res.,* 6, 217, 1972.

Skidmore, J. F., Respiration and osmoregulation in rainbow trout with gills damaged by zinc sulphate, *J. Exp. Biol.,* 52, 481, 1970.

Smart, G. R., The effect of ammonia exposure on gill structure of the rainbow trout *(Salmo gairdneri),* *J. Fish Biol.,* 8, 471, 1976.

Smart, G., Investigations of the toxic mechanisms of ammonia to fish–gas exchange in rainbow trout *(Salmo gairdneri),* exposed to acutely lethal concentrations, *J. Fish. Biol.,* 12, 93, 1978.

Smith, C. E. and Piper, R. G., Pathological effects in formalin-treated rainbow trout *(Salmo gairdneri),* *J. Fish. Res. Bd. Can.,* 29, 328, 1972.

Snieszko, S. F., Bacterial gill disease of freshwater fishes, in, *Fish Disease Leaflet,* 62, U.S. Fish and Wildlife Service, Division of Fishery Ecology Research, Washington, D.C., 1980.

Sorensen, D. M. B., Henry, R. E. and Ramirez-Mitchell, R., Arsenic accumulation, tissue distribution and cytotoxicity in teleosts following indirect aqueous exposures, *Bull. Environ. Cont. Toxicol.,*21, 162, 1979.

Spry, D. J. and Wood, C. M., Acid-base, plasma ion and blood gas changes in rainbow trout during short term toxic zinc exposure, *J. Comp. Physiol. B,* 154, 149, 1984.

Stromberg, P. C., Ferrante, J. G. and Carter, S., Pathology of lethal and sublethal exposure of fathead minnows; *Pimephales promelas* to cadmium, *J. Toxicol. Environ. Health,* 11, 247, 1983.

Thomas, R. E. and Rice, S. D., The effect of exposure temperatures on oxygen consumption and opercular breathing rates of pink salmon fry exposed to toluene, naphthalene, and water-soluble fractions of Cook Inlet crude oil and No. 2 fuel oil, in, *Marine Pollution; Functional Responses,* Vernberg, W. B., Thurberg, F. P., Calabrese, A. and Vernberg, F. J., Eds., Academic Press, New York, 1979, 39.

Tort, L. and Madsen. L., The effects of the heavy metals cadmium and zinc on the contraction of ventricular fibres in fish, *Comp. Biochem. Physiol.,* 99C, 353, 1991.

Tortorelli, M., Hernandez, D., Vazquez, G. and Salibian, A., Effects of paraquat on mortality and cardiorespiratory function of catfish fry *Plecostomus commersoni, Arch. Environ. Contam. Toxicol.,* 19, 523, 1990.

Travis, T. W. and Heath, A. G., Some physiological responses of rainbow trout *(Salmo gairdneri)* to intermittent monochloramine exposure, *Water Res.,* 15, 977, 1981.

Turner, J. D., Wood, C. M. and Clark, D., Lactate and proton dynamics in the rainbow trout *(Salmo gairdneri),* *J. Exp. Biol.,* 104, 247, 1983.

Tuurala, H. and Soivio, A., Structural and circulatory changes in the secondary lamellae of *Salmo gairdneri* gills after sublethal exposures to dehydroabietic acid and zinc, *Aquat. Toxicol.,* 2, 21, 1982.

Ultsch, G. R. and Gros, G., Mucus as a diffusion barrier to oxygen: possible role in oxygen uptake at low pH in carp *(Cyprinus carpio)* gills, *Comp. Biochem. Physiol.,* 62A, 685, 1979.

van der Putte, I., Laurier, M. B. H. M. and van Eijk, G. J. M., Respiration and osmoregulation in rainbow trout *(Salmo gairdneri)* exposed to hexavalent chromium at different pH values, *Aquat. Toxicol.,* 2, 99, 1982.

Wedemeyer, G. A., Nelson, N. C. and Yasutake, W. T., Physiological aspects of ozone toxicity to rainbow trout *(Salmo gairdneri),* *J. Fish. Res. Bd. Can.,* 36, 605,1979.

Wilson, R. and Taylor, E., The physiological responses of freshwater rainbow trout, *Oncorhynchus mykiss,* during acutely lethal copper exposure, *J. Comp. Physiol. B.,* 163, 38, 1993.

Wobeser, G., Acute toxicity of methyl mercury chloride and mercuric chloride for rainbow trout *(Salmo gairdneri)* fry and fingerlings, *J. Fish. Res. Bd. Can.,* 32, 2005, 1975.

Chapter 4

Hematology

I. INTRODUCTION

Hematology is defined as the study of blood and the blood-forming tissues. Blood is composed of the liquid plasma and the blood cells (sometimes called formed elements). Serum is the fluid left after blood has clotted; it has essentially the same chemical makeup as plasma, except for the absence of some of the clotting factors. While many workers include the chemistry of plasma as a part of hematology (Hawkins and Mawdesley-Thomas, 1972; Folmar, 1993), they will be kept separate here, because the composition of plasma is a reflection of the function of numerous organs (e.g., gills, liver, kidney, muscle, etc.) that are considered in other parts of this book. Thus, this chapter will be limited to a review of the "toxicophysiology" of blood cells only.

II. FISH BLOOD CELLS AND THEIR MEASUREMENT

The blood cells of teleost fish are produced in the hematopoietic tissue which is located primarily in the spleen and kidney (Satchell, 1991; Fange, 1992). In contrast to mammals, there is no bone marrow in fish, nor do they have lymph nodes. Within the blood cell-forming tissues, the hemocytoblast is the cell that gives rise to all the others, both erythrocytes and leukocytes. The latter are divided into the lymphocytes, granulocytes (of which the neutrophils and eosinophils are the most common), and thrombocytes. Lymphocytes are the most numerous of all the leukocytes. Fish do not have platelets as in mammals, instead the thrombocytes contain the clotting factors that are normally found in platelets.

An overall reduction in leukocytes (leukocytopenia) in response to stress is characteristic of all sorts of vertebrates. It is a non-specific response to any type of inner or environmental stressor and is mediated by the corticosteroid hormones. It is a fundamental mechanism in the increased susceptibility of fish to disease organisms when they are suffering from the stress of pollution (see Chapter 11).

The erythrocytes (red blood cells) of fish are nucleated and similar in size to the leukocytes, in contrast to mammals in which the latter are the much larger cell. Erythrocytes contain hemoglobin, which enables the blood to hold a far greater amount of oxygen than it otherwise could (50- to 100-fold, depending on the hemoglobin concentration).

The term anemia is applied to any condition where the concentration of the hemoglobin in the blood is abnormally low, whether this is due to a reduction in number of erythrocytes or to an inadequate amount of hemoglobin in the cells.

For the measurement of hematological factors in fish, techniques similar to those used in human and veterinary clinical laboratories are generally used, with minor modifications. Blaxhall and Daisley (1973), Wedemeyer and Yasutake (1977), Ellis (1977), and especially Houston (1990) provide practical guides to these methods adapted for use on fish blood. The following are the principle factors measured in work on the action of pollutants.

A. HEMATOCRIT

Hematrocrit is sometimes called the "packed cell volume" and is determined by spinning a sample of whole blood (usually contained in a sealed capillary tube) in a centrifuge and the relative volume of the packed cells is then measured. Hypoxia (*in vitro*) causes the cells to swell, thereby increasing the hematocrit. Because erythrocytes of fish have a high intrinsic rate of oxygen consumption (Eddy, 1977), delays in spinning down the hematocrit tubes can yield excessively high values due to cell swelling induced by their utilization of the available oxygen in the microhematocrit tube.

B. BLOOD CELL COUNT

Fish erythrocytes are easier to count than mammalian ones because they are larger and fewer in number. Both these and the leukocytes can be counted in the same hemocytometers used for human blood or with the aid of automatic electronic counters found in many clinical laboratories. (The latter may not distinguish leukocytes because in fish they are approximately the same size as the erythrocytes.)

C. HEMOGLOBIN CONCENTRATION

For this purpose, whole blood is analyzed after the cells have been hemolyzed to release the hemoglobin into solution. This requires less blood than does a hematocrit so when working with small fish where volumes are severely limited, this may be preferable to use.

There is generally a good correlation between the hematocrit, hemoglobin concentration, and RBC count, but if all three are measured, it is useful to calculate the following:

$$MCV = \text{Mean corpuscular volume in } \mu m^3/\text{cell}$$

$$= \text{Hematocrit ratio} \times 1000/\text{RBC count}$$

$$MCH = \text{Mean corpuscular hemoglobin in } \mu g/\text{cell}$$

$$= \frac{\text{Hb conc. in } g/1000 \text{ ml}}{\text{RBC count}}$$

In both cases, the RBC count is expressed in the unit millions/mm^3. With these calculated data, one may be able to detect the presence of a physiological lesion in the hemoglobin-forming process as opposed to the rate of blood cell mitosis.

D. BLOOD CELL DIFFERENTIAL COUNT

In order to perform the counts necessary for calculating these, fresh blood smears are made and stained for microscopic examination. The relative numbers of lymphocytes, granulocytes, and thrombocytes are then enumerated. Some workers also count different developmental stages of the erythrocytes.

Houston et al. (1993) use the term "erythron organization" to refer to the relative numbers of karyorrhetic cells (those with disintegrated nuclei), dividing blood cells, and changes in cytomorphology. They maintain this is more sensitive to pollutant stressors than the more traditional hematological measures. Moreover, the blood smears can be prepared in the field, an obvious advantage where natural populations are being sampled.

The location in the fish from which blood is sampled can influence the hematological values recorded. Soivio et al. (1981) have discovered that blood coming from the gills (i.e., in the dorsal aorta) has a higher hematocrit than blood going into the gills (i.e., in the ventral aorta). This is due to "plasma skimming" by the gills. The skimmed off plasma is recycled back to the heart via the venous system so blood taken from the severed caudal peduncle (a common practice) will have a slightly higher hematocrit and hemoglobin concentration than that taken from the heart (another common practice).

The estimation of normal values for hematocrit, hemoglobin, and RBC count can be a challenging problem (Miller et al., 1983). Variables that can influence one or more of the factors include season (Denton and Yousef, 1975; Bidwell and Heath, 1993), disease (Barham et al., 1980), stress of capture (Lowe-Jinde and Niimi, 1983), and chemical pollutants, which are the topic of this chapter.

Increases or decreases in hematological factors in response to environmental stressors, chemical or otherwise, may be due to water loss or gain in the blood (see Chapter 7). Thus osmoregulatory dysfunction may cause an apparent anemia or its opposite, polycythemia, even though the blood-forming machinery is unaffected by the pollutant. Consequently, it is useful to have measures of electrolytes, osmolality, and/or plasma protein concentration as additional information when doing hematological studies of fish. Finally, if repeated blood sampling from the same specimen is contemplated, whether from a catheter or by heart puncture, the matter of blood loss needs to be considered (Hoffman and Lommel, 1984).

Exposure to chemical pollutants or environmental hypoxia can induce either increases or decreases in the different hematological measures; the trend is almost always the same in all three traditional measures. We will discuss first the conditions that produce anemia and then consider the opposite physiological response. Within groups, the mechanisms responsible for the changes differ somewhat with the chemical.

III. CHEMICALS THAT CAUSE ANEMIA

A. CADMIUM

Cadmium has been reported to cause anemia in a variety of fish species at low and high concentrations (e.g., Larsson et al., 1976; Newman and McLean, 1974; Larsson et al., 1985; Gill and Pant, 1985; Houston et al., 1993). It is also well known for its ability to cause anemia in mammals (Friberg et al., 1974). The mechanism involved in these animals is reportedly to be due, largely, to a reduction in absorption of iron from the gut for the synthesis of hemoglobin. It is not known if this mechanism also occurs in fish. If so, it is certainly not the only one, as an anemic condition is produced in flounders exposed to cadmium at 1 ppm for 15 days during which time they were not fed (Larsson et al., 1976), thus, there was no opportunity for iron uptake from the gut in either control or experimental animals.

Cadmium also causes an abnormally large number of malformed erythrocytes (Newman and McLean, 1974; Houston et al., 1993) which implies a lesion in the machinery for forming blood cells. Measurements of the enzyme delta-aminolevulinic acid dehydratase (ALA-D; see section on lead below) in the kidney have shown a compensatory increase as a result of exposure to cadmium (Johansson-Sjobeck and Larsson, 1978). This indicates that the first steps of hemoglobin synthesis are not blocked by cadmium, but in spite of the increased enzyme activity, hemoglobin synthesis still goes down. The net effect is a slowing of erythrocyte maturation (Houston et al., 1993). In addition to the reduction in capacity to produce normal blood cells, some workers have also noted increased rates of erythrocyte breakdown (Houston et al., 1993). Palace et al., (1993) reported that erythrocytes from rainbow trout exposed to cadmium for 181 days had a tendency for lysis. This increased cell fragility was attributed to lipid peroxidation of the erythrocyte membranes which affects membrane fluidity. They present other evidence of stimulated antioxidant defenses which are mobilized to counteract increased peroxidation.

After an induction of anemia by cadmium in flounder, Larsson et al. (1985) found that a year's recovery in non-contaminated water abolished the anemia. This recovery is interesting in light of the finding by the same research team that carbohydrate metabolism was still disturbed, possibly due to permanent damage to the insulin-secreting cells in the pancreas (see Chapter 8). From this, it appears that some cadmium-induced lesions are far more permanent than others.

Cadmium has also been shown to alter the leukocyte differential count, but the direction of change has not been consistent between different studies. Newman and MacLean (1974) reported a dose-dependent increase (threefold) in neutrophils and a dose-dependent decrease (also nearly threefold) in lymphocytes in the cunner (*Tautogolabrus adspersus*). This contrasts with the "markedly enhanced lymphocyte count" observed by Larsson et al. (1985) in perch from a river polluted with cadmium. The contradictory results could be due to a different pattern of disease organisms present in the two studies. Differential leukocyte counts are especially susceptible to this variable.

B. LEAD

Chronic exposure of the cyprinid *Barbus* to lead for up to 60 days at 47 µg/L resulted in severe reductions (12 to 31%) in erythrocyte count, hematocrit, hemoglobin concentration, and MCV (Tewari et al., 1987). This element has been shown to cause anemia in mammals by inhibiting hemoglobin synthesis and shortening the lifespan of circulating erythrocytes (Hernberg, 1976). Lead inhibits the enzyme ALA-D which is required in the early stages of hemoglobin synthesis in the hemopoietic tissue. The inhibition of this enzyme by lead is quite specific; other metals such as cadmium have no effect on it in mammals. Such specificity of inhibition is rare, so measurement of ALA-D activity in erythrocytes of humans has been useful for diagnosing lead poisoning (Hernberg et al., 1970), and, whereas it is stimulated by cadmium in the kidney of fish (discussed above), erythrocytic ALA-D is unaffected by this element (Hodson et al., 1984).

Trout, exposed to waterborne lead at concentrations of 10, 75, and 300 µg/L for 30 days, exhibited anemia only at the highest concentration. On the other hand, there was a dose-dependent inhibition of ALA-D in the erythrocytes and spleen (Johansson-Sjobeck and Larsson, 1979). At the highest dose, an 86% inhibition of erythrocytic ALA-D was observed while the lower lead concentrations induced an inhibition of 74 and 21%, respectively. No effect on hemoglobin synthesis was noted at these lower levels of lead implying the enzyme has a large reserve capacity, because essentially normal function was possible with only 25% of the usual enzymatic activity.

The recovery of ALA-D activity is very slow when fish are allowed to reside in lead-free water following an exposure. Johansson-Sjobeck and Larsson (1979) found that after 7 weeks of recovery there was only a slight improvement in enzyme activity. It is not clear whether this is due to the continued presence of lead in the tissues (it was not measured), or a slow rate of enzyme synthesis. Intuitively, the former seems far more likely as lead is retained in the kidney of fish for at least a month following its uptake from the environment (Reichert et al., 1979). From a practical standpoint, the continued inhibition of ALA-D following exposure to lead may add to its usefulness for diagnosing lead poisoning in fish taken from polluted waterways. The assay of ALA-D activity is presumably easier than measuring lead in the tissues.

C. MERCURY

Both winter flounder and striped bass exhibited a distinct anemia when exposed to mercury (inorganic) for 60 days (Dawson, 1979, 1982). The magnitude of the change was smaller in the flounder even though a higher mercury concentration was used with that species. Anemia was also reported in the plaice with higher and shorter duration exposures (Fletcher and White, 1986). Mercury tends to concentrate in the kidney of teleosts (Penreath, 1976) where it probably inhibits uroporphyrinogen I synthetase, a heme biosynthetic enzyme (Tephly et al., 1978). It has also been shown to reduce the deformability of the erythrocyte membranes which could contribute to their early destruction (Brouwer and Brower-Hoexum, 1985).

Methylmercury appears to be less "effective" at producing anemia than is inorganic mercury even though the former is more toxic (Lock et al., 1981). When rainbow trout were exposed to 15 µg/L waterborne methylmercury for 75–119 days, they accumulated whole-body residue levels of 3–12 mg/kg, but no discernible hematological effects were found (Niimi and Lowe-Jinde, 1984). In addition, the effect on the hematocrit from methylmercury in the food is equivocal; there have been increases and decreases reported (Rogers and Beamish, 1982; Bidwell and Heath, 1993).

One final point about mercury in fish should be mentioned. Methylmercury binds reversibly with the hemoglobin in erythrocytes. This has been shown both *in vitro* and *in vivo*; the reaction is with the –SH groups of the hemoglobin molecule (Massaro, 1974). Thus, it could have an effect on the binding of hemoglobin with oxygen, although this seemingly is not known.

D. CHLORAMINE

When chlorine (as a gas or as hypochlorite) is added to water as a disinfectant or for antifouling purposes, it immediately forms hypochlorous acid, which is commonly called "free chlorine". If ammonia is present, the chlorine will unite with it and become a chloramine (sometimes called "combined chlorine"). There are similarities and differences in the physiological actions of free and combined chlorine (Travis and Heath, 1981). Buckley et al. (1976) found hemolytic anemia in salmon exposed for 12 weeks to municipal wastewater containing chloramines. Free chlorine, on the other hand, has been shown to cause an increase in hemoglobin concentration (Bass and Heath, 1977). This increase is probably an adaptation to the internal hypoxia induced by the damage to the gill tissue. Internal hypoxia does not occur with chloramines (Travis and Heath, 1981) except perhaps at unrealistically high doses.

Exposure of fish to chloramine, in addition to causing anemia, results in oxidation of part of the hemoglobin to methemoglobin, which does not transport oxygen (Grothe and Eaton, 1975). This is in contrast to free chlorine which causes little methemoglobin formation in trout (Bass and Heath, 1977) or even when the erythrocytes are exposed to the chemical *in vitro* (Buckley, 1981). Thus, both forms of chlorine reduce the available oxygen to the tissue. Chloramine lowers the oxygen-carrying capacity of the blood while free chlorine lowers the oxygen tension in the arterial blood by causing an impairment of oxygen uptake by the gills (Chapter 3). From the standpoint of the fish, chloramine will probably be more debilitating because, due to the oxygen dissociation characteristics of the blood, a considerable decrease in arterial oxygen tension is required before a reduction in actual oxygen availability occurs. Free chlorine affects only the tension whereas chloramine reduces the amount of oxygen on the hemoglobin molecule.

E. NITRITE

Nitrite, from the bacterial reduction of nitrate, enters the gills of fish via the chloride-transporting mechanism. It can accumulate in the blood plasma against a concentration gradient (Williams and Eddy, 1988). As with chloramine discussed above, nitrite oxidizes hemoglobin to methemoglobin; the extent of this conversion is directly proportional to the plasma nitrite concentration which, in turn, is proportional to environmental concentration (Tomasso, 1986). Along with reducing the oxygen-carrying capacity of the blood by oxidizing the hemoglobin, nitrite also seems to reduce oxygen affinity, as was shown with *in vitro* studies of carp blood where the P_{50} was increased (indicating reduced affinity) by almost threefold (Williams et al., 1993). This could have a considerable effect on the ability of the fish to tolerate low dissolved oxygen (see Chapter 2). The conversion of hemoglobin to methemoglobin causes hemolytic anemia which can cause the blood plasma to acquire a reddish-brown color (Williams and Eddy, 1988).

Nitrite poisoning of hemoglobin also affects the temperature tolerance. Watenpaugh et al. (1985) exposed channel catfish to several sublethal concentrations of nitrite for 24 h and then tested the critical thermal maximum. This was inversely related to the nitrite concentration with mean lethal temperatures ranging from 38°C for controls to 35.9°C for those exposed to 1.4 mg/L nitrite.

F. PULPMILL EFFLUENT

Long-term (25 days) exposure to the effluent from kraft pulpmills has been found to cause anemia in salmon (McLeay, 1973). The composition of this effluent is extremely variable between different mills and at various times at a single one (Davis, 1976). There are several toxic components in the effluent, one of which is dehydroabietic acid, one of the naturally occurring resin acids extracted from softwood trees. When coho salmon were exposed for up to 4 days to a concentration of this chemical approximating half the 96-h LC50, there were no changes found in the hematology. However, there was a significant increase in clotting time in blood from exposed animals (Iwama et al., 1976). Because there was also a decrease in total white cell count, it is probable that thrombocytes were down, although they were not counted separately. Such a decrease in leukocytes is a typical response to virtually any environmental stress (Ellis, 1981). However, a decrease in clotting time is the usual response to stress, instead of an increase as was seen here. Thus, this effect on clotting time could be a rather specific action of dehydroabietic acid.

In an *in vitro* study using rainbow trout erythrocytes, Bushnell et al. (1985) found the resin acid caused a decrease in cellular ATP and oxygen consumption and marked increase in hemolysis. In other work they cite, jaundice has been noted in fish exposed to some pulpmill effluents; the hemolytic anemia could be the primary reason for that condition.

Lehtinen et al. (1990) used hematological measurements (along with those for osmo-regulation and mixed function oxidase) to compare different bleaching processes in kraft mills. The exposures were for 7 weeks at concentrations of 400 and 2000 times dilution. Little effect was seen on hematology except with the effluent from conventional chlorine bleaching which caused a considerable anemia. This approach of comparing different industrial processes using several biomarkers in fish is one that should be utilized more extensively.

G. MISCELLANEOUS

A variety of chemicals are probably capable of causing varying degrees of anemia in fish. These include organochlorine pesticides (Venkateshwarlu et al., 1990), the fungicide chlorothalonil (Davies, 1987), and cobalt (Pamila et al., 1991). Cobalt is somewhat interesting in that it inhibits the enzyme 5-aminolevulinate synthetase, one of the enzymes involved in heme synthesis. When *Sarotherodon mossambicus*, an Indian freshwater teleost, was exposed to cobalt for up to 15 days, a marked increase in erythrocyte count occurred within 5 days, although total hemoglobin content of the blood was declining. It appeared the fish were overcompensating for the inhibited hemoglobin synthesis.

H. BIOLOGICAL SIGNIFICANCE OF ANEMIA

In evaluating the chemicals listed above and others that may be found, it must be kept in mind that a mild condition of anemia, such as was found in most of the studies mentioned here, is probably not particularly debilitating. There is some evidence to suggest that fish are able to tolerate a far greater degree of anemia than can mammals. Holeton (1971) subjected rainbow trout to carbon monoxide thereby completely inactivating their hemoglobin and found the fish were able to survive this (at rest), providing temperatures did not exceed 12°C. Anthony (1961) did the same thing with goldfish and reported that they survived over 24 h at 30°C and indefinitely at lower temperatures. More importantly, routine activity in the goldfish did not appear to depend on the presence of hemoglobin; and, of course, there are ice fish living around Antarctica which lack hemoglobin, but have compensatory adaptations, such as a large cardiac output (Holeton, 1970).

Where a low hemoglobin level may be most important is in fish subjected to severe environmental hypoxia (Holeton, 1972) (also see Chapter 2), or forced to swim at a high speed for long distances. Jones (1971) induced a condition of hemolytic anemia in rainbow trout by injection of phenylhydrazine and then tested their ability to swim in a water tunnel. By this means, he determined that a reduction of the hematocrit to one third of normal caused a 34 to 40% reduction in maximum sustained swimming speed. Thus, this could be an especially important debilitation for salmon during the upstream spawning migration.

IV. CHEMICALS CAUSING AN INCREASE IN HEMATOLOGICAL VARIABLES

An increase in the hematocrit alone has often been observed in fish subjected to a non-specific stressor (such as being lifted out of the water), and this can occur within minutes of the onset of stress (Casillas and Smith, 1977). Part, if not all, of this change is due to swelling of the erythrocytes which occurs whenever fish blood cells are exposed to a hypoxic environment (Soivio and Nikinmaa, 1981). It apparently does not matter whether the blood cells are exposed to the hypoxia in vitro or in vivo (Soivio and Nikinmaa, 1981; Swift and Lloyd, 1974). Thus, virtually any dose of pollutant that results in gill damage, and a subsequent internal hypoxia (this includes many chemicals, see Chapter 3) can be expected to also cause an increase in hematocrit.

Casillas and Smith (1977) further found that the hematocrit may return to normal within 60 min of cessation of the stress (in their case it was handling). These findings argue for the measuring of more than just the hematocrit when doing hematological studies; and transient increases in response to an acute exposure dose mean very little from a hematological standpoint.

A. ACID

It is quite common to have an increase in hematocrit, hemoglobin, and/or RBC count in fish residing in acidic water (see Wood and McDonald, 1982 for review). At an acutely low pH, this is largely due to hemoconcentration and swelling of blood cells brought on by a failure to regulate plasma electrolytes (see Chapter 10). There is also a release of erythrocytes from the spleen (Milligan and Wood, 1982). When the exposure to a low pH is of a more chronic type, the hematological increases could be a response to an impaired oxygen transport by the blood. Due to the well-known Bohr effect, the blood will have less affinity for oxygen at a lower pH so stimulation of erythrocyte production would then increase the oxygen capacity and thus partially compensate for this.

B. PESTICIDES

Among organic pesticides, it appears that some chlorinated hydrocarbons stimulate erythropoiesis; aldrin, chlordane, and pentachlorophenol are all in this group (Dhillon and Gupta, 1983; Davies, 1987; Iwama et al., 1986). Aldrin caused a dose-dependent increase in RBC count and total hemoglobin concentration, but the mean corpuscular hemoglobin actually went down (Dhillon and Gupta, 1983). This suggests a large increase in cell formation; so much so that hemoglobin synthesis did not keep up with it. The mechanism of this stimulatory effect of chlorinated hydrocarbons insecticides is unclear. In part, it may be due to an impairment in oxygen transfer at the gills, but that is only a speculation.

Organophosphate insecticides mostly cause increases in hematological variables (Natarajan, 1984; Lal et al., 1986). Evidence is beginning to accumulate supporting the idea that the effect is due to histological damage to gill tissues which produces an internal hypoxia and subsequent stimulation of erythropoiesis (Areechon and Plumb, 1990).

C. COPPER

McKim et al. (1970) observed that copper increased the hematocrit, hemoglobin, and RBC count in trout when exposed for 6 days to a concentration far below the LC50. By 21 days of continued exposure the hematocrit had returned to normal and by the 11th month, hemoglobin and blood cell counts had also returned to normal. Waiwood (1980) also reported increased hematocrits in trout exposed to copper but he was able to account for the entire change as being due to a shift of water from the plasma to the muscle cells, thereby producing hemoconcentration, as has also been observed by Wilson and Taylor (1993) in trout exposed to lethal copper concentrations. The McKim et al. (1970) chronic study found just the opposite; a slight increase in plasma water. Because copper is required for hemoglobin synthesis, a mild excess may be stimulatory, particularly if the fish were marginally deficient in copper at the start of an experiment.

In mammals, excess copper can cause hemolytic anemia by inhibiting glycolysis in the erythrocytes, denaturing the hemoglobin, and oxidizing glutathione (Fairbanks, 1967). Because fish erythrocytes have a greater aerobic capacity than mammalian ones (Eddy, 1977), perhaps they are not as sensitive to this element.

Other chemicals that stimulate blood cell/hemoglobin production, such as ozone (Wedemeyer et al., 1979), nickel (Ghazaly, 1992), zinc (Mishra and Srivastava, 1979; Hilmy et al., 1987), and hexavalent chromium (van der Putte et al., 1982) generally induce a hypoxic condition in the fish. Pentachlorophenol, on the other hand, may increase the metabolic demand by the organism for oxygen (Holmberg et al., 1972). Thus, these observed increases in oxygen capacity of the blood can be viewed as an adaptation to an altered respiratory homeostasis caused by the pollutant and not a toxic or direct stimulatory action of the chemical on the blood cell-forming tissues. A similar thing happens during acclimation to environmental hypoxia (see Chapter 2), and, in any situation where acute stress is imposed on the animal, adrenergic stimulation of the spleen can cause it to contract and release stored erythrocytes into the circulation Nilsson and Grove, 1974.

V. CONCLUDING REMARKS

Some final caveats are in order regarding the presumed stimulatory or inhibitory effect of some chemical on hematological variables in teleosts. It is not unheard of for a substance to produce one type of effect when exposure is long term but cause an exact opposite effect if the exposure is highly acute (e.g., Dhillon and Gupta, 1983). Where histological damage to gills reduces the transfer of oxygen into the blood, this can stimulate erythropoiesis while the chemical may, at the same time, be inhibiting some step in the formation of hemoglobin. The net effect on hematological variables will then be some kind of algebraic sum of these opposite effects. Finally, recent work (Bollard et al., 1993) has shown that artificially elevating the cortisol level can cause a drop in mean cellular hemoglobin content. Cortisol is frequently elevated in fish under a variety of stressors, physical, chemical or psychological, so this might be a further mechanism producing an anemia. Thus, considerable caution must be exercised when designating a chemical as being anemia-causing, or the reverse.

REFERENCES

Anthony, E. H., Survival of goldfish in presence of carbon monoxide, *J. Exp. Biol.*, 38, 109, 1961.

Areechon, N. and Plumb, J., Sublethal effects of malathion on channel catfish, *Ictalurus punctatus, Bull. Environ. Contam. Toxicol.*, 44, 435, 1990.

Barham, W. T., Smit, G. L. and Schoonbee, H. J., The haematological assessment of bacterial infection in rainbow trout (*Salmo gairdneri*), *J. Fish Biol.*, 17, 275, 1980.

Bass, M. L. and Heath, A. G., Cardiovascular and respiratory changes in rainbow trout (*Salmo gairdneri*), exposed intermittently to chlorine, *Water Res.*, 11, 497, 1977.

Bidwell, J. and Heath, A., An *in situ* study of rock bass (*Ambloplites rupestris*) physiology: effect of season and mercury contamination, *Hydrobiologia*, 264, 137, 1993.

Blaxhall, P. C. and Daisley, K. W., Routine haematological methods for use with fish blood, *J. Fish Biol.*, 5, 771, 1973.

Bollard, B., Pankhurst, N. and Wells, R., Effects of artificially elevated plasma cortisol levels on blood parameters in the teleost fish *Pagrus auratus* (Sparidae), *Comp. Biochem. Physiol.*, 106A, 157, 1993.

Brouwer, M. and Brouwer-Hoexum, T., Cupric and mercuric ions decrease red cell deformability, *Environs*, February 3, 1985.

Buckley, J. A., Intoxication of trout erythrocytes from hypochlorous acid and monochloramine *in vitro*: evidence for different modes of action, *Comp. Biochem. Physiol.*, 69C, 133, 1981.

Buckley, J. A., Whitmore, C. M. and Matsuda, R., Changes in blood chemistry and blood cell morphology in coho salmon (*Oncorhynchus kisutch*), following exposure to sublethal levels of total residual chlorine in municipal wastewater, *J. Fish. Res. Bd. Can.*, 33, 776, 1976.

Bushell, P., Nikinmaa, M. and Oikari, A., Metabolic effects of dehydroabietic acid on rainbow trout erythrocytes, *Comp. Biochem. Physiol.*, 81C, 391, 1985.

Casillas, E. and Smith, L. S., Effect of stress on blood coagulation and hematology in rainbow trout (*Salmo gairdneri*), *J. Fish Biol.*, 10, 481, 1977.

Davies, P., Physiological, anatomic and behavioural changes in the respiratory system of *Salmo gairdneri* on acute and chronic exposure to chlorothalonil, *Comp. Biochem. Physiol.*, 88C, 113, 1987.

Davis, J. C., Progress in sublethal effect studies with kraft pulpmill effluent and salmonids, *J. Fish. Res. Bd. Can.*, 33, 2031, 1976.

Dawson, M. A., Hematological effects of long-term mercury exposure and subsequent periods of recovery on the winter flounder, *Pseudopleuronectes americanus*, in, *Marine Pollution: Functional Responses*, Vernberg, W. B., Thurberg, F. P., Calabrese, A. and Vernberg, F. J., Eds., Academic Press, New York, 1979, 171.

Dawson, M. A., Effects of long-term mercury exposure on hematology of striped bass, *Morone saxatilis*, *Fish. Bull.*, 80, 389, 1982.

Denton, J. E. and Yousef, M. K., Seasonal changes in hematology of rainbow trout (*Salmo gairdneri*), *Comp. Biochem. Physiol.*, 51A, 151, 1975.

Dhillon, S. S. and Gupta, A. K., A clinical approach to study the pollutants intoxication in a freshwater teleost *Clarias batrachus*, *Water, Air, Soil Pollut.*, 20, 63, 1983.

Eddy, F. B., Oxygen uptake by rainbow trout (*Salmo gairdneri*) blood, *J. Fish Biol.*, 10, 87, 1977.

Ellis, A. E., The leucocytes of fish: a review, *J. Fish Biol.*, 11, 453, 1977.

Ellis, A. E., Stress and modulation of defense mechanisms in fish, in, *Stress and Fish*, Pickering, A.D., Ed., Academic Press, New York, 1981, chap. 7.

Fairbanks, U. F., Copper sulfate-induced hemolytic anemia, *Arch. Intern. Med.*, 120, 428, 1967.

Fange, R., Fish blood cells, in, *Fish Physiology*, Volume 12B, Hoar, W. S. and Randall, D. J., Eds., Academic Press, New York, 1992, chap. 1.

Fletcher, T. and White, A., Nephrotoxic and haematological effects of mercuric chloride in the plaice (*Pleuronectes platessa*), *Aquat. Toxicol.*, 8, 77, 1986.

76

Folmar, L., Effects of chemical contaminants on blood chemistry of teleost fish: a bibliography and synopsis of selected effects, *Environ. Toxicol. Chem.*, 12, 337, 1993.

Friberg, L. T., Piscator, M., Nordberg, G. and Kjellstrom, T., *Cadmium in the Environment*, 2nd ed., CRC Press, Boca Raton, FL, 1974.

Ghazaly, K., Sublethal effects of nickel on carbohydrate metabolism, blood and mineral contents of *Tilapia nilotica, Water, Air, Soil Pollut.*, 64, 525, 1992.

Gill, T. and Pant, J., Erythrocytic and leucocytic responses to cadmium poisoning in a freshwater fish, *Puntius conchonius, Environ. Res.*, 36, 327, 1985.

Grothe, D. and Eaton, J., Chlorine-induced mortality in fish, *Trans. Am. Fish. Soc.*, 104, 800, 1975,

Hawkins, R. E. and Mawdesley-Thomas, L. E., Fish haematology — a bibliography, *J. Fish Biol.*, 4, 193, 1972.

Hernberg, S., Biochemical, subclinical, and clinical responses to lead and their relation to different exposure levels as indicated by concentration of lead in blood, in, *Effects and Dose-Response Relationships of Toxic Metals*, Norberg, G. F., Ed., Elsevier, Amsterdam, 1976, 404.

Hernberg, S., Nikkanen, J., Mellin, G. and Lilius, H., Delta-aminolevulinic acid dehydratase as a measure of lead exposure, *Arch. Environ. Health*, 21, 140, 1970.

Hilmy, A., Domiaty, N., Daabees, A. and Latife, H., Some physiological and biochemical indices of zinc toxicity in two freshwater fishes, *Clarias lazera* and *Tilapia zilli, Comp. Biochem. Physiol.*, 87C, 297, 1987.

Hodson, P. V., Blunt, B. R. and Whittle, D. M., Monitoring lead exposure of fish, in *Contaminant Effects on Fisheries*, Cairns, V. W., Hodson, P. V. and Nriagu, J. O., Eds., John Wiley and Sons, New York, 1984, chap. 8.

Hoffman, R. and Lommel, R., Effects of repeated blood sampling on some blood parameters in freshwater fish, *J. Fish Biol.*, 24, 245, 1984.

Holeton, G. F., Oxygen uptake and transport by the rainbow trout during exposure to carbon monoxide, *J. Exp. Biol.*, 54, 239, 1971.

Holeton, G. F., Oxygen uptake and circulation by a hemoglobinless Antarctic fish (*Chaenopcephalus aceratus* Lonnberg) compared with three red-blood Antarctic fish, *Comp. Biochem. Physiol.*, 34, 457, 1970.

Holeton, G. F., Gas exchange in fish with and without hemoglobin, *Respir. Physiol.*, 14, 142, 1972.

Holmberg, B., Jensen, S., Larsson, A., Lewander, K. and Olsson, M., Metabolic effects of technical pentachlorophenol (PCP) on the eel *Anguilla anguilla* L., *Comp. Biochem. Physiol.*, 43B, 171, 1972.

Houston, A., Blood and circulation, in, *Methods for Fish Biology*, Schreck, C. and Moyle, P., Eds., American Fisheries Society, Bethesda, MD, 1990, chap. 9.

Houston, A., Blahut, S., Murad, A. and Amikrtharaj, P., Changes in erythron organization during prolonged cadmium exposure: an indicator of heavy metal stress?, *Can. J. Fish. Aquat. Sci.*, 50, 217, 1993.

Iwama, G. K., Greer, G. L. and Larkin, P. A., Changes in some hematological characteristics of coho salmon (*Oncorhynchus kisutch*) in response to acute exposure to dehydroabietic acid (DHAA) at different exercise levels, *J. Fish. Res. Bd. Can.*, 33, 285, 1976.

Iwama, G., Greer, G. and Randall, D., Changes in selected haematological parameters in juvenile chinook salmon subjected to a bacterial challenge and a toxicant, *J. Fish Biol.* 28, 563, 1986.

Johansson-Sjobeck, M. and Larsson, A., Effects of inorganic lead on delta-aminolevulinic acid dehydratase activity and hematological variables in the rainbow trout (*Salmo gairdneri*), *Arch. Environ. Contam. Toxicol.*, 8, 419, 1979.

Johansson-Sjobeck, M. and Larsson, A., The effect of cadmium on the hematology and on the activity of delta aminolevulinic acid dehydratase (ALA-D) in blood and hematopoietic tissues of the flounder, *Pleuronectes flesus* L., *Environ. Res.*, 17, 191, 1978.

Jones, D. R., The effect of hypoxia and anaemia on the swimming performance of rainbow trout (*Salmo gairdneri*), *J. Exp. Biol.*, 55, 541, 1971.

Lal, B., Singh, A., Kumari, A. and Sinha, N., Biochemical and haematological changes following malathion treatment in the freshwater catfish *Heteropneustes fossilis, Environ. Pollut. Ser. A*, 42, 151, 1986.

Larsson, A., Bengtsson, B. and Svanberg, O., Some haematological and biochemical effects of cadmium on fish, in, *Effects of Pollutants on Aquatic Organisms*, Lockwood, A. P. M., Ed., Cambridge University Press, Cambridge, 1976, 35.

Larsson, A., Haux, C. and Sjobeck, M., Fish physiology and metal pollution: results and experiences from laboratory and field studies, *Ecotoxicol. Environ. Safety*, 9, 250, 1985.

Lehtinen, K., Kierkegaard, A., Jakobsson, E. and Wandell, A., Physiological effects in fish exposed to effluents from mills with six different bleaching processes, *Ecotoxicol. Environ. Safety*, 19, 33, 1990.

Lock, R. A. C., Cruijsen, P. M. and van Overbeeke, A. P., Effects of mercuric chloride and methylmercuric chloride on the osmoregulatory function of the gills in rainbow trout, *Salmo gairdneri*, *Comp. Biochem. Physiol.*, 68C, 151, 1981.

Lowe-Jinde, L. and Niimi, A. J., Influence of sampling on the interpretation of haematological measurements of rainbow trout (*Salmo gairdneri*), *Can. J. Zool.*, 61, 396, 1983.

Massaro, E. J., Pharmacokinetics of toxic elements in rainbow trout, U.S. Environmental Protection Agency Ecological Research Series, EPA-660/3-74-027, Washington, D.C., 1974.

McKim, J. M., Christensen, G. M. and Hunt, E. P., Changes in the blood of brook trout (*Salvelinus fontinalis*) after short- and long-term exposure to copper, *J. Fish. Res. Bd. Can.*, 27, 1883, 1970.

McLeay, D. J., Effects of a 12-hr and 25-day exposure to kraft pulp mill effluent on the blood and tissues of juvenile coho salmon (*Oncorhynchus kisutch*), *J. Fish. Res. Bd. Can.*, 30, 395, 1973.

Miller, W. R., Hendricks, A. C. and Cairns, J., Normal ranges for diagnostically important hematological and blood chemistry characteristics of rainbow trout (*Salmo gairdneri*), *Can. J. Fish. Aquat. Sci.*, 40, 420, 1983.

Milligan, C. L. and Wood, C. M., Disturbances in haematology, fluid volume distribution and circulatory function associated with low environmental pH in the rainbow trout, *Salmo gairdneri*, *J. Exp. Biol.*, 99, 397, 1982.

Mishra, J. and Srivastava, A. K., Malathion induced hematological and biochemical changes in the Indian catfish *Heteropneustes fossilis*, *Environ. Res.*, 30, 393, 1983.

Natarajan, G. M., Effect of sublethal concentration of metasystox on selected oxidative enzymes, tissue respiration, and hematology of the freshwater air-breathing fish, *Channa striatus* (Bleeker), *Pest. Biochem. Physiol.*, 21, 194, 1984.

Newman, M. W. and MacLean, S. A., Physiological response of the cunner, *Tautogolabrus adspersus*, to cadmium. VI. Histopathology, *National Oceanic and Atmospheric Administration Technical Report*, NMFS SSRF-681, 27, 1974.

Niimi, A. J. and Lowe-Jinde, L. L., Differential blood cell ratios of rainbow trout (*Salmo gairdneri*) exposed to methylmercury and chlorobenzenes, *Arch. Environ. Contam. Toxicol.*, 13, 303,1984.

Nilsson, S. and Grove, D. J., Adrenergic and cholinergic innervation of the spleen of the cod: *Gadus morhua*, *Eur. J. Pharmacol.*, 28, 135, 1974.

Palace, V. P., Majewski, H. S. and Klaverkamp, J. F., Interactions among antioxidant defenses in liver of rainbow trout (*Oncorhynchus mykiss*) exposed to cadmium, *Can. J. Fish. Aquat. Sci.*, 50, 156, 1993.

Pamila, D., Subbiayan, P. and Ramashwamy, M., Toxic effects of chromium and cobalt on *Sarotherodon mossambicus*, *Indian J. Environ. Health*, 33, 218, 1991.

Penreath, R. J., The accumulation of inorganic mercury from the seawater by the plaice, *Pleuronectes platessa*, *J. Exp. Mar. Biol. Ecol.*, 24, 103, 1976.

Reichert, W. L., Federighi, D. A. and Malins, D. C., Uptake and metabolism of lead and cadmium in coho salmon (*Oncorhynchus kisutch*), *Comp. Biochem. Physiol.*, 63C, 229, 1979.

Rogers, D. W. and Beamish, F. W. H., Dynamics of dietary methylmercury in rainbow trout (*Salmo gairdneri*), *Aquat. Toxicol.*, 2, 271, 1982.

Satchell, G., *Physiology and Form of Fish Circulation*, Cambridge University Press, New York, 1991.

Soivio, A., Nyholm, K. and Westman, K., Notes on haematocrit determinations on rainbow trout (*Salmo gairdneri*), *Aquaculture*, 2, 31, 1973.

Soivio, A., Nikinmaa, M. and Westman, K., The role of gills in the responses of *Salmo gairdneri* during moderate hypoxia, *Comp. Biochem. Physiol.*, 70A, 133, 1981.

Soivio, A. and Nikinmaa, M., The swelling of erythrocytes in relation to the oxygen affinity of the blood of rainbow trout (*Salmo gairdneri*) Richardson, in, *Fish and Stress*, Pickering, A. D., Ed., Academic Press, New York, 1981, chap. 5.

Swift, D. J. and Lloyd, R., Changes in urine flow rate and haematocrit value of rainbow trout (*Salmo gairdneri*) exposed to hypoxia, *J. Fish Biol.*, 6, 379, 1974.

Tephly, T. R., Wagner, G., Sedman, R. and Piper, W., Effects of metals on heme biosynthesis and metabolism, *Fed. Proc.*, 37, 35, 1978.

Tewari, H., Gill, T. and Pant. J., Impact of chronic lead poisoning on the hematological and biochemical profiles of fish, *Barbus conchonius, Bull. Environ. Contam. Toxicol.,* 38, 748, 1987.

Thomas, P. and Neff, J. M., Plasma corticosteroid and glucose response to pollutants in striped mullet: different effects of naphthalene, benzo(a)pyrene, and cadmium exposure, in, *Marine Pollution and Physiology: Recent Advances,* Vernberg, F. J., Thurberg, F. P., Calabrese, A. and Vernberg, W., Eds., University of South Carolina Press, Columbia, 1985, 63.

Tomasso, J. R., Comparative toxicity of nitrite to freshwater fishes, *Aquat. Toxicol.,* 8, 129, 1986.

Travis, T. W. and Heath, A. G., Some physiological responses of rainbow trout (*Salmo gairdneri*) to intermittent monochloramine exposure, *Water Res.,* 15, 977, 1981.

van der Putte, L., Laurier, M. B. H. M. and van Eijk, G. J. M., Respiration and osmoregulation in rainbow trout (*Salmo gairdneri*) exposed to hexavalent chromium at different pH values, *Aquat. Toxicol.,* 2, 99, 1982.

Venkateshwarlu, P., Rani, V., Janaiah, C. and Prasad, M., Effect of endosulfan and kelthane on haematology and serum biochemical parameters of the teleost *Clarias batrachus, Indian J. Comp. Anim. Physiol.,* 8, 8, 1990.

Waiwood, K. G., Changes in hematocrit of rainbow trout exposed to various combinations of water hardness, pH, and copper, *Trans. Am. Fish. Soc.,* 109, 461, 1980.

Watenpaugh, D., Beitinger, T. and Huey, D., Temperature tolerance of nitrite-exposed channel catfish, *Trans. Am. Fish. Soc.,* 114, 274, 1985.

Wedemeyer, G. A. and Yasutake, W. T., Clinical methods for the assessment of the effects of environmental stress on fish health, *Technical Paper 89,* U.S. Fish and Wildlife Service, U.S. Department of Interior, Washington D.C., 1977.

Wedemeyer, G. A., Nelson, N. C. and Yasutake, W. T., Physiological and biochemical aspects of ozone toxicity to rainbow trout (*Salmo gairdneri*), *J. Fish. Res. Bd. Can.,* 36, 605, 1979.

Williams, E. and Eddy. F. B., Anion transport, chloride cell number and nitrite-induced methaemoglobinaemia in rainbow trout (*Salmo gairdneri*) and carp (*Cyprinus carpio*), *Aquat. Toxicol.,* 13, 29, 1988.

Williams, E., Glass, M. and Heisler, N., Effects of nitrite-induced methaemoglobinaemia on oxygen affinity of carp blood, *Environ. Biol. Fishes,* 37, 407, 1993.

Wilson, R. and Taylor, E., The physiological responses of freshwater rainbow trout, *Oncorhynchus mykiss,* during acutely lethal copper exposure, *J. Comp. Physiol., B,* 163, 38, 1993.

Wood, C. M. and McDonald, D. G., Physiological mechanisms of acid toxicity to fish, in *Acid Rain/ Fisheries,* Johnson, R. E., Ed., American Fisheries Society, Bethesda, MD, 1982, 197.

Chapter 5

Uptake, Accumulation, Biotransformation, and Excretion of Xenobiotics

I. INTRODUCTION

Investigations of uptake and accumulation are fundamental in helping to understand the effects these chemicals have on specific organ systems in fish. Figure 1 outlines in a general way the possible fates of a contaminant coming in contact with a fish either by way of the water or the food of the fish. The chemical, once it is absorbed via the gills or gut, is usually bound to a protein and then transported by the blood to either a storage point, such as fat, or to the liver for transformation and/or storage. If transformed by the liver, it may be stored there, excreted in the bile, or passed back into the blood for possible excretion by the kidney or gills, or stored in extrahepatic tissues such as fat. Thus, the concentration found in different organs after environmental exposure for a particular time depends on several simultaneous dynamic processes. The total body burden of a foreign chemical will be a weighted average of all the tissues, which may differ from each other in concentration of this chemical by orders of magnitude.

A word about terminology: the expressions "bioaccumulation" and "bioconcentration" are used rather synonymously in the literature. They refer to the process of a fish acquiring a body burden of some chemical that is at a higher concentration than that in the water. The biological concentration factor (BCF) describes the extent to which something accumulates in an aquatic organism. It is a unitless value obtained by dividing the concentration in one or more of the tissues by the average concentration in the water. Veith et al. (1979) suggest the term bioconcentration be limited to the accumulation of a chemical directly from the water but excluding that obtained via the food, while bioaccumulation would refer to accumulation from both food and water, a suggestion that will be followed here.

"Biomagnification" is best reserved for the process of acquiring a greater body burden from being at a higher trophic level. Of course, a fish may be "practicing" both bioconcentration directly from the water and biomagnification from eating lower on the food chain. It might be noted here that not all substances lend themselves to biomagnification. For example, mercury biomagnifies easily whereas cadmium and lead seemingly do not; nor do the body concentrations of the latter two metals increase with age or size of the fish as does mercury (Spry and Wiener, 1991).

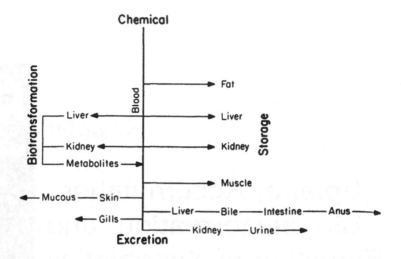

Figure 1 Schematic diagram of the possible directions of movement and fates of a pollutant after it has been absorbed into the bloodstream of a fish.

In the literature, several workers have used the term "uptake" when referring to the accumulation of some pollutant chemical in the fish. For clarity, it would be better to use this term to describe the process(es) by which the substance is actually taken into the body of the fish. The amount that is actually accumulated will depend on the balance between uptake rate, metabolism of the chemical, and excretion rate.

II. UPTAKE FROM THE ENVIRONMENT

The routes by which a foreign chemical is taken into a terrestrial animal for the most part do not apply to fish. This is because the aquatic environment imposes several constraints on the animals in it. One of the most critical is the matter of obtaining sufficient oxygen. A fish must breathe roughly 20 times more of its respiratory medium (i.e., water) than a terrestrial animal in order to obtain an equivalent amount of oxygen. For example, a 250-g trout at rest will pass approximately 48 L of water over its gills each hour (Reid and McDonald, 1991). Therefore, the gill tissue (which is the main point of entry for dissolved substances) will be exposed to a far greater amount of selected pollutants than a terrestrial animal breathing air, even allowing for a lower oxygen consumption in the water-dwelling poikilotherm. Furthermore, the gills possess a countercurrent blood/water flow system, very thin epithelial membranes, and large surface areas; these are all features that facilitate the uptake of materials from the water and their transfer to blood.

The matter of drinking water will be different in fish from mammals, but in a somewhat peculiar way. Contrary to popular belief, only marine fish "drink like a fish". Freshwater fish drink almost nothing except what is swallowed with food.

There are four possible routes for a substance to enter a fish: gills, food, drinking water, and skin. Gills and food vary greatly as routes of exposure, depending mostly on the availability of the substance in question (i.e., whether it is concentrated in the food or dissolved in the water). For example, Spry and Wiener (1991) conclude after reviewing the literature that over 90% of the mercury accumulated in wild fish is acquired directly from the food. The importance of skin varies with the size of the fish and toxicant. For large fish (about 1 kg), it accounts for less than 10% of total absorbed dose (McKim and Nichols, 1991), but for small fish of 4 g or less, due to the large surface to volume ratio, skin can account for up to one half of the absorbed dose (Liem and McKim, 1993).

The distribution into various organs is influenced by the route of uptake. This is because xenobiotics that enter the circulation via the gills can be rapidly distributed throughout the body to all organs. Xenobiotics that are absorbed via the intestine will, on the other hand, pass first to the liver, via the hepatic portal system, where some metabolism or sequestering may occur.

Assimilation by the intestine influences the relative importance of intestinal uptake vs. gill. For example, at least some polycyclic aromatic hydrocarbons are not absorbed well by the digestive tract and yet are taken up very readily by the gills (Kennedy and Law, 1990).

A. METALS

When metals enter natural waters there are several things that may happen to them which can greatly affect their bioavailability (Figure 2). If there is a considerable amount of organic material or suspended solids, the actual amount of dissolved metal available to be absorbed by the fish will be greatly reduced. This tendency to form complexes with organic and inorganic ligands (primarily chloride, carbonate, and hydroxide) varies with the metal (Sprague, 1985). For example, copper binds to organics far more readily than does either cadmium or silver (Engel et al., 1981). Consequently, it is important that the dissolved metal be measured, rather than just the total concentration in the water, in order to assess the amount of metal actually available for absorption by the fish.

It is assumed that most metals are absorbed by fish in the ionic form, although methylmercury is certainly an exception, and studies of cultured gill epithelial cells suggest that some cadmium chloride can be taken up along with the cadmium ion (Block and Part, 1992). Because the outer surface of the gill tissue has a negative charge, it will attract metallic ions (Reid and McDonald, 1991), but recently, it has been found that not all metals bind to the surface equally well. Reid and McDonald (1991) report that affinity of the gill for metals tested was La > Ca = Cd > Cu. The affinity of the gill for metals

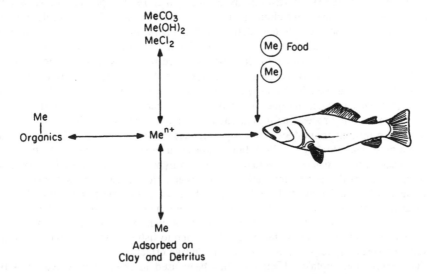

$MeCO_3$
$Me(OH)_2$
$MeCl_2$

(Me) Food

(Me)

Me
|
Organics ⟷ Me^{n+} ⟶

Me

Adsorbed on
Clay and Detritus

Figure 2 Conceptual model of the different forms of metal (Me) present following its addition to fresh, estuarine, or marine water. The double-ended arrows indicate that the metal is in a dynamic equilibrium with the various components of the water column, and that a change in any one aspect of the system can shift the interrelationships. (Based on Engel et al., *Biological Monitoring of Marine Pollutants,* Vernberg, J., Calabrese, A., Thurberg, F. and Vernberg, W. B., Academic Press, New York, 1981, 127.)

is determined by the microenvironment of the gill surface which is complex, because it includes the epithelial membranes as well as the mucus layer with its mixture of glyco-proteins, mucopolysaccharides, assorted low molecular weight compounds, and water. Contrary to what might first be expected, the absorption of a metal into the cells of the gills is inversely proportional to the surface-binding affinity. Thus, copper, which has a low affinity, is absorbed into the cells much more readily than either calcium or cadmium (Reid and McDonald, 1991).

The fact that binding to the outer surface of the gill epithelium does not facilitate further uptake is especially noteworthy with aluminum. In acidic soft water (where aluminum contamination is most prevalent), much of it precipitates due to a rise in pH from excreted ammonia as the water passes over the gill and this is sloughed off with the mucus. The rest remains as a charged form bound to the gill surface (Playle and Wood, 1991). As a result of these effects of the gill microenvironment, relatively little dissolved aluminum actually gets into the fish (Spry and Wiener, 1991; Handy and Eddy, 1990).

The mechanism of metal uptake through the gills has been assumed to be simple diffusion (Bryan, 1979), although carrier mediation or endocytosis have not been elimi-nated as possibilities (Luoma, 1983). Using an isolated perfused head preparation, Spry and Wood (1988) showed that zinc is not absorbed in a simple concentration-dependent manner. Rather, the kinetics suggest a saturable uptake mechanism, which they point out is not necessarily active or carrier mediated but could be. They also present interesting indirect evidence that much of the zinc is actually absorbed via the chloride cells in the lamellar epithelium.

Mucus on the surface of the gill has a considerable influence on the accumulation by the gills of metals (Handy and Eddy, 1990). Starved fish accumulate zinc on their gill surface more rapidly than those fed due to a change in composition of the mucus during starvation (Handy and Eddy, 1990). Apparently, there are more metal-binding ligands in the mucus from starved fish.

Part and Lock (1983) reported that trout mucus in an *in vitro* preparation slowed the diffusion rate of mercury and cadmium, but had no effect on the diffusion of calcium. This difference was probably due to the relative binding characteristics of the metals with the mucus. Overall, it appears that mucus reduces the uptake of metals into fish.

The rate of metal uptake into the gill tissue correlates with the weight-specific metabolic rate, thus small fish accumulate it more rapidly than large ones (Anderson and Spear, 1980). Because the higher flow rate of water over the gills of small fish apparently results in greater uptake, the question arises as to what effect environmental hypoxia would have on this process, as it also causes a greater rate of ventilatory water flow. Exposure to mild hypoxia does cause a greater accumulation by the gills of cadmium, chromium, or lead (Freeman, 1980), but Hughes and Flos (1978) found that the rate of zinc uptake into gill tissue was actually slower in hypoxic fish compared to normoxic controls. Just why this is remains unclear. Hughes and Flos (1978) suggest the possibility that the histological damage to the gills caused by the zinc may be greater under hypoxia than under more normal conditions. This may have the effect of making the gills less permeable to zinc, as it does to oxygen. A point not mentioned by them is that hypoxia alone has been shown to cause histological damage to fish gills (see Chapter 2), so this could be a sort of synergism, or at least, an additive effect.

The expression "uptake efficiency" has been used to refer to how readily a fish removes a contaminant from the water via the gills (Boddington et al., 1979; McKim and Goeden, 1982). Fish respiratory physiologists have a long tradition of using the terms "utilization efficiency" or "extraction coefficient" for the percent of the oxygen in the inhalant water that is removed as it passes over the gills (Truchot and DeJours, 1989). The parallel with uptake efficiency for pollutants is obvious. For these measures, one must determine the concentration of some chemical (or oxygen) in the exhalant water, which is not an easy thing to do, for it must be kept from mixing with the inhalant water. The fish must be restrained in a chamber that is divided by a flexible rubber membrane into front and back. The fish's head is inserted part way through the membrane and sutured just in front of the operculae. In this way, the fish in the process of breathing pumps the water from the front to the back chamber, both of which are arranged to avoid either a positive or negative pressure gradient for water flow. This design was originated by Van Dam (1938) and such devices are frequently called "Van Dam chambers".

Figure 3 illustrates a slightly more sophisticated form of Van Dam chamber. In this design, there is a third chamber separating the back portion of the body from the gill exhalant water. Analyses of the water from the rear chamber indicate the extent of excretion or uptake of a chemical through the skin. Also by having a third chamber, the size of the rear chamber is reduced so the turnover time is faster and a shorter sample interval can be used. Note the urinary and fecal catheters enabling measures of volume and composition of these excretory modes. This system also makes possible the determination of the volume of water passing over the gills in a given time interval.

Obviously, the process of obtaining data from a fish in this manner is quite tedious. Unfortunately, the simpler method of cannulation of the opercular cavity in order to obtain samples of the exhalant water yields misleading results, at least when it is applied to the estimation of oxygen utilization efficiency (Davis and Waters, 1970).

Boddington et al. (1979) used an apparatus somewhat similar to that shown in Figure 3 (except that it was two-chambered) to measure the uptake efficiency of waterborne methylmercury. Radiolabeled methylmercury was taken up with an average efficiency of 7.1%, which is about 0.13 of the extraction coefficient for oxygen in the same fish. It would be interesting to compare different metals in this regard.

Rogers and Beamish (1983) use the term uptake efficiency in a slightly different manner. For their work, they measured the rate of methylmercury uptake and related this to oxygen consumption (not extraction efficiency). The equation of this is $E = [\Delta P/(P)/Q/(O_2)]$ Delta P is the rate of methylmercury uptake in nanograms per hour. (P) is the average mercury concentration in the water, Q is the rate of oxygen consumption (mg/h),

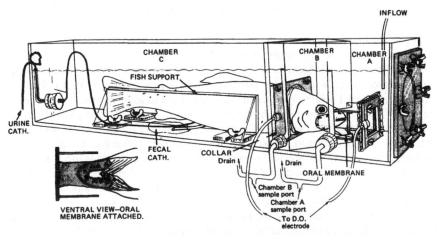

Figure 3 Schematic view of respirometer-metabolism chamber containing a fish and emphasizing the design and placement of the oral membrane. (From McKim, J. M. and Goeden, H., *Comp. Biochem. Physiol.*, 72C, 65, 1982. With permission.)

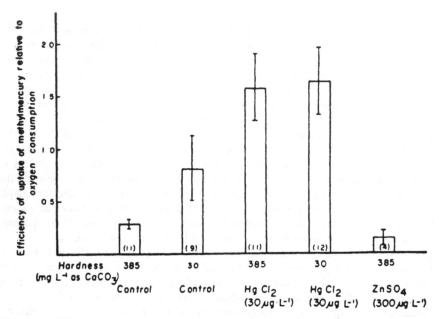

Figure 4 Efficiency of uptake of methylmercury relative to oxygen of rainbow trout, *Salmo gairdneri,* and showing effects of water hardness, inorganic mercury, and zinc, Bars indicate mean ± 95% confidence limits. Sample sizes in parentheses. (From Rogers, D. W. and Beamish, F. W. H., *Can. J. Fish. Aquat. Sci.,* 40, 824, 1983. With permission.)

and O_2 is the mean oxygen concentration in milligrams per liter. The main questions asked in their study were whether hardness and the presence of other pollutants affected the uptake rate of methylmercury. The data in Figure 4 show that these factors do indeed have a considerable effect. Increasing the hardness decreased the rate of mercury uptake. A similar relationship between water calcium concentration and uptake of waterborne lead has been noted (Varanasi and Gmur, 1978). These findings may help explain the often repeated observation that acute toxicity of heavy metals is less in hard water (Sprague, 1985).

Water hardness affects the gill permeability to water and ions so that the harder the ambient water, the less permeable the tissue (Hunn, 1985). The calcium ion, which is the major cation responsible for what we call "hardness", also causes the electrical charge on the outside of the gills to be more positive (McWilliams and Potts, 1978). This would further tend to repel positively charged molecules such as toxic metals. It clearly reduces the rate of uptake of these metals by the gills (Everall et al., 1989; Wicklund and Runn, 1988). The zinc flux rates in hard vs. soft water can differ by as much as 40-fold (Spry and Wood, 1988). Spry and Wood (1988) have further shown that the effect of water hardness is not due exclusively to the presence of external calcium, for fish that have been acclimated to hard water retain their lower zinc uptake rate when later tested in soft water. Thus, the hard water must cause some changes in the gill epithelial cells or the junctions between them that inhibits the flux of metals.

In the aforementioned study on the uptake of mercury, the presence of inorganic mercury actually increased the uptake of methylmercury (but canceled out the hardness effect; Figure 4). Rogers and Beamish (1983) speculate the increased uptake of methylmercury was due to an altered structure of the gill mucus caused by the inorganic mercury. However, simply increasing the mucus as should occur with exposure to zinc, actually decreased the uptake efficiency of methylmercury. The authors suggest this may represent zinc competing for sites of methylmercury uptake on the surface of the gill.

Ramamoorthy and Blumhagen (1984) found that zinc in soft water actually stimulated the uptake of inorganic mercury by trout. They did not test hard water nor methylmercury, so it is not possible to determine at this time the relative importance of these factors. Finally, we must note that low levels of cadmium can inhibit the uptake of zinc from the water, but zinc seems to have no effect on cadmium uptake (Bentley, 1992). Clearly, the interactions of various metals and water hardness on metal uptake by fish is complex.

Two additional environmental factors that can influence the rate of metal uptake are water pH and temperature. At a given concentration of calcium in the ambient water, lowering the pH decreases the rate of zinc uptake into gill tissue of rainbow trout (Bradley and Sprague, 1985). This generalization holds, however, only for high zinc concentrations and gill tissue (the primary target of toxic action at high doses). Using much lower concentrations and measuring total body burden, Bentley (1992) found that lowering the pH increased zinc uptake, so sublethal effects might be greater at lower pH.

Copper uptake by the gills of fathead minnows was slower at lower pHs, but only in hard water (Playle et al., 1992). In soft water, there was no difference in copper accumulation by gills between pH 4.8 and 6.3. The authors conclude that calcium and hydrogen ions compete with copper for gill-binding sites.

Based on toxicity data, cadmium appears to be taken up in direct correlation with pH. However, lead exhibits exactly the opposite; lowering the pH increases uptake (Campbell and Stokes, 1985).

The effects of water pH on gill function are made complicated by the fact that carbon dioxide and ammonia are excreted by the gill which alters the microenvironment of the gill lamellae. The net effect of these substances is to decrease the water pH if the inhalant water has a pH of 6 or above while increasing the pH of the interlamellar water if the inhalant water has a pH of less than 5 (Randall et al., 1991). Thus, the exhalant water will tend not to vary in pH as much as the inhalant water (Playle et al., 1992). Still, measurement of the exhalant water does not actually represent the true pH for the entire length of a gill lamellum, because according to Randall et al. (1991) this changes in a nearly linear fashion as it passes the lamellum. Thus, the epithelium of the lamellae is exposed to a different pH at the two ends.

When rainbow trout were tested for aluminum accumulation on gill tissue at inspired pHs of 5.1, 4.7, or 4.1, Playle and Wood (1991) found accumulation greatest at an inspired pH of 5.1 and lowest at 4.1. As expected, the expired pH was 0.2–0.7 pH units higher than inspired pH. They also noted that only about 10% of the aluminum extracted from the water by the gills remained bound there; most of the remainder was sloughed off by the mucus.

Laboratory data on the effect of pH on the rate of uptake of methylmercury are not in agreement. Drummond et al. (1974) found a faster rate of accumulation into the gills and blood cells of brook trout at pH 6 than at pH 9. In contrast, Rogers et al. (1987) reported that pH over the range of 5–7 had no effect on methylmercury accumulation by walleyes and rainbow trout. Neither of these studies measured direct uptake by the gill so relative rates of depuration and other factors such as food may have contributed to the differences. Also, the pH range and concentration of mercury tested may be important variables. Ponce and Bloom (1991) exposed rainbow trout fingerlings to a very low (1.38 ng/L) concentration of methylmercury for 2 months. They used four different pH levels (5.8, 6.3, 7.0, 8.2) and found that uptake was the same in the upper three pHs. However, the rate of uptake was approximately twice as fast at pH 5.8 as 6.3, which suggests a pH threshold. Because many natural waters that are contaminated with methylmercury have pH levels below 6, this finding may have considerable practical significance (Spry and Wiener, 1991).

One pollutant may have a considerable effect on the uptake rate of another one. With the aid of an isolated gill preparation, Part et al. (1985) tested the effect of the detergents linear alkylaryl sulfonate (LAS) and nonylphenol ethoxylate (NP-10EO) on the gill uptake of cadmium. They found that LAS stimulated cadmium uptake, and because LAS has a low affinity for the metal, the effect must be directly on the gill tissue instead of the formation of a lipid soluble complex with LAS. The other detergent had no effect on cadmium uptake, which indicates the mechanism is not merely due to a reduction of surface tension. Instead, the authors suggest it involves an action of LAS on the proteins of the gill epithelium.

In summary, we see that the particular metal in question and various environmental factors can influence the rate of uptake of waterborne metals by the gills of fish. The bioavailability will depend on the degree of complexation with ligands in the water (the greater the tendency to form complexes the less the bioavailability), the microenvironment of the gill lamellae of the gills, and changes in the gill epithelium.

Uptake of metals via the food can be quite important in nature. As a general rule, invertebrates often accumulate higher levels of metals than do fish under similar conditions (Waldichuck, 1974). Thus, predators on these invertebrates may obtain a considerable body burden from this source. Younger fish are reported to take up zinc via the food more rapidly than older ones (Patrick and Loutit, 1978). This is to be expected as they have higher rates of metabolism (per gram of tissue).

Phillips and Buhler (1978) exposed rainbow trout to methylmercury in the food, water, or both and found the body burden increased linearly for at least 24 days irregardless of the route of uptake, and methylmercury accumulated from one source had no effect on the rate of uptake from the other. In other words, they were additive. The efficiency of mercury removal from the medium was different, however. Approximately 70% of the methylmercury ingested was absorbed whereas the gills took only 10% of the chemical out of the water.

The concentration of mercury in the food can affect the assimilation efficiency by the gut. Rogers and Beamish (1982) found a high assimilation efficiency for methylmercury when the dose was low, but when the fish were fed a much higher dose of mercury in the ration, the assimilation efficiency declined to less than 50%. This implied to the authors a saturation of the uptake process in the gut and/or an induction of some sort of specific block to the assimilation of methylmercury.

Handy (1993) recently tested the assimilation of aluminum in rainbow trout and found a very low rate of less than 1%. In light of the study mentioned above on methylmercury in which it was found that assimilation decreased with concentration in the food, it should be noted that the aluminum study was done with very high concentrations. Thus, aluminum assimilation might be better at lower doses. Even so, it appears that aluminum is not taken up by fish very well through either the gills or gut, but it does cause a considerable effect on gill function (see Chapters 7 and 10).

Uptake via the drinking water effectively is no different than that via the food. However, it must be realized that marine fish drink approximately 0.5% of their body weight per hour, so a considerable amount of water enters the gut compared to the freshwater fish, which drink very little water (see Chapter 7). Thus, dissolved substances may enter by way of the gut to a much greater extent in marine fish, but careful studies are needed to partition these routes of entry. Somero et al. (1977) reported that the estuarine teleost *Gillichthys mirabilis* accumulates lead at a rate that is proportional to the water salinity. This can be interpreted as being due to a greater drinking rate in saline water.

There may be a distinct difference between marine and freshwater fish as to the assimilation of dietary lead. According to Hodson et al. (1978) dietary lead is not absorbed by rainbow trout, but more recently, Juedes and Thomas (1984) observed that marine croakers can assimilate it.

B. ORGANICS

As with metals, there is no evidence for active transport of organics across the gills into the blood. In spite of this, some remarkably high uptake efficiencies have been found for a few organics while for others, the gill tissue is nearly impermeable. There are several physicochemical mechanisms that influence the movement of chemicals across biological membranes. These include molecular weight, charge on the molecule, molecular volume, concentration in the water, and lipid solubility. Furthermore, the bioavailability of organics is greatly reduced by the presence of various adsorbents (e.g., humic acids, sediments, and suspended solids) (Spacie et al., 1982).

It has been shown (Veith et al., 1979) that the accumulation of organic chemicals is correlated with their octanol/water partition coefficients (log K_{ow}), which is a measure of the lipid solubility of a chemical. However, more recent work indicates the relationship is not linear because of the effect of blood and water flows. McKim and Erickson (1991) summarized several studies of the direct uptake by gills of rainbow trout of 14 different organic chemicals and found the relationship varies considerably with the log K_{ow} (Figure 5). The exchange coefficient represents uptake efficiency as was described above for metals. Thus, the uptake efficiency was around 7% for those chemicals with log K_{ow} below 1. For those with coefficents between 1 and 3, there was an approximately direct relationship between uptake efficiency and the log K_{ow}, but then between log 3 and 6, there was no further increase in uptake efficiency and a considerable decline in efficiency above log K_{ow} 6. Hansch and Clayton (1973) have suggested that the low fat solubility of chemicals with low log K_{ow} prevents their entry through membranes (except through pores) while those with high lipid solubilities may become bound to the lipid membrane and thus would be slow to move on into the blood.

It is interesting that the flux of organic chemicals in the opposite direction (i.e., from blood to water) across the gill is also influenced by lipid solubility. However, those with log K_{ow} of 1–4 move more rapidly than those with log K_{ow} of 4–6 (Thomas and Rice, 1981). This is, of course, the reverse of what was found with the inward movement.

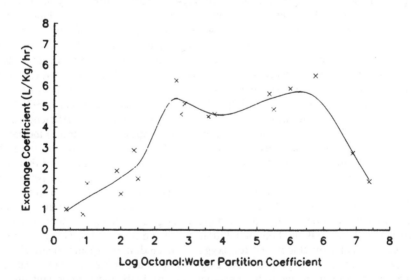

Figure 5 Relationship between exchange coefficient in the gill and log octanol:water partition coefficient of test chemicals. (Data from McKim, J. M. and Erickson, R. J., *Physiol. Zool.*, 64, 39, 1991.)

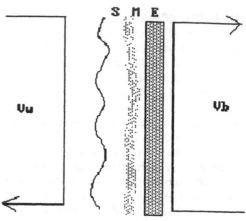

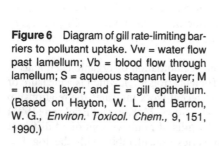

Figure 6 Diagram of gill rate-limiting barriers to pollutant uptake. Vw = water flow past lamellum; Vb = blood flow through lamellum; S = aqueous stagnant layer; M = mucus layer; and E = gill epithelium. (Based on Hayton, W. L. and Barron, W. G., *Environ. Toxicol. Chem.*, 9, 151, 1990.)

Lipid solubility is only one of the factors that affect the transfer of chemicals across the gill into the blood. Hayton and Barron (1990) have discussed several other barriers that may be rate limiting. These include in addition to the epithelial cells the aqueous stagnant layer next to the epithelium and the water and blood flows (Figure 6). Hayton and Barron (1990) did not include the layer of mucus that usually covers the epithelium (Hughes, 1979), which may have different diffusion properties than either the water or the membranes. They make the point that, "In general, for any particular chemical, only one of these barriers is operative with the resistance offered by the others being negligible" (p. 151).

Hayton and Barron (1990) predicted that uptake efficiency of chemicals with low log P would be limited by blood flow while those with high log P would be limited by water flow over the gills. This was confirmed in an elegant experiment by Schmieder and Weber (1992). Using a McKim and Goeden (1982) chamber they altered ventilation of the fish and gill perfusion by changing the dissolved oxygen concentration. They found that the uptake of a hydrophobic chemical (decanol) with high log P was ventilation limited whereas the hydrophilic chemical butanol, with a low log P, was blood flow limited. The mechanism for this distinction is probably that hydrophilic compounds will go primarily into the aqueous phase of the blood and will thus quickly reach equilibrium with the blood in the gill. Consequently, they become limited by blood flow. Compounds with higher log P bind to blood proteins better so the limit then becomes the rate at which the water brings the material to the "blood" at the gill epithelium (Schmieder and Weber, 1992).

It is interesting that endrin is taken up from the water at a higher efficiency than is oxygen, and this is seemingly independent of the concentration of endrin in the ambient water (Figure 7). McKim and Goeden (1982) further found the ventilation volume of their fish was 65–75 ml/min for a 0.5-kg fish. Gill blood perfusion for a trout of that size is approximately 8–9 ml/min (Daxboeck et al., 1982). Because the blood perfusion is considerably less than the ventilation volume, the concentration of endrin in the blood coming from the gills must be far greater than that in the water. This illustrates strikingly the importance of blood proteins which maintain a large diffusion gradient by binding the chemical and thereby taking it out of solution (Streit and Sire, 1993). This is, of course, analogous to what hemoglobin does for oxygen. The binding of organic chemicals to the blood proteins increases in proportion to Log K_{ow} up to about 3 or 4. Above that point, organic complexation in the water becomes progressively more important and this reduces bioavailability of the chemical (McKim and Erickson, 1991).

Uptake efficiency of oxygen and endrin (an organochlorine insecticide) decreases with increasing ventilation volume caused by hypoxia (Figure 7) (McKim and Goeden, 1982). In spite of the decreased efficiency, the actual amount of endrin taken up per day rose as

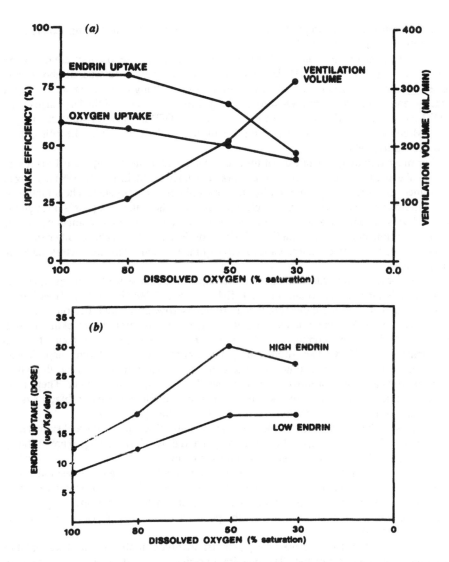

Figure 7 (a) Uptake efficiency of endrin and of oxygen in the gills of brook trout. Note that endrin has a greater uptake efficiency than does oxygen, but efficiency of endrin uptake declines more rapidly with increased ventilation volume produced by environmental hypoxia. (b) Effect of dissolved oxygen concentration on the total endrin taken up by brook trout exposed to a high (0.072 µg/L) and low (0.046 µg/L) endrin concentration. (Redrawn from McKim, J. M. and Goeden, H., *Comp. Biochem. Physiol.*, 72C, 65, 1982.)

the dissolved oxygen was decreased so the effective dose in the fish was higher during hypoxia. This indicates that the uptake of endrin is water flow limited up to some threshold, above which diffusion limitation becomes dominant.

The rate of oxygen consumption by a fish is profoundly affected by temperature with $Q10$ of 2 or higher common (Brett, 1972). Thus, the ventilation volume will also be temperature dependent. So it is not surprising that Black et al. (1991) found that toxicant uptake is directly related to temperature and correlates well with oxygen consumption. The effect is not, however, explained by ventilation alone, because lower temperatures resulted in declines in uptake efficiency. The direct effect of temperature on permeability of gill membranes to hydrophobic compounds has been confirmed using isolated gill preparations (Sijm et al., 1993).

Two other factors that can greatly affect uptake of organic xenobiotics are dissolved organic matter (DOM) and pH. DOM complexes with metals (discussed above) and organic pollutants. The binding affinity of an organic with high molecular weight DOM greatly reduces the rate of passage through the gill tissues because of the large size of the DOM molecules and the presence of polar groups on DOM. Thus, for some chemicals, a DOM concentration of as little as 3 mg/L can nearly block uptake by the gills (Black and McCarthy, 1988).

For most xenobiotic chemicals (if they are weak acids), increasing the pH decreases the uptake because they become more ionized (McKim and Erickson, 1991). Ionized substances have relatively little ability to cross epithelial membranes. The situation is made more complex, however, by the changes in pH of the water as it passes through the gills (discussed above) and shifts in speciation of the xenobiotic compound in the microenvironment of the gill lamellae. McKim and Erickson (1991) have rather successfully modeled the effects these various environmental and chemical factors have on uptake rates.

Waterborne organics can certainly be taken up by the skin as well as the gills, but the skin route is far more important in small fish (Liem and McKim, 1993). Tovell et al. (1975) obtained evidence indicating that as much as 20% of the total detergent sodium lauryl sulfate (SLS) accumulated from the water entered through the skin of adult goldfish. Naphthalene accumulates in the skin of trout exposed to the compound for 24 h (Varanasi et al., 1978). Part of this may have come from bloodborne naphthalene absorbed via the gills, but the time course of changes in concentration suggests there may be a significant amount of direct uptake by the skin from the environment. There is an extensive microcirculation in the skin that is part of the fish's secondary blood system (Satchell, 1991), and these blood vessels might facilitate uptake there.

In general, chemicals up to a log K_{ow} of 3 are mainly taken up by gills and skin. Those with log K_{ow} 3–6 are taken up by both gill and gut and those above log K_{ow} 6 are probably all gut uptake (James McKim, personal communication).

Because of the great diversity of morphology evident in fish gastrointestinal tracts, there is a great deal to learn about the mechanisms involved in xenobiotic uptake across the intestinal wall. For example, many fish lack a stomach and intestinal lengths vary considerably in different species, as does the presence or absence of pyloric ceca (Fange and Grove, 1979).

Van Veld (1990) has reviewed the processes involved in absorption of xenobiotics from the digestive tracts of fish. The proximal portion of the intestine is where most of the food absorption occurs and the pyloric ceca, if present, add absorptive surface area. Lipophilic toxicants are assimilated in much the same way as dietary fat. The lipids (and toxicants) are first digested by pancreatic lipase and bile salts yielding a fine suspension of micelles which are colloidal particles of fatty elements clustered together with bile salts in such a way as to facilitate their diffusion through the aqueous phase. The presence and digestion of dietary lipids actually facilitates the accumulation of toxicants (Van Veld, 1990), perhaps by stimulation of bile and/or lipase secretion. Passage into the intestinal enterocytes occurs presumably by diffusion of the fatty materials from the micelle, but absorption of some materials such as fatty acids may be facilitated by a protein carrier.

Following absorption into the enterocyte, the fat digestion products are used to resynthesize triglycerides. They then enter the circulation in the form of low-density lipoproteins (Sire et al., 1981).

Compounds with molecular weights above 600 are poorly absorbed by the digestive tract, but because most organic contaminants in waterways have molecular weights less than this, that does not appear to be a problem. Octanol-water partition coefficients (log

K_{ow}) of chemicals do not appear to affect their absorption by the gut. Rather, the triglyceride solubility seems to show a good correlation with absorption efficiency. The absorption of hydrophobic organochlorines appears to be by simple diffusion rather than lipid cotransport (Gobas et al., 1993). Overall, it is common for lipophilic toxicants to have assimilation efficiencies in fish of 50% or more (Van Veld, 1990).

Assimilation efficiency can be influenced by the amount of food in the diet. When guppies were fed four different chlorinated hydrocarbons in the food, the rate of accumulation was proportional to the concentration of xenobiotic in the food (i.e., assimilation efficiency was constant) (Clark and Mackay, 1991). However, if the "dose" was increased by feeding more contaminated food (keeping the concentration in food constant), the rate of uptake was not proportional to dose. This suggested the assimilation efficiency was reduced by increasing the amount of food, some of which apparently was egested. The authors note that changing the fat content of the food might also have interesting effects on assimilation efficiency, a point well taken in light of the observations mentioned above relative to the parallel digestion of lipophilic xenobiotics and lipids.

III. TRANSPORT WITHIN THE FISH OF METALS AND ORGANICS

Most if not all metals are probably carried by the blood bound to protein (Roesijadi and Robinson, 1994). For example, the plasma protein ceruloplasmin binds copper in mammals. Copper and zinc-binding proteins have also been found in fish (Fletcher and Fletcher, 1980). There may be a different protein for each essential trace metal, and presumably non-essential metals use one of the already existing proteins. Some metals may also bind to amino acids (Stagg and Shuttleworth, 1982). Methylmercury is rather unique in that it binds to the protein of hemoglobin inside the erythrocytes (Massaro, 1974). What effect, if any, the mercury has on hemoglobin function is apparently unknown.

Insecticides such as dieldrin, carbaryl, and parathion bind to both albumin and lipoproteins in the plasma (Denison and Yarbrough, 1985). By having a chemical bind loosely to a protein, it effectively removes it from the blood side of the gill, thus maintaining a water-blood diffusion gradient even when the ambient concentration is quite low. As was mentioned in Section II, this is a very important mechanism for bioaccumulation of both inorganic and organic substances from the water environment. The importance of blood proteins in transport and accumulation of lipophilic contaminants has recently been modeled by Streit and Sire (1993).

Organic contaminants that have been absorbed by the intestine are also largely associated with plasma albumin and lipoproteins, however, the situation is a bit more complex. The intestinal cells have the capability of metabolizing many organic substances so the products of that metabolism may be what are actually entering the circulation. For example, hydroxylated metabolites of polyaromatic hydrocarbons (PAH) bind primarily to albumin whereas the unmetabolized PAH associate more with plasma lipoprotein (Van Veld, 1990). Based on work largely done in mammals, this partitioning of xenobiotics in the plasma would appear to have an effect on the subsequent uptake by tissues. Endothelial cells in certain organs (e.g., liver) have receptors for the albumin-ligand complex so they are preferentially absorbed in these organs. Lipoproteins, on the other hand, merely diffuse into endothelial cells of essentially all tissues and thus will exhibit less specificity.

IV. ACCUMULATION OF METALS IN DIFFERENT ORGANS

As mentioned earlier, the biological concentration factor is the ratio of the concentration of a chemical inside an organ (or whole animal) to the environmental concentration. For

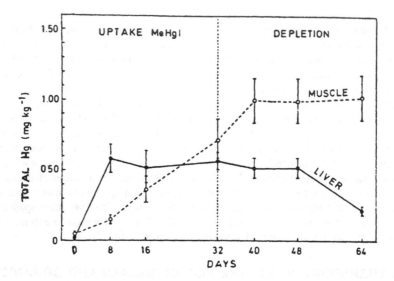

Figure 8 Uptake of mercury in the muscle and liver of cod fed methylmercury and the effect of a subsequent depletion period. (From Julshamm, K., Ringdal, O. and Braekkan, O. R., *Bull. Environ. Contam. Toxicol.*, 29, 544, 1982. With permission.)

nearly all metals, this is a positive number (Waldichuck, 1974) because of protein binding in the tissues. The concentration factor in a given tissue changes over time of exposure so it is not an absolute value for any metal or organ. For example, the data illustrated in Figure 8 show how the methylmercury in the food of the fish moved first into the liver and more gradually into the muscle. After day 32 in the study illustrated, the fish were fed non-contaminated food, but note how the concentration continued to rise in the muscle. The authors (Julshamm et al., 1982) suggest this was due to movement of the mercury from the liver to the muscle. Similar time-course changes in the ratio of liver:muscle mercury concentrations were reported by Olson et al. (1978) in rainbow trout. Clearly, one would observe a different concentration factor for these two tissues depending on when the measurements were made. Massaro (1974) reported that 50% of the total mercury dose given intragastrically to trout was in the muscle when sampled at 100 days, even though higher concentrations were found in organs having large blood volumes, such as the spleen and kidney. This points out the importance of the relatively large muscle mass for sequestering some pollutant chemicals, such as methylmercury.

Mercury in tissues from wild-caught fish is mostly in the methylated form. Fish cannot methylate mercury directly, although bacteria in the gut can (Rudd et al., 1980), and there have been suggestions that slow demethylation may occur (see Bryan, 1979 and Rogers and Beamish, 1982 for review). While methylation in the gut would make inorganic mercury more bioavailable, according to Julshamm et al. (1982), dietary methylmercury is accumulated about 10 times more readily than dietary inorganic mercury into the tissues. Also, inorganic mercury, if taken up, is eliminated from the fish more rapidly than is methylmercury (Spry and Wiener, 1991).

Histochemical observation of liver and kidney shows that methylmercury accumulates mostly in the glomeruli of the kidney and in the nucleus and endoplasmic reticulum of liver cells (Baatrup et al., 1986). Because a number of marine teleost species have kidneys which lack glomeruli, it would be interesting to see where the mercury accumulates in them.

Mercury reaches fairly high concentrations in large predacious fish. This is due to a combination of biomagnification up the food chain and old age (i.e., they have longer to acquire a body burden). Moreover, fish eliminate methylmercury very slowly compared to uptake rate (McKim et al., 1976). Overt toxicity probably does not occur until whole-body concentrations exceed 10–30 µg/g wet weight (Spry and Wiener, 1991).

Depending on the species, cadmium can differ considerably from mercury in its distribution within the body of a fish. In contrast to mercury, neither the skeletal muscle or brain of spot (*Leiostomus xanthurus*) accumulated cadmium (Hawkins et al., 1980). Exposure to waterborne cadmium for 48 h yielded the highest accumulation in liver, followed by gut, kidney, and gill in that order. The spot is a marine species so it is interesting to compare it with rainbow trout which accumulated 99% of its total body burden of cadmium in liver, kidney, and gills (Thomas et al., 1983). There was a virtual absence of cadmium in the gut of the trout. Its high concentration in the gut of the spot shows the importance of the marine environment, where the fish must drink rather large volumes of water to replace that lost osmotically.

More recently, Harrison and Klaverkamp (1989) assessed the accumulation of cadmium in various tissues in rainbow trout and lake whitefish (*Coregonus clupeaformis*) exposed to either waterborne or foodborne cadmium. Figure 9 illustrates the dynamics that occurred during the 72-day exposures and subsequent 56-day depuration. Most notable is the tendency for gills to accumulate the most cadmium, even in those exposed to cadmium via the food. Gut and kidney also accumulate considerable amounts of the metal when it is in the food, but not when in the water. Finally, depuration appears most rapid for gut and gill. The kidney, however, shows little depuration of cadmium, perhaps due to transfer of cadmium from other tissues to the kidney. The percentage of the dose accumulated was almost 10-fold greater from the food than from the water.

Rainbow trout are considerably more sensitive to cadmium than are roach (*Rutilus rutilus*) and stone loach (*Noemacheilus barbatulus*), as reflected in lower LC50 values for the trout. Norey et al., (1990) showed that this difference is correlated with the relative rate of accumulation of cadmium into tissues. As has been repeatedly emphasized in this book, excretion could in part account for this difference. Thus, it is interesting that they showed that there was no species difference in their depuration rates. Indeed, over a period of 132–170 days in cadmium-free water, there was little loss of cadmium in any of the three species tested, so the differences in toxicity must be related to uptake rate and/or differences in sensitivity of target organ.

The metal chromium was not accumulated above the ambient concentration in trout following exposure for 22 days (Buhler et al., 1977). However, when the exposure was prolonged for months, some bioaccumulation of the element occurred. The authors suggest there exists a "fast-turnover pool" (body fluids) for short-term exposures and a "slow-turnover pool" for chronic exposures. The latter may involve synthesis of a metallothionein-type of protein for binding. In spite of this, it appears that chromium has relatively little bioaccumulation potential in trout. Other fish species, however, such as bluegill (*Lepomis macrochirus*) can accumulate chromium quite well (Freeman, 1980).

The distribution of lead into various organs appears to vary greatly depending on the fish species. According to Spry and Wiener, 1991), lead accumulates in freshwater fish mostly in scales, bone, kidney, gill, and liver. The order of tendency (from greatest to least) for accumulation of lead in the organs of an estuarine fish was spleen, gill, fin, and intestine. Surprisingly, liver and muscle accumulated almost no lead (Somero, et al., 1977). This can be compared with the order of lead residual concentration found in brook trout after exposures of 2–38 weeks (Holcombe et al., 1976). In this species, the order was kidney, gill, liver, and spleen with virtually none in the muscle (intestine, scales, and bone were not measured). Binding to mucus may be an especially important mechanism for lead accumulation and subsequent depuration (Somero et al., 1977) and gill and intestine have an abundance of this material.

The above-mentioned studies all conclude that muscle does not accumulate lead, thus the findings of Kumar and Mathur (1991) are interesting. They exposed the Indian

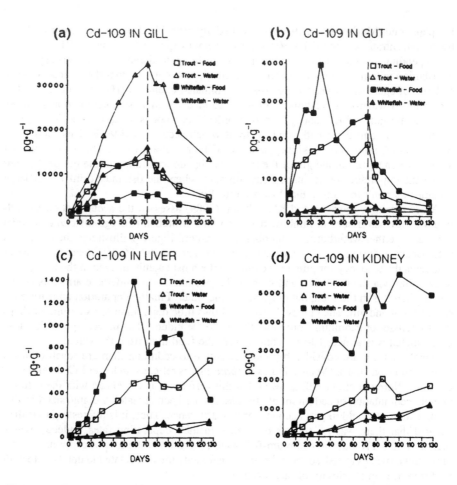

(a) Cd–109 IN GILL

(b) Cd–109 IN GUT

(c) Cd–109 IN LIVER

(d) Cd–109 IN KIDNEY

Figure 9 Cadmium concentrations (pg/g wet wt.) in gill, gut, liver, and kidney tissue on each sample day for food (30.7 µg/kg) and water (1.25 ng/L) exposed rainbow trout and whitefish. Vertical dashed line marks end of exposure period. Note that the scale on the Y axis is different for each tissue. (From Harrison, S. E. and Klaverkamp, J. F., *Environ. Toxicol. Chem.*, 8, 87, 1989. With permission.)

freshwater teleost *Colisa fasciatus* to lead for up to 24 days and measured lead accumulation in gill, liver, and muscle. Surprisingly, gill and muscle both exhibited approximately the same bioconcentration factor (7–8) while liver had a factor of 0.9. Unfortunately, no other organs were measured nor do the authors seem to be aware of the contradiction with earlier studies.

When bluegill were exposed to 0.5 ppm of waterborne lead, chromium, or cadmium for 7 days the gills had higher concentrations of all three metals than did the liver (Freeman, 1980). Moreover, the differences were large (i.e., order of magnitude greater). In gills, the chromium and lead reached higher concentrations than did cadmium, whereas all three metals were about the same in liver.

With copper, the liver has the highest concentration factor for almost any exposure time (Brungs et al., 1973; Buckley et al., 1982; Felts and Heath, 1984; Stagg and Shuttleworth, 1982). However, Felts and Heath (1984) observed that when bluegill sunfish were exposed to a sublethal level of copper, only the gills exhibited significant increases at 3 days. By 7 days of exposure, the liver exhibited large increases in copper concentration. At least some of the elevated copper concentration in the gills could be due to the element complexing with the mucus, which is impossible to completely remove from between the lamellae before the tissue is prepared for analysis. This, incidentally, is a point that should be appreciated when doing analysis of any substance in the gill tissue and may help explain seemingly high concentrations of various metals there.

In a time-course investigation involving chronic exposure of coho salmon to copper, the plasma copper rose on day one, but when the fish were sampled at 2 weeks and longer times, it was not elevated. During this time the liver copper was steadily increasing (Buckley et al., 1982). Thus, it appears the copper was being rapidly removed from the plasma by the liver and/or there was a reduction in the uptake rate by the gills during the first few days of copper exposure. If this phenomenon can be verified, it could be quite significant as it would provide a mechanism for regulation of copper accumulation other than excretion.

Accumulation of copper by liver can be greatly influenced by the nutritional level of the experimental fish. When fed and starved yearling roach (*R. rutilus*) were exposed to waterborne copper for 7 days, only the starved fish accumulated copper in the liver (Segner, 1987). Because one of the major functions of bile is digestion, it is suggested that starved fish were unable to dump copper from the liver into the bile due to reduced bile production.

In contrast to some other metals, present evidence suggests there is relatively little tendency for copper to accumulate in the brain (Felts and Heath, 1984) or kidney (Brungs et al., 1973; Buckley et al., 1982) of fish.

Zinc shows the greatest bioconcentration factor in skin and bone (Mount, 1964), although liver, gill, and kidney also accumulate it to a considerable extent (Holcombe et al., 1979). The ratio of the concentration of zinc in the gill to bone (actually opercular flap) has been proposed as an autopsy technique for zinc-caused mortality (Mount, 1964).

Dietary zinc may show a different distribution pattern from the waterborne metal. Hardy et al. (1987) gave rainbow trout a single feeding of isotopic zinc and measured levels in various organs and tissues after 72 h. Blood concentrations of zinc were higher than that of gill, liver, kidney, and spleen, whereas all other tissues (including bone) had lower concentrations than did the blood.

When pregnant guppies (*Poecilia reticulata*) are exposed to waterborne zinc they take it up and apparently actively transfer it to the developing embryo *in utero*. This results in fry with high body zinc concentrations. As the fry grow, the levels of body zinc decline (Pierson, 1981). Several other pollutants have been found to transfer from female fish into developing oocytes and be deposited with the eggs (see Chapter 13).

The largest concentration factor for selenium is in the liver following chronic waterborne exposure (Hodson et al., 1980). This element also accumulates to some extent in the digestive organs and kidney. Dietary selenium is more toxic than the waterborne metal and Hodson (1988) has suggested this is related to its metabolism by the fish. Waterborne uptake and excretion is dependent exclusively on diffusion whereas dietary selenium is accumulated independent of dietary concentration. Also, at high dietary concentrations, the half-life for excretion decreases so more is retained. This perhaps illustrates a saturation of excretory pathways.

Tributyltin, sometimes called organotin, accumulates rapidly from the water in marine fish. Davies and McKie (1987) exposed Atlantic salmon to doses of tributyltin for 26 days and reported that those receiving the 1.0 µg/L dose had organ concentrations ranging from 0.24 mg/kg in ceca to 1.62 mg/kg in liver. Muscle, gonad, and gill were only slightly higher in concentration than ceca, so this compound clearly shows an affinity for liver.

A factor that can have a considerable impact on total body burden of a metal is the size of the fish. Most metabolic processes in animals are inversely proportional to body size so weight-specific rate functions generally exhibit an exponent of 0.75 (Schmidt-Nielsen, 1984) (i.e., small animals of the same species operate at a higher rate than larger ones). Not surprisingly, mosquitofish (*Gambusia*) show decreasing zinc accumulation with increasing body size. Small fish also depurated zinc more rapidly (Newman and Mitz, 1988).

All of the above observations on accumulation of metals actually are viewing the result of both uptake and excretion. This is dramatically seen in the study of Eisler and Gardner (1973) where they noted that dead fish accumulated more copper and zinc than live ones when exposed to the waterborne elements for 48 h. The dead fish, of course, could not excrete metals, and evidently the skin is remarkably permeable to them. (Another possible explanation is that the metals adsorbed to the mucus on the dead fish, whereas live fish would slough off this material.)

V. REGULATION OF METAL CONCENTRATION

By regulation, we are actually referring primarily to the ability to excrete a metal. The rate of uptake of essential trace elements, such as iron, copper, and zinc may be controlled, although the mechanisms involved in fish are largely unknown. Acute levels of several different metals (e.g., cadmium, lead, zinc, copper) stimulate mucus secretion. This could serve as a way of reducing uptake rate by chelation of metals and inhibition of diffusion (Miller and MacKay, 1982; Handy and Eddy, 1990; Part and Lock, 1983). During chronic exposures there is little or no mucus accumulation so this variable declines in importance.

The term depuration is frequently used to refer to the ridding from the body of the fish of some chemical after the animal is put in non-contaminated water, or after it has been fed the test chemical. Different tissues depurate at different rates. For example, Figure 8 shows the slow rate at which the muscle lost mercury compared to liver. Whole-body depuration rate is inversely proportional to body size so small fish are able to do this more rapidly than large ones (Sharpe et al., 1977; Newman and Mitz, 1988).

Fish have different routes for possible excretion of harmful chemicals from mammals; these include the gills, bile (via feces), kidney, and skin. When the metal is present in the water, depending on the concentration, gills and skin may tend to take it up rather than excrete it.

Fish can excrete, at least to some extent, copper, chromium, cadmium, mercury, selenium, arsenic, lead, and zinc (Bryan, 1976; Hodson et al., 1980; Hardy et al., 1987; Rogers and Beamish, 1982; Somero et al., 1977; Sorenson et al., 1979). Under continuous exposure to a waterborne chemical, the excretory processes may take several hours or days to become activated. Thus, the body burden can rise rapidly and then actually decline somewhat with continued constant exposure.

The mechanisms of metal excretion in fish are only beginning to be understood. The liver is the main organ for metal homeostasis in mammals (Klassen, 1976) and at least to some extent seems to serve a similar function in fish. In mammals, many metals are excreted via the bile which is formed by the liver (Klassen, 1976). Several metals have been found at elevated concentrations in the bile of fish during or following their ingestion or waterborne exposure. These include chromium (Buhler et al., 1977), arsenic (Sorenson et al., 1979), triethyltin (Stroganov et al., 1973), and copper (Felts and Heath, 1984). In time-course measurements on bluegill exposed to low doses of copper (Felts and Heath, 1984), it was found that the concentration of copper rose first in the bile (i.e., in 3 days) and then in the liver at 9 days. An interpretation of this is that the liver of the bluegill picked up copper from the blood and immediately "dumped" it into the bile. When the amount of copper exceeded the ability to empty it into the bile, it then began to store it in the liver. It appears that mercury (Massaro, 1974) and zinc (Hardy et al.,

1987) are excreted to a small extent in fish bile. Evidence presented by Hardy et al. (1987) for dietary zinc implicates the gill as the major excretory route for zinc.

Excretion of a metal in the bile does not necessarily rid the body of the metal in question. The bile flows from the gallbladder through the bile duct into the anterior end of the intestine where it then mixes with the foodstuffs that are being digested. In mammals, there is often a significant uptake of the bile constituents, including harmful chemicals, back into the blood from the intestine (Klassen, 1976). This process is referred to as enterohepatic circulation because a chemical could, theoretically, be recycled more than once. How important this enterohepatic circulation is in fish is unknown. The marked diversity of anatomical and functional features of teleost digestive tracts (Fange and Grove, 1979) suggests that it may vary considerably between different species.

The excretion of metals via the urine of teleost fish has received little attention, although it is generally assumed that the kidney excretes some metals (Reichert et al., 1979; Rogers and Beamish, 1982), but apparently not dietary zinc (Hardy et al., 1987). As mentioned above, the kidney frequently accumulates metals to rather high concentrations, but this may be merely a location for sequestering them and does not necessarily indicate excretion. Reichert et al. (1979) report (p. 233) that "Salmonids exhibit a strong tendency to retain lead and cadmium in the kidney for at least a month after the termination of exposure."

The structure and function of marine and freshwater fish kidneys differ considerably. Many of the former lack glomeruli so must function exclusively as a secretory kidney (Smith, 1982) and probably would have little ability to excrete metals via this route. Thus the capability for excretion of metals in the urine may differ depending on whether it is a marine or freshwater species.

There is some evidence that methylmercury must be demethylated before excretion, and a major mode of exit from the body may be the gills (Olson et al., 1978), as has already been mentioned for zinc. Rates of methylmercury elimination increase with continued feeding of the material in the diet. Rogers and Beamish (1982) suggest this may indicate an induction of enzymes involved in this process.

Exposure to waterborne zinc, copper, or cadmium has been shown to stimulate the development of more chloride cells in the gills (Baker, 1969; Matthieson and Brafield, 1973; Oronsaye and Brafield, 1984). The response is quite rapid, requiring only a few hours of exposure to the element. Oronsaye and Brafield (1984) speculate these cells are excreting the metals, although they also recognize this could be an osmoregulatory response, which seems more likely (see Chapter 7).

With the use of a Van Dam chamber for the test fish, Oladimeji et al. (1984) found that 40% of an ingested dose of arsenic was excreted via the gills in 7 days. Urinary excretion in this time accounted for 15% of the dose. The authors suggest that with further clearance time, the urine would account for only an additional 5% of the total taken in. Presumably the remainder would be partitioned between gills and bile.

Loss of metals via the skin and gills probably involves the mucus. This proteinaceous material is constantly secreted and sloughed off by these tissues. Significant amounts of lead and cadmium which have been injected into the fish subsequently appear in the skin mucus (Varanasi and Markey, 1978). This suggests that metals taken in by the food may be excreted by this route, among others.

VI. GLUTATHIONE AND METAL DETOXIFICATION

The tripeptide glutathione performs a variety of critical functions in cells, including metabolism of peroxides and free radicals. It plays a key role in the detoxification of organic xenobiotics in a wide variety of animals (see Section IX). Because of its sulfhydryl groups, it also binds to heavy metals (Luckey and Venugpal, 1976) and is involved in excretion of these via the bile in mammals (Ellinder and Pannone, 1979). As will be seen below, however, such a generalization cannot be made for fish.

Evidence now exists for the involvement of glutathione in the metabolism of metals by fish. Work in Peter Thomas' laboratory on two marine species has shown that chronic exposures to lead, cadmium, and mercury cause increases in hepatic glutathione concentrations (Thomas and Juedes, 1992). Figure 10 illustrates this in mullet exposed to cadmium. Clearly, there was a dose- and time-dependent relationship. Similar changes occurred in the posterior kidney, but there were no significant alterations in glutathione concentration in the brain of the fish, possibly because the blood-brain barrier prevented much metal accumulation there. Measured concentrations of cadmium in the tissues of the mullet correlated with those of glutathione. A similar relationship between metal accumulation and glutathione increase was seen with lead exposure.

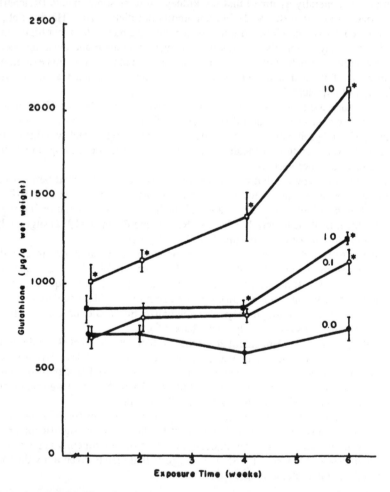

Figure 10 Hepatic glutathione concentration in mullett (*Mugil cephalus*) exposed to 0, 0.1, or 10 mg/L cadmium for up to six weeks. (Vertical bars = SEM, N = 6 for each point.) Asterisks denote means significantly different from controls at alpha = 0.05. (From Thomas, P. and Wofford, H., in *Physiological Mechanisms of Marine Pollutant Toxicity*, Vernberg, W. B., Calabrese, A., Thurberg, F. P. and Vernberg, F. J., Eds., Academic Press, New York, 1982, 109. With permission.)

Thomas and Juedes (1992) have shown that for lead, at least, the increase in glutathione is due to a stimulation of synthesis, rather than a decrease in utilization of the tripeptide. Normally, glutathione acts in a negative feedback manner on its own synthesis, so the elevations in rate of synthesis caused by xenobiotics may reflect a change in setpoint (Thomas and Juedes, 1992).

Although lead and glutathione have been found to accumulate in the cytosol of liver cells of fish, the lead did not bind to the glutathione, as it does in mammals. Instead, the elevation in glutathione may help protect the cell from lipid peroxidation and aid in maintenance of mitochondrial calcium content (see Thomas and Juedes, 1992, for review).

A highly acute dose of waterborne cadmium can produce a depression in glutathione. Chatterjee and Bhattacharya (1984) exposed climbing perch to a concentration of 166 mg/L and found that within a day the liver glutathione in these fish dropped by 33% and stayed at that level on day 2; but when allowed to recover in non-treated water, the glutathione concentration was elevated by 50% above controls in 15 days. In this study, one of the enzymes (glutathione-S-transferase) that utilizes glutathione as a substrate was assayed. By day 2 of exposure the activity of this enzyme had increased 3.5-fold. This may help explain the marked depression in glutathione observed as the transferase may be "using" the glutathione in a conjugation reaction.

The involvement of glutathione in the metabolism of mercury is somewhat unclear. Increases in kidney and liver glutathione from mercury exposure have been reported (Thomas and Wofford, 1984; Heisinger and Wait, 1989). However, in a combined laboratory and field study, Bidwell and Heath (1993) found a severe decrease in liver glutathione in rock bass (a freshwater teleost) with exposure to either methylmercury or inorganic mercury. A large amount of mercury was found in the bile which suggests the glutathione may have been involved in that excretory process. Ballatori and Boyer (1986) found that two different elasmobranchs handled mercury rather differently. The skate excreted the mercury in the bile largely bound to glutathione whereas the dogfish shark excreted it much more slowly and with less involvement of glutathione. Thus, there may be some significant species differences in how various fish utilize glutathione in the metabolism of mercury.

Finally, mention should be made of the distribution of glutathione in tissues other than liver. Recently, it was found that levels of glutathione in gills and kidney of channel catfish (*Ictalurus punctatus*) were about half that of the liver, indicating considerable detoxification capacity in these extrahepatic tissues (Gallagher and DiGiulo, 1992). The activity of the associated enzyme glutathione-S-transferase paralleled the levels of glutathione. However, there may be considerable species differences in the distribution of glutathione and its associated enzymes. Leaver et al. (1992) measured glutathione-S-transferase activity in various tissues of the marine flatfish, the plaice (*Pleuronectes platessa*). In this species, the gills had only 1% of the liver-specific activity of this enzyme and the kidney had 6%. These low levels would suggest that the gills and kidney of this species might be less able to handle contaminants than was seen in the catfish.

VII. INVOLVEMENT OF METALLOTHIONEIN IN METAL ACCUMULATION AND ACCLIMATION TO METALS

Metallothioneins (MTs) are low molecular weight polypeptides (sometimes called proteins) with many sulfhydryl groups due to the large amount of cysteine in the molecule. MTs have been found in a very wide variety of organisms including prokaryotes, plants, and animals. At least 34 species of elasmobranchs and teleosts have been reported to have them (Roesijadi, 1992). Because of the numerous sulfhydryl groups in the molecule, MTs bind a wide variety of metals, both essential and non-essential. For the essential metals, copper and zinc, MTs sequester the metals or donate them for biochemical reactions depending on demand for and concentration of a metal in the cell. Non-essential metals, such as mercury, are sequestered by MTs thereby reducing toxic interactions with other cellular proteins. The "spill over" hypothesis of Brown and Parsons (1978) proposed the idea that if the capacity of MTs to sequester a toxic metal is exceeded, then the metal will spill over and bind to other cellular proteins such as enzymes and produce toxic effects. It is possibly an oversimplified view of the kinetics of cellular toxicity, but it still retains considerable support, especially where exposures are chronic (Brown et al., 1990; see Roch and McCarter, 1984a). In fish, MTs have been found in gills, intestine, kidney, and liver with the latter receiving the bulk of attention.

MTs exist generally in a metal-saturated form (Hodson, 1988). Thus, in order for them to detoxify additional metal, there must be synthesis of additional MTs (or other metal-binding proteins) or displacement of one metal by another. All three of these mechanisms come into play at times.

It is well established that a variety of metals induce the synthesis of additional MTs (Hamer, 1986). It has also been found that acclimation to low levels of zinc (Hobson and Birge, 1989), copper, (Buckley et al., 1982; Dixon and Sprague, 1981), or cadmium (Klavercamp and Duncan,1987) can result in elevated LC50 values (reduced toxicity) for these metals. This increased tolerance has been associated with elevated levels of MTs in one or more tissues. Elevated levels of MTs in gills or intestine may reduce the rate of metal uptake into the blood (Petering et al., 1990), although the major mechanism of increased tolerance is presumed to be the enhanced ability to sequester the accumulated metal. Since a primary target organ from acute exposure of several metals is gill, the MTs there may aid the increased tolerance. As Lauren and McDonald (1987) have pointed out, for acute toxicity, there would be little benefit in protecting the liver with increased MTs as there is no evidence that it is harmed (by copper at least) at these doses.

The ability of fish to acclimate to a metal varies with the particular metal. Chapman (1985) reviewed the literature and ranked them in decreasing order: Zn > Cu > Cd > Cr. This is the reverse of their binding affinities to MTs which is: Hg > Cd > Cu > Zn (Eaton, 1985), so the relationship of MTs to acclimation may not be as obvious as it appears. Roch and McCarter (1984b) found elevated levels of MTs in field-exposed fish that did not translate into higher metal tolerance. Part of the uncertainty regarding the relationship of MTs to metal acclimation may be due to the fact that there are several other metal-binding proteins in cells that may act in a manner similar to MTs. Hodson (1988) has emphasized that the role of MTs will not be understood until we better understand these other proteins.

Mercury has an especially strong affinity for MTs, so much so that it displaces copper and zinc from the molecule. According to Brown and Parsons (1978), this has the effect of decreasing the normal tissue concentrations of these two essential elements when salmon are chronically exposed to mercuric chloride.

In contrast to inorganic mercury, methylmercury apparently does not bind to MTs (Weis, 1984), although it may bind to other cellular proteins that could provide some capacity for sequestering it. Thus the form of the mercury to which a fish is exposed could have a considerable bearing on the metabolism of other trace elements.

The kinetics of MT formation has been investigated by McCarter and Roch (1984). With continuous copper exposure, hepatic MT rose steadily and then stabilized at 4 weeks at a level dependent on the copper concentration in the water. Thus, the rate of MT synthesis was controlled by the rate of copper uptake. Transfer of the acclimated fish to water not contaminated with copper resulted in no significant loss in MTs after 4 weeks, so once it is formed, MTs appear rather stable.

There is a widely held assumption that metals, in general, induce MT synthesis. Evidently, however, not all metals are capable of doing this. Lead has been shown to not induce MTs when fed orally for a week to croakers, even though glutathione was induced in the liver (Juedes and Thomas, 1984).

There has been considerable interest in using MT induction as a biomarker of metal stress. The reasoning is that induction would indicate actual metal bioavailability, although measurements of the concentration of individual metals in the tissues might do essentially the same thing with less analytical effort. Also, in the field numerous stresses that are not metal related can cause alterations in MTs. Thus, for fish it is generally considered premature to utilize MT induction as a biomarker for metal contamination until their normal physiological functions and controls are better understood (Stegeman et al., 1992).

VIII. BIOCONCENTRATION OF ORGANIC POLLUTANTS

Very broadly speaking, the organic pollutants of concern for fish physiology fall into the following groups: PAHs (mostly associated with petroleum), polychlorinated dibenzo-*p*-dioxins, dibenzofurans, polyhalogenated biphenyls, and various insecticides and herbicides.

As was mentioned at the beginning of this chapter, the route of uptake of a specific chemical is going to depend on whether it is primarily in the food, dissolved in the water, or both. Many organic pollutants that happen to be in the food (e.g., DDT, PCBs) are taken up in direct proportion to the dietary level, which will depend on the quantity of food consumed and the concentration in the food (Spigarelli et al., 1983; Hilton et al., 1983). Factors that influence the metabolic rate of the fish, such as temperature or simultaneous exposure to other pollutants (see Chapter 8), will alter the food consumption and thereby the rate of accumulation of an organic contaminant.

During the 1970s, models were developed to approximate the bioconcentration factor of organic chemicals that are dissolved in the water (Ernst, 1977; Krzeminski et al., 1977; Vieth et al., 1979). These are based on the assumption that uptake and elimination rates can be estimated by first-order kinetic expressions (i.e., they are in some manner proportional to the concentration of the chemical in the environment and that in the fish). To quote Veith et al. (1979, p. 1041): "Measurements of BCF can be made by exposing organisms for a period sufficiently long that the steady state residue is observed, or relying on the accuracy of a given kinetic model to measure appropriate rate constants and calculate the BCF assuming the chemical behaved according to the model." They go on to make the point that the first method requires that the fish be small and the BCF must not be too large, otherwise the time to reach equilibrium in a laboratory regime becomes excessive. The kinetic model, however, can yield misleading results as uptake and/or depuration rates may change during the exposure. As will be seen later in this section, biotransformation processes are inducible so it is to be expected that depuration will indeed change over time of exposure.

The bioconcentration factor for 30 common organic chemicals was determined in fathead minnows by exposing them for 32 days in the laboratory (Vieth et al., 1979). The BCF ranged from 2.7 to 194,000. Part of the aim of this study was to determine if the BCF was related to the *n*-octanol/water partition coefficient (P), which is a measure of the lipid solubility of a molecule. It is well known that permeability of cell membranes by organic molecules is highly dependent on their lipid solubility. With some exceptions, the BCF was related to the partition coefficient.

More recently, however, Nichols et al. (1991; p. 388) "...have attempted with only limited success to relate chemical log K_{ow} to chemical accumulation in blood and tissues." They go on to remark that other information such as triglyceride partitioning or the physical and chemical properties of molar volume, dipolarity/polarizability, and hydrogen bonding might be better for predicting partitioning in various tissues.

Laboratory derived estimates of residue accumulation from water neglect uptake via the food, so they may greatly underestimate what would be found in fish taken from the natural environment. They should also be looked upon as relative, rather than absolute, as a host of things influence the accumulation of organics. These include size of fish, lipid stores, temperature, age of the fish, and the processes of biotransformation and excretion.

Limited data (Vieth et al., 1979) suggest that within a species, age and body size has little effect on BCF. However, an increase in temperature increases the BCF approximately twofold for each 10°C difference, and an oscillating temperature causes more uptake than one which is constant. Spigarelli et al. (1983) exposed brown trout (*Salmo trutta*) to PCBs from natural food and water under three thermal regimes: (1) a daily temperature cycle ranging from 7–18 with a mean of 12.5°C; (2) a constant temperature of 12.9°C; and (3) natural fluctuations of ambient inshore temperatures (4°–11°C) with a mean of 7.7°C. Direct uptake from the water accounted for only about 10% of the total PCB load, so most came from the food. The greatest PCB accumulation occurred in fish from thermal regime number 1, and the fish exposed to the cycling temperatures fed and grew at the rate they would have if acclimated to the maximum temperature experienced. The authors conclude the temperature effect on PCB accumulation was because the temperature controlled food consumption, growth, and lipid content of the fish.

In field studies, Gossett et al. (1983) analyzed for 27 organic compounds in livers of fish taken in the vicinity of the Los Angeles, CA, wastewater treatment plant. They found that the tissue concentrations of the compounds were correlated with the log P of the compound ($r = 0.63–0.75$), but were not correlated with the environmental concentration. While the overall correlation with lipid solubility was good, there were some notable exceptions which the authors attribute to metabolism of the parent compound.

The molecular weight of a pesticide has been shown to affect the BCF (Kanazawa, 1982). When log BCF (Y) is plotted against the log molecular weight (X) of pesticides of all sorts ranging in weight from 187–412, there is a correlation coefficient of 0.846. There is a linear relationship between the lipid solubility of these same pesticides and their BCFs, so the cause and effect here is not clear, but it is interesting that the larger the pesticide molecule (at least up to a molecular weight of 412), the more it accumulates in fish tissues. A similar relationship between molecular chain length and accumulation has been found for linear alkylbenzene sulfonate (LAS), an anionic surfactant used in detergents (Comotto et al., 1979).

Thus far in this discussion of accumulation of organics, the fish has been considered as a single compartment, when it is, of course, made up of different types of tissues. In comparison to the amount of information available on total body burden, the relative accumulation of organic xenobiotics by different tissues is lacking. Pentachlorophenol will now be used as a type of contaminant as there is considerable information about it, thus a few generalizations begin to become evident.

Examination of Figure 11 illustrates how different tissues may accumulate markedly different levels of an organic contaminant, and even compounds superficially similar may accumulate in rather dissimilar ways. In this study, trout were exposed to either pentachlorophenol (PCP) or pentachloroanisole (PCA) at a concentration similar to that later found by others in contaminated natural waters (Delfino, 1979), but well below the lethal concentration (Niimi and McFadden, 1982). PCP accumulated most in the liver whereas PCA was accumulated far more in the fat. Neither compound showed much affinity for muscle.

Glickman and Melancon (1982) reported BCF values after 48-h exposures for radiolabeled PCP and 2.4.5-trichlorophenol (TCP) of approximately 50× for muscle, 100× for blood, 1000× for liver, and 100,000× for bile. The test concentration in the water was fairly high (0.05 mg/L), but the partitioning into different organs is especially striking.

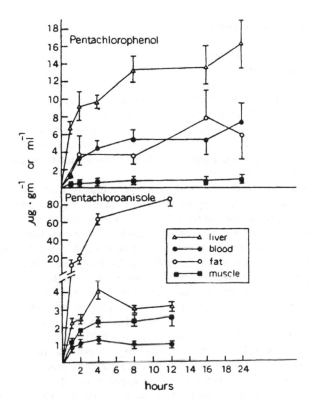

Figure 11 Uptake of ¹⁴C-labeled PCP and PCA in tissues of rainbow trout exposed to 0.025 μg/L in the water. Concentrations are calculated as μg/g wet wt. Each point represents the mean ± the SE from at least six fish. (From Glickman, A. H., Statham, C. N. and Lech, J. J., *Toxicol. Appl. Pharmacol.*, 41, 649, 1977. With permission.)

The elimination of PCP may be relatively fast in that the half-life for fat is 24 h while blood, liver, muscle, gills, and heart range from 6.2–10.3 h (Glickman et al., 1977). Of course, the highest concentration is in the liver so the greatest amount was depurated from there. This is to be expected since the liver is the primary organ for metabolism of PCP (Lech et al., 1978). One final note on depuration of PCP: whereas the first half-life is quite short, the data on this goes only out to 24 h (Glickman et al., 1977) and there may be a low residue that remains for some unknown period of time. Indeed, Niimi and Cho (1983) report a half-life for depuration of PCP in whole trout as 7 days, which is probably due to longer retention time in some tissues than others.

The accumulation and elimination of PCP is greatly affected by the ambient salinity. Freshwater acclimated killifish exhibited a BCF of 1680 (whole body) while seawater acclimated killifish had a BCF of only 370 (Tachikawa et al., 1991). This is somewhat surprising as seawater fish are about 10× more permeable to salt ions and water than are those in freshwater (see Chapter 7). Furthermore, they drink a significant amount of seawater to replace water lost by osmosis, but according to Tachikawa et al., (1991) the amount of PCP taken in by the drinking of seawater-acclimated fish contributed only "slightly" to the uptake. They attribute the reduced accumulation rate in seawater to a slower rate of uptake and especially to a more rapid metabolism and excretion. The mechanisms by which these processes would change with salinity acclimation are currently obscure.

The BCF for specific organs of organochlorine pesticide residues has been shown to vary with the dose. Under acutely toxic doses, the liver, spleen, and fatty tissues have much higher concentrations than either muscle or gill, but when the exposure is chronic, concentrations in the liver decline to those found in muscle or gill while spleen and fatty tissue have much higher levels (Holden, 1966). The lower levels in the liver with chronic exposure are probably due to metabolism of the pesticide by that organ.

Fish captured in the field that are contaminated with organochlorine pesticides may exhibit a seasonal difference in partitioning into different tissues. Trout and whitefish captured from freshwaters in Russia showed highest concentrations of these pesticides in the brain during the spring and summer; the liver accumulated them during late summer and fall (Chernyaev, 1991).

Organophosphorus compounds accumulate very rapidly from the water reaching a steady state in small fish such as guppies within 12–48 h. A biphasic relationship was seen between accumulation rate and log K_{ow} (De Bruijn and Hermens, 1991). Elimination rates generally correlate with log K_{ow} with some exceptions. These exceptions seem to be related to different rates of biotransformation of parent compounds. This can also cause lower bioconcentration factors than would be predicted. De Bruijn and Hermens (1991) review several studies that show the impact that biotransformation can have on the accumulation of various compounds. Welling and De Vries (1992) found that the organophosphorus insecticide chlorpyrifos is eliminated rapidly from guppies but nearly all of the parent compound is biotransformed before elimination. They go on to note that prolonged exposures to the pesticide caused the elimination rate to actually decrease, which is the reverse of what might be expected. Biotransformation enzymes are inducible (see Section IX) so prolonged exposure would be expected to enhance excretory processes.

Fish accumulate aromatic petroleum hydrocarbons rapidly, so concentrations may reach a maximum within a few hours of exposure (Anderson, 1979). As the number of aromatic rings in the molecule increases, the accumulation and retention also increases. Different petroleum hydrocarbons are accumulated to different equilibrium levels so the concentrations in the fish will not necessarily represent the relative concentrations of the different components in the environment. Longer exposures often produce lower tissue concentrations of the parent compounds because of induction of biotransformation enzymes in the liver (Anderson, 1979).

Petroleum hydrocarbons are not partitioned into the tissues uniformly. In Table 1 data are derived from work on dolly varden char (a salmonid) that had received a stomach dose of either radioactive-labled toluene or naphthalene and then the tissues sampled 24 h later. Thomas and Rice (1981) provided the percentage of administered label that appeared in each organ and the average organ weights. In order to get a measure of relative concentration in the different organs, I divided the average percentage of the total label in an organ by the organ weight. It should be kept in mind here that these data reflect both the parent compound and the metabolites. The relative percentage of these two forms varies with the tissue. Several things then become evident in the data presented in Table 2:

1. The naphthalene has a greater potential for accumulation than does toluene, which might be predicted as it is a two-ring compound whereas toluene is a one-ring molecule (later we will see that toluene is excreted more rapidly than naphthalene).
2. Seawater acclimation caused naphthalene to be accumulated more than when the fish were in freshwater, but this environmental variable had no affect on toluene accumulation. The authors present other evidence to the effect that naphthalene is metabolized more rapidly in freshwater-adapted fish, which provides a mechanism for the slower accumulation of this chemical when the fish are in freshwater. Why the salinity would affect this process seems obscure. It may be recalled that the opposite occurred with the

metabolism of PCP in seawater and freshwater acclimated killifish. Seawater acclimation stimulated metabolism and elimination of PCP (Tachikawa et al., 1991).

3. The concentrations in the various tissues differed by orders of magnitude. The bile (called gallbladder by Thomas and Rice) had by far the largest concentrations, and gill had remarkably high concentrations, even though the exposure was internal rather than external. The high hydrocarbon levels in these two organs are probably due to their central role in the excretion of both the parent compounds and their metabolites.

A problem often occurs when studying petroleum hydrocarbons in aquatic organs and their environments in that many of these chemicals are "naturally occurring" in the ambient waters, sediments, and fish tissues. Thus, there has been some interest in the

Table 1   Relative Concentration of Radioisotope in Tissues 24 h After Being Force-Fed Labeled Toluene or Naphthalene Dose

Tissue	Toluene FW	Toluene SW	Naphthalene FW	Naphthalene SW
Muscle	0.02	0.03	0.04	0.31
Gill	0.51	0.80	0.68	9.32
Bile	2.12	3.12	10.50	1.50
Liver	0.15	0.10	0.30	0.52
Brain	0.06	0.18	0.06	0.22

Note: See text for method of calculation. FW = Freshwater-acclimated fish. SW = Seawater-acclimated fish.

Data calculated from Thomas, R.E. and Rice, S.D., 1981.

Table 2   Physiological Parameters Used in a Physiological-Based Toxicokinetic Model for Adult Rainbow Trout[a]

Body weight (kg)	1.0
Ventilation volume (L H_2O/h)	10.6
Effective respiratory vol. (L H_2O/h)	7.2
Oxygen consumption rate (mg O_2/h)	63.0
Cardiac output (L blood/h)	2.07
Arterial blood flow	
To liver	0.06
To fat	0.176
Poorly perfused compartment	1.242
Richly perfused compartment	0.476
Kidney	0.116
Tissue group volumes (L)	
Liver	0.012
Fat	0.098
Poorly perfused compartment	0.818
Richly perfused compartment	0.063
Kidney	0.009

[a]   From Nichols et al. (1990) using a trout at 12°C.

alkane squalane which is apparently correlated only with hydrocarbon sources such as fossil fuels and other petroleum by-products (Matsumoto and Hanya, 1981). This material has a high (about 40%) assimilation efficiency in trout (Cravedi and Tulliez, 1986). The partitioning of this compound into various tissues is, however, rather different than other hydrocarbons. In spite of its high lipophilicity, very little is accumulated in fat where other hydrocarbons usually go (Cravedi and Tulliez, 1982). Instead, most accumulation occurs in the liver, kidney, heart, and spleen. The authors (Cravedi and Tulliez) speculate this may be due to uptake by the reticuloendothelial cells in these organs, although no direct evidence is provided.

The detergents, LAS, and SLS, are accumulated rapidly in fish tissues (Comotto et al., 1979; Tovell et al., 1975). Muscle and brain generally exhibit the lowest BCF and the highest values are in gallbladder, liver, and gut. Apparently, metabolism of and excretion of detergents are fairly rapid as a steady state is achieved at almost any dose after 3 days, and depuration of over 85% of the accumulated material occurs within 4 days of being placed in clean water.

In recent years there has developed an interest in creation of physiologically based toxicokinetic models for fish exposed to waterborne toxicants (Nichols et al., 1990, 1991). The more traditional approach to modeling accumulation of a contaminant treats the fish as a two or more compartment model and is based on measures of accumulation. The physiologically based model uses a list of physiological variables to predict the time course of contaminant accumulation. This theoretically then permits extrapolation to other species and exposure conditions. Table 2 presents the physiological parameters used in such a model. The actual values are from the literature. Other non-physiological parameters used include concentrations of the test substance in the inspired water and some characteristics of the chemical. One point of explanation is in order. In Table 2, the arterial blood flow to a poorly perfused compartment may seem high as there is not much blood flow there. However, in the fish this compartment is mostly white muscle which makes up over 50% of the body, so even though the perfusion is limited, the total bulk allows a considerable blood flow.

The physiologically based toxicokinetic model has been tested with pentachloroethane, hexachloroethane, and tetrachloroethane (Nichols et al., 1991) using cannulated trout in chambers similar to the one shown in Figure 3. The model proved reasonably good at predicting the dynamics of accumulation of these chemicals. One of the outgrowths of this type of study is how it illustrates the effects that chemical affinity and blood perfusion have on accumulation dynamics. For example, fat tissue has a high affinity for the test chemicals but is poorly perfused. As a result, uptake is slow into this compartment so that its concentration may continue to rise long after an equilibrium concentration has been reached in the blood.

While the physiologically based models have considerable potential, species differences will be an interesting challenge. For example, when the distribution of a single oral dose of tetrachlorodibenzo-p-dioxin was compared in cod and in seawater-adapted rainbow trout, the highest concentrations in the cod were found in the liver and central nervous systems. In the trout, on the other hand, the highest concentrations were in the visceral and extravisceral fat depots and relatively little in the liver and central nervous system (Hektoen et al., 1992). These differences are probably explained by differences in physiology, which makes understanding comparative physiology a key issue for different species.

IX. BIOTRANSFORMATION OF ORGANIC CONTAMINANTS

Biotransformation processes in fish have been reviewed extensively (e.g., Lech et al., 1982; Buhler and Williams, 1988; Goksoyr and Forlin, 1992; Andersson and Forlin,

1992; Stegeman and Hahn, 1994). This is a very large and fast moving field so only a brief presentation of some basic concepts will be attempted here.

The biotransformation reactions tend to make the relatively lipophilic parent compound into one that is more hydrophilic (i.e., water soluble) and more polar. This enhances the tendency for the substance to be excreted by the gills, bile, and kidney while at the same time it reduces its affinity for plasma and tissue proteins and for adipose tissue. Hydrophilic compounds are also less permeable to cell membranes so are less likely to be accumulated in various tissues.

It is usually assumed that biotransformations are detoxification reactions. While that may be true for a great majority of xenobiotics, there are examples in fish where the transformation makes the substance more toxic or carcinogenic; the latter has serious implications for human health but is clearly beyond the scope of this book. Increased toxicity to the fish is seen with the classical case of the organophosphate insecticide parathion. This insecticide must actually be converted to paraoxone by mixed function oxidate (MFO) before it becomes an inhibitor of cholinesterase in the nervous system (Ludke et al., 1972).

The biotransformation reactions are broadly divided into asynthetic (Phase I) and synthetic (Phase II). The former group includes oxidation, reduction, and hydrolysis. The synthetic reactions usually involve the conjugation of endogenous compounds such as glutathione with the metabolites produced in the Phase I steps. However, in some cases the conjugation is of the parent compound so it is not always necessary that a Phase I reaction precede a Phase II reaction.

Most of the Phase I reactions are oxidations catalyzed by the cytochrome P450 system referred to as the MFO, monooxygenase (MO), or CYP1A1 system. For convenience, it will be referred to here as the MFO system. It is a coupled electron transport system with several accessory enzymes which acts to insert one atom of oxygen into the substrate and reduces the second atom of oxygen to form water. The system is located predominantly in the endoplasmic reticulum of the cells (microsomal fraction when separated in an ultracentrifuge).

Phase I hydrolysis reactions are catalyzed by epoxide hydrolases which hydrolyze various epoxides and diols. These enzymes are located both in the ribosomes and in the cytosol of the cell (De Bruin, 1976).

Considerable species variability undoubtedly exists at the quantitative level in these Phase I processes. The enzymes of the MFO system have very broad substrate specificities due to the presence of multiple isozymes (Andersson and Forlin, 1992). They have also been found in the embryonic stages of several species so are functional at a very early stage in the life cycle. Induction of MFO activity in larvae can occur via direct exposure of the egg to crude oil (Goksoyr et al., 1991). Indeed, xenobiotics in the gametes of lake trout have been shown to induce MFO activity in the livers of the offspring derived from those gametes (Binder and Lech, 1984).

Increased activity of MFO enzymes is induced by exposure to xenobiotics. This has been extensively studied in mammals and in fish. Classically, there were two major classes of MFO-inducing chemicals referred to as the 3-methylcholanthrene (3-MC-like) and phenobarbital-like. Now chemicals are classed by the specific family of P450 genes that are activated. Over a dozen P450 forms have been purified from fish tissues; the most extensively studied is from rainbow trout and is designated P4501A1 (Andersson and Forlin, 1992). The inductive response is usually detected by measuring P450 catalytic activity by assaying the ethoxyresorufin-O-deethylase (EROD) reaction which is low or sometimes undetectable in non-contaminated fish.

The marine and freshwater environments are loaded with thousands of different chemical effluents. It has been recognized since the mid-1970s that induction of MFO might be a useful biomarker for environmental contamination as it would yield an integrated signal for bioavailability of unknown xenobiotics and the defense responses by the resident fish. Thus far, there has been considerable success in the utilization of this biomarker, reviewed in Payne et al. (1987); Goksoyr and Forlin (1992); Collier et al. (1992). Useful guidelines for the application of such procedures are found in Stegeman et al. (1992).

The MFO system is not the "Holy Grail" of biomarkers, for it does have some problems. To begin with, while a wide variety of xenobiotics induce increased MFO activity in fish, the phenobarbital group of substances does not (Goksoyr and Forlin, 1992). The chlorinated hydrocarbon DDT falls in this category and fails to induce MFO activity when fed to trout (Addison, 1977). Fortunately, most of the organic pollutants of concern are inducers of this system.

The presence of cadmium (and possibly some other metals?) can lower MFO activity (Forlin et al., 1986), so this might cancel out the effect of organic induction of MFO activity. The sex of the fish affects basal levels; male trout have over twice the levels of females (Williams et al., 1986), although sockeye salmon during the spawning migration showed no sexual difference (Kennish et al., 1992). It appears that the sex hormones in both sexes can have a considerable influence on MFO, but not in the same direction. Andersson and Forlin (1992) review several studies which show that testosterone seems to upregulate MFO whereas estradiol downregulates the same system. The situation in the female during spawning season when both hormones rise is quite complicated and varies with the species.

Another steroid interaction with the MFO system has recently been revealed using an *in vitro* preparation of trout hepatocytes. Devaux et al. (1992) found that cortisol- or dexamethasone-induced MFO activity, thus elevated cortisol levels from environmental stress unrelated to pollutant contamination might produce high levels of MFO. This deserves further study in intact fish.

Tributyltin is a common metal contaminant in harbors and has been shown to inhibit MFO activity and the induction response (Fent and Stegeman, 1993). This could have serious implications for fish exposed to mixed effluents which, of course, is the usual situation in harbors and bays.

The process of induction of MFO enzymes can be looked on as a form of adaptation to a contaminated environment. However, it also may have some adverse side effects. This is because a function of the MFO system in non-contaminated fish is the metabolism of steroid hormones. Lech et al. (1982) speculated that an enhanced rate of metabolism of the sex hormones from pollutant-induced MFO activity could have harmful effects on reproduction. Since then, several investigations have provided indirect confirmation that elevated MFO can disrupt reproductive functions (e.g., Thomas, 1990), although direct cause and effect mechanisms are not clear.

Because fish are poikilotherms, temperature might have a notable effect on MFO function, however, current findings are in conflict on this matter. The elimination rate of benzo[a]pyrene by bluegill (*L. macrochirus*) was increased threefold over a 10°C rise in acclimation temperature and enzyme profiles indicated that biotransformation was similarly affected (Jimenez et al., 1987). However, trout have been reported to exhibit ideal temperature compensation for MFO activity (Blanck et al., 1989; Koivusaari, 1983) (i.e., the same rate is obtained at all acclimation temperatures if incubation temperatures for the tests are run at the acclimation temperature). Members of the genus *Lepomis* generally show "good" temperature compensation in their glycolytic and Kreb's cycle enzyme systems (Shaklee et al., 1977) so the lack of temperature compensation in the MFO system of the bluegill deserves further investigation.

There are still a good many questions remaining concerning the MFO system in fishes. Goksoyr and Forlin (1992) note that there is almost no information on the effect of photoperiod, salinity, handling, etc., on this enzyme group. Of course, the extreme diversity of fish species presents a tremendous challenge to our further understanding of this remarkable system.

The Phase II, or conjugation reactions involved in xenobiotic transformation, include conjugations with glucuronide, sulfate, glutathione, or taurine. This has been referred to as a protective synthesis (De Bruin, 1976) whereby both natural substances (e.g., steroids) and foreign xenobiotics are coupled to endogenous substrates such as glutathione. There are specific enzymes (e.g., glutathione transferases) which catalyze the transfer of the conjugating agent from its bound coenzyme form to the foreign molecule. Obviously, this is a greatly simplified summary of some rather complex biochemistry, much of which has been learned from work on mammals. The interested reader is referred to the reviews of Lech and Bend (1980), Smith et al. (1983), and George (1994).

Conjugated metabolites of a variety of organic chemicals have been found in fish (Stehly and Hayton, 1989). The phenolics have probably received the most attention and there are species differences in the conjugation of phenol and PCP. For example, the goldfish and sheepshead conjugated a greater percentage of PCP with sulfate whereas rainbow trout, fathead minnows, and firemouth utilized the glucuronide conjugation more (Stehly and Hayton, 1989). The dose can also influence the conjugation mode. Nagel (1983) found that when the phenol dose was increased the percent glucuronylated increased while the percent sulfated decreased. Evidently, as has been seen in mammals, the sulfate conjugation process is easily saturated, and when phenol is injected, as opposed to putting it in the water, the material is rapidly transported to the gills where the nonconjugated phenol is eliminated (Nagel and Ulrick, 1980).

Glutathione is seemingly a "molecule for all seasons". Earlier in this chapter it was mentioned that it plays a role, at present poorly understood, in the metabolic handling of metals. In mammals, and probably fish, it serves a variety of functions in both natural substrate and xenobiotic metabolism. The cell utilizes glutathione along with ascorbic acid and vitamin E in the process of scavenging free radicals by superoxide dismutase and catalase. The conjugation of glutathione to xenobiotics, or their Phase I products, is an active field of study. The concentration of glutathione depends on the activity of several enzymes, some of which catalyze its conjugation with other substrates, others catalyze its synthesis.

Several investigators have reported elevated concentrations of glutathione in liver or gill from exposures to various organic chemicals (Thomas and Wofford, 1984; Lindstrom-Seppa and Oikari, 1991; Gallagher et al., 1992). Gallagher et al. (1992) showed that the fungicide chlorothalonil induced an increased activity in hepatic gammaglutamylcysteine synthetase which is a key enzyme in the synthesis of glutathione. Whether other chemicals cause the rise in glutathione by this mechanism remains to be determined.

The importance of glutathione in counteracting toxicity was clearly shown by Gallagher et al. (1992) when they depleted the liver and gill of glutathione with a drug. This caused the toxicity of chlorothalonil to increase threefold.

Exposure of fish to either phenol or ammonia rapidly induced (i.e., within a day) a two- to threefold increase in the activity of liver glutathione-S-transferase activity. This enzyme catalyzes conjugation of glutathione with the xenobiotic. At the same time, there was a transient rise in glutathione concentration (Chatterjee and Bhattacharya, 1984). The concentration of glutathione will undoubtedly depend on the rate of synthesis minus the rate of conjugation with xenobiotic.

Most of the work that is done on the transformation of xenobiotics by the liver has been performed on fish sacrificed at particular times during an exposure regime, or following an injection. In an attempt to get at some of the dynamics of the process, Forlin and Andersson (1981) developed a system to use an *in vitro* perfused rainbow trout liver. The isolated livers were kept functional for at least several hours which made it possible to add a chemical to the "blood" going into the liver and measure the removal of the material and the formation of metabolites as they appear in the blood and/or bile. This sounds easier than it really is, and it is necessary to use large fish (>400 g) so work with small species would be precluded.

It has been long assumed that biotransformation of xenobiotic organic chemicals occurs mostly in the liver. It is now appreciated that other organs, most especially the intestine and gills, also have considerable capacity as well (Van Veld, 1990; Miller et al., 1989). Because the gills and intestine are the main modes of uptake for pollutants, having the capacity for some metabolism of these compounds there has obvious advantages. Van Veld et al. (1988) have developed a technique whereby the portal vein is catheterized so the blood coming from the intestine can be sampled for metabolites produced by the intestine. With this system, they found that most of the metabolites appearing in the blood from the ingestion of benzo[a]pyrene were derived from the intestine instead of the liver. There is probably considerable species variability in extrahepatic metabolism of xenobiotics (Lindstrom-Seppa et al., 1981).

Intestinal microorganisms may transform organics present in the food (Addison, 1976). This can cause problems in determining the location where a metabolite was formed, and for this reason, in laboratory studies the xenobiotic may be injected rather than included in the food. (The injection mode of entry also precludes problems with incomplete assimilation.)

The metabolic transformation of xenobiotic chemicals that occurs in fish has important implications for monitoring programs. When doing residue analysis, if analysis for only the suspected parent compound is done, misleading results may be obtained should part or all of the chemical be transformed (Giesy et al., 1983). This can happen within a matter of hours of exposure. In general, the bile will contain the conjugated products of the transformations. The fact that these metabolites concentrate in the liver and bile argues for analysis of these select tissues separate from muscle when doing residue analyses. Biotransformations also make predictions of steady-state tissue concentrations based on chemical characteristics quite suspect unless it is shown that the chemical is not significantly metabolized during the period of exposure.

X. EXCRETION OF ORGANIC CONTAMINANTS

Depuration of some foreign chemical is usually measured by allowing the contaminated fish to reside in non-contaminated water for some specified period of time and then residue analysis is performed on the whole body or specific tissues. The latter sort of analyses are far more informative as the rate of depuration varies greatly between different tissues and this is related to their relative abilities for metabolic transformation of the chemical. Depuration measurements give a general picture of the elimination of a specific parent chemical which has practical application when there is concern for how long a group of fish may remain contaminated following a chemical spill. Our interest here is primarily with the physiological routes by which this is accomplished. (The excretion of metals was reviewed earlier in this chapter.)

As with inorganic pollutants such as the metals, a fish has several possible routes for excretion of an organic pollutant. These include the gills, skin, mucus, bile, feces, and urine, which are the same as those "used" by metals, only the primary routes differ considerably between different chemicals.

PCP is rapidly absorbed by fish from the water and it is rapidly conjugated with sulfate and glucuronic acid. During exposure of goldfish to 0.1 ppm PCP, the conjugated PCP began to appear in the bile within an hour and further accumulation there was linear for up to at least 48 h, by which time the concentration factor was 12,000. Only the glucuronide conjugate and a small amount of free PCP was detected in the bile (Kobayashi, 1978). In spite of the high concentrations of PCP conjugate in the bile, work in several species has shown that over 60% of the PCP is excreted mostly via the gills, but with some through the urine (Stehly and Hayton, 1989; McKim et al., 1986).

Looking for metabolites in the surrounding water following exposure of fish to a pollutant has caused some confusion in the literature as to what metabolites are actually formed from phenol and PCP (Nagel, 1983). Perhaps the delay in releasing material from the gallbladder may be a major problem here. This structure is in part a digestive organ which contracts to empty its contents into the intestine when food is present. Because the fish used in xenobiotic metabolism studies are generally postabsorptive, there may be a tendency for material to stay longer in the gallbladder than would occur under "natural" conditions where the fish is feeding.

A wide variety of organic materials and/or their metabolites have been found to concentrate in fish bile, and this can occur very rapidly. Mention has already been made of the accumulation of conjugated PCP there. Miyamato et al. (1979) reported considerable accumulation of fenitrothion (an insecticide) in the bile of trout after only 6 h of exposure. Peterson and Guiney (1979) noted that PCB metabolites are excreted via the bile of fish. Using nine very different isotopically labeled organic substances added to the water for 24 h, Statham et al. (1976) obtained bioconcentration factors of 11–10,000 in the bile of trout. In most cases, the material in the bile was actually the conjugated metabolite. It is interesting that the lowest concentration factors in the bile are associated with compounds with the highest lipid solubility. Recall that these have the high octanol/water partition coefficients and therefore tend to be accumulated by fish to the greatest degree. Evidently, they have a relatively low rate of metabolism or conjugation by the liver, a characteristic that would contribute to their tendency to accumulate.

Finding a high concentration of some pollutant in the bile does not necessarily mean this is the only, or even the dominant mode for excretion. In order to partition the excretory modes it is necessary to use specially designed chambers similar to the one shown in Figure 3. Using a chamber such as this, Thomas and Rice (1981, 1982) have investigated the excretion of radiolabeled aromatic hydrocarbons associated with petroleum. The test chemical was inserted into the stomach in a gelatin capsule and the fish allowed 24 h to absorb and excrete it. The gills turned out to be an especially important organ for excretion of aromatic hydrocarbons. The rate of excretion of individual hydrocarbons by the gills was shown to be inversely proportional to the size of the molecule and the log K_{ow}. More recent work by Le Bon et al. (1987) has shown that the isoprenoid hydrocarbon, pristane, is metabolized first before being excreted via the gills.

The total isotope from naphthalene and toluene in the rear chamber, urine, and gallbladder was less than 20% of that excreted via the gills (Thomas and Rice, 1981) indicating that the latter organ is the primary mode of excretion of these two highly toxic aromatic hydrocarbons. Even though high concentrations of these hydrocarbons were found in the gallbladder and urine, the total volumes of fluid involved were so small that they did not account for a very large amount of the overall hydrocarbon excretion compared to that excreted via the gills. This points out again the fact that finding a high concentration of some compound in the urine or bile does not necessarily establish it as an important mode of excretion for that compound.

For larger organic molecules, the gills may not be particularly useful as an excretory organ. Thomas and Rice (1982) concluded that excretion of anthracene and benzo[a]pyrene via the gills was of little significance compared to the bile, although even that mode was slow.

Phenol and cresol are rapidly excreted by the gills, bile, urine, and skin with the greatest percentage coming out through the skin and bile (Thomas and Rice, 1981). The skin and bile has also been shown to be an effective mode for excretion of naphthalene metabolites in rainbow trout (Varanasi, 1979). In these fish, the bile contained little of the hydrocarbon material over the first 24 h of exposure, but it became an important method of excretion in the long run so it apparently takes some time for the excretory machinery in the liver to get "geared up".

Anthracene and benzo[a]pyrene are excreted very slowly compared to the aromatic hydrocarbons (Thomas and Rice, 1982). In 24 h only somewhat less than 4% of the total dose was excreted compared to 66% for phenol, 30% for toluene, and 10.8% for naphthalene. This delay is apparently due to the requirement that anthracene and benzo[a]pyrene be metabolized before excretion. Anthracene depuration has been shown to be most rapid at night (Lindstadt and Bergman, 1984). Such a periodic depuration could be characteristic of a variety of compounds, but it seemingly has not been looked for in other studies.

It is well known that seawater-adapted fish excrete little urine volume compared to the freshwater-adapted forms (even of the same species), and the osmotic flow of water through the gills is inward in freshwater while outward in seawater. Thus, it is conceivable that a greater amount of material might be excreted via the kidneys of freshwater fish. In spite of these theoretical considerations, from very limited data it appears that the salinity has little effect. The excretion of naphthalene and toluene via the urine was little changed when the dolly varden used in the studies discussed above were adapted to freshwater (Thomas and Rice, 1981). This represents only one species and two aromatic hydrocarbons so the generalization should be treated as very tentative.

The detergent SLS is apparently excreted exclusively via the urine and bile. Using radiolabled SLS Tovell et al. (1975) reported finding no isotope in the "gill" water of their experimental goldfish. (Their fish had the rear half of the body enclosed in a plastic bag which collected both urine and cloacal excretions.) Analysis of the bile showed very high concentrations there of both the metabolized and non-metabolized form of the compound, so it is evident that it is not necessary that SLS be metabolized before excretion.

Using catheterized channel catfish in a split chamber and LAS, another detergent, Schmidt and Kimerle (1981) were able to partition the excretion more completely. Ninety-six hours following injection of the isotope, 42% percent of the radioactivity had been excreted. Of that, 60% appeared in the urine, 29% in the anal chamber (presumably from bile), and 9.5% in the gill chamber.

Carboxymethyltartonate (CMT), a potential partial replacement for phosphates in detergents was also tested. All of it was excreted in 96 h with 75% appearing in the urine, 16% in the anal chamber, and 8% was excreted via the gills. Thus, for this group of organic compounds, little gill excretion occurs in freshwater fish, but instead they exit the fish largely by way of the urine and to a lesser extent the bile.

The excretion of the lamprey poison Bayer 73 by teleosts occurs by both the urine and bile in roughly equal amounts. Some may be lost across the gill tissue to the water but this likely is relatively insignificant compared to the other two routes (Allen et al., 1979).

Ethyl-m-aminobenzoate (commonly called MS 222) is widely used as a fish anesthetic by adding it to the water. One of its virtues as an anesthetic is the rapid recovery of the fish following their being placed in anesthetic-free water. This must be in part due to the rapid clearance from the body of the fish. Stenger and Maren (1974) reported a 95% clearance via the gills in 2 h. Renal excretion of the injected drug was less than 5% of the total lost. An interesting sidelight to this study was the finding that IV injection failed to produce anesthesia. Materials injected IV in fish must pass the gills before reaching the brain, so evidently most of the drug was lost to the water upon one passage of the blood through the gills (Hunn and Allen, 1974).

The commonly used carbamate pesticide, aldicarb, is rapidly distributed in the fish following either oral administration or IP injection (Schlenk et al., 1992). It is also excreted rapidly, mostly as the unmetabolized molecule via the gills. Its rapid exit through the gills is probably due to it being a rather water-soluble molecule.

The phenoxyacetic acid herbicides (e.g., 2,4-D), the primary metabolite of DDT (DDA), and the metabolites of benzo[a]pyrene are all easily excreted by the teleost kidney. Indeed, the kidney actively secretes them against a concentration gradient so the urine contains a far higher concentration of the first two compounds than was present in the blood (Pritchard and Bend, 1984). (The metabolites of BP were only slightly concentrated in the urine.) Pritchard and Bend (1984) note that a high lipid solubility tends to reduce the ability of the kidney to excrete an organic xenobiotic. They also point out that fish kidneys may be able to handle some compounds better than mammalian kidneys because of the extensive renal portal system in the fish, which accentuates the role of tubular transport, and the fact that the xenobiotics have a higher binding affinity to plasma proteins in mammals.

It is difficult, if not impossible, to draw many generalizations regarding the excretion of foreign chemicals in fish because these compounds are so diverse and the fish is endowed with so many excretory modes. It does seem safe to conclude that the lipid-soluble xenobiotics are usually metabolized into a water-soluble form before excretion, and the products of that metabolism are then excreted largely by way of the bile and/or urine. The gills can be an important mode of exit for small compounds that require little if any biotransformation before excretion, and this is presumably a passive diffusion process. Finally, the skin is especially important for small fishes.

REFERENCES

Addison, R. F., Organochlorine compounds in aquatic organisms: their distribution, transport and physiological significance, in, *Effects of Pollutants on Aquatic Organisms,* Lockwood, A. P. M., Ed., Cambridge University Press, New York, 1976, 127.

Allen, J. L., Dawson, V. K. and Hunn, J. B., Excretion of the lampricide Bayer 73 by rainbow trout, in, *Aquatic Toxicology,* ASTM STP 667, Marking, L. L. and Kimerle, R. A., Eds., American Society for Testing and Materials, Philadelphia, 1979, 52.

Anderson, P. D. and Spear, P. A., Copper pharmacokinetics in fish gills. 1. Kinetics in pumpkinseed sunfish, *Lepomis gibbosus,* of different body sizes, *Water Res.,* 14, 1101, 1980.

Anderson, J. W., An assessment of knowledge concerning the fate and effects of petroleum hydrocarbons in the marine environment, in, *Marine Pollution: Functional Responses,* Vernberg, W. B., Calabrese, A., Thurberg, F. P. and Vernberg, F. J., Eds., Academic Press, New York, 1979, 3.

Anderson, J. W., Neff, J. M., Cox, B. A., Tatem, H. E. and Hightower, G. M., The effects of oil on estuarine animals: toxicity, uptake and depuration, respiration, in, *Pollution and Physiology of Marine Organisms,* Vernberg, F. J. and Vernberg, W. B., Eds., Academic Press, New York, 1974, 285.

Andersson, T. and Forlin, L., Regulation of the cytochrome P450 enzyme system in fish, *Aquat. Toxicol.,* 24, 1, 1992.

Baatrup, E., Nielsen, M. and Danscher, G., Histochemical demonstration of two mercury pools in trout tissues: mercury in kidney and liver after mercuric chloride exposure, *Ecotoxicol. Environ. Safety,* 12, 267, 1986.

Baker, J. T. P., Histological and electron microscopical observations on copper poisoning in the winter flounder (*Pseudopleuronectes americanus*), *J. Fish. Res. Bd. Can.,* 26, 2785, 1969.

Ballatori, N. and Boyer, J. L., Slow biliary elimination of methyl mercury in the marine elasmobranchs *Raja erinacea* and *Squalus acanthias, Toxicol. Appl. Pharmacol.,* 85, 407, 1986.

Benoit, D., Leonard, E., Christensen, G. and Fiandt, J., Toxic effects of cadmium on three generations of brook trout (*Salvelinus fontinalis*), *Trans. Am. Fish. Soc.,* 105, 550, 1976.

Bentley, P. J., Influx of zinc by channel catfish (*Ictalurus punctatus*): uptake from external environmental solutions, *Comp. Biochem. Physiol.,* 101C, 215, 1992.

Bidwell, J. R. and Heath, A. G., An *in situ* study of rock bass, *Ambloplites rupestris,* physiology: effect of season and mercury contamination, *Hydrobiologia,* 264, 137, 1993.

Binder, R. L. and Stegeman, J. J., Basal levels and induction of hepatic aryl hydrocarbon hydroxylase activity during the embryonic period of development in brook trout, *Biochem. Pharmacol.,* 32, 1983.

114

Binder, R. L. and Lech, J. J., Xenobiotics in gametes of Lake Michigan lake trout, *Salvelinus namaycush*, induce hepatic monooxygenase activity in their offspring, *Fundam. Appl. Toxicol.*, 4, 1042, 1984.

Black, M. C. and McCarthy, J. F., Dissolved organic macromolecules reduce the uptake of hydrophobic organic contaminants by the gills of the rainbow trout *(Salmo gairdneri)*, *Environ. Toxicol. Chem.*, 7, 593, 1988.

Black, M. C., Millsap, D. S. and McCarthy, J. F., Effects of acute temperature change on respiration and toxicant uptake by rainbow trout, *Salmo gairdneri* (Richardson), *Physiol. Zool.*, 64, 145, 1991.

Blanck, J., Londstrom-Seppa, P., Agren, J., Hanninen, O., Rein, H. and Ruckpaul, K., Temperature compensation of hepatic microsomal cytochrome P-450 activity in rainbow trout. I. Thermodynamic regulation during water cooling in autumn, *Comp. Biochem. Physiol.*, 93C, 55, 1989.

Block, M. and Part, P., Uptake of ^{109}Cd by cultured epithelial cells from rainbow trout *(Oncorhynchus mykiss)*, *Aquat. Toxicol.*, 23, 137, 1992.

Boddington, M. J., MacKenzie, B. A. and DeFreitas, A. S. W., A respirometer to measure the uptake efficiency of waterborne contaminants in fish, *Ecotoxicol. Environ. Safety*, 3, 383, 1979.

Bradley, R. W. and Sprague, J. B., Accumulation of zinc by rainbow trout as influenced by pH, water hardness and fish size, *Environ. Toxicol. Chem.*, 4, 685, 1985.

Brett, J. R., The metabolic demand for oxygen in fish, particularly salmonids, and a comparison with other vertebrates, *Respir. Physiol.* 14, 151, 1972.

Brown, D. A., Bay, S. M. and Hershelman, G. P., Exposure of scorpionfish *(Scorpaena guttata)* to cadmium: effects of acute and chronic exposures on the cytosolic distribution of cadmium, copper and zinc, *Aquat. Toxicol.*, 16, 295, 1990.

Brown, D. A. and Parsons, T. R., Relationship between cytoplasmic distribution of mercury and toxic effects to zooplankton and chum salmon *(Oncorhynchus keta)* exposed to mercury in a controlled ecosystem, *J. Fish. Res. Bd. Can.*, 35, 880, 1978.

Brungs, W. A., Leonard, E. N. and McKim, J. M., Acute and long-term accumulation of copper by the brown bullhead, *Ictalurus nebulosus, J. Fish. Res. Bd. Can.*, 30, 583, 1973.

Bryan, G. W., Some aspects of heavy metal tolerance in aquatic organisms, in, *Effects of Pollutants on Aquatic Organisms*, Lockwood, A. P. M., Ed., Cambridge University Press, New York, 1976, 7.

Bryan, G. W., Bioaccumulation of marine pollutants, *Philos. Trans. Res. Soc. London* B, 286, 483, 1979.

Buckley, J. T., Roch, M., McCarter, J. A., Rendell, C. A. and Matheson, A. T., Chronic exposure of coho salmon to sublethal concentrations of copper. I. Effect on growth, on accumulation, and distribution of copper, and on copper tolerance, *Comp. Biochem. Physiol.*, 72C, 15, 1982.

Buhler, D. R., Stokes, R. M. and Caldwell, R. S., Tissue accumulation and enzymatic effects of hexavalent chromium in rainbow trout *(Salmo gairdneri)*, *J. Fish. Res. Bd. Can.*, 34, 9, 1977.

Buhler, D. R. and Williams, D. E., The role of biotransformation in the toxicity of chemicals, *Aquat. Toxicol.*, 11, 19, 1988.

Campbell. P. G. C. and Stokes, P. M., Acidification and toxicity of metals to aquatic biota, *Can. J. Fish. Aquat. Sci.*, 42, 2034, 1985.

Chapman, G. A., Acclimation as a factor influencing metal criteria, in, *Aquatic Toxicology and Hazard Assessment: Eighth Symposium*, ASTM STP 891, 119, 1985.

Chatterjee, S. and Bhattacharya, S., Detoxification of industrial pollutants by the glutathione gluthione-S-transferase system in the liver of *Anabas testudineus* (Bloch), *Toxicol. Lett.*, 22, 187, 1984.

Chernyaev, Z. A. and Strekozov, V. P., Biomonitoring of ecotoxicological situation in preserves by detecting organochlorine pesticides in fish tissues, *Biol. Nauki (Moscow)* (11), 114, 1991. (BIOSIS abstract BA 94-114878).

Clark, K. E. and Mackay, D., Dietary uptake and biomagnification of four chlorinated hydrocarbons by guppies, *Environ. Toxicol. Chem.*, 10, 1205, 1991.

Collier, T. K., Thomas, L. C. and Malins, D. C., Influence of environmental temperature on disposition of dietary naphthalene in coho salmon: isolation and identification of individual metabolites, *Comp. Biochem. Physiol.*, 61C, 23, 1978.

Collier, T. K., Connor, S. D., Eberhart, B. L., Anulacion, B. F., Goksoyr, A. and Varanasi, U., Using cytochrome P450 to monitor the aquatic environment: initial results from regional and national surveys, *Mar. Environ. Res.*, 34, 195, 1992.

Comotto, R. M., Kimerle, R. A. and Swisher, R. D., Bioconcentration and metabolism of linear alkylbenzene sulfonate by daphnids and fathead minnows, in, *Aquatic Toxicology*, Marking, L. L. and Kimerle, R. A., Eds., ASTM STP 667, American Society Testing and Materials, Philadelphia, 1979.

Cook, C. H. and Moore, J. C., Determination of malathion, malaxon, and mono- and dicarboxylic acids of malathion in fish, oyster, and shrimp tissue, *Agric. Food Chem.*, 24, 631, 1976.

Cravedi, J. P. and Tulliez, J., Accumulation, distribution and depuration in trout of naphthenic and isoprenoid hydrocarbons (dodecycyclohexane and pristane), *Bull. Environ. Contam. Toxicol.*, 28, 154, 1982.

Cravedi, J. P. and Tulliez, J., Fate of hydrocarbon pollution indicator in fish: absorption, deposition and depuration of squalane in *Salmo gairdneri*, *Environ. Pollut.*, 42A, 247, 1986.

Davies, I. M. and McKie, J. C., Accumulation of total tin disposition and toxicity in fish, *Aquat. Toxicol.*, 11, 3, 1988.

Davis, J. C. and Watters, K., Evaluation of opercular catheterization as a method for sampling water expired by fish, *J. Fish. Res. Bd. Can.*, 27, 1627, 1970.

Daxboeck, C., Davie, P. S., Perry, S. F. and Randall, D. J., Oxygen uptake in a spontaneously ventilating, blood-perfused trout preparation, *J. Exp. Biol.*, 101, 35, 1982.

De Bruijn, J. and Hermens, J., Uptake and elimination kinetics of organophosphorous pesticides in the guppy (*Poecilia reticulata*): correlations with the octanol/water partition coefficient, *Environ. Toxicol. Chem.*, 10, 791, 1991.

De Bruin, A., *Biochemical Toxicology of Environmental Agents*, Elsevier Biomedical Press, Amsterdam, 1976.

Delfino, J. J., Toxic substances in the Great Lakes, *Environ. Sci. Tech.*, 513, 1462, 1979.

Denison, M. S. and Yarbrough, J. D., Binding of insecticides to serum proteins in mosquitofish (*Gambusia affinis*), *Comp. Biochem. Physiol.*, 81C, 105, 1985.

Devaux, A., Pesonen, M., Monod, G. and Andersson, T., Glucocorticoid-mediated potentiation of P450 induction in primary culture of rainbow trout hepatocytes, *Biochem. Pharmacol.*, 43, 898, 1992.

Dixon, D. G. and Sprague, J. B., Copper bioaccumulation and hepatoprotein synthesis during acclimation to copper by juvenile rainbow trout, *Aquat. Toxicol.*, 1, 69, 1981.

Drummond, R. A., Olson, G. F. and Batterman, A. R., Cough response and uptake of mercury by brook trout, *Salvelinus fontinalis*, exposed to mercuric compounds at different hydrogen-ion concentrations, *Trans. Am. Fish. Soc.*, 103, 244, 1974.

Eaton, D. I., Effect of various trace metals on the binding of cadmium to rat hepatic metallothionein determined by the Cd/hemoglobin affinity assay, *Toxicol. Appl. Pharmacol.*, 78, 158.

Eisele, T. A., Coulombe, R. A., Pawlowski, N. E. and Nixon, J. E., The effects of route of exposure and combined exposure of mixed function oxidase inducers and suppressors on hepatic parameters in rainbow trout (*Salmo gairdneri*), *Aquat. Toxicol.*, 5, 211, 1984.

Eisler, R. and Gardner, G. R., Acute toxicology to an estuarine teleost of mixtures of cadmium, copper and zinc salts, *J. Fish Biol.*, 5, 131, 1973.

Elinder, C. G. and Pannone, M., Biliary excretion of cadmium, *Environ. Health Perspect.*, 28, 123, 1979.

Engel, D. W., Sunda, W. G. and Fowler, B. A., Factors affecting trace metal uptake and toxicity to estuarine organisms. I. Environmental parameters, in, *Biological Monitoring of Marine Pollutants*, Vernberg, J., Calabrese, A., Thurberg, F. and Vernberg, W. B., Eds., Academic Press, New York, 1981, 127.

Ernst, W., Determination of the concentration potential of marine organisms — a steady-state approach, *Chemosphere*, 11, 731, 1977.

Everall, N. C., Macfarlane, N. A. A. and Sedgwick, R. W., The effects of water hardness upon the uptake, accumulation and excretion of zinc in the brown trout, *Salmo trutta* L., *J. Fish. Biol.*, 35, 881, 1989.

Fange, R. and Grove, D., Digestion, in, *Fish Physiology*, Vol. 8, Hoar, W. S., Randall, D. J. and Brett, J. R., Eds., Academic Press, New York, 1979, chap. 4.

Felts, P. A. and Heath, A. G., Interactions of temperature and sublethal environmental copper exposure on the energy metabolism of bluegill, *Lepomis macrochirus*, *J. Fish Biol.*, 25, 445, 1984.

Fent, K. and Stegeman, J. J., Effects of tributyltin in vivo on hepatic cytochrome P450 forms in marine fish, *Aquat. Toxicol.*, 24, 219, 1993.

Fletcher, C. R., Osmotic and ionic regulation in the cod (*Gadus callarias* L.), *J. Comp. Physiol.*, 124, 157, 1978.

Fletcher, P. E. and Fletcher, G. L., Zinc and copper-binding proteins in the plasma of winter flounder (*Pseudopleuronectes americanus*), *Can. J. Zool.*, 58, 609, 1980.

Forlin, L., Haux, C., Karlsson-Norrgren, L., Runn, P. and Larsson, A., Biotransformation enzyme activities and histopathology in rainbow trout, *Salmo gairdneri*, treated with cadmium, *Aquat. Toxicol.*, 8, 51, 1986.

Forlin, L. and Andersson, T., Effects of clophen A50 on the metabolism of paranitroanisole in an *in vitro* perfused rainbow trout liver, *Comp. Biochem. Physiol.*, 68C, 239, 1981.

Freeman, B. J., Accumulation of Cadmium, Chromium, and Lead by Bluegill Sunfish *(Lepomis macrochirus)* Under Temperature and Oxygen Stress, Ph.D. dissertation, University of Georgia, Athens, 1980.

Fukami, J. L., Shishido, T., Fukunaga, K. and Casida, J. E., Oxidative metabolism of rotenone in mammals, fish and insects and its relation to selective toxicity, *J. Agric. Food Chem.*, 17, 1217, 1969.

Gallagher, E. P. and DiGiulio, R. T., A comparison of glutathione-dependent enzymes in liver, gills and posterior kidney of channel catfish *(Ictalurus punctatus)*, *Comp. Biochem. Physiol.* 102C, 543, 1992.

Gallagher, E. P., Canada, A. T. and DiGiulio, R. T., The protective role of glutathione in chlorothalonil-induced toxicity to channel catfish, *Aquat. Toxicol.*, 23, 1992.

George, S. G., Enzymology and molecular biology of phase II xenobiotic-conjugating enzymes in fish, in, *Aquatic Toxicology: Molecular, Biochemical, and Cellular Perspectives*, Malins, D. C. and Ostrander, G. K., Eds., Lewis Publishers, Boca Raton, FL, 1994, chap. 3.

Giesy, J. P., Bartell, S. M., Landrum, P. F., Leversee, G. J. and Bowling, J. W., Fates and biological effects of polycyclic aromatic hydrocarbons in aquatic systems, *U.S.E.P.A. Research and Development Project Summary*, EPA-600-S3-83-053, U.S. Environmental Protection Agency, Washington, D.C., 1983.

Glickman, A. H., Statham, C. N. and Lech, J. J., Studies on the uptake, metabolism and disposition of pentachlorophenol and pentachloroanisole in rainbow trout, *Toxicol. Appl. Pharmacol.*, 41, 649, 1977.

Glickman, A. H. and Melancon, M. J., The accumulation, distribution, metabolism and elimination of phenol and several substituted phenols by rainbow trout, paper presented at the 3rd Annual SETAC meeting, Arlington, VA, November 1982.

Gobas, F. A., McCorquodale, J. R. and Haffner, G. D., Intestinal absorption and biomagnification of organochlorines, *Environ. Toxicol. Chem.*, 12, 567, 1993.

Goksoyr, A. and Forlin, L., The cytochrome P450 system in fish, aquatic toxicology and environmental monitoring, *Aquat. Toxicol.*, 22, 287, 1992.

Goksoyr, A., Solberg, T. and Serigstad, B., Immunochemical detection of cytochrome P450IA1 induction in cod larvae and juveniles exposed to a water soluble fraction of North sea crude oil, *Mar. Pollut. Bull.* 22, 122, 1991.

Gossett, R. W., Brown, D. A. and Young, D. R., Predicting the bioaccumulation of organic compounds in marine organisms using octanol/water partition coefficients, *Mar. Pollut. Bull.*, 14, 387, 1983.

Hamer, D. H., Metallothioneins, *Annu. Rev. Biochem.*, 55, 913, 1986.

Handy, R. D., The accumulation of dietary aluminum by rainbow trout, *Oncorhynchus mykiss*, at high exposure concentrations, *J. Fish Biol.*, 42, 603, 1993.

Handy, R. D. and Eddy, F. B., The interactions between the surface of rainbow trout, *Oncorhynchus mykiss*, and waterborne metal toxicants, *Funct. Ecol.*, 4, 385, 1990.

Hansch, C. and Clayton, J. M., Lipophilic character and biological activity of drugs. II. The parabolic case, *J. Pharmacol. Sci.*, 62, 1, 1973.

Hansen, L. G., Kapoor, L. P. and Metcalf, R., Biochemistry of selective toxicity and biodegradability: comparative *O*-dealkylation by aquatic organisms, *Comp. Gen. Pharmacol.*, 3, 339, 1972.

Hardy, R. W., Sullivan, C. V. and Koziol, A. M., Absorption, body distribution, and excretion of dietary zinc by rainbow trout, *Fish Physiol. Biochem.*, 3, 133, 1987.

Harrison, S. E. and Klaverkamp, J. F., Uptake, elimination and tissue distribution of dietary and aqueous cadmium by rainbow trout *(Salmo gairdneri* Richardson) and lake white fish *(Coregonus clupeaformis* Mitchell), *Environ. Toxicol. Chem.*, 8, 87, 1989.

Hawkins, W. E., Tate, L. G. and Sharpie, T. G., Acute effects of cadmium on the spot *Leiostomus xanthurus* (Teleostei): tissue distribution and renal ultrastructure, *J. Toxicol. Environ. Health*, 6, 283, 1980.

Hayton, W. L. and Barron, M. G., Rate-limiting barriers to xenobiotic uptake by the gill, *Environ. Toxicol. Chem.*, 9, 151, 1990.

Heisinger, J. F. and Wait, E., The effects of mercuric chloride and sodium selenite on glutathione and total nonprotein sulfhydryls in the kidney of the black bullhead *Ictalurus melas, Comp. Biochem. Physiol.,* 94C, 139, 1989.

Hektoen, H., Ingebrigtsen, K., Magne, E. and Oehme, M., Interspecies differences in tissue distribution of 2,3,7,8-tetrachlorodibenzo-*p*-dioxin, between cod, *Gadus morhua,* and rainbow trout, *Oncorhynchus mykiss, Chemosphere,* 24, 581, 1992.

Hilton, J. W., Hodson, P. V., Braun, H. E., Leatherland, J. L. and Slinger, S. J., Contaminant accumulation and physiological response in rainbow trout *(Salmo gairdneri)* reared on naturally contaminated diets, *Can. J. Fish. Aquat. Sci.,* 40, 1987, 1983.

Hobson, J. F. and Birge, W. J., Acclimation-induced changes in toxicity and induction of metallothionein-like proteins in the fathead minnow following sublethal exposure to zinc, *Environ. Toxicol. Chem.,* 8, 157, 1989.

Hodson, P. V., The effect of metal metabolism on uptake, and tributyltin in muscle tissue of farmed Atlantic salmon, *Mar. Pollut. Bull.,* 18, 405, 1987.

Hodson, P. V., The effect of metal metabolism on uptake, disposition and toxicity in fish, *Aquat. Toxicol.,* 11, 3, 1988.

Hodson, P. V., Blunt, B. R. and Spry, D. J., Chronic toxicity of water-borne and dietary lead to rainbow trout *(Salmo gairdneri)* in Lake Ontario water, *Water Res.,* 12, 869, 1978.

Hodson, P. V., Spry, D. J. and Blunt, B. R., Effects on rainbow trout *(Salmo gairdneri)* of a chronic exposure to waterborne selenium, *Can. J. Fish. Aquat. Sci.,* 37, 233, 1980.

Holcombe, G. W., Benoit, D. A., Leonard, E. N. and McKim, J. M., Long-term effects of lead exposure on three generations of brook trout *(Salvelinus fontinalis), J. Fish. Res. Bd. Can.,* 33, 1731, 1976.

Holcombe, G. W., Benoit, D. A. and Leonard, E. N., Long-term effects of zinc exposures on brook trout *(Salvelinus fontinalis), Trans. Am. Fish. Soc.,* 108, 76, 1979.

Holden, A. V., Organochlorine insecticide residues in salmonid fish, *J. Appl. Ecol.,* 3(Suppl.), 45, 1966.

Hughes, G. M. and Flos, R., Zinc content of the gills of rainbow trout *(Salmo gairdneri)* after treatment with zinc solutions under normoxic and hypoxic conditions, *J. Fish Biol.,* 13, 717, 1978.

Hughes, G. M., Scanning electron microscopy of the respiratory surfaces of trout gills, *J. Zool. London,* 188, 443, 1979.

Hunn, J. B., Role of calcium in gill function in freshwater fishes, *Comp. Biochem. Physiol.,* 82A, 543, 1985.

Hunn, J. B. and Allen, J. L., Movement of drugs across the gills of fishes, *Annu. Rev. Pharmacol.,* 14, 47, 1974.

Hunn, J. B., Schoettger, R. A. and Willford, W. A., Turnover and urinary excretion of free and acetylated MS 222 by rainbow trout, *Salmo gairdneri, J. Fish. Res. Bd. Can.,* 25, 25, 1968.

Jimenez, B. D., Cirmo, C. P. and McCarthy, J. F., Effects of feeding and temperature on uptake, elimination and metabolism of benzo[a]pyrene in the bluegill sunfish *(Lepomis macrochirus), Aquat. Toxicol.,* 10, 41, 1987.

Juedes, M. J. and Thomas, P., Effect of lead on the acid-soluble thiol content of croaker, *Micropogonias undulatus, Am. Zool.,* 24, 137A, 1984.

Julshamm, K., Ringdal, O. and Braekkan, O. R., Mercury concentration in liver and muscle of cod *(Gadus morhua)* as an evidence of migration between waters with different levels of mercury, *Bull. Environ. Contam. Toxicol.,* 29, 544, 1982.

Kanazawa, J., Relationship between molecular weights of pesticides and their bioconcentration factors by fish, *Experientia,* 38, 1045, 1982.

Karara, A. H. and McFarland, V. A., A pharmacokinetic analysis of the uptake of polychlorinated biphenyls, PCBs, by golden shiners, *Environ. Toxicol. Chem.,* 11, 315, 1992.

Kennedy, C. J. and Law, F. C. P., Toxicokinetics of selected polycyclic aromatic hydrocarbons in rainbow trout following different routes of exposure, *Environ. Toxicol. Chem.,* 9, 133, 1990.

Kennish, J., Bolinger, R., Chambers, K. and Russel, M., Xenobiotic metabolizing enzyme activity in sockeye salmon *(Oncorhynchus nerka)* during spawning migration, *Mar. Environ. Res.,* 34, 293, 1992.

Klassen, C. D., Biliary excretion of metals, *Drug Metab. Rev.,* 5, 165, 1976.

Klaverkamp, J. F. and Duncan, D. A., Acclimation to cadmium toxicity by white suckers: cadmium binding capacity and metal distribution in gill and liver cytosol, *Environ. Toxicol. Chem.,* 6, 275, 1987.

Kobayashi, K., Metabolism of pentachlorophenol in fishes, in, *Pentachlorophenol*, Rao, K. R., Ed., Plenum Press, New York, 1978, 89.

Kobayashi, K. and Nakamura, N., Isolation and identification of a conjugated PCP excreted in the urine of goldfish, *Bull. Jpn. Soc. Sci. Fish.,* 45, 1001, 1979.

Koivusaari, U., Thermal acclimation of hepatic polysubstrate monooxygenase and UDP-glucuronsyltransferase of mature rainbow trout *(Salmo gairdneri), J. Exp. Zool.,* 227, 35, 1983.

Krzeminski, S. F., Gilbert, J. T. and Ritts, J. A., A pharmacokinetic model for predicting pesticide residues in fish, *Arch. Environ. Contam. Toxicol.,* 5, 157, 1977.

Kumar, A. and Mathur, R., Bioaccumulation kinetics and organ distribution of lead in a freshwater teleost, *Colisa fasciatus, Environ. Technol.,* 12, 731, 1991.

Kurelec, B., Kezic, N., Singh, H. and Zahn, R. K., Mixed-function oxidases in fish: their role in adaptation to pollution, *Mar. Environ. Res.,* 14, 409, 1984.

Lauren, D. J. and McDonald, D., Acclimation to copper by rainbow trout, *Salmo gairdneri*: biochemistry, *Can. J. Fish. Aquat. Sci.,* 44, 105, 1987.

Le Bon, A. M., Cravedi, J. D. and Tulliez, J., Fate of the isoprenoid hydrocarbon, pristane, in rainbow trout, *Ecotoxicol. Environ. Safety,* 13, 274, 1987.

Leaver, M. J., Scott, K. and George, S., Expression and tissue distribution of plaice glutathione-*S*-transferase A, *Marine Environ. Res.,* 34, 237, 1992.

Lech, J., Glickman, J. and Statham, C. N., Studies on the uptake, disposition and metabolism of pentachlorophenol and pentachloroanisole in rainbow trout *(Salmo gairdneri),* in, *Pentachlorophenol,* Rao, K. R., Ed., Plenum Press, 1978, 107.

Lech, J. J. and Bend, J. R., Relationship between biotransformation and the toxicity and fate of xenobiotic chemicals in fish, *Environ. Health Perspect.,* 34, 115, 1980.

Lech, J. J., Vodicnik, M. J. and Elcombe, C. R., Induction of monooxygenase activity in fish, in, *Aquatic Toxicology,* Weber, L. J., Ed., 1982, 107.

Lee, R. F., Sauerheber, R. and Dobbs, G. H., Uptake, metabolism and discharge of polycyclic aromatic hydrocarbons by marine fish, *Mar. Biol.,* 17, 201, 1972.

Liem, G. J. and McKim, J. M., Predicting branchial and cutaneous uptake of 2,2",5,5"-tetrachlorobiphenyl in fathead minnows *(Pimephales promelas)* and Japanese medaka *(Oryzias latipes)*: rate limiting factors, *Aquat. Toxicol.,* 27, 15, 1993.

Lindstadt, G. and Bergman, H. L., Periodic depuration of anthracene metabolites by rainbow trout, *Trans. Am. Fish. Soc.,* 113, 513, 1984.

Lindstrom-Seppa, P., Koivusaari, U. and Hanninen, O., Extrahepatic xenobiotic metabolism in north-European freshwater fish, *Comp. Biochem. Physiol.,* 69C, 259, 1981.

Lindstrom-Seppa, P. and Oikari, A., Biotransformation activities of feral fish in waters receiving bleached pulp mill effluents, *Environ. Toxicol. Chem.,* 9, 1415, 1991.

Luckey, T. D. and Venugpal, B., Metal toxicity in mammals, in, *Physiologic and Chemical Basis for Metal Toxicity,* Plenum Press, New York, 1976, 103.

Ludke, J. L., Gibson, J. R. and Lusk, C., Mixed function oxidase activity in freshwater fishes: aldrin epoxidation and parathion activation, *Toxicol. Appl. Pharmacol.,* 21, 89, 1972.

Luoma, S. N., Bioavailability of trace metals to aquatic organisms — a review, *Sci. Total Environ.,* 28, 1, 1983.

MacLeod, J. C. and Pessah, E., Temperature effects on mercury accumulation, toxicity and metabolic rate in rainbow trout *(Salmo gairdneri), J. Fish. Res. Bd. Can.,* 30, 485, 1973.

Marafante, E., Binding of mercury and zinc to cadmium-binding protein in liver and kidney of goldfish *(Carassius auratus), Experientia,* 32, 149, 1976.

Maren, T. H., Broder, L. E. and Stenger, V. G., Metabolism of ethyl m-aminobenzoate (MS 222) in the dogfish, *Squalus acanthias, Bull. Mt. Desert Isl. Biol. Lab.,* 8, 39, 1968.

Massaro, E. J., Pharmacokinetics of toxic elements in rainbow trout, *U.S.E.P.A. Ecological Research Series,* No. EPA-660/3-74-027, 30p, U.S. Environmental Protection Agency, Washington, D.C., 1974.

Matsumoto, G. and Hanya, T., Comparative study on organic constituents of polluted and unpolluted inland aquatic environments. I. Features of hydrocarbons for polluted and unpolluted waters, *Water Res.,* 15, 217, 1981.

Matthiessen, P. and Brafield, A., Uptake and loss of dissolved zinc by the stickleback, *Gasterosteus aculeatus, J. Fish Biol.,* 10, 399, 1977.

McCarter, J. A. and Roch, M., Hepatic metallothionein and resistance to copper in juvenile coho salmon, *Comp. Biochem. Physiol.,* 74C, 133, 1983.

McCarter, J. A. and Roch, M., Chronic exposure of coho salmon to sublethal concentrations of copper. III. Kinetics of metabolism of metallothionein, *Comp. Biochem. Physiol.*, 77C, 83, 1984.

McKim, J., Olson, G., Holcombe, G. and Hunt, E., Long-term effects of methylmercuric chloride on three generations of brook trout *(Salvelinus fontinalis)*: toxicity, accumulation, distribution and elimination, *J. Fish. Res. Bd. Can.*, 33, 2726, 1976.

McKim, J. M. and Goeden, H., A direct measure of the uptake efficiency of a xenobiotic chemical across the gills of brook trout *(Salvelinus fontinalis)* under normoxic and hypoxic conditions, *Comp. Biochem. Physiol.*, 72C, 65, 1982.

McKim, J. M. and Heath, E. M., Dose determinations for waterborne 2,5,2',5-tetrachlorobiphenyl and related pharmacokinetics in two species of trout *(Salmo gairdneri* and *Salvelinus fontinalis)*: a mass-balance approach, *Toxicol. Appl. Pharmacol.*, 68, 177, 1983.

McKim, J., Schmieder, P. and Veith, G., Absorption dynamics of organic chemical transport across trout gills as related to octanol-water partition coefficient, *Toxicol. Appl. Pharmacol.*, 77, 1, 1985.

McKim, J. M., Schmieder, P. K. and Erickson, R. J., Toxicokinetic modeling of pentachlorophenol in the rainbow trout *(Salmo gairdneri)*, *Aquat. Toxicol.*, 9, 59, 1986.

McKim, J. M. and Erickson, R. J., Environmental impacts on the physiological mechanisms controlling xenobiotic transfer across fish gills, *Physiol. Zool.*, 64, 39, 1991.

McKim, J. M. and Nichols, J. W., Physiologically based toxicokinetic modeling of the dermal uptake of three waterborne chlorethanes in rainbow trout *(Oncorhynchus mykiss)*, *Abstr. Toxicologist*, 11, 35, 1991.

McWilliams, P. C. and Potts, W. T. W., The effects of pH and calcium concentrations on gill potentials in the brown trout, *Salmo trutta, J. Comp. Physiol.*, 126, 277, 1978.

Meister, A., Selective modification of glutathione metabolism, *Science*, 220, 470, 1983.

Melancon, M. J. and Lech, J. J., Distribution and elimination of naphthalene and 2-methylnaphthalene in rainbow trout during short and long term exposures, *Arch. Environ. Contam. Toxicol.*, 7, 207, 1978.

Melancon, M. J. and Lech, J. J., Distribution and biliary excretion products of Di-2-ethylhexyl phthalate in rainbow trout, *Drug Metab. Disp.*, 2, 112, 1976.

Miller, T. G. and Mackay, W. C., Relationship of secreted mucus to copper and acid toxicity in rainbow trout, *Bull. Environ. Contam. Toxicol.*, 28, 68, 1982.

Miller, M. R., Hinton, D. E. and Stegeman, J. J., Cytochrome P-450E induction and localization in gill pillar (endothelial) cells of scup and rainbow trout, *Aquat. Toxicol.*, 14, 307, 1989.

Miyamoto, J., Takimoto, Y. and Mihara, K., Metabolism of organophosphorus insecticides in aquatic organisms with special emphasis on fenitrothion, in, *Pesticide and Xenobiotic Metabolism in Aquatic Organisms*, Kahn, M. A. Q., Lech, J. J. and Menn, J. J., Eds., (ACS Symposium Series 99), American Chemical Society, Washington, D.C., 1979, 3.

Mount, D., An autopsy technique for zinc-caused fish mortality, *Trans. Am. Fish. Soc.*, 93, 174, 1964.

Nagel, R., Species differences, influence of dose and application on biotransformation of phenol in fish, *Xenobiotica*, 13, 101, 1983.

Nagel, R. and Urich, K., Kinetic studies on the elimination of different substituted phenols by goldfish *(Carassius auratus)*, *Bull. Environ. Contam. Toxicol.*, 24, 374, 1980.

Neff, J. M., *Polycyclic Aromatic Hydrocarbons in the Aquatic Environment*, Allied Science Publishers, London, 1979.

Newman, M. C. and Mitz, S., Size dependence of zinc elimination and uptake from water by mosquitofish *(Gambusia affinis)*, *Aquat. Toxicol.*, 12, 17, 1988.

Nichols, J. W., McKim, K. M., Anderson, M. E., Gardas, M. L., Clewell, H. J. and Erickson, R. J., A physiologically based toxicokinetic model for the uptake and disposition of waterborne organic chemicals in fish, *Toxicol. Appl. Pharmacol.*, 106, 433, 1990.

Nichols, J. W., McKim, J. M., Lien, G. J., Hoffman, A. D. and Bertelsen, S. L., Physiologically based toxicokinetic modeling of three waterborne chloroethanes in rainbow trout *(Oncorhynchus mykiss)*, *Toxicol. Appl. Pharmacol.*, 110, 374, 1991.

Niimi, A. and Oliver, V. G., Influence of molecular weight and molecular volume on dietary absorption efficiency of chemicals by fishes, *Can. J. Fish. Aquat. Sci.*, 45, 222, 1988.

Niimi, A. J. and McFadden, C. A., Uptake of sodium pentachlorophenate (NaPCP) from water by rainbow trout *(Salmo gairdneri)*, exposed to concentrations in the ng/l range, *Bull. Environ. Contam. Toxicol.*, 28, 11, 1982.

Niimi, A. J. and Cho, C. Y., Laboratory and field analysis of pentachlorophenol (PCP) accumulation by salmonids, *Water Res.*, 17, 1791, 1983.

Norey, C. G., Brown, M. W., Cryer, A. and Kay, J., A comparison of the accumulation, tissue distribution and secretion of cadmium in different species of freshwater fish, *Comp. Biochem. Physiol.*, 96C, 181, 1990.

Oladimeji, A. A., Quadri, S. U. and deFreitas, A. S. W., Measuring the elimination of arsenic by the gills of rainbow trout (*Salmo gairdneri*) by using a two compartment respirometer, *Bull. Environ. Contam. Toxicol.*, 32, 661, 1984.

Olafson, R. W. and Thompson, J. A. J., Isolation of heavy metal binding proteins from marine vertebrates, *Mar. Biol.*, 28, 83, 1974.

Olson, K. R., Squibb, K. S. and Cousins, R. J., Tissue uptake, subcellular distribution, and metabolism of CHJHgCI and 14CH3 203 HgCI by rainbow trout, *Salmo gairdneri, J. Fish. Res. Bd. Can.*, 35, 381, 1978.

Olson, L. E., Allen, J. L. and Hogan, J. W., Biotransformation and elimination of herbicide dinitramine in carp, *J. Agric. Food Chem.*, 25, 554, 1977.

Opperhuizen, A., Bioconcentration and biomagnification: is a distinction necessary?, in, *Bioaccumulation in Aquatic Systems,* Nagel, R. and Loskill, R., Eds., VCH Publishers, New York, 1991, chap. 6.

Oronsaye, J. A. O. and Brafield, A. E., The effect of dissolved cadmium on the chloride cells of the gills of the stickleback, *Gasterosteus aculeatus* L., *J. Fish Biol.*, 25, 253, 1984.

Overnell, J. and Coombs, T. L., Purification and properties of plaice metallothionein, a cadmium binding protein from the liver of the plaice (*Pleuronectes platessa), Biochem. J.*, 183, 277, 1979.

Part, P. and Lock, R., Diffusion of calcium, cadmium and mercury in a mucus solution from rainbow trout, *Comp. Biochem. Physiol.*, 76C, 254, 1983.

Part, P., Svanberg, O. and Bergstrom, E., The influence of surfactants on gill physiology and cadmium uptake in perfused rainbow trout gills, *Ecotoxicol. Environ. Safety,* 9, 135, 1985.

Patrick, F. M. and Loutit, M. W., Passage of metals to freshwater fish from their foods, *Water Res.*, 12, 395, 1978.

Payne, J. F., Fancey, L. L., Rahimtula, A. D. and Porter, E., Review and perspective on the use of mixed-function oxygenase enzymes in biological monitoring, *Comp. Pharmacol. Physiol.*, 86C, 233, 1987.

Penreath, R. J., The accumulation and retention of Zn and Mn by the plaice, *Pleuronectes platessa* L., *J. Exp. Mar. Biol. Ecol.*, 12, 1, 1973.

Petering, D. H., Goodrich, W. H., Krezoski, S., Weber, D., Shaw, C., Spieler, R. and Zettergren, L., Metal-binding proteins and peptides for the detection of heavy metals in aquatic organisms, in, *Biomarkers of Environmental Contamination,* McCarthy, J. and Shugart, L., Eds., Lewis Publishers, Boca Raton, FL, 1990, 239.

Peterson, R. E. and Guiney, P. D., Disposition of polychlorinated biphenyls in fish, in, *Pesticides and Xenobiotic Metabolism in Aquatic Organisms,* Kahn, M. A. Q., Lech, J. J. and Menn, J. J., Eds., 1979.

Phillips, G. R. and Buhler, D. R., The relative contributions of methylmercury from food or water to rainbow trout (*Salmo gairdneri*) in a controlled laboratory environment, *Trans. Am. Fish. Soc.*, 1978.

Pierson, K. B., Effects of chronic zinc exposure on the growth, sexual maturity, reproduction, and bioaccumulation of the guppy, *Poecilia reticulata, Can. J. Fish. Aquat. Sci.*, 38, 23, 1981.

Playle, R. C. and Wood, C. M., Mechanisms of aluminum extraction and accumulation at the gills of rainbow trout, *Oncorhynchus mykiss* (Walbaum), in acidic soft water, *J. Fish Biol.*, 38, 791, 1991.

Playle, R. C., Gensemer, R. W. and Dixon. D. G., Copper accumulation on gills of fathead minnows: influence of water hardness, complexation and pH of the gill micro-environment, *Environ. Toxicol. Chem.*, 11, 381, 1992.

Ponce, R. and Bloom, N., Effect of pH on the bioaccumulation of low level, dissolved methylmercury by rainbow trout (*Oncorhynchus mykiss*), *Water, Air, Soil Pollut.*, 56, 631, 1991.

Pritchard, J. B. and Bend, J. R., Mechanisms controlling the renal excretion of xenobiotics in fish: effects of chemical structure, *Drug Metab. Rev.*, 15, 655, 1984.

Pritchard, J. B. and Bend, J. R., Relative roles of metabolism and renal excretory mechanisms in xenobiotic elimination by fish, *Environ. Health Perspect.*, 90, 85, 1991.

Ramamoorthy, S. and Blumhagen, K., Uptake of Zn, Cd, and Hg by fish in the presence of competing compartments, *Can. J. Fish Aquat. Sci.*, 41, 750, 1984.

Randall, D., Lin, H. and Wright, P., Gill water flow and the chemistry of the boundary layer, *Physiol. Zool.*, 64, 26, 1991.

Reichert, W. L., Federighi, D. A. and Malins, D. C., Uptake and metabolism of lead and cadmium in coho salmon (*Oncorhynchus kisutch*), *Comp. Biochem. Physiol.*, 63C, 229, 1979.

Reid, S. and McDonald, D., Metal binding activity of the gills of rainbow trout (*Oncorhynchus mykiss*), *Can. J. Fish. Aquat. Sci.*, 48, 1061, 1991.

Rice, S. D., Short, J. W. and Karinen, J. F., Comparative oil toxicity and comparative animal sensitivity, in, *Fate and Effects of Petroleum Hydrocarbons in Marine Ecosystems and Organisms*, Wolfe, D. A., Ed., Pergamon Press, New York, 1977, 78.

Roch, M. and McCarter, J. A., Hepatic metallothionein production and resistance to heavy metals by rainbow trout (*Salmo gairdneri*). I. Exposed to an artificial mixture of zinc, copper and cadmium, *Comp. Biochem. Physiol.*, 77C, 71, 1984a.

Roch, M. and McCarter, J. A., Hepatic metallothionein production and resistance to heavy metals by rainbow trout (*Salmo gairdneri*). II. Held in a series of contaminated lakes, *Comp. Biochem. Physiol.*, 77C, 77, 1984b.

Roesijadi, G., Metallothioneins in metal regulation and toxicity in aquatic animals, *Aquat. Toxicol.*, 22, 81, 1992.

Roesijadi, G. and Robinson, W. E., Metal regulation in aquatic animals: mechanisms of uptake, accumulation and release, in, *Aquatic Toxicology: Molecular, Biochemical, and Cellular Perspectives*, Malins, D. C. and Ostrander, G. K., Eds., Lewis Publishers, Boca Raton, FL, 1994, chap. 9.

Rogers, C. A. and Stalling, D. L., Dynamics of an ester of 2,4-D in organs of three fish species, *Weed Sci.*, 20, 101, 1972.

Rogers, D. W. and Beamish, F. W. H., Dynamics of dietary methylmercury in rainbow trout, *Salmo gairdneri*, *Aquat. Toxicol.*, 2, 271, 1982.

Rogers, D. W. and Beamish, F. W. H., Water quality modifies uptake of waterborne methylmercury by rainbow trout, *Salmo gairdneri*, *Can. J. Fish Aquat. Sci.*, 40, 824, 1983.

Rogers, D., Watson, T., Langan, J. and Wheaton, T., Effects of pH and feeding regime on methylmercury accumulation within aquatic microcosms, *Environ. Pollut.*, 45, 261, 1987.

Rudd, J. W. M., Furutani, A. and Turner, M. A., Mercury methylation by fish intestinal contents, *Appl. Environ. Microbiol.* 40, 777, 1980.

Satchell, G. H., *Physiology and Form of Fish Circulation*, Cambridge University Press, Cambridge, 1991

Schlenk, D., Erickson, D., Lech, J. and Buhler, D., The distribution, elimination, and *in vivo* biotransformation of aldicarb in the rainbow trout, (*Oncorhynchus mykiss*), *Fund. Appl. Toxicol.*, 18, 131, 1992.

Schmidt, E. J. and Kimerle, R. A., New design and use of a fish metabolism chamber, in, *Aquatic Toxicology and Hazard Assessment: Fourth Conference*, ASTM STP 737, Branson, D. R. and Dickson, K. L., Eds., American Society for Testing and Materials, Philadelphia, 1981, 436.

Schmidt-Nielsen, K., *Scaling, Why is Animal Size So Important*, Cambridge University Press, Cambridge, 1984.

Schmieder, P. and Weber, L., Blood and water flow limitations on gill uptake of organic chemicals in the rainbow trout (*Oncorhynchus mykiss*), *Aquat. Toxicol.*, 24, 193, 1992.

Segner, H., Response of fed and starved roach, *Rutilus rutilus*, to sublethal copper contamination, *J. Fish Biol.*, 30, 423, 1987.

Shaklee, J. B., Christiansen, J. A., Sidel, B. D., Prosser, C. L. and Whitt, G. S., Molecular aspects of temperature acclimation in fish: contributions of changes in enzyme activities and isozyme patterns to metabolic reorganization in the green sunfish, *J. Exp. Zool.*, 201, 1, 1977.

Sharpe, M. A., deFreitas, A. S. W. and McKinnon, A. E., The effect of body size on methylmercury clearance by goldfish (*Carassius auratus*), *Environ. Biol. Fish.*, 2, 177, 1977.

Sijm, D., Part, P. and Opperhuizen, A., The influence of temperature on the uptake rate constants of hydrophobic compounds determined by the isolated perfused gills of rainbow trout (*Oncorhynchus mykiss*), *Aquat. Toxicol.*, 25, 1, 1993.

Sire, M. F., Lutton, C. and Vernier, J. M., New views on intestinal absorption of lipids in teleostean fishes: an ultrastructural and biochemical study in the rainbow trout, *J. Lipid Res.*, 22, 81, 1981.

Smith, L. S., *Introduction to Fish Physiology*, T. F. H. Publications, Neptune, NJ, 1982.

Smith, C. V., Hughes, H., Lauterburg, B. H. and Mitchell, J. R., Chemical nature of reactive metabolites determines their biological interactions with glutathione, in, *Functions of Glutathione: Biochemical, Physiological, Toxicological and Clinical Aspects*, Larsson, A., Orrenius, A., Holmgren, H. and Mannervik, B., Eds., Raven Press, New York, 1983, 125.

Somero, G. N., Chow, T. J., Yancey, P. H. and Snyder, C. B., Lead accumulation rates in tissues of the estuarine teleost fish *Gillichthys mirabilis:* salinity and temperature effects, *Bull. Environ. Contam. Toxicol.,* 6, 337, 1977.

Sorensen, E. M. B., Henry, R. E. and Ramirez-Mitchell, R., Arsenic accumulation, tissue distribution and cytotoxicity in teleosts following indirect aqueous exposures, *Bull. Environ. Contam. Toxicol.,* 21, 162, 1979.

Spacie, A. and Hamelink, J. L., Alternative models for describing the bioconcentration of organics in fish, *Environ. Toxicol. Chem.,* 1, 309, 1982.

Spigarelli, S. A., Thommes, M. M. and Prepejchal, W., Thermal and metabolic factors affecting PCB uptake by adult brown trout, *Environ. Sci. Technol.,* 17, 88, 1983.

Sprague, J. B., Factors that modify toxicity, in, *Fundamentals of Aquatic Toxicology,* Rand, G. M. and Petrocelli, S. R., Eds., Hemisphere Publishing, Washington, D.C., 1985, chap. 6.

Spry, D. and Wiener, J., Metal bioavailability and toxicity to fish in low alkalinity lakes: a critical review, *Environ. Pollut.,* 71, 243, 1991.

Spry, D. J. and Wood, C. M., Zinc influx across the isolated, perfused head preparation of the rainbow trout *(Salmo gairdneri)* in hard water, *Can. J. Fish. Aquat. Sci.,* 45, 2206, 1988.

Stagg, R. M. and Shuttleworth, T. J., The accumulation of copper in *Platichthys nesus* L. and its effects on plasma electrolyte concentrations, *J. Fish Biol.,* 20, 491, 1982.

Statham, C. N., Melancon, M. J. and Lech, J. J., Bioconcentration of xenobiotics in trout bile: a proposed monitoring aid for some waterborne chemicals, *Science,* 193, 680, 1976.

Stegeman, J., Brouwer, M., DiGiulo, R., Forlin, L., Fowler, B., Sanders, B. and Van Veld, P., Enzyme and protein synthesis as indicators of contaminant exposure and effect, in, *Biomarkers; Biochemical, Physiological, and Histological Markers of Anthropogenic Stress,* Huggett, R., Kimele, R., Mehrle, P. and Bergman, H., Eds., Lewis Publishers, Boca Raton, FL, 1992, chap. 6.

Stegeman, J. J. and Hahn, M. E., Biochemistry and molecular biology of monooxygenases: current perspectives on forms, functions, and regulation of cytochrome P450 in aquatic species, in, *Aquatic Toxicology, Molecular, Biochemical, and Cellular Perspectives,* Malins, D. and Ostrander, G., Eds., Lewis Publishers, Boca Raton, FL, 1994, chap. 3.

Stehly, G. R. and Hayton, W. L., Metabolism of pentachlorophenol by fish, *Xenobiotica,* 19, 75, 1989.

Stenger, V. G. and Maren, T. H., The pharmacology of MS 222 (Ethyl-m-aminobenozoate) in *Squalus acanthias, Comp. Gen. Pharmacol.,* 5, 23, 1974.

Streit, B. and Sire, E., On the role of blood proteins for uptake, distribution, and clearance of waterborne lipophilic xenobiotics by fish: a linear system analysis, *Chemosphere,* 26, 1031, 1993.

Stroganov, N. S., Parina, O. V. and Sorvachev, K. F., Uptake of triethyl tin chloride from the water and its distribution in the organs and tissue of the carp, *Gidrobiol. Zhurn,* 6, 59, 1973.

Sudershan, P. and Kahn, M. A. Q., Metabolic fate of (14C)Endrin in bluegill fish, *Pest. Biochem. Physiol.,* 14, 5, 1980.

Tachikawa, M., Sawamura, R., Okada, S. and Hamada, A., Differences between freshwater and seawater killifish *(Oryzias latipes)* in the accumulation and elimination of pentachlorophenol, *Arch. Environ. Contam. Toxicol.,* 21, 146, 1991.

Thomas, D. G., Cryer, A., Solbie, J. D. and Kay, J., A comparison of the accumulation and protein binding of environmental cadmium in the gills, kidney and liver of rainbow trout *(Salmo gairdneri* Richardson), *Comp. Biochem. Physiol.,* 76C, 241, 1983.

Thomas, D. G., Solbe, J. F., Kay, J. and Cryer, A., Environmental cadmium is not sequestered by metallothionein in rainbow trout, *Biochem. Biophys. Res. Commun.,* 110, 584, 1983.

Thomas, P., Teleost model for studying the effects of chemicals on female reproductive endocrine function, *J. Exp. Zool.,* Suppl. 4, 126, 1990.

Thomas, P. and Juedes, M. J., Influence of lead on the glutathione status of Atlantic croaker tissues, *Aquat. Toxicol.,* 23, 1992.

Thomas, P. and Wofford, H. W., Effects of metals and organic compounds on hepatic glutathione, cysteine, and acid-soluble thiol levels in mullet *(Mugil cephalus), Toxicol. Appl. Pharmacol.,* 76, 172, 1984.

Thomas, P. and Wofford, H. W., Effect of cadmium on glutathione content of mullet *(Mugil cephalus)* tissues, in, *Physiological Mechanisms of Marine Pollutant Toxicity,* Vernberg, W. B., Calabrese, A., Thurberg, F. P. and Vernberg, F. J., Eds., Academic Press, New York, 1982. 109.

Thomas, R. E. and Rice, S. D., Excretion of aromatic hydrocarbons and their metabolites by freshwater and seawater dolly varden char, in, *Biological Monitoring of Marine Pollutants,* Vernberg, F. J., Calabrese, A., Thurberg, F. P. and Vernberg, W. B., Eds., Academic Press, New York, 1981, 425.

Thomas, R. E. and Rice, S. D., Metabolism and clearance of phenolic and mono-di, and polynuclear aromatic hydrocarbons by dolly varden char, in, *Physiological Mechanisms of Marine Pollutant Toxicity,* Vernberg, W. B., Calabrese, A., Thurberg, F. P. and Vernberg, F. J., Eds., Academic Press, New York, 1982, 161.

Tovell, P. W. A., Howes, D. and Newsome, C. S., Absorption, metabolism and excretion by gold-fish of the anionic detergent sodium lauryl sulphate, *Toxicology,* 4, 17, 1975.

Truchot, J. and Dejours, P., Comparative respiratory physiology — quantities, dimensions and units, in, *Techniques in Comparative Respiratory Physiology, An Experimental Approach,* Cambridge University Press, New York, 1989, 3.

Van Dam, L., On the utilization of oxygen and regulation of breathing in some aquatic animals, Doctoral dissertation, University of Groningen, Groningen, The Netherlands, 1938.

Van Veld, P., Absorption and metabolism of dietary xenobiotics by the intestine of fish, *Rev. Aquat. Sci.,* 2, 185, 1990.

Van Veld, P., Patton, J. S. and Lee, R. F., Effect of preexposure to dietary benzo[a]pyrene (BP) on the first-pass metabolism of BP by the intestine of toadfish (*Opsanus tau*): *in vivo* studies using portal vein-catherized fish, *Toxicol. Appl. Pharmacol.,* 92, 255, 1988.

Varanasi, U. and Markey, D., Uptake and release of lead and cadmium in skin and mucus of coho salmon (Oncorhynchus kisutch), *Comp. Biochem. Physiol.,* 60C, 187, 1978.

Varanasi, U. and Gmur, D. J., Influence of water-borne and dietary calcium on uptake and retention of lead by coho salmon (*Oncorhynchus kisutch*), *Toxicol. Appl. Pharmacol.,* 46, 65, 1978.

Varanasi, U., Uhler, M. and Stranahan, S., Uptake and release of naphthalene and its metabolites in skin and epidermal mucus of salmonids, *Toxicol. Appl. Pharmacol.,* 44, 277, 1978.

Varanasi, U., Gmur, D. J. and Treseler, P. A. P., Influence of time and mode of exposure on biotransformation of naphthalene by juvenile starry flounder (*Platichthys stellatus*) and rock sole (*Lepidopsetta bilineata*), *Arch. Environ. Contam. Toxicol.,* 8, 673, 1979.

Veith, G. D., DeFoe, D. L. and Bergstedt, B. V., Measuring and estimating the bioconcentration factor of chemicals in fish, *J. Fish. Res. Bd. Can.,* 36, 1040, 1979.

Vetter, R. D., Carey, M. C. and Patton, J. S., Coassimilation of dietary fat and benzo[a]pyrene in the small intestine: an absorption model using the killifish, *J. Lipid Res.,* 26, 428, 1985.

Waldichuck, M., Some biological concerns in heavy metals pollution, in, *Pollution and Physiology of Marine Organisms,* Vernberg, F. J. and Vernberg, W. B., Eds., Academic Press, New York, 1974, 1.

Weis, P., Metallothionein and mercury tolerance in the killifish, *Fundulus heteroclitus, Mar. Environ. Res.,* 14, 153, 1984.

Welling, W. and De Vries, J. W., Bioconcentration kinetics of the organophosphorus insecticide chlorpyrifos in guppies (*Poecillia reticulata*), *Ecotoxicol. Environ. Safety,* 23, 64, 1992.

Wicklund, A. and Runn, P., Calcium effects on cadmium uptake, redistribution, and elimination in minnows, *Phoxinus phoxinus,* acclimated to different calcium concentrations, *Aquat. Toxicol.,* 13, 109, 1988.

Williams, D. E., Regiospecific hydroxylation of lauric acid at the (Omega 1) position by hepatic and kidney microsomal cytochrome-P-450 from rainbow trout, *Arch. Biochem.,* 231, 503, 1984.

Williams, D. E., Masters, B., Lech, J. and Buhler, D., Sex differences in cytochrome P-450 isozyme composition and activity in kidney microsomes of mature rainbow trout, *Biochem. Pharm.,* 35, 2017, 1986.

Winge, D. R., Premakumar, R. and Rajagopalan, K. V., Metal-induced formation of metallothionein in rat liver, *Arch. Biochem. Biophys.,* 170, 242, 1975.

Yarbrough, J. D. and Chambers, J. E., The disposition and biotransformation of organochlorine insecticides in insecticide-resistant and susceptible mosquitofish, in, *Pesticide and Xenobiotic Metabolism in Aquatic Organisms,* Khan, M. A. Q., Lech, J. J. and Menn, J. J., Eds., (ACS Symposium Series 99), American Chemical Society, Washington, D.C., 1979, 145.

Chapter 6

Liver

I. INTRODUCTION

The liver is of key importance when considering the action of toxic chemicals on fish. It is the primary organ for biotransformations of organic xenobiotics, and probably also for the excretion of harmful trace metals. Because many of these metals and organics tend to accumulate to high concentrations in the liver (see Chapter 5), the cells there are exposed to far higher (often several orders of magnitude) levels of harmful chemicals than may be present in the environment, or in other organs of the fish. The liver serves a number of functions related to other physiological activities, such as interconversions of foodstuffs (Chapter 8) and metabolism of sex hormones (Chapter 13), so the potential effects of xenobiotics on the liver are numerous, but often seen only indirectly.

II. STRUCTURE OF THE LIVER

The gross and histological organization of the fish liver has been described in considerable detail by Gingerich (1982) and Hinton et al. (1987). While they were able to derive a number of generalizations, it must be emphasized that there is a fair amount of diversity among the various fish groups and this variability could influence the toxicological and physiological responses to pollutants.

The well-defined (histologically) lobular structure of the mammalian liver is apparently not characteristic of fish. Instead, the hepatocytes are arranged as tubules of cells. A tubule viewed in cross section is usually surrounded by five to seven hepatocytes with their apices directed toward the central bile canaliculus and/or bile preductule. These tubules are surrounded by the sinusoids which take the place of capillaries in the liver. Blood enters the sinusoids from the hepatic portal system and hepatic artery and the hepatocytes remove nutrients and xenobiotics from that blood. The bile is secreted into the central bile canaliculus from which it ultimately flows into the gallbladder.

There seems to be some debate among histochemists as to the extent of specialization (metabolic zonation) among hepatocytes in fish livers. Most seem to conclude that the distribution of cellular specialization based on enzyme activity is relatively more uniform in fish than in mammals (reviewed in Gingerich, 1982). However, Schar et al. (1985)

conclude that rather extensive zonation is present in rainbow trout livers, although still less than that found in mammals.

Parenchymal cells (hepatocytes) are the dominant type of cell present in liver. These range in shape from oval to irregular polygons; their ultrastructure has been described by Leland (1983). As would be expected from the functions of the liver, these cells are well endowed with secretory and biosynthetic structures, such as Golgi bodies and rough endoplasmic reticulum. Juvenile rainbow trout appear to have a near absence of these cellular structures, which develop as the fish matures (Chapman, 1981; Leland, 1983). This implies that biotransformation and excretion of xenobiotics may be severely limited in younger fish. However, some induction of the biotransformation enzymes is possible by xenobiotic exposure even in the embryos (Binder and Stegeman, 1983).

At any one time, some 12–15% of the liver mass of fish is blood (Stevens, 1968). This is over twice the amount of most other organs. From a practical standpoint, it means that when a crude homogenate of liver tissue is prepared for biochemical analysis, a significant fraction of the content of that homogenate will actually be blood, rather than the homogenized liver tissue.

In spite of there being a large amount of blood in the liver, the rate of blood flow through the organ in fish is small in comparison with that in mammals. Thus, Gingerich (1982) points out that clearance from the blood by the liver of endogenous and exogenous chemicals is slower in fish than mammals, in large part due to this comparative lack of blood perfusion.

Although the fish liver receives blood from the hepatic artery, it apparently lacks a well-developed arterial blood supply in many species. This could imply a rather low level of oxygen for liver cells, as they are perfused primarily by the portal vein which carries venous blood from the intestine. Measurements of adenylate energy charge in this tissue does indeed suggest a relatively low level of oxygen in these cells compared to those from other organs (van den Thillart et al., 1980; Heath, 1984).

III. ALTERATIONS IN LIVER/SOMATIC INDEX

The ratio of liver to body weight differs greatly between and within various species of fish. The liver/somatic index is usually expressed as the percentage of the wet weight of the liver is to the whole body and is often also referred to as the hepatosomatic index (HSI). The nutritional state of the organism at the time of capture has a considerable influence on liver size. Indeed, the HSI is useful (among other measures) in assessing the general condition of fish captured in the field (Goede and Barton, 1990). These transitory changes in liver size are largely due to variations in glycogen and fat content. Because fat is an important repository for many xenobiotics, the amount of such a chemical deposited in the liver may depend to some extent on the nutritional state of the fish, and variations between species are important; for example, flatfish (Pleuronectidae) and cods (Gadidae) have especially large stores of lipid in comparison with most other teleost groups (Gingerich, 1982).

Decreases in HSI are frequently seen in fish under stress (Lee et al., 1983; Ram and Singh, 1988). This presumably occurs when a chronic stress imposes an energy drain on the fish and is often correlated with loss of energy stores such as liver glycogen. The cause of the decline in liver mass can actually be due to depressed feeding (Barton, 1988) and is not necessarily a direct effect of some contaminant. As discussed in Chapter 8, fish that are under stress often reduce or even cease feeding.

Increases in liver size are commonly seen in fish that have been exposed for long periods of time to organic contaminants (especially petroleum hydrocarbons) in the laboratory or field (Yarbrough et al., 1976; Poels et al., 1980; Fletcher et al., 1982; Oikari and Nakari, 1982; Slooff et al, 1983; Vignier et al., 1992; Everaarts et al., 1993). The increase in HSI can be due to hyperplasia (increased cell number) and/or hypertrophy (increased cell size). It may be associated with increased capacity to metabolize xenobiotics so it could be considered an adaptation to the presence of pollution, rather than a dysfunction. The hypothesis might be proposed that fish that are exposed to pollutant stress may show decreases in HSI if food is limited or if the exposure is of short duration. However, with chronic exposures where food is not limited and feeding rate is normal, a rise in HSI may be the expected response.

IV. HISTOPATHOLOGICAL EFFECTS OF POLLUTANTS

There have been numerous reports of histopathological changes occurring in fish tissues from exposure to pollutants and it is evident that a wide variety of chemicals cause lesions in the livers of fish (Gingerich, 1982; Patton and Couch, 1984; Hinton et al., 1987, 1992). These chemicals include the metals arsenic, cadmium, copper, and mercury; assorted industrial wastes such as ammonia, Aroclor, chlorine, and phenol, both organophosphate and organochlorine pesticides such as endrin, dieldrin, Dursban®, and diazinon; the carbamate insecticide carbofuran and petroleum hydrocarbons. It should be noted that species differences can be considerable. For example, Bass et al. (1977) found histological lesions in the livers of bluegill exposed intermittently to fairly high doses of chlorine, whereas trout showed no changes in liver histology (at the level of light microscopy) except for some loss of glycogen, even when the organ was examined from moribund fish.

The histological changes observed in various studies on livers taken from fish exposed to pollutants include increased vacuoles in the cytoplasm, enlarged lysosomes, changes in nuclear shapes, focal necrosis (death of cells in a localized area), ischemia (blockage of capillary circulation), hepatocellular shrinkage, regression of hepatocytic microvilli at the bile canaliculi, fatty degeneration, and loss of glycogen. The latter two changes can be due to depressed feeding and/or elevated levels of the stress hormones cortisol and adrenalin, and therefore are not necessarily caused by the chemical acting directly on liver cells.

In general, the changes observed in liver tissue are rather non-specific so it is not possible to identify the chemical causing the lesion (Gingerich, 1982; Patton and Couch, 1984). At the ultrastructural level, the smooth and rough endoplasmic reticulum (RER) frequently become reorganized and proliferated in response to PCBs, DDT, or mixtures of organic xenobiotics (Weis, 1974; Hacking et al., 1978; Kohler, 1989). It seems reasonable to assume that this is associated with induction of xenobiotic biotransformation enzymes (see Chapter 5). Just such a correspondence between structural change and function was found in mullet receiving injections of 3-methylcholanthrene (Schoor and Couch, 1979). Thus, this could be interpreted as an adaptation rather than a pathological change. Even so, it may come at a cost of resources to the fish.

Copper at 0.05 or 0.1 mg/L for 4–7 days caused a variety of alterations in cellular structure revealed by transmission electron microscopy in livers of the snake-headed fish (*Channa punctatus*; Khangarot, 1992). These included extensive proliferation of the smooth endoplasmic reticulum, dilation of the RER, loss of mitochondrial structure, proliferation of lysosomes, and accumulation of electron dense bodies. The most dominant features were dilation of the RER, and reduction and shortening of mitochondria. The RER changes are probably associated with detoxification while the mitochondrial changes are clearly pathological.

Fairly acute exposures of tench to copper for 12 days caused accumulation of large amounts of hemoglobinemic pigment in Kuppfer cells (Roncero et al., 1992). The function of Kuppfer cells is removal of damaged erythrocytes so the accumulation of hemoglobin in these cells from copper exposure suggests hemolysis of the blood cells. In mammals, excess copper can cause hemolytic anemia by denaturing hemoglobin and oxidizing glutathione (Fairbanks, 1967), although anemia in fish from copper exposure is not common (Chapter 4).

Exposing brown trout to a concentration of copper, which was below that which inhibited spawning, produced some rather subtle changes in hepatocyte ultrastructure (Leland, 1983). These included slight contraction of mitochondria and enlargement of nuclei. At somewhat higher concentrations of copper, there was evidence of cell swelling, and electron dense granules became prevalent. The latter may be areas of copper deposition, but could also be points of cellular damage. Zinc caused similar changes so the lesions are not specific for copper. At doses causing 50% mortality of juveniles in 42 days, Leland (1983) notes (p. 353): "...loss in integrity of mitochondrial membranes, rupturing of plasma and nuclear membranes, separation of granular and fibrillar nuclear components, fragmentation of endoplasmic reticulum, and extensive autophagic vacuolization were significant features of hepatocytes of surviving juvenile rainbow trout." He further makes the cogent point that aside from the possible sequestering of the metals, the ultrastructural changes observed may be largely a non-specific response to stress.

Cytoplasmic inclusions seem to be a fairly common occurrence with fish that have been exposed to either metals or organic xenobiotics (Patton and Couch, 1984; Hinton et al., 1987). For example, trout reared on diets containing elevated amounts of copper had electron-dense granules in the cytoplasm of their hepatocytes (Lanno et al., 1987). With the aid of electron microprobe analysis, these were demonstrated to contain copper. The authors suggest the granules may be involved in the process of excretion of copper into the bile, or they may also be a mechanism for sequestering harmful chemicals.

Histopathological techniques are increasingly being used in field studies to assess the impact of suspected contaminants (Kohler, 1990; Hinton et al., 1992; Meyers et al., 1992; Young et al., 1994). Because the liver commonly accumulates xenobiotics to very high concentrations, it exhibits exceptional sensitivity to these in the field. Good correlations have been seen between the concentrations of various persistent chlorinated hydrocarbons and metals and liver lesions (Kohler, 1990; Myers, 1992). Kohler (1989) examined the question in flounder of how well the liver lesions regenerate following contaminant exposure. She transferred contaminated flounder into a contaminant-free environment and found partial or complete liver regeneration in 50% of the fish after 20 days and 70% after 40 days. Interestingly, some of the flounder failed to show any regeneration at all and these individuals retained their high levels of contaminants. While the field use of histopathology certainly has promise, some biochemical changes have greater sensitivity to pollutant stress and these will be discussed below.

V. MAJOR FUNCTIONS OF THE LIVER

A. INTERCONVERSION OF FOODSTUFFS

Because the liver is "down stream" from the intestine, foodstuffs absorbed from the diet are often immediately acted upon by the liver. Conversions from one form to another occur primarily in this organ (e.g., from protein to carbohydrate). Some of the key enzymes involved in these conversions are sensitive to pollutants (see Chapter 9) and these same enzyme systems are under control of several major hormones (Chapter 8, Section X). One of the effects of this interconversion of foodstuffs is the production of large amounts of ammonia from the deamination of amino acids (Wood, 1993).

B. STORAGE OF GLYCOGEN

The largest store of carbohydrate in the fish's body is in the liver and this is the primary source for blood glucose in the postabsorptive animal. Changes in liver glycogen concentration are almost always seen in fish exposed to a variety of chemical and physical stressors (Chapter 8).

C. REMOVAL AND METABOLISM OF FOREIGN CHEMICALS IN THE BLOOD

This is discussed at some length in Chapter 5. Metals and organic xenobiotics are often sequestered there so cellular concentrations can become quite high.

D. FORMATION OF BILE

Bile has a number of things in it including the bile salts, which are necessary for normal fat digestion; bile pigments, which are breakdown products of hemoglobin; and xenobiotics or their metabolites which have been removed from the blood (Chapter 5). Harmful metals are also excreted via this route.

E. SYNTHESIS OF MANY PLASMA PROTEINS

The complex mixture of proteins found in the blood largely originates from the liver. It is broadly divided into the fibrinogens and albumins. (The immunoglobulin proteins come from B lymphocytes so are not of liver origin.) The fibrinogens are required for blood clotting, which incidentally, is quite rapid in fish compared to mammals. Fish under stress exhibit an accelerated clotting, but this is associated with an increase in number of thrombocytes rather than fibrinogen (Casillas and Smith, 1977). The albumins are important in maintaining normal plasma osmotic pressure, blood pH buffering, as an amino acid source, and as a transport molecule for hormones and exogenous chemicals such as metals and organics.

F. SYNTHESIS OF CHOLESTEROL FOR USE IN STEROID HORMONES AND CELL MEMBRANES

Cholesterol is the most abundant sterol in vertebrates. It functions to stabilize cell membranes, plasma lipoproteins, and myelin in the nervous system. It also is a precursor for bile acids and steroid hormones (Myant, 1981). Cholesterol is synthesized by the liver, however, in mammals at least, the intestine and skin also synthesize significant amounts of this compound.

G. EXOCRINE PANCREATIC SECRETION

The term "exocrine" means the portion of the pancreas devoted to the secretion of digestive enzymes. This tissue is histologically observed surrounding the hepatic portal veins inside the liver (Groman, 1982). A bony fish also has pancreatic tissue scattered among the mesenteries so it is usually not a discrete organ as is typical of mammals. The exocrine portion of the pancreas secretes a variety of digestive enzymes, but it is not clear how important this pancreatic tissue within the liver is in digestion.

H. METABOLISM OF HORMONES

Many of the hormones of a vertebrate are metabolized by the liver. This avoids the buildup in the blood of such things as estrogen, one of the key sex hormones, and cortisol.

VI. EFFECTS OF POLLUTANTS ON LIVER FUNCTION

Measurements of liver function per se are mostly based on clinical tests developed for mammals (Gingerich and Weber, 1979). These are of three general types: (1) determining the rate at which the liver clears some chemical (usually a dye) from the blood; (2) monitoring the activity of plasma enzymes that are normally concentrated in liver cells, but get into the blood when hepatocytes are damaged; and (3) determining changes in the plasma concentrations of chemicals produced (e.g., albumin and cholesterol) or eliminated (e.g., bilirubin) by the liver. From the brief listing above of the multitude of activities associated with the liver, it is also safe to assume that a pollutant's effect on liver function may be reflected indirectly in such diverse physiological activities as sex drive and digestion.

A. PLASMA CLEARANCE BY THE LIVER OF SPECIFIC DYES

An important function of the liver is the removal of hormones and potentially harmful substances from the plasma and subsequent excretion in the bile. This process involves several steps including uptake by the parenchymal cells, biotransformation of the substance by enzyme systems into a more polar product, and conjugation with another molecule (e.g., gluconate) in preparation for excretion.

The primary interest of fish toxicologists has been the mechanisms for accumulation and biotransformation of the xenobiotics which are discussed at some length in Chapter 5. The question has also been asked as to whether some substances may inhibit these processes in the liver. In order to answer this, the dye clearance test developed for mammalian systems has been applied to fish. Fortunately, the dye sulfobromophthalein (BSP) is cleared from the blood in fish, at least in elasmobranchs (Boyer et al., 1976) and trout (Gingerich et al., 1977), the same as it is in mammals.

The rate of clearance of BSP is measured by injecting the dye into a vein (usually the caudal vein) and then sacrificing the fish for blood removal at intervals up to 2 h postinjection (Gingerich et al., 1978). The concentrations of BSP in the plasma samples is then determined colorimetrically. Considerable attention must be paid to establishing an optimum dye dose and the nutritional state of the animal can influence the findings rather greatly, thus the procedure is not as easy as it may appear. Because many fish alter their feeding rate when in the presence of a pollutant, the application of dye clearance tests to fish chronically exposed to some pollutant may not be justified.

Two chemicals, carbon tetrachloride and monochlorobenzene, are model hepatotoxic agents in mammals. Intraperitoneal administration of carbon tetrachloride to rainbow trout resulted in a dose-dependent plasma retention of BSP (Gingerich et al., 1978). In other words, the clearance rate for BSP by the liver was reduced 24 h after receiving the toxicant. The process of dye clearance from the blood involves several steps, any one of which could be limiting. Thus this measure reveals an overall effect, but does not resolve the mechanisms. Treatment with a chemical that specifically blocks formation of bile in mammals also reduces the clearance of BSP in trout (Gingerich, 1982), but other toxicants may act on different processes.

B. PLASMA ENZYMES OF LIVER ORIGIN

Clinical laboratories have long used the measurement of the activity of certain enzymes in the blood plasma (sometimes called serum) as a diagnostic tool for assessing liver damage in humans. Enzymes such as glutamic oxaloacetic transaminase (GOT), glutamic pyruvic transaminase (GPT), and sorbitol dehydrogenase (SDH) are normally found in a low concentration in blood, so if liver cells are damaged, they may leak these into the plasma causing an increase in their catalytic activity there. Liver cells are particularly rich in transaminases because it is the major organ for interconversions of foodstuffs. Although GOT and GPT may become elevated from damage to other organs and are not specific for liver, SDH apparently is quite specific.

When the hepatotoxic chemical carbon tetrachloride is administered to either freshwater or marine fish, increases in plasma GOT (PGOT) and plasma GPT (PGPT) are usually observed (reviewed by Gingerich, 1982). Peak increases in enzyme activity are seen 2–6 h after injection, which is even more rapid than the mammalian response to this agent. However, the degree of change is generally not nearly as large as that seen in mammals. A significant amount of PGOT can come from erythrocytes, so if hemolysis occurs either in the fish, or in the process of blood collection, a false elevation may be evident. Another source of these enzymes is the kidney, so increases may reflect damage to that organ, or even heart tissue. The most convincing evidence that CCl4 causes release of GOT and GPT from liver exclusively is found in the work of Inui. (1969). He showed that eels with the liver removed exhibit no elevation in the enzyme activities when injected with CCl4, whereas sham operated controls do.

In an investigation by Casillas et al. (1983), histological lesions in the liver were correlated with changes in plasma enzymes. Serum alanine aminotransferase (ALAT) and asparate aminotransferase (ASAT) were elevated 40–60 times that of controls following injection of CCl4. These are much greater increases than GOT or GPT ever reach. Serum albumin, total protein, bilirubin, and urea nitrogen did not change in relation to liver lesions induced by CCl4 injections, thus the plasma enzymes would seem to have better diagnostic value for this type of liver damage than the other plasma constituents.

There has been some interest in using the changes in plasma enzyme activity as a general measure of biochemical alterations induced by pollutants. McKim et al. (1970) exposed brook trout chronically to 32–39 µg/L copper and found a +60% increase in plasma GOT during the first 21 days, but when measured at 11 months of exposure, the PGOT was actually 30% below controls. The initial increase could have been due to heart damage, but because the liver accumulates copper so much more readily than the heart, that seems unlikely. The reduction observed at 11 months conceivably was due to a direct inhibitory action of copper on the enzyme following copper accumulation in the liver. Copper is a strong inhibitor of fish PGOT when the blood is exposed to the element *in vitro* (Christensen, 1971).

Chronic exposure of a marine fish to mercury caused peak increases in PGOT and PGPT within 4–8 days. The extent of change (>2×) was greater with PGOT than PGPT (Hilmy et al., 1981). In the same study, it was reported that alkaline phosphatase exhibited a 50% decrease in activity by 8 days of exposure. The authors think this latter effect may be due to the mercury displacing the cofactor magnesium from the enzyme. Clearly there is some direct inhibition of the enzyme in the plasma by mercury, but what the mechanism is remains unclear. It may be worthwhile to note that Jackim et al. (1970) found that alkaline phosphatase extracted from fish liver was weakly inhibited by mercury when the exposure was *in vitro,* but the enzyme activity was actually stimulated when the exposure was *in vivo.* In other words, mercury is inhibitory of alkaline phosphatase, but seemingly induces synthesis of the molecule in the intact liver which more than compensates for any inhibition.

When added *in vitro,* essentially all the transition metals cause inhibition of LDH and GOT from white sucker plasma, although there is a fair amount of variation between different metals as to the degree of inhibition (Christensen et al., 1977). The significance of these *in vitro* findings is unclear because there are no data on the concentrations the different metals may reach in the blood. Without that information, the finding of inhibition at some concentration *in vitro* means very little because the enzyme may never be exposed to that concentration in the blood.

Exposure of pike for 8 days to phenol caused elevations in both GOT and GPT (Kristoffersson et al., 1974), and Wieser and Hinterleitner (1980) found elevated levels of GOT and GPT in plasma from rainbow trout taken near a sewage outfall (composition apparently unknown).

Perhaps the best plasma enzyme for assessing liver damage in fish is SDH. Dixon et al. (1987) has shown that this enzyme has activity in liver that is an order of magnitude above nine other tissues examined in rainbow trout, thus its elevation in plasma is quite specific for liver damage. They tested the effects of phenol, *p*-chlorophenol, *p*-phenoxyphenol, carbon tetrachloride, cyanide, and copper on the level of this enzyme in plasma. Both direct injections and additions to water were used as methods of exposure. A dose-dependent marked rise was seen with all toxicants tested except cyanide. The plasma activity peaked after around 48 h of exposure. When injections were used as the mode of administration, there was a rise in SDH before histopathology was detectable in the liver. Thus, the biochemical measurement in the plasma was even more sensitive than histopathology.

As with the rainbow trout, SDH activity was found to be much higher in liver than in six other tissues of mullet, a marine species (Ozretic and Ozretic, 1993). In the same study, they found activity of glutamate dehydrogenase (a mitochondrial enzyme) is also higher in liver than in other tissues, however, kidney has a level about one third that of liver. Both enzymes exhibited marked increases in the plasma following injection of either phenol or carbon tetrachloride.

Two other enzymes that rise in the plasma as a result of tissue damage are acid phosphatase and N-acetyl-B-D-glucosaminidase (NAG). Both of these come from lysosomes that break down in the cells. Versteeg and Giesy (1986) found that chronic exposure of bluegill to cadmium for up to 32 days produced increases in these two enzymes in the plasma. Asparte and alanine transaminase enzymes were unaffected. The changes in acid phosphatase and NAG were associated with a decrease in stability of liver lysosome membranes (Versteeg and Giesy, 1985) which suggested a direct effect of the metal on these organelles. It is interesting that no histological changes were seen in the liver but decreases in growth rate of the fish were seen, so the biochemical alterations were predictive of a higher order response to the pollutant.

Lysosomes accumulate a variety of metals and organic compounds (Moore, 1985). As previously mentioned, Versteeg and Giesy (1985) showed that the membranes of these organelles become less stable during chronic cadmium exposure. Kohler (1991) used this measure to assess pollution in contaminated estuaries and found decreased lysosomal stability in a gradient that correlated with the level of pollution. Her data suggest a two-step response in that activation of the lysosomal system occurs first with increases in number and size of lysosomes as they accumulate the contaminants. If this adaptive response is overcome by the pollutant load, then the membrane stability of the lysosomes decreases releasing their contained enzymes into cells and ultimately into the bloodstream.

In general, it appears that many toxicants will cause increases in at least some fish plasma enzyme activities. A word of caution is in order here. We have found in this laboratory that the enzyme activities in trout plasma are quite variable in control fish when measured using an automated clinical analyzer. Part of this variability may be due to the fact that these analyzers are designed for mammalian work and operate at 25°C or higher. Trout enzymes are adapted to function at temperatures considerably below that level, so one may be well beyond the temperature for optimum catalytic activity, especially for salmonids.

C. CHANGES IN PLASMA CONCENTRATION OF CHEMICALS PRODUCED OR EXCRETED BY LIVER

The concentration of cholesterol in the blood is moderately sensitive to the presence of pollutants, but the direction of movement seems to depend on the concentration of the pollutant and on whether it affected osmoregulation of the fish. Gluth and Hanke (1985) exposed carp to a variety of organic pesticide chemicals at sublethal concentrations for 72 h and noted decreases in cholesterol in all cases. Because plasma protein also decreased at the same time and to approximately the same extent, they attribute the changes to water uptake resulting from osmoregulatory failure (see Chapter 7). Decreases in blood cholesterol were also seen with 96-h exposures of *Clarias* to endrin or toluene (Ghazaly, 1991). Because protein remained unchanged in this study, the decrease is probably real and not due to water dilution. When the same species of fish was exposed to mercury (both inorganic and organic) for periods up to 180 days, reductions in free and esterified cholesterol were seen (Kirubagaran and Joy, 1992). This was also associated with a reduction in several other lipids so it could reflect a general depletion of energy stores. The inorganic form of the mercury was more effective than the methylmercury in lowering lipid levels, which is the reverse of what usually occurs with these pollutants. Lead also caused a large decrease in cholesterol in blood, liver, ovary, and testes of rainbow trout with chronic exposures of 30 or 60 days (Tewari et al., 1987). Decreases in cholesterol may indicate impairment of the liver synthesis of this chemical, but there is no direct evidence for this in the literature. Another possible cause would be increased utilization for corticosteroid synthesis.

Marked increases in cholesterol have also been observed in some studies. In apparent contrast to the above work on mercury, Dutta and Haghigh (1986) reported an increase in plasma cholesterol over a 3-day period of exposure to methylmercuric chloride. Copper has also been found to cause a steady increase in cholesterol with the levels reaching almost twice that of controls after 30 days of continuous exposure (Singh and Reddy, 1990). Plasma protein and blood cell count actually went down so hemoconcentration was certainly not the explanation for the rise in cholesterol.

DDT and endrin (another organochlorine insecticide), caused a 50% elevation in serum cholesterol when eels and mullet were exposed to the insecticides for 96 h (Hilmy et al., 1983). Interestingly, there was no dose effect over a 10-fold range in pesticide concentration. Elevations in cholesterol could be explained by an enhanced production by the liver (and other organs), or release of cholesterol from damaged cell membranes. Regarding the first possibility, the hormone cortisol inhibits cholesterol synthesis and this hormone becomes elevated in fish under stress (Chapter 8). Thus, a stimulation of cholesterol synthesis does not seem likley in those fish under severe stress, however, fish that are chronically exposed may not have this problem. Interestingly, it was found fairly recently that chronic exposure to phenol (and perhaps some other substances as well?) inhibits the conversion of cholesterol into steroid sex hormones (Mukherjee et al., 1991). Phenol also caused increased cholesterol in the liver and blood, reflecting perhaps a combination of "backing up" of cholesterol in the blood and stimulation of its synthesis in the liver. It therefore appears that there are a number of ways that cholesterol metabolism can be altered by toxic chemicals and the dose used may have a considerable impact on the character of the effect.

It is well known that the liver is responsible for the elimination, via the bile, of the products of heme breakdown, primarily bilirubin. If this process is impaired, a condition of jaundice may develop and be reflected in a rise of bilirubin concentration in the plasma (Mattsoff and Oikari, 1987). Oikari and Nakari (1982) found that exposure of trout for 11 days to kraft pulpmill effluent at a dose equivalent to 0.15×96 h LC50 produced a marked elevation in plasma bilirubin. A concomitant inhibition of the enzyme UDP-glucuronyltransferase in the liver tissue was also noted. Because this enzyme is required for the normal conjugation reactions by which bilirubin is prepared for excretion into the bile, its inhibition may be the primary cause of the jaundice. More recently, Rabergh et al. (1992) found that the resin acids in pulpmill effluent also inhibit the uptake of bile acid and bilirubin by isolated trout hepatocytes.

Liver dysfunction has been implicated in a study of high mortality of striped bass in San Francisco Bay (Brown et al., 1987; Young et al., 1994). Moribund fish had elevated levels of plasma bilirubin and uric acid was 80-fold in excess of the levels found in reference fish. The plasma of the moribund fish was also yellow-orange in color. The osmoregulatory picture was affected but only moderately so blood water percentage changes were apparently minimal in these fish that were obviously under severe stress. A mixture of agricultural, urban, and industrial contaminants were found in very high concentrations in the livers of the moribund fish, but only low levels of these compounds were found in reference fish from the Pacific Ocean (Cashman et al., 1992).

It is often said, but without much evidence presented, that nutrition can affect the response of a fish to some stress. Dixon and Hilton (1981) have shown that a high carbohydrate diet in trout reduces their tolerance to waterborne copper (i.e., lowers the incipient LC50). What is perhaps more interesting is that Dixon and Hilton provided a physiological mechanism: the high carbohydrate diet caused a greater accumulation of copper by the liver. This is undoubtedly due to a reduced capacity for copper excretion into the bile by the liver in fish with a high carbohydrate diet, although just why that would be is obscure. A further aspect of nutrition will be presented next when discussing ascorbic acid.

VII. ASCORBIC ACID AND POLLUTANT EXPOSURE

Ascorbic acid is better known as vitamin C. It plays a number of known biochemical roles and undoubtedly several that are currently unknown. It is essential for the normal synthesis of collagen. Other functions less well studied include involvement in the metabolism of steroids by the liver, oxidation of tyrosine, and in conjunction with superoxide dismutase and catalase, it helps prevent the buildup of free radicals in cells (Lehninger, 1979). Ascorbic acid is discussed here primarily because most of the work in fish has been on liver ascorbic acid, even though it is found in other tissues as well. Species differences are striking; for example, Thomas and Neff (1984) found high levels in the brain of mullet while Lopez-Torres et al. (1993) report that it was non-detectable in the brain of trout.

It is not clear to what extent various teleosts have a dietary requirement for this vitamin. Because a dietary requirement for ascorbic acid varies considerably among different higher vertebrates, it would not be surprising to find that similar variability was present among the teleosts. Salmonids, channel catfish, and mullet clearly do require it in the food (Halver et al., 1975; Thomas et al., 1982), but the cypriniformes, which include carp, possess the enzyme L-gluconolactone oxidase which is necessary for ascorbic acid synthesis (Yamamoto et al., 1978), and therefore they presumably do not require this vitamin in the diet. Even in those species capable of synthesizing the vitamin, the requirement may exceed that capability when stressed (Mayer et al., 1978).

In the tissues of fish, there appears to be two forms of ascorbate: L-ascorbic acid (C1) and L-ascorbic acid-2-sulfate (C2) (Benitez and Halver, 1982). The latter is generally used as a dietary supplement as it is the more stable form (Halver et al., 1975). Benitez and Halver (1982) suggested it may actually be the dominant form in which it is stored in the tissues, but Thomas et al. (1982), working on mullet, found very little of C2 in gill, brain, kidney, or liver, whereas there were considerable quantities of C1 in these tissues. The enzyme C2 sulfatase catalyzes the removal of the sulfate from C2 yielding C1. Benitez and Halver (1982) have proposed that this enzyme serves as a sort of modulator of the cellular levels of vitamin C in fish, in that as the C1 becomes utilized, the enzyme is derepressed and increases in activity to promote replenishment of C1. This mechanism may work in those species with large amounts of C2, but if there is not much C2 available, as in the mullet, such a function seems of little value.

In a rather interesting series of experiments, Mayer et al. (1978) exposed channel catfish to various amounts of toxaphene (an organochlorine insecticide) in the diet for 150 days. They noted a dose-dependent depression of backbone collagen and spinal deformities, but when ascorbate was added to the diet at several different doses, the extent of collagen depression and the incidence of deformities in the fish exposed to toxaphene was decreased roughly in proportion to the ascorbate dose. Toxaphene was shown to cause a depletion of vitamin C in the spine but not in the liver. Because vitamin C supplements also reduced the whole-body residues of toxaphene, the authors hypothesized that this molecule is used by the liver in detoxifying the toxaphene but at the expense of that in the collagen.

Exposure of mullet to a rather acute dose of pentachlorophenol (PCP) produced an elevation in liver ascorbic acid in either 24 h (100 ppb PCP) or 48 h (200 ppb PCP). There were corresponding increases in plasma cortisol which indicate interrenal activation (Thomas et al., 1981). Because stress has been shown in trout to cause ascorbic acid depletion in interrenal tissue (Wedemeyer, 1969), the possibility exists that ascorbate was transported from the interrenal to the hepatic cells. Other hypotheses mentioned by Thomas et al. include a reduction in ascorbic acid breakdown, or a stimulated synthesis, caused by cortisol. The latter seems especially unlikely as mullet have virtually no ability to synthesize ascorbic acid.

PCP is rapidly conjugated to glucuronid in goldfish (see Chapter 5) and this process requires ascorbate in mammals, so the process of detoxification would probably cause depletion of hepatic ascorbate. Indeed, in the mullet, at 120 h of exposure (when the experiments were terminated), the liver ascorbate had returned to near "normal" (Thomas et al., 1981). A more prolonged chronic exposure might eventually cause a lowering of ascorbate as is seen in those fish exposed to waterborne metals (discussed next). However, it is interesting that perch taken from coastal waters polluted by kraft mill effluents had elevated liver ascorbate levels (Anderson et al., 1988). These types of effluents have several types of chlorinated phenolics including PCP (Oikari et al., 1985).

Mullet exposed to cadmium (10 mg/L) exhibited a brief 50% drop in liver ascorbate at 6 h, but then by the third day the ascorbate became elevated. Then, continued exposure produced a more-or-less steady depletion in hepatic ascorbate (Thomas et al., 1982). Brain, kidney, and gill showed a somewhat similar pattern of change although the percent depletion was not as great as that seen in the liver.

It is noteworthy that cadmium produced no initial (other than a very transient burst) increase in liver ascorbate, as was seen in the PCP exposed animals. The hypothesis that a rise in ascorbate is the result of increases in cortisol from the interrenal gland (as proposed by Thomas et al., 1981) is indirectly supported by these cadmium data. This is because work by Schreck and Lorz (1978) on salmon has shown that cadmium is rather unique in that exposure to doses that are high enough to even cause death fail to activate the interrenal gland, the way most other stressors do.

The mechanism responsible for the depletion of tissue ascorbate following cadmium exposure is not known but must involve a stimulated rate of utilization of this vitamin. Copper in the water has been shown to increase the dietary requirement for vitamin C in trout (Yamamato et al., 1981), and supplementing the diet with vitamin C reduced the accumulation of copper in the liver and gills. It could be this latter effect is due to an enhanced excretion rate of copper produced by elevated vitamin C, but that is only speculation. One cannot help but wonder if dietary supplementation with this vitamin would also enhance the depuration of other metals and even organic compounds (Mayer et al., 1978). Providing fish with high doses of vitamin C in the diet has been shown to decrease the effect of waterborne thiotox and malathion on several enzymes in the gills, liver and brain (Verma et al., 1982).

The depletion of tissue stores of ascorbate during chronic exposure to pollutants (except chlorinated phenolics) may turn out to be common depending, of course, on the ability of the species of fish in question to synthesize this vitamin. The extent of change can be sizable. Six-month exposures to carbofuran caused a 44% decline in Indian catfish (Ram and Singh, 1988), and a week of exposure to the watersoluble fraction of No. 2 fuel oil caused a 45% loss of liver ascorbate in mullet (Thomas and Neff, 1984). In the latter study it was also found that liver ascorbate was not affected much by "natural" stressors such as changes in temperature, salinity, or handling, so measurements of this trace nutrient in the liver may have potential for specifically indicating chemical pollution.

From the standpoint of the fish, depletion of ascorbate can produce a variety of pathological conditions including delayed wound healing, anemia, skin lesions, fin necrosis, scoliosis, etc. (Hilton et al., 1977; Thomas et al., 1982). Lipid peroxidation of cell membrane lipids is a primary mechanism of cell injury by xenobiotics (Recknagel and Glende, 1977) and has been seen in several studies of fish (see Thomas, 1990 for references). Because one of the functions of ascorbic acid is prevention of peroxidation in cells (Winston and Diguilo, 1991), a reduction in its concentration from various chemical or physical stresses has obvious implications for possible cell injury.

REFERENCES

Anderson, T., Forlin, L., Hardig, J. and Larsson, A., Biochemical and physiological disturbances in fish inhabiting coastal waters polluted with bleached kraft mill effluents, *Mar. Environ. Res.*, 24, 233, 1988.

Barton, B., Schreck, C. and Fowler, L., Fasting and diet content affect stress-induced changes in plasma glucose and cortisol in juvenile chinook salmon, *Prog. Fish Cult.*, 50, 16, 1988.

Bass, M. I., Berry, C. R. and Heath, A. G., Histopathological effects of intermittent chlorine exposure on bluegill (*Lepomis macrochirus*) and rainbow trout (*Salmo gairdneri*), *Water Res.*, 11, 731, 1977.

Benitez, L. V. and Halver, J. E., Ascorbic acid sulfate sulfohydrolase (C2 sulfatase): the modulator of cellular levels of L-ascorbic acid in rainbow trout, *Proc. Natl. Acad. Sci. U.S.A.*, 79, 5445, 1982.

Binder, R. L. and Stegeman, J. J., Basal levels and induction of hepatic aryl hydrocarbon hydroxylase activity during the embryonic period of development in brook trout, *Biochem. Pharmacol.*, 32, 1324, 1983.

Boyer, J. L., Schwarz, J. and Smith, N., Selective hepatic uptake and biliary excretion of S-sulfobromophthalein in marine elasmobranchs, *Gastroenterology*, 70, 254, 1976.

Brown, C., Young, G., Nishioka, R. and Bern, H., Preliminary report on the physiological status of striped bass in the Carquinez Strait die-off, *Fish. Res.*, 6, 5, 1987.

Cashman, J., Maltby, D., Nishioka, R., Bern, H., Gee, S. and Hammock, B., Chemical contamination and the annual summer die-off of striped bass (*Morone saxatillis*) in the Sacramento-San Joaquin delta, *Chem. Res. Toxicol.*, 5, 100, 1992.

Casillas, E. and Smith, L. S., Effect of stress on blood coagulation and haematology in rainbow trout (*Salmo gairdneri*), *J. Fish Biol.*, 10, 481, 1977.

Casillas, E., Myers, M. and Ames, W. E., Relationship of serum chemistry values to liver and kidney histopathology in English sole (*Parophrys vetulus*) after acute exposure to carbon tetrachloride, *Aquat. Toxicol.*, 3, 61, 1983.

Chapman, G. B., Ultrastructure of the liver of the fingerling rainbow trout, *Salmo gairdneri* Richardson, *J. Fish Biol.*, 18, 553, 1981.

Christensen, G. M., Effects of metal cations and other chemicals upon the *in vitro* activity of two enzymes in the blood plasma of the white sucker, *Catostomus commersoni* (Lacepede), *Chem. Biol. Interact.*, 4, 351, 1971–72.

Christensen, G., Hunt, E. and Fiandt, J., The effect of methylmercuric chloride, cadmium chloride and lead nitrate on six biochemical factors of the brook trout (*Salvelinus fontinalis*), *Toxicol. Appl. Pharmacol.*, 42, 523, 1977.

Dixon, D. G. and Hilton, J. W., Influence of available dietary carbohydrate content on tolerance of waterborne copper by rainbow trout, *J. Fish Biol.*, 19, 509, 1981.

Dixon, D. G., Hodson, P. V. and Kaiser, K. L., Serum sorbitol dehydrogenase activity as an indicator of chemically induced liver damage in rainbow trout, *Environ. Toxicol. Chem.*, 6, 685, 1987.

Dutta, H. and Haghigh, A., Methylmercuric chloride and serum cholesterol level in the bluegill (*Lepomis macrochirus*), *Bull. Environ. Contam. Toxicol.*, 36, 181, 1986.

Everaarts, J., Shugart, L., Gustin, M., Hawkins, W. and Walker, W., Biological markers in fish: DNA integrity, hematological parameters and liver somatic index, *Mar. Environ. Res.*, 35, 101, 1993.

Fairbanks, U. F., Copper sulfate-induced hemolytic anemia, *Arch. Intern. Med.*, 120, 428, 1967.

Fletcher, G., King, M., Kiceniuk, J. and Addison, R., Liver hypertrophy in winter flounder following exposure to experimentally oiled sediments, *Comp. Biochem. Physiol.*, 73C, 457, 1982.

Ghazaly, K. S., Physiological alterations in *Clarias lazera* induced by two different pollutants, *Water, Air, Soil Pollut.*, 60, 181, 1991.

Gingerich, W. H., Hepatic toxicology of fishes, in, *Aquatic Toxicology,* Weber, L., Ed., Raven Press, New York, 1982, 55.

Gingerich, W. and Weber, L., Assessment of clinical procedures to evaluate liver intoxication in fish, *Ecol. Res. Ser.,* EPA, 600 3-79-088, 1979.

Gingerich, W. H., Weber, L. J. and Larson, R. E., Hepatic accumulation, metabolism and biliary excretion of sulfobromophthalein in rainbow trout *(Salmo gairdneri), Comp. Biochem. Physiol.,* 58C, 113, 1977.

Gingerich, W. H., Weber, L. J. and Larson, R. E., The effect of carbon tetrachloride on hepatic accumulation, metabolism, and biliary excretion of sulfobromophthalein in rainbow trout, *Toxicol. Appl. Pharmacol.,* 43, 159, 1978.

Gluth, G. and Hanke, W., A comparison of physiological changes in carp, *Cyprinus carpio,* induced by several pollutants at sublethal concentration, *Ecotoxicol. Environ. Safety,* 9, 179, 1985.

Goede, R. and Barton, B., Organismic indices and an autopsy-based assessment as indicators of health and condition of fish, *Am. Fish. Soc. Symp.,* 8, 93, 1990.

Groman, D. B., *Histology of The Striped Bass,* American Fisheries Society Monograph No. 3, 1982.

Hacking, M. A., Budd, J. and Hodson, K., The ultrastructure of the liver of the rainbow trout: normal structure and modifications after chronic administration of a polychlorinated biphenyl Aroclor 1254, *Can. J. Zool.,* 56, 477, 1978.

Halver, J. E., Smith, R. R., Tolbert, B. M. and Baker, E. M., Utilization of ascorbic acid in fish, *Ann. N.Y. Acad. Sci.,* 258, 81, 1975.

Heath, A. G., Changes in tissue adenylates and water content of bluegill, *Lepomis macrochirus,* exposed to copper, *J. Fish Biol.,* 24, 299, 1984.

Hilmy, A. M., Shabana, M. B. and Said, M. M., The role of serum transaminases (SGO-T and SGP-T) and alkaline phosphatase in relation to inorganic phosphorus with respect to mercury poisoning in *Aphanius dispar* (Teleostei) of the red sea, *Comp. Biochem. Physiol.,* 68C, 69, 1981.

Hilmy, A. M., Badawi, H. K. and Shabana, M. B., Physiological mechanism of toxic action of DDT and Endrin in two euryhaline freshwater fishes, *Anguilla vulgaris* and *Mugil cephalus, Comp. Biochem. Physiol.,* 76C, 173, 1983.

Hilton, J. W., Cho, C. Y. and Slinger, S. J., Evaluation of the ascorbic acid status of rainbow trout *(Salmo gairdneri), J. Fish. Res. Bd. Can.,* 34, 2207, 1977.

Hinton, D. E., Lantz, R. C., Hampton, J., McCuskey, P. and McCuskey, R., Normal versus abnormal structure: considerations in morphologic responses of teleosts to pollutants, *Environ. Health Perspect.,* 71, 139, 1987.

Hinton, D. E., Baumann, P., Gardner, G., Hawkins, W., Hendricks, J., Murchelano, R. and Okihiro, M., Histopathologic biomarkers, in, *Biomarkers, Biochemical, Physiological, and Histological Markers of Anthropogenic Stress,* Hugett, R., Kimerle, R., Mehrle, P. and Bergman, H., Eds., Lewis Publishers Boca Raton, FL, 1992, chap. 4.

Inui, Y., Mechanism of the increase of plasma glutamic oxalacetic transaminase and plasma glutamic pyruvic transaminase activities in acute hepatitis of the eel, *Bull. Freshw. Fish. Res. Lab.,* 19, 25, 1969.

Jackim, E., Hamlin, J. M. and Sonis, A., Effects of metal poisoning on five liver enzymes in the killifish *(Fundulus heteroclitus), J. Fish. Res. Bd. Can.,* 27, 383, 1970.

Khangarot, B. S., Copper-induced hepatic ultrastructural alterations in the snake-headed fish, *Channa punctatus, Ecotoxicol. Environ. Safety,* 23, 282, 1992.

Kiceniuk, J. and Khan, R., Effect of petroleum hydrocarbons on Atlantic cod, *Gadius morhua,* following chronic exposure, *Can. J. Zool.,* 65, 490, 1987.

Kirubagaran, R. and Joy, K., Toxic effects of mercury on testicular activity in the freshwater teleost, *Clarias batrachus* L., *J. Fish Biol.,* 41, 305, 1992.

Kohler, A., Regeneration of contaminant-induced liver lesions in flounder — experimental studies toward the identification of cause-effect relationships, *Aquat. Toxicol.,* 14, 203, 1989.

Kohler, A., Identification of contaminant-induced cellular and subcellular lesions in the liver of flounder *(Platichthys flesus)* caught at differently polluted estuaries, *Aquat. Toxicol.,* 16, 271, 1990.

Kohler, A., Lysosomal perturbations in fish liver as indicators for toxic effects of environmental pollution, *Comp. Biochem. Physiol.,* 100C, 123, 1991.

138

Kristoffersson, R., Broberg, S., Oikari, A. and Pekkarinen, M., Effect of a sublethal concentration of phenol on some blood plasma enzyme activities in the pike *(Esox lusius)* in brackish water, *Ann. Zool. Fennici*, 11, 220, 1974.

Lanno, R., Hicks, B. and Hilton, J., Histological observations on intrahepatocytic copper-containing granules in rainbow trout reared on diets containing elevated levels of copper, *Aquat. Toxicol.*, 10, 251, 1987.

Lee, R., Gerking, S. and Jezierska, B., Electrolyte balance and energy mobilization in acid-stressed rainbow trout, *Salmo gairdneri*, and their relation to reproductive success, *Environ. Biol. Fishes*, 8, 115, 1983.

Lehninger, A. L., *Biochemistry*, Second ed., Worth Publishers, New York, 1979.

Leland, H. V., Ultrastructural changes in the hepatocytes of juvenile rainbow trout and mature brown trout exposed to copper or zinc, *Environ. Toxicol. Chem.*, 2, 353, 1983.

Lopez-Torres, M., Perez-Campo, R., Cadenas, S., Rojas, C. and Barja, G., A comparative study of free radicals in vertebrates. II. Non-enzymatic antioxidants and oxidative stress, *Comp. Biochem. Physiol.*, 105, 757, 1993.

Mattsoff, L. and Oikari, A., Acute hyperbilirubinaemia in rainbow trout *(Salmo gairdneri)* caused by resin acids, *Comp. Biochem. Physiol.*, 88C, 263, 1987.

Mayer, F. L., Mehrle, P. M. and Crutcher, P. L., Interactions of toxaphene and vitamin C in channel catfish, *Trans. Am. Fish. Soc.*, 107, 326, 1978.

McKim, J. M., Christensen, G. M. and Hunt, E. P., Changes in the blood of brook trout *(Salvelinus fontinalis)* after short-term and long-term exposure to copper, *J. Fish. Res. Bd. Can.*, 27, 1883, 1970.

Meyers, M., Olson, O., Johnson, L., Stehr, C., Hom, T. and Varanasi, U., Hepatic lesions other than neoplasms in subadult flatfish from Puget Sound, Washington: relationships with indices of contaminant exposure, *Mar. Environ. Res.*, 34, 45, 1992.

Moore, M. N., Cellular responses to pollutants, *Mar. Pollut. Bull.*, 16, 134, 1985.

Murkherjee, D., Guha, D., Kumar, V. and Chakrabarty, S., Impairment of steroidogenesis and reproduction in sexually mature *Cyprinus carpio* by phenol and sulfide under laboratory conditions, *Aquat. Toxicol.*, 21, 29, 1991.

Myant, N. B., *The Biology of Cholesterol and Related Steroids*, W. Heinemann Medicine Books, London, 1981.

Oikari, A. and Nakari, T., Kraft pulp mill effluent components cause liver dysfunction in trout, *Bull. Environ. Toxicol. Chem.*, 28, 266, 1982.

Oikari, A., Holmbom, B., Nas, E., Millunpalo, M., Kruzynski, G. and Castren, M., Ecotoxicological aspects of pulp and paper mill effluents, discharged to an inland water system: distribution in water, toxicant residues, and physiological effects in caged fish *(Salmo gairdneri), Aquat. Toxicol.*, 6, 219, 1985.

Ozretic, B. and Krajnovic-Ozretic, M., Plasma sorbitol dehydrogenase, glutamate dehydrogenase, and alkaline phosphatase as potential indicators of liver intoxication in grey mullett *(Mugil auratus), Bull. Environ. Contam. Toxicol.*, 50, 586, 1993.

Patton, J. and Couch, J., Can tissue anomalies that occur in marine fish implicate specific pollutant chemicals?, in, *Concepts in Marine Pollution Measurements*, White, H., Ed., University of Maryland, 1984, 511.

Poels, C., van der Gaag, M. and van de Kerkhoff, J., An investigation into the long-term effects of Rhine water on rainbow trout, *Water Res.*, 14, 1029, 1980.

Rabergh, C., Isomaa, B. and Eriksson, J., The resin acids dehydroabietic acid and isopimaric acid inhibit bile acid uptake and perturb potassium transport in isolated hepatocytes from rainbow trout *(Oncorhynchus mykiss), Aquat. Toxicol.*, 23, 169, 1992.

Ram, R. and Singh, R. S., Carbofuran-induced histopathological and biochemical changes in liver of the teleost fish, *Channa punctatus* (Boch), *Ecotoxicol. Environ. Safety*, 16, 194, 1988.

Recknagel, R. and Glende, E., Lipid peroxidation: a specific form of cellular injury, in, *Handbook of Physiology: Reactions to Environmental Agents*, Section 9, Lee, D. H. K., Ed., American Physiological Society, Bethesda, MD, 1977, 591.

Roncero, V., Duran, E., Soler, F., Masot, J. and Gomez, L., Morphometric, structural, and ultrastructural studies of tench, *Tinca tinca*, hepatocytes after copper sulfate administration, *Environ. Res.*, 57, 45, 1992.

Schar, M., Maly, I. and Sasse, D., Histochemical studies on metabolic zonation of the liver in the trout *(Salmo gairdneri), Histochemistry*, 83, 147, 1985.

Schoor, W. P. and Couch, J. A., Correlation of mixed function oxidase activity with ultrastructural changes in the liver of a marine fish, *Cancer Biochem. Biophys.*, 4, 95, 1979.

Schreck, C. B. and Lorz, H. W., Stress response of coho salmon *(Oncorhynchus kisutch)* elicited by cadmium and copper and potential use of cortisol as an indicator of stress, *J. Fish. Res. Bd. Can.*, 35, 1124, 1978.

Singh, H. S. and Reddy, T., Effect of copper sulfate on hematology, blood chemistry and hepato-somatic index of an Indian catfish, *Heteropneustes fossilis*, and its recovery, *Ecotoxicol. Environ. Safety*, 20, 30, 1990.

Slooff, W., van Kreijl, C. and Baars, A., Relative liver weights and xenobiotic-metabolizing enzymes of fish from polluted surface waters in The Netherlands, *Aquat. Toxicol.*, 4, 1, 1983.

Stevens, E. D., The effect of exercise on the distribution of blood to various organs in rainbow trout, *Comp. Biochem. Physiol.*, 25, 614, 1968.

Tewari, H., Gill, T. and Pant, J., Impact of chronic lead poisoning on the hematological and biochemical profiles of a fish, *Barbus conchonius, Bull. Environ. Contam. Toxicol.*, 38, 748, 1987.

Thomas, P., Molecular and biochemical responses of fish to stressors and their potential use in enviromental monitoring, *Am. Fish. Soc. Symp.*, 8, 145, 1990.

Thomas, P., Carr, R. S. and Neff, J. M., Biochemical stress responses of mullet *Mugil cephalus* and polychaete worms *Neanthes virens* to pentachlorophenol, in, *Biological Monitoring of Marine Pollutants*, Vernberg, J., Calabrese, A., Thurberg, F. P. and Vernberg, W. B., Eds., Academic Press, New York, 1981, 73.

Thomas, P., Bally, M. and Neff, J. M., Ascorbic acid status of mullet, *Mugil cephalus* Linn., exposed to cadmium, *J. Fish Biol.*, 20, 183, 1982.

Thomas, P. and Neff, J. M., Effects of pollutant and other environmental variables on the ascorbic acid content of fish tissues, *Mar. Environ. Res.*, 14, 489, 1984.

van den Thillart, G., Kesbeke, F. and van Waarde, A., Anaerobic energy-metabolism of goldfish, *Carassius auratus* (L): influence of hypoxia and anoxia on phosphorylated compounds and glycogen, *J. Comp. Physiol.*, 136, 45, 1980.

Verma, S. R., Tonk, I. P. and Dalela, R. C., Effects of a few xenobiotics on three phosphatases of *Saccobranchus fossilis* and the role of ascorbic acid in their toxicity, *Tox. Lett.*, 10, 287, 1982.

Versteeg, D. J. and Giesy, J. P., Lysosomal enzyme release in the bluegill sunfish *(Lepomis macrochirus)* exposed to cadmium, *Arch. Environ. Contam. Toxicol.*, 14, 631, 1985.

Versteeg, D. J. and Giesy, J. P., The histological and biochemical effects of cadmium exposure in the bluegill sunfish *(Lepomis macrochirus)*, *Ecotoxicol. Environ. Safety*, 11, 31, 1986.

Vignier, V., Vandermeulen, J. and Fraser, A., Growth and food conversion by Atlantic salmon parr during 40 days exposure to crude oil, *Trans. Am. Fish. Soc.*, 121, 322, 1992.

Wedemeyer, G., Stress-induced ascorbic acid depletion and cortisol production in two salmonid fishes, *Comp. Biochem. Physiol.*, 29A, 1247, 1969.

Weis, P., Ultrastructural changes induced by low concentrations of DDT in the livers of zebrafish and guppy, *Chem. Biol. Interact.*, 8, 25, 1974.

Wieser, W. and Hinterleitner, S., Serum enzymes in rainbow trout as tools in the diagnosis of water quality, *Bull. Environ. Contam. Toxicol.*, 25, 188, 1980.

Winston, G. and Diguilo, R., Prooxidant and antioxidant mechanisms in aquatic organisms, *Aquat. Toxicol.*, 19, 137, 1991.

Wood, C. M., Ammonia and urea metabolism and excretion, in, *The Physiology of Fishes*, Evans, D. H., Ed., CRC Press, Boca Raton, FL, 1993, chap. 13.

Yamamoto, Y., Hayama, K. and Ikeda, S., Effect of dietary ascorbic acid on the copper poisoning in rainbow trout, *Bull. Jpn. Soc. Sci. Fish.*, 47, 1085, 1981.

Yamamoto, Y., Sato, M. and Ikeda, S., Existence of L-gluconolactone oxidase in some teleosts, *Bull. Jpn. Soc. Sci. Fish.*, 44, 775, 1978.

Yarbrough, J. D., Neitz, J. R. and Chambers, J. O., Physiological effects of crude oil exposure in the striped mullet *Mugil cephalus*, *Life Sci.*, 19, 775, 1976.

Young, G., Brown, C. L., Nishioka, R. S., Folmar, L. C., Andrews, M., Cashman, J. R. and Bern, H. A., Histopathology, blood chemistry, and physiological status of normal and moribund striped bass *(Morone saxatilis)* involved in summer mortality ("die-off") in the Sacramento-San Joaquin Delta of California, *J. Fish Biol.*, 44, 491, 1994.

Chapter 7

Osmotic and Ionic Regulation

I. OVERVIEW OF SOME GENERAL PRINCIPLES

Osmoregulation refers to the processes by which the osmotic pressure of the body fluids and the water volume in an animal are held relatively constant. The osmotic pressure of fish blood is mostly provided by inorganic salts and is approximately one third that of seawater, with marine fish having slightly more salt in their blood than freshwater species. Because gills are permeable to water as well as oxygen, CO_2, etc., there is an osmotic flow of water out of the marine fish because the salt concentration of its blood is less than that of the ocean, and into the freshwater fish because of an opposite osmotic gradient. The diffusion gradient for sodium chloride (the primary salt) is in the direction opposite to the osmotic diffusion of water across the gill epithelium. Figure 1 illustrates these major ion and water fluxes in typical teleost fish.

The water lost by diffusion from a marine teleost is replaced by the animal drinking an equal amount of seawater. In order for the swallowed seawater to actually enter the blood from the gut lumen, there must be active transport of NaCl from the lumen into the intracellular space which sets up a local osmotic gradient for water movement into the blood (Kirschner, 1991). Due to this uptake of salt from the gut and the passive diffusion of salt into the animal, marine fish must constantly excrete a considerable amount of salt. This is done through the gill and opercular epithelia by way of special "chloride cells" (Payan and Girard, 1984). These cells are well endowed with mitochondria to generate ATP and the enzyme Na,K-activated ATPase. The movement of sodium and chloride out of marine fish by way of these cells is a complex process involving a combination of electrical potential (inside positive) and an Na/K exchange. Sodium and chloride ions are transported separately: sodium goes through the leaky paracellular junctions, chloride directly through the chloride cells (Evans, 1993).

Freshwater fish have the opposite problem to their marine cousins in that excess water gained from osmosis must be excreted. This is done by producing a relatively large volume (5 ml/kg/h) of dilute urine. Because of a loss of salt from diffusion across the gills and skin and a small loss via the urine, these fish must actively transport salt from the water into the blood. This is presumably done by chloride cells in the gills (but not the operculum), and possibly other gill epithelial cells as well, again utilizing energy from ATP and catalyzed by Na,K-activated ATPase. The chloride cells in freshwater fish

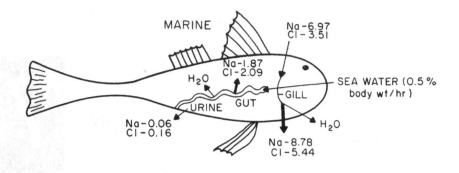

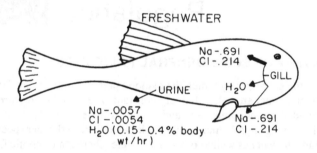

Figure 1 Comparison of ion and water fluxes in a marine and a freshwater fish. Ion fluxes are in mM/kg/h. Thin arrows indicate passive processes, thick ones active transport. (Data from Fletcher, C. R., *J. Comp. Physiol.*, 124, 157, 1978; Eddy, F. B. and Bath, R. N., *J. Exp. Biol.*, 83, 181, 1979).

transport sodium and chloride independently of each other. These electrolyte ions in the water are exchanged for ammonia, hydrogen ions, and bicarbonate in the blood thus aiding the excretion of these waste products (Payan and Girard, 1984).

Recently, it has been demonstrated that the adsorption of sodium to the gill surface plays an important role in presenting sodium ions to the chloride cells (Handy and Eddy, 1991a). At low external sodium concentrations, as would be found in freshwater conditions, calcium inhibits this adsorption process.

The ionic fluxes for marine fish are approximately an order of magnitude greater than those for freshwater species (Figure 1). This reflects the greater permeability of the marine fish to these ions. In order to achieve the high rate of salt excretion required by such large fluxes, they have more chloride cells than do their freshwater relatives. It has, however, been reported (Lauren and McDonald, 1985) that trout acclimated to soft water have more chloride cells than those acclimated to hard (i.e., high calcium) water. There are also marked species differences even among the freshwater fishes in ability to transport electrolytes.

The differences in flux rates between marine and freshwater fish may actually be greater for some species. Most comparisons are based on trout as the freshwater species, whereas recently it was reported that gill permeabilities for electrolytes and water are greater in trout than in other less active species such as yellow perch (*Perca flavescens*) and smallmouth bass (*Micropterus dolomieui*) (McDonald et al., 1991a). These greater fluxes in the trout are facilitated by a greater density of chloride cells and the individual cells have a greater transport capacity when compared with a species such as the carp (*Cyprinus carpio*) (Williams and Eddy, 1988a).

Euryhaline fish, such as the well investigated *Fundulus heteroclitus,* have the ability to reverse the active movement of sodium and chloride across the gill tissue, to change the amount and composition of urine excreted, and to alter epithelial permeability, all within a matter of hours. Anadromous and catadromous species, such as eels and salmon, also do this reversal "act" but at a slower pace (i.e., days instead of hours).

The parr-smolt transformation in salmonids is a form of metamorphosis involving physiological changes that produce an increased salinity tolerance (i.e., conversion into a marine form). This is done in "anticipation of seaward migration" and is keyed to photoperiod, environmental temperature, and perhaps an endogenous "clock" mechanism (reviewed by Boeuf, 1993).

Acclimation temperature seemingly has little effect on blood electrolyte or osmolality setpoints in either marine or freshwater fish (Burton, 1986) except for a tendency for the setpoint to rise at temperatures near freezing in marine forms and fall in freshwater species. This may, at least in some cases, represent a form of osmoregulatory failure at temperatures near the extremes of the normal ranges. Rapid temperature changes, and especially a sudden chill, can produce decreased plasma electrolyte levels in freshwater species (see Eddy, 1981 for review).

Regulation of the various osmoregulatory processes is under hormonal control and is only beginning to be understood (Bern and Madsen, 1992). Cortisol and prolactin are probably the most important of the hormones controlling osmoregulation. In general, cortisol promotes the gain of salt (and retention) in freshwater and its loss in seawater (see Eddy, 1981). It achieves this by acting on Na^+/K^+ ATPase in the kidney, intestine, and urinary bladder as well as the gills (Madsen, 1990). Prolactin acts to reduce membrane permeability to ions and water. It also inhibits chloride secretion in marine species so it is important in the adjustment to freshwater by anadromous fishes (Foskett et al., 1983). In euryhaline teleosts, prolactin is the key hormone for adjusting to freshwater, whereas cortisol is the main one involved in adjustment to seawater (Bern and Madsen, 1992).

Adrenalin has been shown to stimulate uptake of sodium from freshwater in an *in vitro* preparation (Payan et al., 1975), but in seawater-adapted trout, it inhibits salt secretion (Girard and Payan, 1980). Adrenalin also increases water permeability in the gills of freshwater fish, presumably as a by-product of the increased oxygen permeability of the gills associated with stress (Isia, 1984). Thus, adrenalin might not be considered as being adaptive for osmotic and ionic regulation as it may actually exacerbate the problem of living in freshwater.

A variety of other hormones have been shown to affect chloride cells. These include somatostatin, urotinsins (from the caudal neurosecretory system), glucagon, and vasoactive intestinal polypeptide. In seawater teleosts, the first three inhibit secretion of chloride while vasoactive intestinal polypeptide and glucagon stimulate it (Foskett et al., 1983). Thyroid hormones (T3 and T4) are instrumental in the physiological preparations of salmonids for moving into seawater during the parr-smolt transformation (Grau et al., 1981). Leatherland (1985) has also found that T4 injection reduces the elevation in plasma sodium seen in rainbow trout abruptly transferred from freshwater to seawater. Other hormones with effects superficially similar to T4 are the atrial natriuretic factor (Arnold-Reed and Balment, 1991), which presumably comes from the heart, and a group of hormones normally associated with regulation of growth: insulin-like growth factor-I, growth hormone, and insulin (McCormick et al., 1991). Some of this latter group of hormones may exert their osmoregulatory effects indirectly through stimulation of cortisol secretion. Growth hormone is especially important in promoting seawater adaptation in salmonids (Bern and Madsen, 1993; Boeuf, 1993).

Aldosterone, a major ion regulatory hormone in terrestrial vertebrates, may have relatively little importance for teleost fish (Reinking, 1983). Interestingly, the renin-angiotensin system, which in terrestrial animals is largely responsible for the regulation of aldosterone levels, stimulates drinking in seawater teleosts and may thus be necessary for normal adaptation to seawater (Carrick and Balment, 1983). It also appears to stimulate urination in freshwater species, perhaps because it elevates blood pressure which would cause increased filtration by the kidney (Arillo et al., 1981; Olson, 1992).

Along with the problem of maintaining internal osmotic pressure and monovalent electrolyte cation ratios, a fish must also regulate divalent cations, most particularly calcium. This element is critical in the maintenance of normal excitability in nerve and muscle cells, stabilization of the branchial permeability, intracellular regulatory processes, and it also plays a roll in reproduction, among other functions (Hunn, 1985). In the ocean, fish are in a calcium-rich environment, while freshwater forms may or may not be in a calcium-poor environment depending on the water hardness.

Higher vertebrates utilize bone as a calcium pool to help modulate plasma calcium changes and this whole process is regulated by hormones from the parathyroid glands. Cyclostomes have no calcified skeleton and Chondrichthyes have only calcified cartilage which is in simple equilibrium with body fluids (Taylor, 1985), whereas the skeleton of teleosts has considerable calcium in it. Taylor (1985) concludes that unlike higher animals, it cannot undergo much remodeling and therefore serves only as a slight buffer of plasma calcium and phosphate concentrations. The calcium pool that can be exchanged with the environment represents only about 3% of the total body calcium content (Hobe et al., 1984).

The rate of calcium transport from the water into the blood of freshwater fish is only about 3–10% of that of sodium and chloride (Hobe et al., 1984). Gills have been considered to be by far the most important site, even for marine fish which drink large amounts of calcium-rich water (Taylor, 1985). Sayer et al. (1991), however, have reported that brown trout fry took up over 90% of their calcium from the intestine, perhaps because they were in soft water and the calcium would have been largely dietary.

As in mammals, the blood calcium concentration is under hormonal regulation, but there is a good bit of variation between different species of fish as to the normal concentration of blood calcium (Hunn, 1985). Indeed, individual fish can tolerate rather large variations in plasma calcium and it is not regulated as tightly as in tetrapods (Bern and Madsen, 1992). The hormone calcitonin from the ultimobranchial bodies causes a lowering of blood calcium by inhibition of Ca uptake in the gills. The Corpuscles of Stannius, located on the surface of the posterior kidneys, produce the hormone stanniocalcin which lowers the calcium concentration in the blood.

Whereas the above-mentioned hormones all cause a decrease in plasma calcium, prolactin from the anterior pituitary stimulates uptake of calcium from the water by the gills (Flick et al., 1984). However, in trout, prolactin is reputed to have little effect. Instead, uptake of calcium from the environment in trout is stimulated by cortisol (Flick and Perry, 1989). The mechanisms by which these hormones exert their effects are at present unclear, but the gills and kidneys are most likely the target organs and bone absorption or release of calcium seems rather unimportant (Taylor, 1985).

Obviously, the hormonal control mechanisms of electrolyte levels in fish are complex and interrelated with other functions. For example, cortisol and adrenalin are important stress hormones which affect a wide variety of functions in addition to osmotic and ionic homeostasis. Thus, changes in hormonal levels under various environmental conditions may be in response to altered blood ionic composition and/or to other homeostatic challenges. Little is known about how the secretion rate of the various osmoregulatory hormones is regulated.

II. EFFECTS OF POLLUTANTS ON OSMOTIC AND IONIC REGULATION

Changes in osmoregulation in fish exposed to some stressor have been generally elucidated by measuring the blood plasma (or serum) electrolytes and/or total osmolality. The mechanisms of any changes in blood plasma composition are investigated by measuring flux rates of electrolytes through gills or kidney using radioactive isotopes and by assaying the rate of activity of the various cellular ATPases. Handling of a fish can, however, cause changes in osmoregulatory parameters independent of environmental pollutants, due to the phenomenon of handling diuresis (i.e., excess urination), and changes in hormonal levels (Eddy, 1981; Ellsaesser and Clem, 1987). These changes may last for several hours or even days. Other laboratory stresses and temperature changes also affect osmoregulation in both marine and freshwater forms (Eddy, 1981, 1982; Sleet and Weber, 1983; Waring et al., 1992). Thus, there is a need for control fish that receive exactly the same treatment as the experimentals, except for exposure to the test pollutants.

Recent reviews on certain effects of pollutants on osmoregulation are Wendelaar Bonga and Lock (1992) and Lauren (1991). Both correctly emphasize effects on gill function including permeability and active ion uptake. The trends that have been observed in osmoregulatory and ionoregulatory function when fish are exposed to a toxicant are summarized in Table 1. What follows are some observations and elaboration of some of the information in that table with an emphasis on the mechanisms involved.

A. COPPER

Copper causes a comparatively large upset in osmoregulation in freshwater fish; exposed fish exhibit a rather rapid decrease in plasma electrolytes and/or osmolality. The cause of death from acute exposure in freshwater fish appears to be a sequence of events which start with the electrolyte loss. If this occurs rapidly enough (as it would with acute exposures), a massive hemoconcentration follows which in turn causes a big increase (ca. 2×) in arterial blood pressure. Finally, the heart apparently fails from having to deal with the viscous blood and excessive pressure head (Wilson and Taylor, 1993a). This sequence of physiological dysfunctions is very similar to that seen in fish dying from acid exposure (Milligan and Wood, 1982, also see Chapter 10).

The mechanism of the copper effect on osmoregulation in freshwater appears to be an inhibition of sodium and chloride uptake by the gills, although at fairly high copper concentrations a stimulation of passive electrolyte efflux also occurs (McDonald et al., 1988). The inhibition of electrolyte uptake is probably due to an inhibition by copper *in vivo* of the enzyme Na,K ATPase in gill tissue (Lorz and McPherson, 1976). The decreased plasma electrolytes in fish exposed to copper also causes a movement of water from the blood into the tissues (Heath, 1984) which would facilitate the hemoconcentration observed.

Decreasing alkalinity, pH, or hardness of the water enhance the toxicity of copper to freshwater fish (Miller and Mackay, 1980). Lauren and McDonald (1986) showed that the water alkalinity had a considerable effect on the osmoregulatory dysfunction induced by copper whereas hardness had little effect in 24-h exposures. However, for chronic exposures, hardness may become far more important because calcium has such a strong influence on gill permeability, but it requires several days for a change in environmental calcium to alter gill permeability (Lauren and McDonald, 1986).

Prolonged exposure to low copper concentrations may result in a decrease in blood electrolyte levels initially, but then an adaptation to the toxicant occurs and the electrolyte levels may recover (McKim et al., 1970; Christensen et al., 1972). Acclimation to sublethal concentrations of copper makes the fish better able to tolerate subsequent higher doses (Buckley et al., 1982). The acclimation process is fairly complex and time dependent (Lauren and McDonald, 1987a). Initially, there appears to be a decrease in permeability of the gills to sodium which reduces the efflux to below that of control fish in non-contaminated water. Later, the rate of sodium influx shows a return toward normal due to a synthesis of more Na,K ATPase in the gills, presumably to compensate for the copper inhibition of this enzyme (Lauren and McDonald, 1987b).

Table 1 Osmoregulatory Effects of Water Pollutants

Substance and exposure	Type of water	Effect observed	Species	Ref.
Copper				
Acute	FW	Decreased plasma osmolality	Golden shiner	Lewis and Lewis, 1971
		Decreased plasma osmolality	Channel catfish	Lewis and Lewis, 1971
		Increased blood volume	Striped bass	Courtois and Meyerhoff, 1975
		Decreased Na, K, Cl uptake	Rainbow trout	Lauren and McDonald, 1985
		Increased tissue water	Bluegill	Heath, 1984
		Decreased plasma Na, Cl	Rainbow trout	Wilson and Taylor, 1993a
	SW	Increased K, Na, Mg, Cl	Sheepshead	Caudelhac et al., 1972
	BR	Increased plasma Na, K, Ca, Cl	Rainbow trout	Wilson and Taylor, 1993b
	SW	No effect plasma ions	Rainbow trout	Wilson and Taylor, 1993
6,21 d	FW	Decrease Cl and osmolality	Brook trout	McKim et al., 1970
337 d	FW	No effect	Brook trout	McKim et al., 1970
7–42 d	FW	Decreased Na, Cl	Flounder	Stagg and Shuttleworth, 1982
	SW	Increase Na, K, Mg, Cl	Flounder	Stagg and Shuttleworth, 1982
14 d	SW	Decreased Ca uptake	Flounder	Dodoo et al., 1992
1–30 d	FW	Increased Na, K	Indian catfish	Singh and Reddy, 1990
30 d	FW	Decreased Cl	Brown bullhead	Christensen et al., 1972
1–172 d	FW	Reduced SW survival and migration	Coho salmon	Lorz and McPherson, 1976
1–172 d	FW	Inhibition gill Na, K-ATPase	Coho salmon	Lorz and McPherson, 1976
Zinc				
Acute	FW	No effect plasma electrolytes	Rainbow trout	Skidmore, 1970
	FW	Decreased Cl, Ca; no effect Na, K	Rainbow trout	Spry and Wood, 1985
	FW	Decreased plasma osmolality	Channel catfish	Lewis and Lewis, 1970
72 h	FW	Decreased Ca; increased electrolyte efflux	Rainbow trout	Spry and Wood, 1985
30 d	FW	No effect osmolality or ions; increased gill Na, K- and Mg-ATPase	Rainbow trout	Watson and Beamish, 1980
Cadmium				
1–178 d	FW	Electrolyte movement into tissues; no effect kidney function	Rainbow trout	Giles, 1984
25–50 d	FW	Decreased Na; increased muscle Na	Goldfish	McCarty and Houston, 1976
2–8 wk	FW	Increased plasma Na	Brook trout	Christensen et al., 1977
4 days posthatch	FW	Decreased Na, K, Ca, water uptake	Atlantic salmon	Rombough and Garside, 1982
20–180 h	FW	Time-dependent decline plasma Ca	Rainbow trout	Roch and Maly, 1979
2–35 d	FW	Transient Ca decrease and Mg increase; no effect Na, K, osmolality	Tilapia	Pratap et al., 1989
Dietary	FW	Degeneration gill pavement cells and chloride cell activity	Tilapia	Pratap et al., 1989

Table 1 (continued) Osmoregulatory Effects of Water Pollutants

Substance and exposure	Type of water	Effect observed	Species	Ref.
2–35 d	FW	Decreased osmolalitys; transient decrease Na, Ca; increase prolactin	Tilapia	Fu et al., 1989
Acute	SW	Increased Na, Cl	Cunner	Thurberg and Dawson, 1974
4–9 wk	BR	No effect NaCl; decreased K, Ca	Flounder	Larsson et al., 1981
Mercury				
4 h	FW	Decreased Cl influx; increased NaCl efflux	Lamprey	Stinson, 1989
1 wk	FW	Decrease Na, Cl; slight increase K	Rainbow trout	Lock et al., 1981
2–8 wk	FW	Increase NaCl	Brook trout	Christensen et al., 1977
45–180 d	FW	Decreased cortisol	Indian Catfish	Kirubagaran and Joy, 1991
60 d	SW	Increase osmolality; decrease Ca	Flounder	Dawson, 1979
Injected	SW	No effect osmolality or ions	Flounder	Schmidt-Nielsen et al., 1977
Chromium				
Acute	FW	Decreased NaCl and osmolality	Coho salmon	Van der Putte et al., 1982
2–4 wk	FW	Reduced survival in SW	Coho salmon	Sugatt, 1980
Lead				
1 mo; 49 d recovery	BR	Increase K; decrease Na; no effect Ca	Rainbow trout	Haux and Larsson, 1982
2–8 wk	FW	Increase NaCl	Brook trout	Christensen et al., 1977
Field	FW	Low Na, No effect K	Whitefish	Haux et al., 1985
Aluminum				
Acute or chronic	FW	Marked ionregulatory dysfunction in soft acid water but not in hard water At pH 7 ionoregulatory dysfunction slight if at all	Mostly salmonids	See text
Tin				
14 d	BR	No effect Na, K, Ca, Mg; increased gill Na, K- and Mg ATPases	Striped bass	Pickney et al., 1989
Iron				
Acute	FW	Little effect Na	Brook charr	Gonzalez et al., 1990
Manganese				
Acute	FW	Decrease Na	Brook charr	Gonzalez, 1990
Detergents				
Acute	FW	Increase gill permeability to water	Rainbow trout	Jackson and Fromm, 1977
	FW	Decrease plasma Na	Rainbow trout	McKeown and March, 1978
	SW	Increase plasma Na	Rainbow trout	McKeown and March, 1978
	SW	Increase Na, Cl, Ca, Mg; no effect K	Flounder	Baklien et al., 1986
Phenol				
Acute	FW	No effect plasma Cl	Rainbow trout	Swift, 1981
1 wk	BR	No effect Na, K, Mg, Ca, Cl	Pike	Kristoffersson et al., 1973
DDT				
In vitro		Decrease intestinal water uptake and Na, K-ATPase	SW eel	Janicki and Kinter, 1971

Table 1 (continued) Osmoregulatory Effects of Water Pollutants

Substance and exposure	Type of water	Effect observed	Species	Ref.
Fed 18 d	BR	No effect Na, K, Ca, PO_4	Flounder	Haux and Larsson, 1979
Fed 2 wk	Var.	Increase osmolality in SW; no effect in FW	Rainbow trout	Leadem et al., 1974
Acute	SW	Increase Na; decrease gut water uptake and Na, K-ATPase	Fundulus	Miller and Kinter, 1977
4 d	SW(?)	Increase Na and K	Eels and mullet	Hilmey et al., 1983
Lindane				
Fed 30 d and 109 d	FW	Decrease plasma Na and muscle Na, K-ATPase	Carp	Demael et al., 1987
Fenvalerate				
Acute	FW	Increase urine Na, K, and osmolality	Rainbow trout	Bradbury et al., 1987
Endosulfan				
24–240 h	FW	Decrease plasma Ca, Mg	Tilapia	Rangaswamy and Naidu, 1989
Nitrite				
Acute	FW	Decrease Na, K, Cl; rapid recovery	Rainbow trout	Williams and Eddy, 1988b
	FW	Inhibition of Cl uptake by gills	Rainbow trout and carp	Williams and Eddy, 1988a
1–12 d	FW	Decreased K; no change Na, Cl	Atlantic salmon	Bowser, 1989
Naphthalene				
Acute	Var.	Transient increase osmolality in SW Increase ions SW; decrease ions FW	Fundulus	Levitan and Taylor, 1979
Petroleum				
Acute	FW	No effect plasma Na	Rainbow trout	McKeown and March, 1978
	FW	Transient decrease plasma Cl	Rainbow trout	Zbanyszek and Smith, 1984
	FW	Decrease NaCl and osmolality	Rainbow trout	Englehardt et al., 1981
	SW	Increase NaCl and osmolality	Rainbow trout	Englehardt et al., 1981
	SW	Decrease gill Na, K-ATPase, no effect plasma osmolality	Sculpin	Boese et al., 1982
Ammonia				
Acute	FW	Increase urine flow rate	Rainbow trout	Lloyd and Swift, 1976
	FW	Increase plasma renin	Rainbow trout	Arillo et al., 1981
	FW	No effect Cl	Rainbow trout	Smart, 1978
	FW	Slight increase osmolality	Striped bass	Oppenborn and Goudie, 1993
	FW	Loss of Na, Cl, K	Trout fry	Paley et al., 1993
	FW	Increase plasma K; no change Na	Rainbow trout	Zeitoun et al., 1977
Chlorine				
Acute	BR	No effect Na, K, or gill ATPase	Morone	Block, 1977
Ozone				
Acute	FW	Decrease plasma NaCl	Rainbow trout	Wedemeyer et al., 1979
1–3 mo	FW	No effect plasma NaCl	Rainbow trout	Wedemeyer et al., 1979
Acid				
Acute and chronic	FW	Large ionoregulatory effect	Many species	See Chapter 10

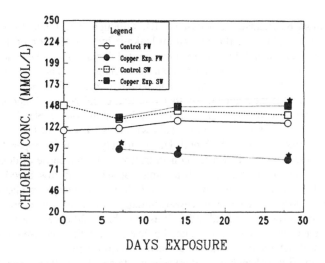

Figure 2 Comparison of effect of waterborne copper on plasma chloride concentration in seawater and freshwater-adapted flounder (*Platichthys flesus*). Copper concentrations were 170 μg/L in seawater and 105 μg/L in freshwater. Asterisks indicate means significantly different from controls ($p < 0.05$). (Data from Stagg, R. M. and Shuttleworth, T. J., *J. Fish Biol.*, 20, 491, 1982.)

There is one report of a rise in plasma sodium and potassium from exposure to copper in a freshwater fish (Singh and Reddy, 1990). This was in an air-breathing catfish species but that should not account for a rise in these electrolytes, which is opposite to what has been found in all other published studies. Unfortunately, the authors do not attempt to explain this, nor do they seem to recognize it as a contradiction.

In seawater, exposure to copper may increase the blood osmolality, but the extent of deviation from controls is not nearly as great as that seen in freshwater fish (Stagg and Shuttleworth, 1982). Figure 2 illustrates how it required 28 days of copper exposure of seawater-acclimated flounders before a significant increase in plasma chloride was seen, whereas only 7 days of exposure of the same species adapted to freshwater caused a significant decrease. This was in spite of the copper concentration being lower during the exposures in freshwater.

The mechanism by which seawater reduces the ionoregulatory dysfunction produced by copper is partially revealed by the work of Wilson and Taylor (1993b). They exposed rainbow trout which had been acclimated to either seawater or brackish water to the same copper concentration (6.3 μmol/L) and observed a rapid decline in plasma electrolytes in those from brackish water but only minor changes occurred in the seawater-acclimated fish. The authors propose the hypothesis that the basolateral location of the gill Na,K ATPase protects it from copper in a seawater environment. Thus, the process of pumping salt out of the fish would be unaffected by the metal. In dilute seawater, ionoregulatory disturbances occur due to increased permeability caused by the copper displacing surface-bound calcium. In full strength seawater, on the other hand, the higher concentration of calcium (and Mg?) helps prevent the calcium displacement and thereby provides some protection.

Exposure of sheepshead in seawater to a massive dose of copper (exceeding the 48-h LC50) was followed by a large increase in all the plasma electrolytes. Cardeilhac et al. (1979) suggest that the greatly elevated plasma potassium may be the cause of death from exposure to copper in seawater because the levels they recorded would probably cause heart stoppage. Because the plasma potassium concentration exceeded that in the seawater, the source must be leakage from tissue cells, although they failed to recognize that. Internal hypoxia could cause this potassium leakage by an inhibition of the Na,K-pump in cellular membranes, and the authors mention gill damage from these doses of copper which certainly could induce internal hypoxia (see Chapter 3). They discount hypoxia as being important, but do not offer evidence on which to reject this hypothesis, nor do they suggest an alternative mechanism. Another one that could be mentioned is the possible action of copper directly on cell membranes by inhibition of Na,K ATPase which provides the power for the Na,K-pump. This would allow some of the potassium in the cells to diffuse passively into the blood. Unfortunately, the high copper doses used are probably not relevant to "real world" concentrations, and the cause of death could be a combination of factors including high plasma potassium and low blood oxygen. It is noteworthy that Stagg and Shuttleworth (1982) found no change in plasma potassium with chronic exposure to copper in seawater-adapted flounders even though there were slight increases in sodium and chloride. Thus, the threshold dose of waterborne copper required to alter these blood ions must be lower than that required for changes in concentration of potassium, at least in a marine environment.

Coho salmon (*Oncorhynchus kisutch*) migrate to sea during their second year of life. At the time of this migration, they become more euryhaline. Chronic exposure of these fish to copper while still in freshwater can cause a large reduction in survival when transferred to seawater, a loss of ability to regulate plasma osmolality, and a concomitant decrease in activity of the gill Na,K-activated ATPase (Lorz and McPherson, 1976). The activity of this enzyme normally increases as the fish move into the seawater. A somewhat similar effect of that seen with copper was observed for salmon exposed for 2 weeks to sodium dichromate (Sugatt, 1980). In contrast to copper, however, dichromate in freshwater did nothing to the serum osmolality. Zinc produced different results than either of the other two metals: chronic exposure at a comparable toxic concentration to that used with copper did not affect either survival in seawater or ATPase activity in the coho salmon (Lorz and McPherson, 1976).

While exposure to waterborne copper or dichromate reduces survival of salmonids transferred to seawater, work on bluegill (*Lepomis macrochrirus*) suggests an opposite effect on this strictly freshwater species (Heath, 1987). After being subjected to copper at a sublethal dose for a week or more, the fish were able to then tolerate a hypertonic (250 mM) sodium chloride solution considerably better than could controls. Analysis of the plasma of those in the hypertonic solution revealed that those which had been exposed to copper were regulating their plasma osmolality and chloride better than the controls. It was hypothesized that chronic exposure to copper reduced the permeability of the gill to sodium chloride thus slowing its passive influx. The acclimation studies mentioned above (Lauren and McDonald, 1987b) lend some support for this although in their case, the change was in passive efflux of NaCl.

One other aspect of copper toxicity that may deserve attention is the effect it has on calcium metabolism. Copper apparently depresses the mechanism for uptake of calcium from the water, even when the fish is in seawater (Dodoo et al., 1992), a phenomenon that deserves further study.

B. ZINC

The results of studies on the effects of zinc exposure on osmoregulation may appear somewhat contradictory (Table 1), but this is probably due to species differences and different dose levels. Lewis and Lewis (1971) found a decrease in plasma osmolality of channel catfish when the exposure was acute. However, it should be noted that it required approximately five times the concentration of zinc to cause the same degree of blood osmolality change as was produced by copper in the same fish species. Skidmore (1970) observed no change in blood ions in trout exposed to a very high concentration of zinc (they died within 8 h). Skidmore's fish were probably dying of internal hypoxia before any changes in blood ions could take place. Within only a few hours, the blood oxygen had dropped to a very low level.

The most definitive studies on zinc toxicity are those of Spry and Wood (1984, 1985) on rainbow trout. They tested a high concentration (but not nearly as high as Skidmore) and one approximating the 96-h LC50. At the high dose, the fish die of hypoxemia, but at lower doses there was no hypoxemia. There was, however, a decreased plasma calcium and stimulation of sodium efflux, which over a longer period of time would undoubtedly cause osmoregulatory dysfunction. It is noteworthy that they observed some mortality even though the plasma sodium chloride and blood oxygen was unchanged. They hypothesize that the lethal mechanisms of the zinc at low environmental concentrations may be at the cellular level and do not necessarily involve either osmoregulatory or respiratory dysfunction.

Spry and Wood (1985) also found that after 48–60 h in water containing a high sublethal concentration of zinc, there was a stimulation of sodium uptake by the gills. This fits in with an earlier finding (Watson and Beamish, 1980) that chronic exposure to zinc induces an increased ATPase activity in gill tissue of trout. It was attributed to a greater passive flux of sodium out of the fish which then necessitated increased active uptake of the electrolyte from the water. This would represent a form of physiological adaptation by the fish to zinc in the water; a finding similar to what happens when fish are acclimated to copper (Lauren and McDonald, 1987b).

It has been reported that *in vitro* exposure of the gill Na,K ATPase enzyme to zinc caused an inhibition in activity, in apparent contradiction to the above finding of a stimulation of activity when the exposure was *in vivo* (Watson and Beamish, 1980). These observations point out the often conflicting results obtained when enzyme activities are measured in tissues from fish exposed to some substance in the water, compared to adding the same material to the incubation medium of an enzyme preparation (see Chapter 9). The latter method yields results that have relatively little meaning, except perhaps as a way of comparing the sensitivity of a given enzyme to more than one type of chemical.

There does not appear to be any information on the effect of zinc on whole-body osmoregulation in fish adapted to seawater. There has, however, been some interesting *in vitro* work. Using an isolated opercular epithelium preparation from seawater-adapted *F. heteroclitus*, the killifish, Crespo and Karnaky (1983) found that copper or zinc at a concentration of $4 \times 10^{-5} M$ inhibited chloride transport when on the blood (serosal) side of the preparation. Exposure of the other side had no effect, which suggests that the metals are interacting with the ATPase molecules on the basal area of the chloride cells. This fits with the hypothesis proposed by Wilson and Taylor (1993b) to explain why seawater adapted fish are much less sensitive to metals like copper. Crespo and Karnaky (1983) also present evidence that not all of the inhibition of chloride transport is caused by inhibition of ATPase, but must also involve the NaCl carrier in the chloride cell.

Overall, zinc appears to cause only mild dysfunction in monovalent cation regulation but the interesting effects on calcium regulation may warrant further study.

C. CADMIUM

This common pollutant appears to cause only moderate alterations in osmotic regulation, but specific ionic regulation, especially that of divalent cations, is quite sensitive to cadmium. First, we will examine what has been seen in studies of the effects of cadmium on osmoregulation per se.

When tilapia were exposed to doses ranging up to 1000 µg/L cadmium, they exhibited a dose-dependent decrease in plasma sodium which became progressively more severe over the test period of 2–35 days (Fu et al., 1989). Those in the highest dose all died in 8 days and had quite low plasma sodium levels, but when the exposure was at a more realistic environmental concentration of 10 μg/L, the change in plasma sodium was only slight and transient.

Goldfish exposed to cadmium experienced a modest decrease in plasma sodium and a shift of sodium, potassium, and water into the muscles. McCarty and Houston (1976) attribute these changes in part to an increased water uptake from the environment with much of that water being localized in the extracellular fluid compartments of the body thereby diluting the blood.

Giles (1984) was able to identify a threshold concentration for cadmium effects in rainbow trout. Chronic exposures for up to 178 days to 3.6 μg/L caused no effect on blood electrolytes while a dose of 6.4 μg/L resulted in a slight drop in plasma sodium, potassium, calcium, and chloride but an elevated magnesium. Measurements of urine flow and concentrations indicated that kidney impairment did not occur. Also, because the urine flow rate was constant, this indicates that permeability did not change during cadmium exposure. He explains his findings as being due to the redistribution of electrolytes that McCarty and Houston noted in the goldfish.

In apparent contrast to McCarty and Houston, and just about everybody else, Christensen et al. (1977) reported a slight, but significant increase in plasma chloride in brook trout exposed to very low levels of cadmium. They also reported elevations in plasma monovalent ions resulting from mercury and lead exposure. Two possible hypotheses for these anomalous findings are presented later in this chapter in the section on mercury. (These authors present no explanations, nor express surprise at the results.)

In seawater-adapted fish, a rather high concentration of cadmium is seemingly required before changes occur in plasma sodium and chloride. Thurberg and Dawson (1974) used several concentrations in 96-h exposures, but only when they reached 48 ppm cadmium, was there a change in plasma ions. This is a concentration an order of magnitude higher than that which was required to alter osmoregulation in freshwater fish (McCarty and Houston, 1976). In marine fish, oxygen consumption of gill tissue was found by the same workers to be depressed at all cadmium concentrations above 3 ppm. Because marine fish rely on gill function to excrete excess sodium chloride, it is interesting that the *in vitro* metabolic rate of this tissue can be decreased by 50% without any effect on the plasma electrolytes (Thurberg and Dawson, 1974). Of course, *in vitro* measures of gill tissue metabolism may or may not have much similarity to the *in vivo* metabolic rate of the same tissue (see Chapter 8). In any case, for marine fish, cadmium seems to alter other functions, such as those of the kidney, more so than osmoregulatory activity by the gills.

Larsson et al. (1981) studying the flounder, *Platichthys flesus,* found that exposure to cadmium of fish adapted to brackish water (which was almost isoosmotic to the blood) produced no changes in plasma sodium or chloride, but the concentrations of other ions were changed. The brackish water environment of the fish had less potassium and calcium but more magnesium than the plasma. Thus, the cadmium-induced changes in blood electrolytes (elevated magnesium but decreased potassium and calcium) resulted from enhanced net diffusion of magnesium in and calcium and potassium ions out. A fish normally excretes excess magnesium via the kidney, so, since it rose in the plasma in response to cadmium, this kidney function must have been inhibited. Histopathological damage to the kidney from cadmium has been seen in euryhaline killifish (Gardner and Yevich, 1970) and cunner (*Tautogolabrus adspersus*); (Newman and MacLean, 1974), so this organ is certainly an important target organ for cadmium.

Cadmium appears to affect calcium metabolism in freshwater fish to a greater extent than in seawater forms. The plasma calcium dropped by approximately 50% in a week of exposure of rainbow trout to 0.3 ppm cadmium (Roch and Maly, 1979). Such a concentration in the brackish water studies mentioned above caused only a slight drop in plasma calcium. This difference is expected, as fish in marine or brackish waters are in a calcium-rich environment which provides less of a diffusion gradient for this element.

When tilapia were exposed to cadmium at 10 μg/L, they exhibited a severe drop in plasma calcium but no significant change in monovalent cations or osmolality (Pratap et al., 1989). In the same laboratory, it was learned that upon continued exposure to this dose of cadmium, there was an increased prolactin secretion rate and recovery of plasma calcium (Fu et al., 1989). Interestingly, as time went on with the fish still in the water contaminated with cadmium, prolactin activity returned to normal which suggests that some secondary mechanism was induced. Pratap et al. (1989) also found that cadmium in the food (but not in water) caused hypocalcemia in much the same way as the waterborne metal, although the effect was somewhat delayed. They also concluded that the effect of cadmium on this freshwater species was on the gills, not on the kidney, which differs from that seen in marine species where the kidney is definitely a target organ.

Decreases in plasma calcium from cadmium exposure are apparently due to inhibition of calcium influx alone, at least at modest exposure concentrations (Reid and McDonald, 1988; Verbost et al., 1987). Sauer and Watabe (1988) suggest the cadmium acts by displacing calcium from protein carriers in the gill epithelium. Changes in calcium efflux by either the gills or kidney are far less important, even though histological damage to kidney occurs from cadmium (Gill et al., 1989). This damage, if produced by modest concentrations of cadmium, seemingly does not result in altered calcium regulation by the kidney in freshwater fish.

The alterations in plasma calcium may help explain the spasms and hyperexcitability seen in some fish exposed to cadmium (Larsson et al., 1981). The altered calcium metabolism does not affect the bone structure, probably because the bone is acellular in many species (Larsson et al. 1981).

The early life stages of fishes are generally presumed to be the most sensitive to cadmium (Benoit et al., 1976; Rombough and Garside, 1982). Thus, the investigation by Rombough and Garside (1984) is particularly relevant. Atlantic salmon, which hatch and undergo early development in freshwater, exhibit an almost linear increase in tissue sodium, potassium, calcium, and water during the first 45 days posthatch (alevin period). The ions increase 2.5- to 4-fold in concentration during this time. (The water content increases by about 20%.) While much of the ion flux is from the yolk, approximately 65% of the sodium, 45% of the potassium, and 75% of the calcium must be taken up from the freshwater (Rombough and Garside, 1984). Cadmium concentrations ranging from 0.47 to 300 μg/L produced a dose-dependent decrease in the rate of uptake of the three ions and water. The greatest effects were on potassium and calcium and the authors propose that this may be a primary cause of mortality from cadmium poisoning in alevins. It is further noteworthy that they detected reduced body potassium and calcium levels at concentrations of cadmium far below that (300 μg/L) required to produce histopathological changes. The exact mechanisms of this reduction in uptake rate of ions and water from the environment is not known. The inhibition of ion-activated ATPase enzymes, increases in membrane permeability (due to low calcium), and loss of superficial mucus are mentioned. Finally, the point is made by Rombough and Garside that the changes seen in the alevins exposed to cadmium are not restricted to this metal alone. Many other pollutants probably produce similar effects on larval fishes, but this has not been explored.

D. MERCURY

Mercury causes altered osmoregulation in both freshwater- and seawater-adapted fish, but in common with other metals, the overall effect may not be as great in the latter (Dawson, 1979). Inorganic mercury caused a slight elevation in plasma osmolality of seawater-adapted flounders. It apparently was not due to an uptake of sodium, chloride, or calcium as these plasma ions were unchanged or actually decreased in concentration. Also, plasma protein concentration was unchanged as well. Increases in magnesium and/ or amino acid content of the blood are possibilities, but these were not measured.

Methylmercury injected into flounders on a daily basis for 13 days produced accumulations of mercury in the gills up to 24 ppm (Schmidt-Nielsen et al., 1977). Still, this treatment caused no effect on either intra- or extracellular electrolytes. There was also no effect on the water content of tissues. The Na,K ATPase activity in gill and intestine remained unchanged in the fish injected with mercury, however, the activity of this enzyme in the bladder increased 2.5- and 1.5-fold in the kidney. Evidently, even though mercury is an inhibitor of this ATPase in seawater flounder when measured *in vitro* (Renfro et al., 1974), there must be a form of overcompensatory synthesis of the enzyme in some tissues under more chronic exposures.

There was an increase in sodium and chloride when brook trout in freshwater were exposed to methylmercury at very low concentrations (0.01–2.93 µg/L (Christensen, et al., 1977). Such an unexpected increase in chloride was also found by the same authors with cadmium exposure, as was cited earlier in this chapter. Although Christensen et al. (1977) propose no explanations, there are at least two possible: (1) increased ATPase activity in the gills or (2) increased rate of urination which would reduce the blood volume and raise its electrolyte concentration. Currently there is no evidence regarding either of these hypotheses other than to note that, under some circumstances, ATPase is stimulated by methylmercury (Schmidt-Nielsen et al., 1977).

Much higher concentrations of mercury (5–200 µg/L) (Lock et al., 1981) decreased plasma osmolality and sodium chloride in rainbow trout exposed for a week to either inorganic mercury or methylmercuric chloride. The depression in osmolality is clearly dose- and time-dependent. Sodium and chloride showed a similar pattern to the osmolality changes except that the percent of alteration was greater for the ions.

It requires 10–20 times more mercuric chloride than methylmercury to produce the same degree of lowering of osmolality and ionic concentrations (Lock et al., 1981). The authors note that this difference is comparable to the differences between the 96-h LC50s for the inorganic and organic forms of mercury. These they found to be 275 µg Hg/L and 24 µg Hg/L respectively. Methylmercuric chloride is more lipophilic and accumulates more rapidly in the tissues, so these results are not unexpected. The relative effectiveness (potency) of the two forms of mercury is apparently due to their relative tendency for accumulation, rather than to the specific action of the molecule. Lock et al. (1981) showed that inorganic mercury is actually more potent than the organic form when tested *in vitro* for effects on the uptake of water by the gills and ATPase activity, which again points out the hazard of drawing conclusions from only *in vitro* work (which these authors are careful to avoid).

One of the contributing causes of death from mercury poisoning in freshwater fish may be osmoregulatory failure. Milligan and Wood (1982) have shown rather conclusively that acid-stressed fish are killed by osmoregulatory failure when the plasma sodium shows a greater than 30% drop, providing this decline is rapid. Trout exposed to the 96-h LC50 of mercury for a week showed a +40% drop in sodium (Lock et al., 1981).

One of the most interesting conclusions to be drawn from the Lock et al. (1981) work on mercury is that both inorganic and organic mercury cause osmoregulatory effects primarily by producing an increase in the permeability of the gills to water. Using an elegant gill perfusion arrangement, they were able to show that mercury compounds stimulated water inflow at lower concentrations than those that resulted in inhibition of gill Na,K ATPase. This enzyme was not affected except at rather acute concentrations of mercury either with the gill perfusion work or with the *in vivo* exposures.

Stinson and Mallatt (1989) have found that methylmercury increases permeability of the gills of lampreys to sodium and chloride so there is a net increase in efflux of these ions. They also, in agreement with Lock et al., 1981, found no inhibition of gill ATPase. Thus, the relative lack of effect of mercury on gill ATPase has now been shown in both seawater- and freshwater-adapted species.

The element selenium has been reported to reduce the toxicity of mercury to fish (Kim et al., 1977). Heisinger and Scott (1985) subsequently showed that the osmoregulatory dysfunction induced by injection of mercuric chloride into freshwater bullhead catfish was completely blocked by simultaneous injection of a sodium selenite solution. A protein-bound complex of mercury and selenium is apparently formed (Burk et al., 1974) which reduces the toxic action of either element.

Finally, it should be mentioned that some of the ionoregulatory changes seen in fish exposed to mercury may, at least in part, be induced by an effect of mercury on blood cortisol levels. Recently, it has been learned that mercury inhibits the secretion rate of cortisol even though the adrenocortical and pituitary ACTH cells exhibit hypertrophy (Kirubagaran and Joy, 1991).

E. CHROMIUM

Sugatt (1980) exposed juvenile coho salmon in freshwater to chromium (as sodium dichromate) for 1–4 weeks at a dose of 0.5 or 0.2 mg Cr/L. The exposed fish were then challenged by transfer to seawater as might occur during the seaward migration. None of the controls died from this challenge, but a high rate of mortality occurred among the exposed fish over the following week. Serum osmolality was unchanged in either chromate exposed or control fish while still in freshwater, however, upon transfer to the seawater, those exposed showed a greater and more prolonged transient increase in osmolality, so the physiological adjustments to seawater were delayed or even prevented by the metal in the water.

The results for chromate are rather similar to those obtained for copper by Lorz and McPherson (1976), which have already been discussed. They illustrate how the challenge of another stressor (in this case salinity) on top of a chronic contaminant exposure can cause physiological changes that might not have been observed with the chronic exposure alone. It seems safe to predict that virtually any chemical pollutant that affects osmoregulation would have an inhibitory effect on the migration of salmon (or other anadromous and catadromous fish species), especially since the pollutants are frequently concentrated in the lower reaches or mouths of rivers where the osmoregulatory adjustments must take place.

F. LEAD

Rainbow trout kept in slightly hypotonic brackish water and exposed to lead (10, 75, or 300 µg Pb/L) for 30 days followed by a recovery of 49 days showed a dose-dependent elevation in plasma potassium (Haux and Larsson, 1982). Given the low concentration of potassium in the water, this element must have come from tissue cells. *In vitro* studies of mammalian erythrocytes show that lead causes leakage of potassium out of the cells (Hasan and Hernberg, 1966). Whether this holds for fish erythrocytes is unknown, but even if it did, that would not limit the source to the blood cells, as high internal potassium concentrations are typical of essentially all animal cells.

A more interesting question is the possible mechanism of this postulated flux of potassium out of body cells. The Na,K ATPase of the cell membranes is responsible for the differential concentration of potassium between body fluids and cytosol. The lead may be inhibiting that enzyme in a wide variety of the body tissues, but if it did in the study by Haux and Larsson (1982), there was not much effect on the regulation of sodium by the gills. This may have been because the fish were in water nearly isotonic to the blood so there was not much of a gradient for sodium either in or out of the fish.

In a study carried out by the same group on fish from lead-contaminated freshwater lakes, the fish exhibited a decrease in plasma sodium but no change in plasma potassium (Haux et al., 1985). The authors make no mention of this apparent contradiction, but one possible explanation is that those fish in the lakes had experienced lead exposure for a considerable period of time, maybe even for several generations, so some physiological adaptations to the lead could have taken place.

G. ALUMINUM

Aluminum is the third most abundant element in the earth's crust so it frequently gets leached from soils experiencing "acid rain", or runoff from land that has been stripmined (Baker, 1982). A variety of studies and reviews have examined the interactions of acid waters and dissolved aluminum on fish (e.g., Baker and Schofield, 1982; Howells et al., 1983; Neville, 1985; McDonald et al., 1991; Potts and McWilliams, 1989). The situation is complex, to say the least, because aluminum affects different physiological functions in the fish depending on the pH and calcium content of the water, and to a lesser extent, the concentration of aluminum.

When exposed acutely to aluminum in soft water, rainbow and brook trout seemingly die from progressive electrolyte loss if the pH is between 4 and 4.5. The initial cause of electrolyte loss is enhanced passive efflux which is then followed in a few days by a reduced NaCl uptake by the gills (Booth et al., 1988). The reduced NaCl uptake is probably caused by inhibition by aluminum of Na, K ATPase in the gills (Staurnes et al., 1984).

As the pH of the water is raised, the effect on blood electrolytes is reduced, but the fish begins to experience an internal hypoxic condition (hypoxemia) which may prove lethal (Neville, 1985; Wood et al., 1988; Malte and Weber, 1988). With trout, the threshold for this transition appears to be around pH 4.5–5.0 so that at pH 4.8, a fish may suffer from both loss of body ions and a decreased blood oxygen.

In acid water, raising the calcium reduces the ionoregulatory dysfunction induced by aluminum (Wood et al., 1988). Finally, in hard acid (pH 5) water, aluminum causes no ionoregulatory dysfunction at all but does cause an acute hypoxia (Malte, 1986). As Wood et al. (1988) have noted, liming as an attempt to raise the pH and calcium levels in lakes or streams has occasionally been associated with fish kills. They go on to point out that such a change in water quality would merely promote respiratory failure while reducing ionoregulatory dysfunction. It might be further mentioned that this is a good example of how fish physiology research can have governmental policy overtones.

One other aspect of calcium needs to be mentioned. Verbost et al. (1992) have shown a dose-dependent inhibition by aluminum of calcium uptake in carp. Calcium efflux was less sensitive than the process of influx to the presence of aluminum. At the pH they were using (5.2) there was little effect on the influx or efflux of sodium except at very high aluminum concentrations. Thus, at higher pHs, aluminum may alter blood calcium levels more than sodium, although this seemingly has not been examined. In any case, alterations in calcium metabolism could have serious implications for reproduction (see Chapter 13).

Trout are able to acclimate to the presence of aluminum and thereby raise the toxic threshold concentration and reduce the physiological dysfunction. Some acclimation in electrolyte balance occurs by four days' exposure. The gills undergo a number of physiological (McDonald et al., 1991b) and histological (Mueller et al., 1991) changes, but perhaps one of the most important is a reduction in aluminum binding to the surface of gill cells (McDonald et al., 1991b). Work by Reid et al. (1991) indicates that during acclimation to aluminum there is an increased affinity of the gills for calcium associated with a decreased binding of aluminum.

Exley et al. (1991) have proposed a cellular mechanism for acute aluminum toxicity involving the binding of aluminum both to the surface of lamellar epithelial cells and subsequently to interior ligands. The sum effect is to increase permeability of the epithelium, interfere in the second messenger system within the cells, and ultimately accelerate cell death.

H. TIN

Tin compounds, generally in the organic form such as tributyltin, are used in antifouling paints on ships, docks, etc. Striped bass, which often live in brackish water, were tested with a variety of concentrations of tributyltin (Pinkney et al., 1989). They found a large stimulation by the tin of Na,K ATPase and Mg ATPase activity in the gills. When tested *in vitro*, however, these two enzymes were inhibited by tributyltin. A similar difference between *in vitro* and *in vivo* effects on ATPase enzymes has also been seen with copper (Stagg and Shuttleworth, 1982) and zinc (Watson and Beamish, 1980) exposures. It may reflect a form of adaptation to the presence of the toxicant in that *in vivo* exposures induced more enzyme synthesis. Indeed, Pinkney et al. (1989) found that the tin compound caused no change in the plasma electrolytes so the adaptation was seemingly successful, although it should be noted that the fish were in brackish water so diffusion gradients between the environment and the fish blood were not large.

I. MANGANESE AND IRON

These two elements generally get into waterways as a result of strip mining. Manganese has a very strong effect on sodium regulation (Gonzalez et al., 1990) when fish are exposed to water with a concentration approaching their 96-h LC50. Body and plasma sodium concentrations declined by 40–50% before death, which suggests this is a primary mode of death in acute doses of manganese. On the other hand, iron at a concentration even twice that of the LC50 caused very little change in plasma sodium so the mode of toxicity of these two metals is markedly different.

Compared to the situation with metals, there is not a great deal known about the action of organic pollutants on osmoregulation or control of electrolytes by fish. The large number and extreme variety of anthropogenic organic chemicals that appear in waterways, however, make this a presumptively important lack of information.

J. DETERGENTS

With the aid of an isolated gill preparation, Jackson and Fromm (1977) demonstrated that acute exposure to a detergent caused the gill to become more permeable to water. The effect was dose dependent (5–100 ppm) and it reached a maximum in about 25 min of exposure. That is probably too fast for much histological damage to take place except at the highest concentrations tested (Abel and Skidmore, 1975). Concentrations below 5 mg/L caused no effect on the permeability of water over the 65-min exposure. This altered permeability of gills to water could provide the mechanism for the decrease in serum sodium which has been observed in freshwater fish exposed to an oil dispersant (McKeown and March, 1978) that presumably acts much like a detergent. An increased inflow of water would dilute the blood and/or result in a greater urine output, thereby carrying some ions out of the body. A change in gill water permeability could also explain the increased serum sodium in the seawater-adapted fish challenged with the same dispersant (McKeown and March, 1978), because an increased permeability to water produces a more rapid osmotic flow of water out of the fish.

Petroleum dispersants may also affect permeability of the gills to inorganic ions as well as affect the ability to excrete an excess amount of these in marine fish. Baklien et al. (1986) reported elevations in all blood ions except potassium when flounders were exposed for four days to a fairly acute dose of dispersant, whereas the changes seen were all statistically significant, they were not especially large. Because there was a fairly high mortality among the exposed fish, the cause of death from this dispersant is probably something other than osmoregulatory dysfunction.

K. PHENOL

Phenol seems to have little or no effect on plasma electrolytes even when exposures are to a concentration one half the 48-h LC50 (Swift, 1981), or when exposed for a week to a concentration approximating the incipient lethal level (Kristofferson et al., 1973). This is in spite of the rather considerable damage that phenol does to gill tissue (Mitrovic et al., 1968). It should be recalled that zinc is also quite damaging to gill tissue (Skidmore and Tovell, 1972), yet does not necessarily alter blood sodium levels when exposure is to acute levels (Skidmore, 1970). The usual type of histological damage to gills from pollutants involves swelling and separation of the epithelium from the basement membrane on the lamellum (see Chapter 3). Evidently then, the physical damage to the gill lamellae does not necessarily result in an immediate failure to maintain osmoregulation, although it certainly reduces respiratory gas exchange (Chapter 3).

L. DDT

DDT is a potent inhibitor of Na,K and Mg ATPase (*in vitro*) in fish tissues (Janicki and Kinter, 1971). Hilmy et al. (1983) reported increases in plasma sodium, potassium, and calcium in mullet and eels exposed to DDT or the other organochlorine insecticide endrin. It is not clear in their paper whether the fish were adapted to freshwater or seawater. Because there were increases in all measured electrolytes as a result of the insecticide exposures, I have assumed they were in seawater. The authors noted a dose-dependent increase in potassium which might indicate an inhibition by DDT of Na,K ATPase in the cell membranes of muscle and other tissues (Demael, 1987). This would permit the intracellular potassium to leak out down its natural concentration gradient. The authors attribute the ionoregulatory dysfunction to liver damage, but the liver serves essentially no such function in fish.

Janicki and Kinter (1971) have found that uptake of water from the gut in seawater-adapted eels is inhibited by DDT, because the process is coupled to active sodium uptake which is dependent on the Na,K ATPase enzyme. Miller and Kinter (1977) subsequently showed that this inhibition of Na transport also impairs the uptake of amino acids from the gut. This finding suggests that exposure to substances which inhibit ATPase, such as copper, in marine fish may cause problems with the absorption of nutrients, which could have more practical implications than a moderate osmoregulatory dysfunction. Freshwater fish, on the other hand, because they drink very little, may not experience as great a problem from this.

DDT in the diet seemingly does not have much osmoregulatory effect on freshwater fish compared to their seawater counterparts (Haux and Larsson, 1979; Leadem et al., 1974). Leadem et al. (1974) fed DDT for 2 weeks to rainbow trout acclimated to freshwater, one third seawater, or 100% seawater. The DDT treated fish in one third seawater and 100% seawater exhibited an elevated serum osmolality and sodium, but the DDT caused no effect on these two parameters when the trout were acclimated to freshwater. The enzyme Na,K ATPase in the gill tissue was inhibited in all the fish fed DDT with the greatest percent of inhibition occurring in those adapted to seawater. This was probably due to the enzyme being located basolaterally in the chloride cells in the seawater-adapted forms. It is noteworthy that even though there was some inhibition of this enzyme in the freshwater trout, there was no alteration in osmoregulation.

M. FENVALERATE

The pyrethroid insecticide fenvalerate is a potent inhibitor of a variety of ATPases (Clark, 1982; Desaiah et al., 1975). When trout were exposed to a highly acute dose, they showed a marked increase in concentration of various cations and osmolality of the urine (Bradbury et al., 1987). The overall excretion rate of these cations also increased suggesting the pesticide was inhibiting the reabsorption function in the kidney so the filtrate flowed straight out. The net effect of this could be a loss of ions from the blood, although they did not measure that parameter.

N. PETROLEUM HYDROCARBONS

It would appear that petroleum and the hydrocarbons that compose it have little effect on osmoregulation. For example, Stickle et al. (1982) found that neither toluene or naphthalene caused much effect on serum osmolality in coho salmon adapted to seawater or freshwater unless the concentrations of the hydrocarbons exceeded the 48-h LC50.

Crude oil emulsions can apparently cause altered osmoregulation, but the concentrations required may be rather high. Using a dose that produced considerable mortality in 7 days (in seawater), Englehardt et al. (1981) observed either depressions in serum monovalent ions in freshwater or elevations of these ions in rainbow trout acclimated to seawater. They also found considerable gill damage as a result of the crude oil emulsions.

Rainbow trout acclimated to freshwater and exposed to massive doses of a synthetic oil mixture (similar to Prudhoe Bay crude oil) died in 1–4 h and the plasma chloride was depressed (Zbanyszek and Smith, 1984). It probably was not, however, the direct cause of death as the extent of chloride depression was not of sufficient severity (Milligan and Wood, 1982). When trout were dosed with an oil mix concentration an order of magnitude less, they experienced a slightly lowered chloride level at 24 h, but by 48 h of continuous exposure it had returned to normal (Zbanyszek and Smith, 1984).

Using the wastewater from a petroleum refinery, Boese et al. (1982) found a 30% inhibition of the gill ATPase, but only at a rather high concentration of waste (20%). In spite of this enzyme inhibition, there was no change in the plasma osmolality. From this observation and those mentioned earlier, it appears that it is possible to have a considerable inhibition of ATPase in the gills without a concomitant alteration in plasma electrolytes. Perhaps, the inhibition is compensated for by other hormonally mediated changes (see summary at end of this chapter).

Salinity may affect the LC50 of various substances, and this is true of naphthalene and other hydrocarbons derived from petroleum. In contrast to most metals, petroleum components are more toxic at higher salinities (Stickle et al., 1982). Levitan and Taylor (1979) attempted to investigate the physiological basis of this differential toxicity by measuring serum osmolality of *Fundulus* exposed to naphthalene. At a concentration which produced 85% mortality in 30 h, there was a transient increase in osmolality, but it returned to normal within 12 h. While the authors think this change in serum osmolality explains the increased toxicity in seawater, the evidence certainly does not support that hypothesis. A more probable cause of the differential mortality is the increased intake of naphthalene produced by increased drinking of seawater. They showed that, indeed, naphthalene is taken up more rapidly when the fish are in hyperosmotic water. Once taken up from the water, it will then concentrate in various tissues and other studies have found that the distribution between tissues differs when the fish are in seawater as compared to when in freshwater (see Chapter 5).

The available evidence seems to suggest that petroleum hydrocarbons have little effect on osmoregulatory processes in either fresh- or seawater fishes. Exceptionally high doses, which could certainly occur in areas of a petroleum spill, may produce severe osmoregulatory dysfunction, but chronic exposures probably affect other physiological processes to a greater extent than those involved with osmoregulation.

Having said that, it should be pointed out that the use of oil dispersants to "clean up" oil spills may cause a marked effect on fish electrolyte/osmoregulation (Baklien et al., 1986), whereas either the oil or dispersant alone has relatively little effect.

O. AMMONIA

Elevated levels of unionized ammonia in the water caused trout to increase their urine production in a concentration-dependent fashion (Lloyd and Swift, 1976). The stimulated urine flow was explained by the authors as being due to increased inflow of water which was then excreted. An alternative explanation later proposed by Arillo et al. (1981) involves the circulatory system. They found that renin activity in the blood was elevated by exposure to ammonia at concentrations and times of exposure that are comparable to those used in the foregoing study.

Renin is released by the kidney and catalyzes the conversion of angiotensinogen to angiotensin in the plasma. This latter protein is a powerful vasoconstrictor which causes an increase in blood pressure (Olson, 1992). Smart (1978) noted a marked increase in blood pressure in trout exposed to ammonia, although he used a very high concentration. A rise in blood pressure would increase the filtration rate by the kidney and thus an increase in urination.

With the theoretical and observational considerations noted above in mind, it appears that ammonia causes little if any effect on the concentration of plasma electrolytes in adult teleosts (Oppenborn and Goudie, 1993). This is so even when the concentration is lethal and the fish are "overturning" (Smart, 1978; Swift, 1981). So, perhaps the compensatory mechanisms are sufficient to prevent the electrolyte loss that would otherwise occur via the increased urine flow.

The Oppenborn and Goudie (1993) study revealed two interesting aspects of ammonia toxicity.

1. Striped bass in freshwater exposed to 0.5 mg/L unionized ammonia for 96 h exhibited a slight rise in plasma osmolality. This suggests the gill regulation of electrolyes was not compromised by ammonia. Instead, the stimulation of urine excretion rate probably lowered the water content of the blood producing hemoconcentration.
2. Striped bass, *Morone saxatilis,* were less sensitive to ammonia than were hybrids (*M. saxatilis X M. chrysops*). The mechanism of this differential sensitivity appears to be that striped bass do not accumulate ammonia in their plasma as much as the hybrids do when the external ammonia level is elevated. Whether this involves superior excretion or less permeability of the gills to ammonia in striped bass is not known. It is a somewhat surprising observation in that the hybrids are considered to be more tolerant of changing environmental conditions.

Although ionic regulation in adult fish may not be especially sensitive to ammonia, larval trout (and other species?) certainly are. Paley et al. (1993) have reported a marked loss of sodium, chloride, and potassium from rainbow trout yolk sac fry when exposed to ammonia for 24 h. The concentrations of unionized ammonia ranged from 7.2–36.2 μmol/L and the effect on electrolytes was greatest for potassium. The loss of sodium was probably due primarily to competitive inhibition of sodium uptake by ammonia. The mechanism of potassium loss is less clear but may be caused by ammonia entry into the cells and subsequent deprotonation which would cause a buildup of H^+ ions there (Paley et al., 1993).

P. NITRITE

Nitrite is formed from nitrate by bacterial reduction. It appears to have relatively modest effects on electrolyte regulation in fish (Williams and Eddy, 1988a,b; Bowser et al., 1989). The effects on hemoglobin function (see Chapter 4) and other cellular activities are probably more important in its toxicity. Nitrite is important in water of low ionic content when nitrite/chloride ratio is high and fish take up nitrite instead of chloride leading to formation of methemoglobin (Eddy and Williams, 1987).

Q. CHLORINE AND OZONE

These are strong oxidizing agents which are quite toxic to fish (Hall et al., 1981). When rainbow trout were subjected to a massive dose of chlorine (causing death within a few hours) an elevation in plasma potassium was noted (Zeitoun et al., 1977). This would have to have come from tissue cells, undoubedly as a result of the acute tissue hypoxia induced by the gill damage (Bass et al., 1977; Bass and Heath, 1977). Unfortunately, the dose in the Zeitoun et al. (1977) study was so high as to have little relevance to a real-world situation. Using a much lower dose and a more prolonged exposure period, Block (1977) found no effect of chlorine on blood electrolytes or gill ATPase on two species of brackish water fishes. The lack of change in electrolytes may have been due to the fact they were in water nearly isoosmotic to their blood, but the lack of an effect on the Na,K ATPase enzyme may be indicative that, at least at low doses, chlorine may not have the potential for much of an effect on osmoregulation.

Ozone is used as a disinfectant and as a substitute for chlorine in the antifouling of steam-generating plants. Wedemeyer et al. (1979) found that acute exposure caused a severe drop in sodium chloride, but chronic exposure caused no osmoregulatory effects, even at concentrations resulting in limited histopathological damage to the gills. The authors attribute death from high doses of ozone to massive gill damage followed by osmoregulatory failure. An equally likely cause of death, perhaps in combination with the osmoregulatory collapse, is internal hypoxia produced by the gill damage, as occurs in acute exposure to chlorine (Bass and Heath, 1977) or zinc (Skidmore, 1970).

III. MUCUS

The importance of mucus in osmoregulation is beginning to be revealed (Shephard, 1982; Handy et al., 1989). Handy and Eddy (1991b) have shown that mucus is usually absent from the gill surfaces of rainbow trout that are free from stress. Thus, while it may aid ionoregulation under some circumstances, it clearly is not required for "normal" function in this species. Whether that is so for other species is not known.

It has been known since at least 1927 that exposure of fish to a variety of metals in the water causes acute production of mucus by the gills, buccal cavity, and skin (Carpenter, 1927). Over the years this observation has been repeated numerous times with a wide variety of chemicals as well as acidic conditions. There now appears to be good evidence both indirect and direct that the mucus helps maintain ionoregulation under these circumstances. When trout were exposed to either methylmercury or mercuric chloride at the same ambient concentration, far more mucus was secreted by those in water with mercuric chloride. (Lock et al., 1981; Lock and Overbeeke, 1981). It was further found that it requires 10–20 times more inorganic mercury in the water to cause the same degree of osmotic dysfunction as that produced by a given concentration of organic mercury. Thus, the mucus may be helping to counteract the increased permeability to water of the gill poisoned by mercury.

Handy and Eddy (1989) have found that mucus of a freshwater fish has a considerably higher concentration of Na and Cl than does the water. This reduces the gradient between the blood and water, thus an increased mucus layer in fish experiencing pollutant stress would reduce the passive diffusive loss of electrolytes. It also has a low permeation coefficient for metals and thus may help protect the tissue from its action on the sensitive gill epithelium (Handy et al., 1989).

Although increased mucus secretion in fish which are experiencing pollutant stress appears to be adaptive regarding electrolyte regulation, it may be maladaptive from a respiratory standpoint. This is because it adds an unstirred layer next to the lamellar surface and therefore increases the diffusion distance for oxygen (see Chapter 3).

IV. CHLORIDE CELL PROLIFERATION

Zinc (Matthiessen and Brafield, 1973), copper (Baker, 1969), cadmium (Oronsaye and Brafield, 1984), and nitrite (Gaino et al., 1984) have been found to stimulate an increase in chloride cell number on the gills of fish. With zinc and cadmium, the authors felt that this is a mechanism to excrete the accumulated metals, although osmoregulatory involvement was not entirely discounted. There seems to be no direct evidence in support or opposition of the hypothesis that these cells are excreting the metals.

Interestingly, as was mentioned above, nitrite is actively taken up from the water by chloride cells (Williams and Eddy, 1988) so their proliferation under nitrite exposure might appear to be maladaptive. Bath and Eddy (1980) have also found that nitrite ions compete with chloride ions in the active uptake of the latter by the chloride cells (Bath and Eddy, 1980). Thus the proliferation of more chloride cells caused by this substance appears to be a compensatory mechanism to maintain normal uptake of chloride from the water. It appears to be successful as nitrite exposure causes little effect on plasma chloride content.

V. SOME SUMMARY COMMENTS REGARDING OSMOREGULATORY AND ELECTROLYTE ALTERATIONS

In seawater-adapted fish, exposure to a wide variety of different pollutants induces an osmoregulatory dysfunction which is reflected in a rise in plasma NaCl and osmolality. In freshwater-adapted fish, the opposite will occur in that there is a loss of NaCl mostly via the gills. Evidently, a large majority of pollutants affect osmoregulation and/or ionoregulation, but there are considerable differences in their potency. Among the metals, copper, aluminum, and methylmercury probably cause the largest amount of dysfunction. Zinc and iron cause little or no effect on osmoregulatory variables, even though zinc, at least, causes marked histopathological damage to the gill epithelium. Thus, physical damage to the gills does not necessarily lead to ionoregulatory dysfunction. Indeed, the hypersecretion of mucus that usually accompanies such histological changes may reduce the ionoregulatory problems.

It appears that the percentage change in blood constituents produced by comparable toxic concentrations of a given chemical may be greater for the freshwater-adapted fish as opposed to marine, but this needs more systematic investigation. The mechanisms for this difference are not obvious, since marine fish have flux rates for water and salts about an order of magnitude greater than their freshwater cousins.

Osmoregulatory disruption by metals and most other chemicals, with the exception of acids, usually involves inhibition of the Na,K ATPase enzymes in gill and perhaps also in the gut. The latter would be of importance primarily in marine habitats where the fish drink large volumes of water and this inhibition may cause a nutritional problem. This is because the uptake of sodium in the gut is coupled to amino acid absorption so inhibition of sodium uptake would inhibit the latter. The ATPase enzymes in the kidney may also be a point of action, particularly in freshwater fish where a good deal of glomerular filtration and reabsorption takes place, but this has not received much attention. Increases in permeability of the gills may or may not take place.

While it has been assumed that altered osmoregulatory function is generally due to effects of some toxicant on peripheral processes, the involvement of the brain and endocrine tissues should not be ignored. For example, it has been reported that the catecholamine content of the hypothalamus changes during acclimation to different salinities (Hegab and Hanke, 1981). Because this area of the brain is affected by a variety of toxicants (Smith, 1984), there may be indirect effects on osmoregulation.

An altered osmotic and ionic makeup of the blood induced by some pollutant should bring about hormonal changes as a compensatory response to restore homeostasis. It is common to observe elevated cortisol levels with exposure to a variety of environmental stressors (Pickering, 1993), and this hormone facilitates the active transport of ions by both marine and freshwater teleosts (Madsen, 1991). Whether the cortisol elevation is part of a generalized stress response or one caused by the electrolyte changes of the blood is not clear. Indeed, it could easily be a bit of both. In any case, if the extent of ion alteration is not too great, recovery of homeostasis may occur even with continued pollutant exposure, and cortisol is probably a key mechanism here as it stimulates increased Na,K ATPase activity in the gills (Madsen, 1991).

A stress severe enough to cause elevated adrenalin/noradrenalin levels may actually accentuate disruption of hydromineral homeostasis because these catecholamines cause a net loss of sodium and chloride (Vermette and Perry, 1987). Thus, from an osmoregulatory standpoint, these hormones appear to be maladaptive.

The hormone prolactin, from the anterior pituitary, has been found to increase markedly during acid stress in tilapia (*Oreochromis mossambicus*); (Wendelaar Bonga et al., 1984). This was associated with a subsequent reduction in water permeability of the gills and restoration of plasma ion levels. A comparable hormonal response may occur with other pollutants as well.

With a wide variety of substances causing osmoregulatory dysfunction, the question arises as to whether this might add to or subtract from the energetic demands of the fish. It is a difficult matter to approach because estimates of the energetic "cost" of osmoregulation vary from negligible to 50% of the resting metabolic rate in freshwater (for review see Febry and Lutz, 1987). Even closely related species may show huge differences. For example, Iwama and Morgan (1990) report a value of 20% and 1% for the rainbow and steelhead trouts, respectively. One thing that they most seem to agree on is that osmoregulation in freshwater requires more energy than it does in seawater. This is intuitively rather suprising as the total fluxes are much greater in the seawater forms. When there is an inhibition of the processes catalyzed by ATPases, as occurs with a variety of pollutants, the demand for ATP might actually go down. Thus, it is conceivable that pollutant exposure might thereby lower the metabolic cost of osmoregulation, but that is pure conjecture at this time.

The calcium concentration in the ambient water critically influences permeability of the gills to both sodium and water. Increasing calcium concentrations decreases the branchial water permeability and rate of sodium loss caused by toxic chemicals.

As was mentioned at the beginning of this chapter, the usual method that has been used to detect effects of pollutants on osmotic and ionic regulation has been to measure the concentrations of Na, K, Ca, and Cl in the plasma after exposures. Recently, Wood (1992) has proposed a major change in approach. He maintains that changes in flux rates are what is actually of interest and these can be assessed by measuring the concentration of these ions in the water in which the fish is residing (providing volumes are small). Such an approach has several advantages such as the fact that since it is non-invasive, changes in a single fish over time can be assessed. The technique is more sensitive to alterations in osmoregulatory processes and can detect changes from toxic chemicals long before they are reflected in internal concentration changes.

Finally, the single pollutant that probably has the greatest effect on electrolyte regulation in fish is acid. Acidic conditions cause an exceptionally large dysfunction in osmotic and ionic regulation which is discussed in some detail in Chapter 10.

REFERENCES

Abel, P. D. and Skidmore, J. F., Toxic effects of an anionic detergent on the gills of rainbow trout, *Water Res.*, 9, 759, 1975.

Arillo, A., Uva, B. and Vallarino, M., Renin activity in rainbow trout (*Salmo gairdneri* Rich.) and effects of environmental ammonia, *Comp. Biochem. Physiol.*, 68A, 307, 1981.

Arnold-Reed, D. E. and Balment, R. J., Atrial natriuretic factor stimulates *in-vivo* and *in-vitro* secretion of cortisol in teleosts, *J. Endocrinol.* 128, R17, 1991.

Baker, J. P., Histological and electron microscopical observations on copper poisoning in the winter flounder *Pseudopleuronectes americanus, J. Fish. Res. Bd. Can.*, 26, 2785, 1969.

Baker, J. P., Effects on fish of metals associated with acidification, in, *Acid Rain/Fisheries*, Johnson, R. E., Ed., American Fisheries Society, Bethesda, MD, 1982, 165.

Baker, J. P. and Schofield, C. L., Aluminum toxicity to fish in acidic waters, *Water, Air, Soil Pollut.*, 18, 289, 1982.

Baklien, A., Lange, R. and Reiersen, L., A comparison between the physiological effects in fish exposed to lethal and sublethal concentrations of a dispersant and dispersed oil, *Mar. Environ. Res.*, 19, 1, 1986.

Bass, M. L., Berry, C. R. and Heath, A. G., Histopathological effects of intermittent chlorine exposure on bluegill (*Lepomis macrochirus*) and rainbow trout (*Salmo gairdneri*), *Water Res.*, 11, 731, 1977.

Bass, M. L. and Heath, A. G., Cardiovascular and respiratory changes in rainbow trout, *Salmo gairdneri*, exposed intermittently to chlorine, *Water Res.*, 11, 497, 1977.

Bath, R. N. and Eddy, F. B., Transport of nitrite across fish gills, *J. Exp. Zool.*, 214, 119, 1980.

Benoit, D. A., Leonard, E. N., Christensen, G. M. and Randt, J. T., Toxic effects of cadmium on three generations of brook trout, *Salvelinus fontinalis, Trans. Am. Fish. Soc.*, 105, 550, 1976.

Bern, H. and Madsen, S., A selective survey of the endocrine system of the rainbow trout (*Oncorhynchus mykiss*) with emphasis on the hormonal regulation of ion balance, *Aquaculture*, 100, 237, 1992.

Block, R. M., Physiological responses of estuarine organisms to chlorine, *Chesapeake Sci.*, 18, 158, 1977.

Boese, B. L., Hohonson, V. G., Chapman, D. E., Ridlington, J. W. and Randall, R., Effects of petroleum refinery waste water on gill ATPase and selected blood parameters in the Pacific staghorn sculpin (*Leptocottus armatus*), *Comp. Biochem. Physiol.*, 71C, 63, 1982.

Boeuf, G., Salmonid smolting: a pre-adaptation to the oceanic environment, in, *Fish Ecophysiology*, Rankin, J. C. and Jensen, F. B., Eds., Chapman and Hall, London, 1993, chap. 4.

Booth, C. E., McDonald, D.G., Simons, B. P. and Wood, C. M., Effects of aluminum and low pH on net ion fluxes and ion balance in the brook trout (*Salvelinus fontinalis*), *Can. J. Fish. Aquat. Sci.*, 45, 1563, 1988.

Bowser, P. R., Wooser, G. A. and Aluisio, A. A., Plasma chemistries of nitrite stressed Atlantic salmon, *Salmo salar, J. World Aquacult. Soc.*, 20, 173, 1989.

Bradbury, S. P., McKim, J. M. and Coats, J. R., Physiological response of rainbow trout (*Salmo gairdneri*) to acute fenvalerate intoxication, *Pest. Biochem. Physiol.*, 27, 275, 1987.

Brown, D. J., Morris, R. and Goldthorpe, S. A., Sublethal effects of acid water, in, *Stress and Fish*, Pickering, A. D., Ed., Academic Press, New York, 1981, 344.

Buckley, J. T., Roch, M., Rendell, C. A. and Matheson, A. T., Chronic exposure of coho salmon to sublethal concentrations of copper. I. Effect on growth, on accumulation and distribution of copper, and on copper tolerance, *Comp. Biochem. Physiol.*, 72C, 15, 1982.

Burk, R. F., Foster, K. A., Greenfield, P. M. and Kiker, K. W., Binding of simultaneously administered inorganic selenium and mercury to a rat plasma protein, *Proc. Soc. Exp. Biol. Med.*, 145, 782, 1974.

Burton, R. F., Ionic regulation in fish; the influence of acclimation temperature on plasma composition and apparent set points, *Comp. Biochem. Physiol.*, 85A, 23, 1986.

Carpenter, K. E., The lethal action of soluble metallic salts on fishes, *Br. J. Exp. Biol.*, 4, 378, 1927.

Carrick, S. and Balment, R. J., The renin-angiotensin system and drinking in the euryhaline flounder, *Platichthys flesus, Gen. Comp. Endocrinol.*, 51, 423, 1983.

Cardeilhac, P. T., Simpson, C. F., Lovelock, R. L., Yosha, S. R., Calderwood, S. F. and Gudat, J. C., Failure of osmoregulation with apparent potassium intoxication in marine teleosts: a primary toxic effect of copper, *Aquaculture*, 17, 231, 1979.

Christensen, G. M., McKim, J. M., Brungs, W. A. and Hunt, E. P., Changes in the blood of the brown bullhead (*Ictalurus nebulosus*, Lesueur) following short and long term exposure to copper (II), *Toxicol. Appl. Pharmacol.*, 23, 417, 1972.

Christensen, G., Hunt, E. and Fiandt, J., The effect of methylmercuric chloride, cadmium chloride, and lead nitrate on six biochemical factors of the brook trout (*Salvelinus fontinalis*), *Toxicol. Appl. Pharmacol.*, 42, 523, 1977.

Clark, J. M. and Matsumura, F., Two different types of inhibitory effects of pyrethroids on nerve Ca⁻ and Ca⁺ Mg-ATPase activity in the squid, *Loligo pealei*, *Pest. Biochem. Physiol.*, 18, 180, 1982.

Courtois, L. A. and Meyerhoff, R. D., Effects of copper exposure on water balance, *Bull. Environ. Contam. Toxicol.*, 14, 221, 1975.

Crespo, S. and Karnaky, K. J., Copper and zinc inhibit chloride transport across the opercular epithelium of seawater-adapted killifish (*Fundulus heteroclitus*), *J. Exp. Biol.*, 102, 337, 1983.

Cutkomp, L. K., Koch, R. B. and Desaiah, D., Inhibition of ATPases by chlorinated hydrocarbons, in, *Insecticide Mode of Action*, Coats, J. R., Ed., Academic Press, New York, 1982, 45.

Dawson, M. A., Hematological effects of long-term mercury exposure and subsequent periods of recovery on the winter flounder, *Pseudopleuronectes americanus*, in, *Marine Pollution: Functional Responses*, Vernberg, W. B., Thurberg, F. P., Calabrese, A. and Vernberg, F. J., Eds., Academic Press, New York, 1979, 171.

Demael, A., Lepot, D., Cossarini-Dunier, M. and Monod, G., Effect of hexachlorocyclohexane (Lindane) on carp (*Cyprinus carpio*). II. Effects of chronic intoxication on blood, liver enzymes and muscle plasmic membrane, *Ecotoxicol. Environ. Safety*, 13, 346, 1987.

Desaiah, D., Cutkomp, K., Vea, E. and Koch, R., The effect of three pyrethroids on ATPase of insects and fish, *Gen. Pharmacol.*, 6, 31, 1975.

Dodoo, D., Engel, D. and Sunda, W., Effect of cupric ion activity on calcium accumulation in juvenile flounder (*Paralichthyes* spp.), *Mar. Environ. Res.*, 33, 101, 1992.

Eddy, F. B., Effects of stress on osmotic and ionic regulation in fish, in, *Stress and Fish*, Pickering, A. D., Ed., Academic Press, New York, 1981, chap. 4.

Eddy, F. B., Osmotic and ionic regulation in captive fish with particular reference to salmonids, *Comp. Biochem. Physiol.*, 73B, 125, 1982.

Eddy, F. B. and Bath, R. N., Ionic regulation in rainbow trout, *Salmo gairdneri*, adapted to freshwater and dilute seawater, *J. Exp. Biol.*, 83, 181, 1979.

Eddy, F. B. and Fraser, J. E., Sialic acid and mucus production in rainbow trout (*Salmo gairdneri*) in response to zinc and seawater, *Comp. Biochem. Physiol.*, 73C, 357, 1982.

Eddy, F. B. and Williams, E. M., Nitrate and freshwater fish, *Chem. Ecol.*, 3, 1, 1987.

Ellsaesser, C. F. and Clem, L. W., Blood serum chemistry measurements of normal and acutely stressed channel catfish, *Comp. Biochem. Physiol.*, 88A, 589, 1987.

Englehardt, F. R., Wong, M. P. and Duey, M. E., Hydromineral balance and gill morphology in rainbow trout *Salmo gairdneri*, acclimated to fresh and seawater, as affected by petroleum exposure, *Aquat. Toxicol.*, 1, 175, 1981.

Evans, D., Osmotic and ionic regulation, in, *The Physiology of Fishes*, CRC Press, Boca Raton, FL, 1993, chap. 11.

Exley, C., Chappel, J. S. and Birchall, J. D., A mechanism for acute aluminum toxicity in fish, *J. Theor. Biol.*, 151, 417, 1991.

Febry, R. and Lutz, P., Energy partitioning in fish: the activity-related cost of osmoregulation in a euryhaline cichlid, *J. Exp. Biol.*, 128, 63, 1987.

Flick, G. and Perry, S., Cortisol stimulates whole body calcium uptake and the branchial calcium pump in freshwater rainbow trout, *J. Endocrinol.*, 120, 75, 1989.

Flick, G., Wendelaar Bonga, S. E. and Fenwick, J. C., Ca^{2+} dependent phosphatase and Ca^{2+} dependent ATPase activities in plasma membranes of eel gill epithelium. III. Stimulation of branchial high-affinity Ca^{2+}-ATPase activity during prolactin-induced hypercalcemia in American eels, *Comp. Biochem. Physiol.*, 79B, 521, 1984.

Fletcher, C. R., Osmotic and ionic regulating the cod (*Gadus callarias* L.), *J. Comp. Physiol.*, 124, 157, 1978.

Foskett, J. K., Bern, H. A., Machen, T. E. and Conner, M., Chloride cells and the hormonal control of teleost fish osmoregulation, *J. Exp. Biol.*, 106, 255, 1983.

Fraser, G. A. and Harvey, H. H., Effects of environmental pH on the ionic composition of the white sucker (*Catostomus commersoni*) and pumpkinseed (*Lepomis gibbosus*), *Can. J. Zool.*, 62, 249, 1984.

Fu, H., Lock, R. A. C. and Wendelaar Bonga, S. E., Effect of cadmium on prolactin cell activity and plasma electrolytes in the freshwater teleost *Oreochromis mossambicus*, *Aquat. Toxicol.*, 14, 295, 1989.

Gaino, E., Arillo, A. and Mensi, P., Involvement of the gill chloride cells of trout under acute nitrite intoxication, *Comp. Biochem. Physiol.,* 77A, 611, 1984.

Gardner, G. R. and Yevich, P. P., Histological and hematological responses of an estuarine teleost to cadmium, *J. Fish. Res. Bd. Can.,* 27, 2185, 1970.

Giles, M. A., Electrolyte and water balance in plasma and urine of rainbow trout (*Salmo gairdneri*) during chronic exposure to cadmium, *Can. J. Fish. Aquat. Sci.,* 41, 1678, 1984.

Gill, T. S., Jagdish, C. P. and Tewari, H., Cadmium nephropathy in a freshwater fish, *Puntius conchonius, Ecotoxicol. Environ. Safety,* 18, 165, 1989.

Girard, J. P. and Payan, P., Ion exchanges through respiratory and chloride cells in freshwater and seawater adapted teleosteans, *Am. J. Physiol.,* 238, R260, 1980.

Gonzalez, R., Grippo, R. S. and Dunson, W.A., The disruption of sodium balance in brook charr, *Salvelinus fontinalis* (Mitchill), by manganese and iron, *J. Fish Biol.,* 37, 765, 1990.

Grau, E. G., Dickhoff, W. W., Nichioka, R. S., Bern, H. A. and Folmar, L. C., Lunar phasing of the thyroxine surge preparatory to seaward migration of salmonid fish, *Science,* 211, 607, 1981.

Hall, L. W., Burton, D. T. and Liden, L. H., An interpretative literature analysis evaluating the effects of power plant chlorination on freshwater organisms, *Crit. Rev. Toxicol.,* 9, 1, 1981.

Handy, R. D. and Eddy, F. B., Effects of inorganic cations on sodium adsorption to the gill and body surface of rainbow trout, *Oncorhynchus mykiss,* in dilute solutions, *Can. J. Fish. Aquat. Sci.,* 48, 1829, 1991a.

Handy, R. D. and Eddy, F. B., The absence of mucus on the secondary lamellae of unstressed rainbow trout, *Oncorhynchus mykiss* (Walbaum), *J. Fish Biol.,* 38, 153, 1991b.

Handy, R. D., Eddy, F. B. and Romain, G., In vitro evidence for the ionoregulatory role of rainbow trout mucus in acid, acid/aluminum and zinc toxicity, *J. Fish Biol.,* 35, 737, 1989.

Hasan, J. and Hernberg, S., Interactions of inorganic lead with human red blood cells, *Arch. Environ. Health,* 2, 26, 1966.

Haux, C. and Larsson, A., Effects of DDT on blood plasma electrolytes in the flounder, Platichthys flesus, in hypotonic brackish water, *Ambio,* 8, 171, 1979.

Haux, C. and Larsson, A., Influence of inorganic lead on the biochemical blood composition in the rainbow trout, *Salmo gairdneri, Ecotoxicol. Environ. Safety,* 6, 28, 1982.

Haux, C., Larsson, A., Lithner, G. and Sjobeck, M., A field study of physiological effects on fish in lead-contaminated lakes, *Environ. Toxicol. Chem.,* 5, 283, 1985.

Heath, A. G., Changes in tissue adenylates and water content of bluegill, *Lepomis macrochirus,* exposed to copper, *J. Fish Biol.,* 24, 299, 1984.

Heath, A. G., Effects of waterborne copper or zinc on the osmoregulatory response of bluegill to a hypertonic NaCl challenge, *Comp. Biochem. Physiol.,* 88C, 307, 1987.

Hegab, S. A. and Hanke, W., Changes in catecholamine content of the hypothalamus during adaptation of fish to changed external salinity, *Gen. Comp. Endocrinol.,* 44, 324, 1981.

Heisinger, J. F. and Scott, L., Selenium prevents mercuric chloride induced acute osmoregulatory failure without glutathione peroxidase involvement in the black bullhead (*Ictalurus melas*), *Comp. Biochem. Physiol.,* 80C, 295, 1985.

Hilmy, A. M., Badawi, H. and Shabana, N., Physiological mechanism of toxic action of DDT and endrin in two euryhaline freshwater fishes, *Anguilla vulgaris* and *Mugil cephalus, Comp. Biochem. Physiol.,* 76C, 173, 1983.

Howells, C. D., Brosn, J. A. and Sadler, K., Effects of acidity, calcium and aluminum on fish survival and productivity — a review, *J. Sci. Food. Agric.,* 34, 559, 1983.

Hunn, J. B., Role of calcium in gill function in freshwater fishes, *Comp. Biochem. Physiol.,* 82A, 543, 1985.

Isaia, J., Water and nonelectric permeation, in, *Fish Physiology,* Vol. XB, Hoar, W.S. and Randall, D. J., Eds., 1984, chap. 1.

Iwama, G. K. and Morgan, J., Metabolic cost of ionic regulation in rainbow and steelhead trout, *Physiologist,* 33, A38, 1990

Jackson, W. F. and Fromm, P. O., Effect of a detergent on flux of tritiated water into isolated perfused gills of rainbow trout, *Comp. Biochem. Physiol.,* 58C, 167, 1977.

Janicki, R. and Kinter, W., DDT inhibits Na, K, Mg ATPase in the intestinal mucosae and the gills of marine teleosts, *Nature,* 223, 148, 1971.

Janicki, R. H. and Kinter, W. B., DDT: disrupted osmoregulatory events in the intestine of the eel *Anguilla rostrata,* adapted to seawater, *Science,* 173, 1146, 1971.

Kim, J. K., Birks, A. and Heisinger, J. F., Protective action of selenium against mercury in northern creek chubs, *Bull. Environ. Contam. Toxicol.,* 17, 132, 1977.

Kirschner, L. B., Water and ions, in, *Environmental and Metabolic Animal Physiology,* Prosser, C. L., Ed., Wiley-Liss, New York, 1991, chap. 2.

Kirubagaran, R. and Joy, K. P., Changes in adrenocortical-pituitary activity in the catfish, *Clarias batrachus* (L.) after mercury treatment, *Ecotoxicol. Environ. Safety,* 22, 36, 1991.

Kristoffersson, R., Broberg, S. and Oikari, A., Physiological effects of a sublethal concentration of phenol in the pike (*Esox lucius* L.) in pure brackish water, *Ann. Zool. Fennici,* 10, 392, 1973.

Larsson, A., Bengtsson, B. E. and Haux, C., Disturbed ion balance in flounder, *Platichthys flesus* L. exposed to sublethal levels of cadmium, *Aquat. Toxicol.,* 1, 19, 1981.

Lauren, D. J., The fish gill: a sensitive target for waterborne pollutants, in *Aquatic Toxicology and Risk Assessment:* Vol. 14, ASTM STP 1124, Mayes, M. A. and Barron, M. G., Eds., American Society for Testing Materials, Philadelphia, 1991, 223.

Lauren, J. and McDonald, D. G., Effects of copper on branchial ionoregulation in the rainbow trout, *Salmo gairdneri* Richardson, *J. Comp. Physiol.,* 155B, 635, 1985.

Lauren, D. J. and McDonald, D. G., Influence of water hardness, pH, and alkalinity on the mechanisms of copper toxicity in juvenile rainbow trout, *Salmo gairdneri, Can. J. Fish. Aquat. Sci.,* 43, 1488, 1986.

Lauren, D. J. and McDonald, D. G., Acclimation to copper by rainbow trout, *Salmo gairdneri:* physiology, *Can. J. Fish. Aquat. Sci.,* 44, 99, 1987a.

Lauren, D. J. and McDonald, D. G., Acclimation to copper by rainbow trout, *Salmo gairdneri:* biochemistry, *Can. J. Fish. Aquat. Sci.,* 44, 105, 1987b.

Leadem, T. P., Campbell, R. D. and Johnson, D. W., Osmoregulatory responses to DDT and varying salinities in *Salmo gairdneri.* I. Gill Na,K-ATPase, *Comp. Biochem. Physiol.,* 49A, 197, 1974.

Leatherland, J. F., Studies of the correlation between stress-response, osmoregulation and thyroid physiology in rainbow trout, *Salmo gairdneri* (Richardson), *Comp. Biochem. Physiol.,* 80A, 523, 1985.

Levitan, W. M. and Taylor, M. H., Physiology of salinity-dependent naphthalene toxicity in *Fundulus heteroclitus, J. Fish. Res. Bd. Can.,* 36, 615, 1979.

Lewis, S. D. and Lewis, W. M., The effect of zinc and copper on the osmolality of blood serum of the channel catfish, *Ictalurus punctatus* Rafinesque, and golden shiner, *Notemigonus crysoleucas* Mitchill, *Trans. Am. Fish. Soc.,* 100, 639, 1971.

Lloyd, R. and Swift, D. J., Some physiological responses by freshwater fish to low dissolved oxygen, high carbon dioxide, ammonia and phenol with particular reference to water balance, in, *Effects of Pollutants on Aquatic Organisms,* Lockwood, A. P. M., Ed., Cambridge University Press, New York, 1976, 47.

Lock, R. A. C., Cruijsen, P. M. and Van Overbeeke, A. P., Effects of mercuric chloride and methylmercuric chloride on the osmoregulatory function of the gills in rainbow trout, *Salmo gairdneri, Comp. Biochem. Physiol.,* 68C, 151, 1981.

Lock, R. A. C. and van Overbeeke, A. P., Effects of mercuric chloride and methylmercuric chloride on mucus secretion in rainbow trout, *Salmo gairdneri* Richardson, *Comp. Biochem. Physiol.,* 69C, 67, 1981.

Lorz, H. W. and McPherson, B. P., Effects of copper or zinc in fresh water on the adaptation to seawater and ATPase activity, and the effects of copper on migratory disposition of coho salmon (*Oncorhynchus kisutch*), *J. Fish. Res. Bd. Can.,* 33, 2023, 1976.

Madsen, S. S., Cortisol treatment improves the development of hypoosmoregulatory mechanisms in the euryhaline rainbow trout, *Salmo gairdneri, Fish Physiol. Biochem.,* 8, 45, 1991.

Malte, H., Effects of aluminum in hard, acid water on metabolic rate, blood gas tensions and ionic status in the rainbow trout, *J. Fish Biol.,* 29, 187, 1986.

Malte, H. and Weber, R. E., Respiratory stress in rainbow trout dying from aluminum exposure in soft, acid water, with or without added sodium chloride, *Fish Physiol. Biochem.,* 5, 249, 1988.

Matthiessen, P. and Brafield, A. E., The effects of dissolved zinc on the gills of the stickleback *Gasterosteus aculeatus* (L), *J. Fish Biol.,* 5, 607, 1973.

Mazik, P. M., Hinman, M. L., Winkelmann, D. A., Klaine, S. J. and Simco, B. A., Influence of nitrite and chloride concentrations on survival and hematological profiles of striped bass, *Trans. Am. Fish. Soc.,* 120, 247, 1991.

McCarty, L. S. and Houston, A. H., Effects of exposure to sublethal levels of cadmium upon water-electrolyte status in the goldfish (*Carassius auratus*), *J. Fish Biol.*, 9, 11, 1976.

McCormick, S. D., Sakamoto, T., Hasegawa, S. and Hirano, T., Osmoregulatory actions of insulin-like growth factor-I in rainbow trout (*Oncorhynchus mykiss*), *J. Endocrinol.*, 130, 87, 1991.

McDonald, D. G., Reader, J. P. and Dalziel, T. K. R., The combined effects of pH and trace metals on fish ionoregulation, in, *Acid Toxicity and Aquatic Animals,* Morris, R., Brown, D. J. A., Taylor, E. W. and Brown, J. A., Eds., Cambridge University Press, Cambridge, 1988.

McDonald, D. G., Cavdek, V. and Ellis, R., Gill design in freshwater fishes: interrelationships among gas exchange, ion regulation, and acid-base regulation, *Physiol. Zool.*, 64, 103, 1991a.

McDonald, D. G., Wood, C. M., Rhem, R. G., Mueller, M. E., Mount, D. R. and Bergman, H. L., Nature and time course of acclimation to aluminum in juvenile brook trout, (*Salvelinus fontinalis*). I. Physiology, *Can. J. Fish. Aquat. Sci.*, 48, 2006, 1991b.

McKim, J. M., Christensen, G. M., and Hunt, E. P., Changes in the blood of brook trout (*Salvelinus fontinalis*) after short-term and long-term exposure to copper, *J. Fish. Res. Bd. Can.*, 27, 1883, 1970.

McKeown, B. A. and March, G. L., The acute effect of bunker C oil and an oil dispersant on serum glucose and gill morphology in both freshwater and seawater acclimated rainbow trout (*Salmo gairdneri*), *Water Res.*, 12, 157, 1978.

Miller, T. and Mackay, V., The effects of hardness, alkalinity and pH of test water on the toxicity of copper to rainbow trout (*Salmo gairdneri*), *Water Res.*, 114, 124, 1980.

Miller, D. S. and Kinter, W. B., DDT inhibits nutrient absorption and osmoregulatory function in *Fundulus heteroclitus*, in, *Physiological Responses of Marine Biota to Pollutants,* Vernberg, F. J., Calabrese, A., Thurberg, F. P. and Vernberg, W. B., Eds., Academic Press, New York, 1977, 63.

Milligan, C. L. and Wood, C. M., Disturbances in haematology, fluid volume distribution and circulatory function associated with low environmental pH in the rainbow trout, *Salmo gairdneri, J. Exp. Biol.*, 99, 397, 1982.

Mitrovic, V. V., Brown, V. M., Shurben, D. G. and Berryman, M. H., Some pathological effects of sub-acute and acute poisoning of rainbow trout by phenol in hard water, *Water Res.*, 2, 249, 1968.

Mueller, M. E., Sanchez, D. A., Bergman, H. L., McDonald, D. G., Rhem, R. G. and Wood, C. M., Nature and time course of acclimation to aluminum in juvenile brook trout (*Salvelinus fontinalis*). II. *Can. J. Fish. Aquat. Sci.*, 48, 2016, 1991.

Murty, A. S., *Toxicity of Pesticides to Fish,* Vol. 2, CRC Press, Boca Raton, FL, 1986.

Neville, C. M., Physiological response of juvenile rainbow trout, *Salmo gairdneri*, to acid and aluminum — prediction of field responses from laboratory data, *Can. J. Fish. Aquat. Sci.*, 42, 2004, 1985.

Newman, M. W. and MacLean, S. A., Histopathology, in, *Physiological Response of the Cunner, Tautogolabrus adspersus, to cadmium,* NOAA Technical Report NMFS SSRF-681, 1974, chap. VI.

Olson, K. R., Blood and extracellular fluid volume regulation: role of renin-angiotensin system, kallikrein-kinin system and atrial natriuretic peptides, in, *Fish Physiology,* Vol. 12, Hoar, W. S. and Randall, D. J., Eds., Academic Press, New York, 1992, chap. 3.

Oppenborn, J. B. and Goudie, C., Acute and sublethal effects of ammonia on striped bass and hybrid striped bass, *J. World Aquacult. Soc.*, 24, 90, 1993.

Oronsaye, J. A. O. and Brafield, A. E., The effect of dissolved cadmium on the chloride cells of the gills of the stickleback, *Gasterosteus aculeatus* L., *J. Fish Biol.*, 25, 253, 1984.

Paley, R., Twitchen, I. D. and Eddy, F. B., Ammonia, Na, K and Cl levels in rainbow trout yolk-sac fry in response to external ammonia, *J. Exp. Biol.*, 180, 273, 1993.

Payan, P. and Girard, J. P., Branchial ion movements in teleosts: the roles of respiratory and chloride cells, in, *Fish Physiology,* Vol. XB, Hoar, W. S. and Randall, D. J., Eds., 1984, chap. 2.

Payan, P., Matty, A. J. and Maetz, J., A study of the sodium pump in the perfused head preparation of the trout *Salmo gairdneri* in freshwater, *J. Comp. Physiol.*, 104, 33, 1975.

Pickering, A. D., Endocrine-induced pathology in stressed salmonid fish, *Fish. Res.*, 17, 35, 1993.

Pinkney, A., Wright, D. A., Jepson, M. A. and Towle, D. W., Effects of tributyltin compounds on ionic regulation and gill ATPase activity in estuarine fish, *Comp. Biochem. Physiol.*, 92C, 125, 1989.

Potts, W. and McWilliams, P., The effects of hydrogen and aluminum ions on fish gills, in, *Acid Toxicity and Aquatic Animals,* Morris, R., Taylor, E., Brown, D. and Brown, J., Eds., Cambridge University Press, New York, 1989, 202.

Pratap, H. B., Fu, H., Lock, R. A. C. and Wendelaar Bonga, S. E., Effect of waterborne and dietary cadmium on plasma ions of the teleost *Oreochromis mossambicus* in relation to water calcium levels, *Arch. Environ. Contam. Toxicol.*, 18, 568, 1989.

Rangaswamy, C. P. and Naidu, B. P., Endosulfan induced changes in the serum calcium and magnesium levels in the food fish, *Tilapia mossambica* (Peters), *J. Environ. Biol.,* 10, 245, 1989.

Reid, S. D. and McDonald, D. G., Effects of cadmium, copper, and low pH on ion fluxes in the rainbow trout, *Salmo gairdneri, Can. J. Fish. Aquat. Sci.,* 45, 244, 1988.

Reid, S. D., McDonald, D. G. and Rhem, R. R., Acclimation to sublethal aluminum: modification of metal-gill surface interactions of juvenile rainbow trout, *Oncorhynchus mykiss, Can. J. Fish. Aquat. Sci.,* 48, 1996, 1991.

Reinking, L. N., Aldosterone response to renin, angiotensin, ACTH, hemorrhage and sodium depletion in a freshwater teleost, *Catostomus macrocheilus, Comp. Biochem. Physiol.,* 74A, 873, 1983.

Renfro, J. L., Schmidt-Nielsen, B., Miller, D., Benos, D. and Allen, J., Methyl mercury and inorganic mercury: uptake, distribution, and effect on osmoregulatory mechanisms in fishes, in, *Pollution and Physiology of Marine Organisms,* Vernberg, F. G. and Vernberg, W. B., Eds., Academic Press, New York, 1974, 101.

Roch, M. and Maly, E. J., Relationship of cadmium-induced hypocalcemia with mortality in rainbow trout *(Salmo gairdneri)* and the influence of temperature on toxicity, *J. Fish. Res. Bd. Can.,* 36, 1297, 1979.

Rombough, P. J. and Garside, E. T., Cadmium toxicity and accumulation in eggs and alevins of Atlantic salmon, *Salmo salar, Can. J. Zool.,* 60, 2006, 1982.

Rombough, P. J. and Garside, E. T., Disturbed ion balance in alevins of Atlantic salmon *Salmo salar* chronically exposed to sublethal concentrations of cadmium, *Can. J. Zool.,* 62, 1443, 1984.

Sauer, G. R. and Watabe, N., The effects of heavy metals and metabolic inhibitors on calcium uptake by gills and scales of *Fundulus heteroclitus* in vitro, *Comp. Biochem. Physiol.,* 91C, 473, 1988.

Sayer, M. D. J., Reader, J. P. and Morris, R., Effects of six trace metals on calcium fluxes in brown trout, *Salmo trutta,* in soft water, *J. Comp. Physiol. B, Biochem. Syst. Environ. Physiol.,* 161, 537, 1991.

Schmidt-Nielsen, B., Sheline, J., Miller, D. S. and Deldonno, M., Effect of methylmercury upon osmoregulation, cellular volume, and ion regulation in winter flounder, *Pseudopleuronectes americanus,* in, *Physiological Responses of Marine Biota to Pollutants,* Vernberg, F.J., Calabrese, A., Thurberg, F. P. and Vernberg, W. B., Eds., Academic Press, New York, 1977, 105.

Shephard, K. L., The influence of mucus on the diffusion of ions across the esophagus of fish, *Physiol. Zool.,* 55, 23, 1982.

Singh, H. S. and Reddy, T. V., Effect of copper sulfate on hematology, blood chemistry, and hepatosomatic index of an Indian catfish, *Heteropneustes fossilis* (Bloch), and its recovery, *Ecotoxicol. Environ. Safety,* 20, 30, 1990.

Skidmore, J. F., Respiration and osmoregulation in rainbow trout with gills damaged by zinc sulphate, *J. Exp. Biol.,* 52, 481, 1970.

Skidmore, J. F. and Tovell, P. W. A., Toxic effects of zinc sulphate on the gills of rainbow trout, *Water Res.,* 6, 217, 1972.

Sleet, R. B. and Weber, L. J., Water and electrolyte imbalances associated with laboratory manipulation of a marine teleost involve the gut, *Can. J. Zool.,* 61, 1202, 1983.

Smart, G., Investigations of the toxic mechanisms of ammonia to fish-gas exchange in rainbow trout *(Salmo gairdneri)* exposed to acutely lethal concentrations, *J. Fish Biol.,* 12, 93, 1978.

Smith, J. R., Fish neurotoxicology, in, *Aquatic Toxicology,* Vol. 2, Weber, L. J., Ed., Raven Press, New York, 1984, 107.

Spry, D. J. and Wood, C. M., Acid-base, plasma ion and blood gas changes in rainbow trout during short term toxic zinc exposure, *J. Comp. Physiol. B.,* 154, 149, 1984.

Spry, D. J. and Wood, C. M., Ion flux rates, acid-base status, and blood gases in rainbow trout, *Salmo gairdneri,* exposed to toxic zinc in natural soft water, *Can. J. Fish. Aquat. Sci.,* 42, 1332, 1985.

Stagg, R. M. and Shuttleworth, T. J., The accumulation of copper in *Platichthys flesus* L. and its effects on plasma electrolyte concentrations, *J. Fish Biol.,* 20, 491, 1982.

Staurnes, M., Sigholt, T. and Reite, O. B., Reduced carbonic anhydrase and Na-K-ATPase activity in gills of salmonids exposed to aluminum-containing acid water, *Experientia,* 40, 226, 1984.

Stickle, W. B., Sabourin, T. D. and Rice, S. D., Sensitivity and osmoregulation of coho salmon *Oncorhynchus kisutch,* exposed to toluene and naphthalene at different salinities, in, *Physiological Mechanisms of Pollutant Toxicity,* Vernberg, W. B., Calabrese, A., Thurberg, F. P. and Vernberg, F. J., Eds., Academic Press, New York, 1982, 331.

Stinson, C. and Mallatt, J., Branchial ion fluxes and toxicant extraction efficiency in lamprey (*Petromyzon marinus*) exposed to methylmercury, *Aquat. Toxicol.*, 15, 237, 1989.

Sugatt, R. H., Effects of sublethal sodium dichromate exposure in freshwater on the salinity tolerance and serum osmolality of juvenile coho salmon, *Oncorhynchus kisutch,* in seawater, *Arch. Environ. Contam. Toxicol.,* 9, 41, 1980.

Swift, D. J., Changes in selected blood component concentrations of rainbow trout, *Salmo gairdneri* Richardson, exposed to hypoxia or sublethal concentrations of phenol or ammonia, *J. Fish Biol.,* 19, 1981.

Taylor, C. W., Calcium regulation in vertebrates: An overview, *Comp. Biochem. Physiol.,* 82A, 249, 1985.

Thurberg, P. and Dawson, M. A., Changes in osmoregulation and oxygen consumption, in, *Physiological Response of the Cunner, Tautogolabrus adspersus,* to cadmium, NOAA Technical Report NMFS SSRF-681, 1974, chap. III.

van der Putte, I., Laurier, M. B. H. M. and van Eijk, G. J. M., Respiration and osmoregulation in rainbow trout (*Salmo gairdneri*) exposed to hexavalent chromium at different pH values, *Aquat. Toxicol.,* 2, 99, 1982.

Verbost, P. M., Flik, G., Lock. R. A. C. and Bonga, S. E. W., Cadmium inhibition of Ca2+ uptake in rainbow trout gills, *Am. J. Physiol.,* 253, R216, 1987.

Verbost, P., Lafeber, F., Spanings, F., Aarden, E. and Wendelaar Bonga, S. E., Inhibition of Ca uptake in freshwater carp, *Cyprinus carpio,* during short-term exposure to aluminum, *J. Exp. Zool.,* 262, 247, 1992.

Vermette, M. and Perry, S., The effects of prolonged epinephrine infusion on the physiology of the rainbow trout *Salmo gairdneri.* II. Branchial solute fluxes, *J. Exp. Biol.,* 128, 255, 1987.

Waring, C. P., Stagg, R. M. and Poxton, M. G., The effects of handling on flounder (*Platichthys flesus*) and Atlantic salmon (*Salmo salar*), *J. Fish Biol.,* 41, 131, 1992.

Watson, T. A. and Beamish, F. W. H., Effects of zinc on branchial ATPase activity in vivo in rainbow trout, *Salmo gairdneri, Comp. Biochem. Physiol.,* 66C, 77, 1980.

Wedemeyer, G. A., Nelson, N. C. and Yasutake, W. T., Physiological and biochemical aspects of ozone toxicity to rainbow trout (*Salmo gairdneri*), *J. Fish. Res. Bd. Can.,* 36, 605, 1979.

Wendelaar Bonga, S. E., van der Meij, J. C. A. and Flik, G., Prolactin and acid stress in the teleost *Oreochromis* (formerly *Sarotherodon*) *mossambicus, Gen. Comp. Endocrinol.,* 55, 323, 1984.

Wendelaar Bonga, S. E. and Lock, R. A., Toxicants and osmoregulation in fish, *Neth. J. Zool.,* 42, 478, 1992.

Williams, E. M. and Eddy, F. B., Anion transport, chloride cell number and nitrite-induced methaemoglobinaemia in rainbow trout (*Salmo gairdneri*) and carp (*Cyprinus carpio*), *Aquat. Toxicol.,* 13, 29, 1988a.

Williams, E. and Eddy, F. B., Regulation of blood haemoglobin and electrolytes in rainbow trout, *Salmo gairdneri* (Richardson) exposed to nitrite, *Aquat. Toxicol.* 13, 13, 1988b.

Wilson, R. and Taylor, E., The physiological responses of freshwater rainbow trout, (*Oncorhynchus mykiss*) during acutely lethal copper exposure, *J. Comp. Physiol. B.,* 163, 38, 1993a.

Wilson, R. and Taylor, E., Differential responses to copper in rainbow trout (*Oncorhynchus mykiss*) acclimated to sea water and brackish water, *J. Comp. Physiol. B.,* 163, 239, 1993b.

Wood, C. M., Flux measurements as indices of H+ and metal effects on freshwater fish, *Aquat. Toxicol.* 22, 239, 1992.

Wood C. M., Playle, R. C., Simons, B. P., Goss, G. G. and McDonald, D. G., Blood gases, acid-base status, ions, and hematology in adult brook trout, *Salvelinus fontinalis,* under acid/aluminum exposure, *Can. J. Fish. Aquat. Sci.,* 45, 1575, 1988.

Zbanyszek, R. and Smith, L. S., The effect of water-soluble aromatic hydrocarbons on some haematological parameters of rainbow trout, *Salmo gairdneri,* Richardson, during acute exposure, *J. Fish Biol.,* 24, 545, 1984.

Zeitoun, I. H., Hughes, L. D. and Ullrey, D. E., Effect of shock exposures of chlorine on the plasma electrolyte concentration of adult rainbow trout (*Salmo gairdneri*), *J. Fish. Res. Bd. Can.,* 34, 1034, 1977.

Chapter

Physiological Energetics

I. INTRODUCTION

This chapter is devoted to certain aspects of the subject of bioenergetics, an immense subject encompassing levels of complexity from the cellular to the ecosystem. Physiological energetics, which is the main thrust of this chapter, includes the study of rates of energy expenditure by individual fish, losses and gains of energy in the body, and transformations and mobilizations of energy "pools" within the fish. Growth, which is inextricably linked to energetics, is also included in this chapter. Indeed, many fish bioenergetics studies are basically studies of growth rate.

After a review of some basic concepts of energetics and methodologies applicable to fish, the effects of pollutants on energy expenditure by individual fish will be considered with implications for growth included where appropriate. The effects of pollutants on growth rate per se of larval and juvenile fish follows. Swimming performance will then be discussed at some length because the presence of toxic chemicals may restrict this highly energy-dependent activity. Finally, a review of the influence of pollutants on body energy stores of carbohydrate, lipid, and protein closes the chapter.

II. GENERAL CONCEPTS

In biological studies, the basic unit of energy measurement has traditionally been the calorie. With the transition to the SI system (Systeme International), the joule (1 cal = 4.1868 J) is becoming the basic measure of energy expenditure, although some continue to use the calorie (or Calorie = kcal). For a good discussion of the SI system and how it relates to animal energetics, the review by Bartholomew (1982) is recommended.

The following simple equation summarizes the principles of energy input and output for any animal, the actual numbers involved become the energy budget.

$$I = G + M + E + F$$

where:

I = gross energy intake
G = tissue production (i.e., growth)
M = respiratory energy expenditure (also called metabolic cost or metabolic rate)
E = energy lost via nitrogenous waste (NH_3)
F = fecal loss (i.e., energy not absorbed by the digestive tract)   (1)

The metabolic rate component can be further partitioned into standard (maintenance), active, and postprandial metabolic rates; and the growth component can be divided into somatic and gonadal (Wieser and Medgyesy, 1991). Bayne et al. (1979) have rearranged Equation 1 slightly:

$$G = A - (M + E)$$ (2)

where A refers to assimilation. They call "G" the "scope for growth" which emphasizes how growth and energy metabolism are intimately related. Thus far, the scope for growth concept has mostly been applied to pollutant effects on mollusks, but there is little reason it could not also be used in fish studies.

Because fish are dynamic systems involving inputs and outputs which are constantly changing, none of the symbols in Equation 1 are constants. A useful way of visualizing what is happening to energy exchanges in a fish is to examine Figure 1.

In this hypothetical fish we start with an intake of 100 cal, 20 of which are lost as feces and 7 as nitrogenous waste. (In fish, incidentally, most of the nitrogenous waste is ammonia which is excreted via the gills, in contrast to terrestrial animals which excrete these wastes via the urine.) The cost of digestion and assimilation of the food is 14 cal

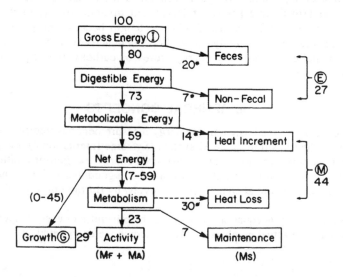

Figure 1 Average partitioning of dietary energy for a carnivorous fish. Numbers refer to calories, letters to the symbols in Equation 1. See text for further explanation. (From Brett, D. J. and Groves, T. D. D., *Fish Physiology*, Vol. 8, Hoar, W. S. and Randall, D. J., Eds., Academic Press, New York, 1979, chap. 6. With permission.)

(Specific Dynamic Action or postprandial) leaving a net energy to be used by the organism of 59 cal. This net energy is partitioned between maintenance (i.e., basal metabolic rate), swimming (active) metabolism, and growth. Clearly then, energy metabolism (which is the same thing as energy expenditure) and growth are competing for the net energy. Thus, if metabolism is elevated because of environmental stressors, growth will be limited unless the intake of food is increased.

The above generalized energy budget applies to a typical carnivorous fish with abundant food. Although 95% of the fish species are carnivorous, some important commercial and aquaculture groups (e.g., cichlids, cyprinids) are herbivorous or omnivorous (Pandian, 1987). Thus, it is interesting to compare generalized budgets for the two different feeding styles (Brett and Groves, 1979):

Carnivores: 100 C = 29 G + 44 M + 7 E + 20 F

Herbivores: 100 C = 20 G + 37 M + 2 E + 41 F

The assimilation efficiency of the herbivores is considerably less than that of carnivores, so a sizable amount of energy is lost via the feces in herbivorous species. Note how the amount of energy lost via ammonia excretion (E) is relatively small compared to other parts of the energy budget.

The percent of food assimilated varies with the diet composition. As a general rule, carbohydrates are poorly assimilated in fish; about 60% is lost in the feces of carnivorous fish. Omnivorous species such as carp do a bit better. Even so, for most fish, protein is considered to be the major source of energy (Pandian, 1987). Having said that, however, it should be further noted that lipids can be utilized by many species and these may reduce the protein requirement. Furthermore, Brown et al. (1990) have shown that in channel catfish (*Ictalurus punctatus*), at feeding levels of maintenance or below, lipids and carbohydrates are utilized more efficiently than protein. As the daily ration is increased above maintenance, protein and lipid are utilized more efficiently.

Protein, in addition to being a major intake energy source, is also a major form in which energy is stored. Diana (1982, p. 395) has noted that: "Proximate composition of fish under various feeding regimes indicated that energy gain or depletion from the body was due to changes in amount of whole body tissue or body protein, rather than specific utilization or storage of lipid."

Tissues that synthesize a lot of protein may use a sizable proportion of their total energy expenditure for protein synthesis. Recently, Pannevis and Houlihan (1992) have found that isolated fish liver cells utilized an average of nearly 80% of their total oxygen consumption for this purpose. This contrasts with a value of 2.8% for the Na,K ATPase activity which maintains the transmembrane electrolyte balance. Thus, changes in standard oxygen consumption of whole animals may at times reflect changes in protein synthesis rates, a point rarely considered in the literature.

The generalizations about the importance of protein as an energy source in fish applies to whole-animal studies. Individual tissues may utilize other substrates preferentially. For example, fish gill (which is a mixture of several cell types) prefers glucose and lactate over amino acids for oxidation (Mommesen, 1984). Even though protein is a major source of energy in fish, the stress of exercise (Driedzic and Hochachka, 1979) or severe environmental hypoxia (Heath and Pritchard, 1965) causes rapid depletion of stores of carbohydrate, primarily liver and muscle glycogen. A considerable amount of work has shown that certain pollutants may do the same thing (discussed later in this chapter).

Since the early 1970s, there have been a number of attempts to compile energy budgets for several species of fish in the laboratory and in the field (see Brafield, 1985; Soofiani and Hawkins, 1985; Adams and Breck, 1990 for review of methodologies used). Hewett and Johnson (1992) have also created a generalized bioenergetics model for fish growth for microcomputers. It enables an experimenter to alter various factors (e.g., oxygen consumption rate or temperature) and follow changes in aspects of the energy budget over time. It is set up to operate with 20 different species including lampreys, several salmonids, centrarchids, and percids.

Most physiological studies of the effects of pollutants on energetics in fish have involved measures of energy expenditure and that will be the emphasis in this chapter. Bartholomew (1982, p. 947) has pointed out that: "The rate of energy metabolism probably integrates more aspects of animal performance than any other single physiological parameter." Because all physiological activities require energy, measurements of the rate of its expenditure by a fish tells us a lot about how the fish is functioning in its environment, and when the variations over time are observed, the relationship of temporal changes in the animal to those in its environment can also be investigated.

III. METHODS OF MEASURING ENERGY EXPENDITURE IN FISH

The most common method of measuring the energy expended by a fish is to measure the rate of oxygen consumption (sometimes called respiration). This is not to be confused with breathing movements which some authors designate by the same name. The frequency of breathing movements is not a good indicator of metabolic rate because gill damage which causes a reduced flow of oxygen from the water to the blood, and thus a decreased consumption of oxygen also causes increased breathing (see Chapter 3).

Attempts have been made to measure heat production (i.e., direct calorimetry) in fish (Brafield, 1985), however, difficult technical problems associated with this approach have prevented its wide application. One can calculate the joules released from oxygen consumption data (19.38 J/ml oxygen consumed). This assumes, however, no anaerobic energy expenditure. If the fish is being forced to swim at burst speed or is under severe hypoxic or pollutant stress, then there will be an anaerobic component. In order to measure this, whole-body lactic acid levels can be determined and then the energy released by anaerobic metabolism computed (0.27 mg lactic acid/4.186 J).

It is also possible with some effort to measure oxygen consumption, and carbon dioxide and ammonia excretion rates all simultaneously (Brafield, 1985). This yields information on substrate utilization but has not been applied much at this writing.

A method to estimate maintenance (i.e., basal metabolism) energy needs is to feed groups of fish at different rates and determine the daily percent energy change in the whole body by bomb calorimetry. When feed rate is plotted against percent energy change per day, a straight line is obtained which passes through the zero energy change (i.e., maintenance level) at some feed rate that can then be used to estimate maintenance feed consumption (Brown et al., 1990).

Oxygen consumption rate is by far the most common way to measure energy expenditure in any animal and fish are no exception. The consumption of oxygen by a fish is usually measured by confining the animal in a chamber through which water slowly flows (Cech, 1990). The difference in concentration of dissolved oxygen between the inflowing and outflowing water multiplied by the flow rate yields the oxygen consumed. Various intermittent flow arrangements have also been devised to avoid problems of lag. In nearly all the literature, oxygen consumption data on fish are expressed as either mg O_2/kg body wt/h, or as ml O_2/h. There has also been a not too successful effort to get investigators to express such data in μmol O_2 (31.251 μmol = 1 mg) (Gnaiger, 1983).

Data on consumption of oxygen (metabolic rate) by fish are classified into three types.

1. *Standard metabolic rate.* This approximates the basal metabolic rate commonly used in human energetics but it is difficult to define with precision in fish. Thus, the fish is

allowed a period of acclimatization to the chamber (e.g., 12–24 h), is free from any external disturbances, has been acclimated to the test temperature for approximately a week, and is in a postabsorptive state. Most importantly for the determination of standard metabolic rate is the complete elimination of body movement. Because this is clearly impossible to do without subjecting the organism to stress, and anesthetics cause undesirable side effects, some quantitative measure of movement is often obtained and the data relating oxygen consumption to body movement are extrapolated to zero body movement (Beamish and Mookherjii, 1964; Cech, 1990). The standard rate is a measure of the cost of homeostatic regulation in the animal. Thus, all the various requirements of energy for pumping blood, regulating ionic composition, nervous conduction, tissue repair, etc., are included in this.

2. *Routine metabolic rate.* This rate of oxygen consumption is similar to the standard rate, however, it is more variable because random body movement is not eliminated as a factor. For this reason, the values will be higher. If the fish is a relatively inactive type and the respirometer chamber is small, it will come close to yielding the standard metabolic rate. By far the majority of "resting" metabolic rate data on fish are of this type and some would argue that it has more environmental realism than does the standard rate.

3. *Active metabolic rate.* In order to obtain this, the fish must be forced to swim at a maximum sustained speed while the consumption of oxygen is determined. Various water tunnels and other devices have been devised for this purpose (Cech, 1990; Fry, 1971).

The difference between the standard and active rate is termed the metabolic scope (Bartholomew, 1982). It is a measure of the maximum rate of energy that a fish can expend above that required to keep the homeostatic machinery functioning. It would probably be more precise to refer to this as aerobic scope, for as fish approach a maximum swimming speed, anaerobic metabolic processes begin to play a part in the total expenditure of energy (Driedzic and Hochachka, 1978). Precisely where the line separating aerobic and anaerobic expenditure of energy lies is not known but undoubtedly varies greatly between species of fish and their swimming habits (e.g., sculpins vs. tunas).

When measuring energy metabolism in fish, it is important to realize there are many variables that may have large effects on metabolic rate. These must be recognized and kept as constant as possible when measuring the effect of some pollutant on this rate, unless one wishes to investigate potentially complicating interactions.

Water temperature probably has the greatest effect of any natural environmental variable on metabolic rate. Because fish (except for some species such as tuna) are ectothermic animals, their bodies are for all intents and purposes at the same temperature as the water. Routine metabolic rate of a species of fish from the temperate zones averages an increase of 2.3-fold for a rise of 10°C (i.e., the $Q10 = 2.3$) in the midpoint of the thermal range for that species (Brett and Groves, 1979). While acclimation for a period of days to a new temperature may reduce the extent of this increase, it will not generally eliminate it, and as we shall see, chronic exposure to a pollutant (e.g., copper) may delay the process of temperature acclimation.

The feeding of fish has a large effect on their rate of energy metabolism. Feeding causes an increase in the rate which is roughly proportional to the size of the ration up to a saturation level. In seven species investigated by Jobling (1981), there was an average doubling of the routine metabolic rate when fed at saturation level. This postprandial elevation in metabolism persists for a few hours to several days. In the plaice, at least, the size of the ration has little influence on the duration of the postprandial elevation. Most workers have made their measurements either 24 or 48 h postfeeding, which seems to be a reasonable compromise.

Because long-term exposures to pollutants may reduce appetite, when doing chronic studies it is probably best to give controls the same ration size as experimental animals receive, even though this may be less than they were getting before. However, excessive food deprivation can cause an abnormally low rate of metabolism (Beamish, 1964).

Bouts of severe exercise can cause an oxygen debt so the oxygen consumption rate after the exercise may remain elevated for 3–12 h, depending on species and extent of the exercise (Heath and Pritchard, 1962; Goolish, 1989). The same phenomenon occurs following severe hypoxia or anoxia (Heath and Pritchard, 1965; Van den Thillart and Verbeek, 1991).

The greater the body size of an animal, the slower its rate of metabolism per gram of tissue. This rule holds for most poikilothermic and homeothermic animals. Therefore, it is best to either group fish by size or use a limited size range thereby eliminating this factor as a variable.

These precautions may seem obvious to the experienced investigator, but unfortunately, in many studies of metabolism adequate attention has not been paid to them. It is hoped this brief review will raise the consciousness of those who plan to pursue these kinds of studies in the future. Still, even with careful elimination of extraneous variables, there is an element of truth in the comment by Wood and McDonald (1982, p. 199): "In our experience, one only has to think evil of a fish to double its rate of oxygen consumption."

Fry (1971) has devised a system of classification for environmental factors based on how they affect the metabolic rate of a fish. A factor that reduces the active oxygen consumption is called a "limiting factor". An example of this would be the concentration of dissolved oxygen. At progressively lower levels of oxygen in the water the active oxygen consumption decreases in proportion to the severity of the hypoxia (cf. Chapter 2). A factor that alters the standard metabolic rate by changes in the flow rates of molecular components through the cellular metabolic pathways is referred to by Fry as a "controlling" factor. Temperature is an example of this.

If an environmental factor influences the metabolic cost of maintenance, this has been designated as "masking" (Fry, 1971). An example is the requirement for energy associated with osmotic and ionic regulation. As the salinity (i.e., the masking factor) changes these costs increase or decrease but not necessarily in proportion to the change in osmotic or ionic gradient (Rao, 1968). The possible interaction of these three categories of factors should not be overlooked. For example, it is well known that low dissolved oxygen has a more limiting effect at high rather than low temperatures (Fry, 1971). On the other hand, acclimation temperature apparently does not affect the metabolic cost of osmoregulation (Rao, 1968). Pollutants may act as one or more of these environmental factors.

IV. EFFECTS OF METALS ON METABOLIC RATE

A. COPPER

We begin this section with a review of the work involving copper as the "pollutant" because there is probably more known about the effects of this on energy metabolism in fish than any other chemical. The observations should not, however, be interpreted as a "typical" response to a metal, as will be seen when the other metals are discussed.

When routine oxygen consumption of bluegill (*Lepomis macrochirus*) was measured while they were exposed to copper for 7 days, there was an initial stimulation followed by inhibition (O'Hara, 1971). Both the stimulation and inhibition were concentration dependent.

The initial stimulatory effect is most likely due to simple irritation of the mucous membranes of the oral cavity which causes fish to move around more. With subsequent adaptation of the receptors, this effect could abate in a day or two. More interesting is the depression of metabolism with prolonged exposure to copper. Felts and Heath (1984) more recently observed that bluegills exposed to 0.21 mg/L copper showed no inhibition in oxygen consumption until sometime between 9 and 30 days of continuous exposure. This concentration of copper is below any of those tested by O'Hara (1971) and because the inhibition required so long to occur, it suggests the extent of inhibition might be dependent on the amount of copper accumulated from the water. If this were the case, individual tissue respiration rate could be affected. To test this hypothesis, the *in vitro* oxygen consumption of liver, brain, and gill tissue was measured (Felts and Heath, 1984). At 30 days of exposure (20°C), only the respiration rate of the liver was different from that of controls, and it was slightly elevated rather than inhibited; thus the hypothesis is rejected. The stimulation of liver energy consumption may reflect an increase in the detoxification activity and associated protein synthesis in the liver. Also, copper has been found to stimulate respiration when added *in vitro* to fish liver mitochondria (Zaba and Harris, 1978).

The explanation for the whole-body metabolic depression produced by copper must then lie with a reduction in muscle metabolism. This could be due to: (1) inhibition of energy metabolism enzymes by copper, (2) decreased muscle tone, or (3) decreased spontaneous activity. The first possibility is probably not important, at least at this concentration of copper, because such an inhibition should have caused reduced respiration in the other tissues measured and that was not observed in the Felts and Heath study. A reduction in spontaneous activity of bluegill exposed to sublethal concentrations of copper has been quantified (Ellgaard and Guillot, 1988). Ellgaard and Guillot suggest this might be a direct effect of the copper on the enzymatic reactions of respiration. However, the *in vitro* tissue respiration work of Felts and Heath (1984) does not support that hypothesis.

In work by Waiwood and Beamish (1978) the effect of body movement was kept constant by measuring the oxygen consumption of fish swimming at controlled speeds following their exposure to copper for 5 days (Figure 2). The concentration of copper used was approximately that which would cause 20% mortality in 240 h. Even at this low concentration, there was a significant effect on metabolism. Copper caused the oxygen consumption to become elevated at all swimming speeds suggesting an increased cost of maintenance (masking effect), contrary to the results discussed above. The metabolic scope was also reduced (limiting effect), which is a separate issue to be considered later. Unfortunately, Waiwood and Beamish (1978) were using trout while the other work mentioned was done on bluegill; therefore there may be a species-specific difference. A more likely explanation is that the copper caused an increased maintenance cost in certain tissues (e.g., liver) while at the same time it reduced spontaneous movement in fish not forced to swim. Recall the discussion at the beginning of this chapter regarding the high cost of protein synthesis in liver.

Support for the hypothesis of increased maintenance cost of copper-exposed fish is found in work on perch growth and food consumption. Collvin (1985) found that growth was suppressed at very low levels of copper even though food consumption remained unchanged. Copper presumably does not cause a reduction in assimilation efficiency in fish (Lett et al., 1976), so the only logical explanation for the reduced growth is a greater utilization of energy by the fish leaving less to be used for growth.

In summary, it appears that exposure of fish to sublethal levels of copper causes metabolism in some non-muscular tissues (e.g., liver and gill) to increase while depressing spontaneous muscular activity. This latter effect may involve some effects of the metal on the central nervous system. The net effect on whole-animal oxygen consumption will depend on whether one measures it in a swimming fish, where muscle metabolism is kept the same in experimental and control animals, and therefore any differences are due to changes in other tissues, or in "resting fish", in which case small variations in muscular activity are probably the dominant influence on metabolic rate.

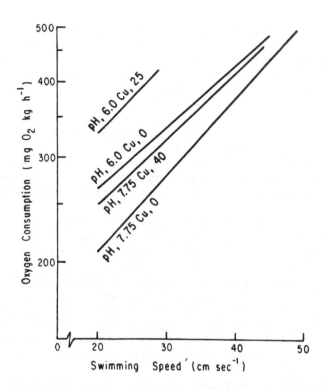

Figure 2 Oxygen consumption rate of rainbow trout *(Salmo gairdneri)* at different swimming speeds measured on fish which were first exposed to waterborne copper for 5 days. The concentrations of copper are in μg/L. (From Waiwood, K. G. and Beamish, F. W. H., *Water Res.*, 12, 611, 1978. With permission.)

With the close coupling between metabolic rate and growth, changes in the latter are to be expected when metabolism is altered. So, it is not surprising that chronic exposure to copper depresses the rate of growth in young trout (Lett et al., 1976). However, over a period of 10–40 days of continued exposure there appears to be an acclimation to the copper and return to normal growth. The speed of return to normal is inversely proportional to the concentration of copper in the water.

Lett et al. (1976) have also noted a virtual cessation of eating upon initial exposure to copper. Then over a period of 5–15 days, the fishes' appetite returns to normal. As with growth, the rate of recovery depends on the concentration of copper. Because the appetite returns to normal more rapidly than the growth rate, it appears that a lack of eating is not the sole cause of the slowed rate of growth, although it must contribute to it.

Collvin (1985) also found that perch acclimated to the copper so that after 20 or more days (depending on copper concentration in the water), the effects on growth were reduced or disappeared. In his study, however, food consumption was constant so it raises the possibility that once the detoxification mechanisms (metallothionein?) are fully induced, the energetic demands are reduced so they are not competing as much with growth for the energy intake.

Suppression of appetite is commonly observed in fish when they are exposed to metals (Lett et al., 1976 and personal observation). The mechanism for this has not been determined, but probably is in part caused by hormonal changes. Fish under almost any type of stress show elevated concentrations of plasma corticosteroid (Donaldson, 1981) and adrenaline (Mazeaud and Mazeaud, 1981). These hormones could cause a direct inhibition of eating or they might even do it indirectly by producing changes in the blood glucose level. Colgan (1973) has presented evidence that appetite in fish is suppressed by high levels of glucose in the blood. The higher the blood glucose, the less the food intake. The above-mentioned hormonal changes cause mobilization of liver glycogen into blood glucose, so in essence the system may be fooled into "accepting" the caloric intake as being more than adequate. The acclimation phenomenon whereby the appetite returns to normal could reflect a return to "normal" levels of the corticosteroid hormones, although no direct evidence is available on this. A literature review on the control of appetite in fish (Fletcher, 1984) emphasizes the multifactorial nature of this process.

At present, changes in the maintenance cost as a result of copper or any other metal is not well explained. Excretion of copper and repair of damaged tissue could contribute to this as could alterations in osmoregulation (see Chapter 7). Waiwood and Beamish (1978) make the interesting suggestion, based on their results and others that the masking effect could be due to copper causing some blockage of nervous transmitters or other neuronal malfunctions of the central nervous system. If this speculation is correct, the net result would be a reduction in coordination and, thereby, swimming efficiency. It is hypothesis that does not appear to have been further tested.

One final interaction of copper with energetics has to do with the question of whether exposure to the metal affects the ability of fish to acclimate to a temperature change. Figure 3 shows how raising the water temperature 10°C while bluegill (*L. macrochirus*) were exposed to a sublethal level of copper caused essentially the same amount of increase in metabolism in controls and exposed fish, but upon continued exposure to the copper and elevated temperature, the controls showed a classical type III (Precht, 1958) partial compensation while those in copper exhibited a greatly delayed acclimation to the new temperature. While increased temperatures generally enhance growth rate (Brett and Groves, 1979), it is evident from these data that copper can also cause the demand for energy to be "excessively elevated" following a temperature rise, and this could compete with growth for the available energy intake.

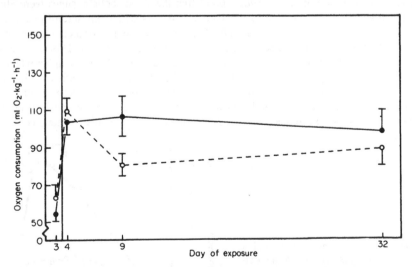

Figure 3 The effect of copper exposure (0.21 mg/L) on metabolic acclimation to a temperature change in bluegill (*L. macrochirus*). Between days 3 and 4 of exposure, the temperature was increased from 20–30°C (represented by the vertical line) and held at 30°C for the remainder of the exposure period. Bars represent 2 SEM above and below the mean. Open circles = controls. Closed circles = copper exposed. (From Felts, P. and Heath, A., *J. Fish Biol.*, 25, 445, 1984. With permission.)

B. MISCELLANEOUS METALS

Mercury is well known for causing neurological toxicity in a variety of animals. From the standpoint of energetics of fish, three studies are noteworthy. MacLeod and Pessah (1973) measured active metabolism in rainbow trout following exposure to inorganic mercury (as $HgCl_2$) for 96 h. The mercury caused a depression of active metabolism and the effect was increased at higher temperatures. At first glance, these results may seem different from those of Waiwood and Beamish (1978) for copper, but recall, however, that copper, while it caused an increased maintenance demand, also resulted in a decrease in the maximum oxygen consumption (i.e., a limiting effect). Active metabolism, as measured by MacLeod and Pessah (1973), is displayed at maximum sustained swimming speed, so the two sets of data are actually similar. Both results may be due to altered nervous function, although other mechanisms such as inhibition of cellular enzyme activities could also be occurring (Brown and Parsons, 1978). As has been emphasized in several places, observed effects may be attributable to more than one cause.

Mercury may inhibit routine as well as active metabolism in some species. Panigrahi and Misra (1978) reported an 84% decrease in "resting" oxygen consumption in *Anabas* exposed for 10 days to 3 mg/L mercuric nitrate. After continued exposure for 45 days these fish were still alive even though the metabolism remained down. Survival at a resting metabolic rate of only 16% of "normal" for this length of time is remarkable. These fish are commonly called the "climbing perch", so they are quite capable of breathing air. Unfortunately, the authors do not adequately describe their methods, but one is left with the feeling the mercury either reduced spontaneous activity, which was probably quite high in their controls (0.19 mg/g/h) and/or caused the fish to utilize more breathing of air, which presumably was not measured. Such a high dose of mercury would probably cause considerable gill damage (Olson et al., 1973) and thus the reduced oxygen consumption from water would reflect this.

The oxygen consumption of larval fish rises dramatically with hatching and as the yolk sac is absorbed. There is then a rapid drop in the oxygen consumption when active feeding begins. Storozhuk and Smirnov (1982) found that these metabolic changes were almost eliminated and the overall rate of metabolism was depressed by exposure of pink salmon larvae to mercury at either 25 or 1 µg Hg/L (Figure. 4). The extent of metabolic inhibition was not dependent on the concentration of mercury in the medium. This is somewhat surprising in that the concentration of mercury in the tissues was 13 times higher when they were kept in the 25 µg Hg/L than when in 1 µg Hg/L. The authors cite earlier work in their laboratory which showed that exposure of salmon larvae to mercury causes a decrease in concentrations of copper, zinc, and iron in the tissues. Consequently, they attribute the metabolic inhibition to a replacement of these elements by mercury in metal-containing enzymes.

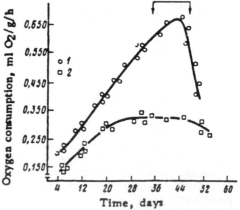

Figure 4 Oxygen consumption by pink salmon *(O. gorguscha)* larvae immediately after hatching. Circles = controls. Squares = exposed to 25 µg Hg/L. Arrows indicate changeover to active feeding. (From Storozhuk, N. G. and Smirnov, B. F., *J. Ichthyol.*, 22, 168, 1982. With permission.)

Aluminum has a variety of physiological effects and apparently, damage to the gills is one of them. Malte (1986) found a marked drop in arterial Po_2 when trout were exposed to 2 mg/L for up to 72 h. The drop in arterial Po_2 was preceded by an elevation in oxygen consumption and this was maintained even in the face of the severely dropping Po_2. The increased rate of oxygen consumption was probably due to the big increase in energy demand of the ventilatory muscles (ventilation frequency increased almost threefold) in response to the falling arterial Po_2.

Chromium, zinc, and lead at sublethal concentrations have all been found to increase oxygen consumption, (Brafield and Mathiesse, 1976; Somero et al., 1977; Pamila et al., 1991). For chromium and zinc, at least, the cause of the increased energy expenditure appears to be increased spontaneous activity. Irritation of skin or mucous membranes clearly causes greater spontaneous locomotor activity in a fish, as it would almost any other animal. Ellgaard et al. (1978) have devised a clever method of quantifying this spontaneous activity in bluegill, which could presumably be used for other species, as well. A 10-gal aquarium is divided into two equal compartments by a transparent partition containing four holes 4.5 cm in diameter. A group of fingerlings is introduced into one side and the number found on the opposite side is counted at time intervals. The data are then treated mathematically as an opposed first-order chemical reaction. With this simple system they were able to show a dose-dependent increase in spontaneous locomotor activity for fish exposed to zinc, cadmium, and chromium (Table 1). These changes reached a plateau after 3 days' exposure and remained at that level for the remainder of the 2-week studies.

The extent of the increase in locomotor activity correlates well with the relative toxicities of the metals tested. For example, cadmium caused greater changes than the other two and it has the lowest LC50. (The decline in locomotor activity in 0.5 ppm cadmium is an artifact due to the fact that considerable mortality was occurring and some of the fish were dying and therefore not moving much.) Because locomotor activity is very much a "driver" of the rate of metabolism, these observations suggest that these metals cause an elevation in routine metabolism, at least in part through this mechanism. In later studies, Ellgaard and colleagues have shown that crude oil (1979) and assorted acids (1984) cause reductions in spontaneous activity of bluegill. Thus, their method holds promise for quantifying both increases and decreases in activity.

The effects of cadmium on metabolic rate has recently been used to model the effects of this metal on growth in bluegill. Sandheinrich and Atchison (1993) successfully

Table 1 Spontaneous Activity of Bluegill
Exposed to a Metal in the Water for 3 Days

Metal added	Concentration	% of controls
None	—	100
Cadmium	0.1 mg/L	148
	0.25	778
	0.5	60
Chromium	0.05	118
	2.4	364
	24.0	650
Zinc	0.1	130
	5.0	380

Note: See text for description of method for quantifying activity.

Data from Ellgaard, E. G., Tusa, J. and Malizia, A., *J. Fish Biol.*, 1, 19, 1978.

predicted 28-day growth rates from intitial measurements of oxygen consumption. It is to be hoped that more of this approach will be used in the future.

The effect of lead may depend on whether the fish is freshwater or marine. Ellgaard and Rudner (1982) observed no effect of lead on spontaneous activity of bluegill, even when tested at lethal concentrations. Somero et al. (1977), however, exposed *Gillichthys mirabilis* (a marine goby) to lead in the water for up to 87 days and found a greatly elevated oxygen consumption throughout this period. There was no gill damage and there was no effect of the lead on *in vitro* respiration of gill tissue even though the gills accumulated high levels of it. They did note (subjectively) a greatly elevated amount of physical activity in the tanks along with the heightened rate of oxygen consumption; so the effect of lead could have been either neurogenic or irritating. Somero et al. (1977) concluded it was the former.

In summary, metals generally cause changes in metabolic rate but the direction of the effect varies with the metal — some stimulate, others inhibit. These changes in routine metabolic rate are generally mediated through effects on motor activity of the fish. Metals can also alter the enzyme systems within cells of liver, and presumably other tissues (see Chapter 9), but what effect this has on whole-animal metabolism cannot be predicted from those findings as changes in the metabolic rate of some tissue such as liver may be masked by changes in motor activity brought about by the metal acting via the nervous system, either by inhibiting or stimulating it, or by irritating sensitive epithelial tissues.

V. GILL TISSUE METABOLISM: EFFECTS OF METALS AND POSSIBLE RELATION OF GILL METABOLISM TO WHOLE-BODY METABOLIC RATE

The rate of oxygen consumption by gill tissue has been of interest in the study of pollution by metals because this tissue is relatively easy to remove and work with from a variety of animals, both finfish and shellfish, and also because gill tissue is one of the first damaged by metal exposure (see Chapter 3). It has also been thought that such a physiological measure might be useful in determining the extent of damage from metal exposure to a sample of organisms drawn from a population in the field suspected of suffering from some effluent. In other words, a biomarker, although all this work was done before the use of the term biomarker came into vogue. The idea may still have merit, however, few measurements of this type have been made in recent years.

It should be first pointed out that measurements of the respiration rate of gill tissue from fish exposed in the laboratory to very high doses means little as the respiration obviously will decrease as the cells are rapidly killed within a few hours by the metal. Also, changes in gill respiration should not be interpreted as representing similar alterations in whole-body respiration rate. The two measurements may or may not be related.

Exposure of cunner (*Tautogolabrus adspersus*) and winter flounder (*Pseudopleuronectes americanus*) to mercury (5–10 ppb) for 30–60 days caused elevated oxygen consumption of gill tissue. This is in contrast to what was found with striped bass (*Morone saxatilis*) by the same authors, in which the gill tissue respiration was depressed by a comparable exposure to mercury (Calabrese et al., 1977). They further found that the bass is actually more tolerant of metal exposure than the two other species, so the gill respiration findings are not indicative of overall metal tolerance by the species.

In apparent contrast to mercury, chronic exposure to cadmium resulted in a 30% depression of gill tissue respiration in cunner, but it is noteworthy that the amount of depression was approximately the same at all concentrations tested (3–48 ppm), so the effect was not concentration dependent (Thurberg and Dawson, 1974). The respiration of gill tissue from striped bass and flounder was also slowed by cadmium exposure (Calabrese et al., 1977).

Tort et al. (1984) exposed dogfish sharks (*Scliorhinus canicula*) to 50 ppm cadmium for 6 days. The 24-h LC50 for this species is 200 ppm so this concentration is sublethal. They observed a severe depression in *in vitro* gill tissue respiration at 2 days' exposure, but by 6 days of continued exposure, the respiration had returned almost to the control level. There was a marked increase in lactate at 2 days and drop in ATP which suggests a reduction in oxygen availability to the gill tissue, probably due to histological changes (see Chapter 3). These variables also returned to control levels by day six of exposure. Thus, the dogfish exhibited a rather good ability to adapt to the presence of cadmium in the water.

These findings might be taken to indicate that cadmium is a general respiratory poison for gill tissue, but such a generalization would have to be limited to teleost fish. For example, exposure of oysters caused an increase in oxygen consumption of the gill tissue. Moreover, the increase tends to be proportional to the amount of cadmium accumulated in the tissue (Engle and Fowler, 1979). Copper causes a somewhat similar result in the oyster. Engle and Fowler attribute the increased rate of respiration to an increase in permeability of cellular and/or mitochondrial membranes. Such a hypothesis was supported by their electron microscopic observations of the tissues which showed damage to these membranes. Zaba et al. (1978) found that copper causes an increase in permeability of fish liver mitochondria to potassium and an increase in respiration at low doses, but an inhibition at high doses.

Studies by Tort et al. (1982) on dogfish exposed to zinc for 4–21 days showed inhibition of gill tissue respiration. The concentrations required to cause inhibition (10 ppm minimum) seem high so the relevance for "real world" situations is questionable. However, since these fish were able to tolerate such seemingly high doses for at least 21 days, perhaps acute gill damage from zinc does not occur in elasmobranchs as it does in teleosts (see Chapter 3).

In bluegill exposed to copper for up to a month, oxygen consumption by isolated gill tissue was unaffected except in one case. It was stimulated by the copper exposure in fish that had also experienced a rise in temperature from 20–30°C on the third day of exposure and held there for 9 days while still receiving copper (Felts and Heath, 1984). The physiological significance of this is that whole-animal oxygen consumption was also elevated above that of the non-exposed fish (see Figure 3), thus copper caused the process of temperature acclimation to be delayed in both whole fish and gill tissues.

The changes observed in gill respiration that have been reviewed here are relevant to research on the percent of the whole-body oxygen requirement consumed by gill tissue. Two studies suggest it is very high. Daxboeck et al. (1982) measured the difference between the amount of oxygen removed from the water and the amount actually appearing in the blood. This was done using spontaneously ventilating, artificially perfused preparations of trout and the results were used to estimate the *in vivo* consumption of oxygen by gill tissue. They concluded that a median of 27% of total whole-body oxygen uptake is utilized by this tissue alone. That seems a remarkably large figure when it is realized that gill tissue represents only 3.9% of the total body weight. Johansen and Pettersson (1981) using perfused head and isolated gill arch preparations, estimated that in cod the contribution of gill to whole-body consumption of oxygen is 6.6%, which is considerably less than that found by Daxboeck et al. (1982) but is still large compared to the percent of body mass.

It may be relevant to note that the Daxboeck et al. (1982) paper is not cited in more recent work (Perry and Walsh, 1989) and reviews on gill metabolism (Mommesen, 1984). Perhaps it is now felt the values are unrealistically high.

Measuring the rate of respiration in isolated tissues requires their removal from the fish, mincing or slicing, and then the measurements of oxygen utilization must be made quickly in a microrespirometer of some sort (LeFavre et al., 1970; Oikawa and Itazawa,

Table 2 Oxygen Consumption of Gill Tissue Measured *In Vitro* or *In Vivo*

Species	Temp.	Gill	Whole-animal	Ref.
			In Vitro	
Bluegill	20	21.56	2.66	Felts and Heath (1984)
Bluegill	20	31.25	11.87	O'Hara (1971)
Pumpkinseed	20	15.02	2.68	Roberts (1967)
Rainbow trout	16	21.25	6.88	Evans et al. (1962)
Carp	20	8.04		Itazawa and Oikawa (1983)
Cunner	20	6.70		Thurberg and Dawson(1974)
Flounder	20	5.35		Calabrese et al. (1975)
Tilapia	25	4.82		Perry and Walsh (1989)
Toadfish	25	13.82		Perry and Walsh (1989)
			In Vivo	
Rainbow trout	7	11.16		Daxboeck et al. (1982)
Cod	15	4.24		Johansen and Pettersson (1981)

Note: All values are expressed as μmol/g wet wt/h. Comparisons with whole-body oxygen consumption are included where these were done by the same author(s).

1983). Unavoidably, the procedures cause breakage of many cells so the question is frequently raised as to how realistic are the rates obtained *in vitro*. Table 2 compares *in vitro* and *in vivo* data for freshwater and marine species.

Several points are presented by this table. Both the *in vitro* and *in vivo* values for the freshwater fish are generally higher than those from marine fish, which is somewhat surprising given the much greater ionic flux in the marine forms (Chapter 7). At first glance the *in vivo* values for gill respiration seem low compared to those *in vitro*. However, the former were measured at lower temperatures. If a Q10 of 2 is assumed, the two rates are remarkably similar. Finally, in all cases, the metabolic rate per gram of tissue weight of gill tissue was several-fold greater than that of the whole animal.

In a more recent study, Perry and Walsh (1989) showed by cellular isolation experiments that the chloride cells have a considerably higher rate of energy metabolism than the other cells of gill tissue. They further noted that the number of chloride cells in the gills is greater in a marine-adapted species, which does not fit the data shown in Table 2 where the marine species have the slower rate of respiration. Still, everybody seems to be in agreement that gill tissue has a very high energy requirement.

It is tempting to speculate that changes in whole-body metabolism of a resting fish are due mostly to changes in the metabolism of tissues other than muscle. Even though muscle makes up some 50–60% of the body mass, it appears to contribute relatively little to the total metabolism except when the fish is swimming. Thus changes seen in fish exposed to pollutants may reflect predominantly effects on non-muscular tissues when measurements are made on resting fish (i.e., standard metabolic rate), but when the fish is forced to swim or when spontaneous activity is altered by the chemical, differences in metabolism between controls and exposed specimens probably reflect effects on muscle metabolism or on the process of muscular coordination and the central nervous system. This latter point is especially noteworthy in fish exposed to pesticides, which will now be discussed.

VI. EFFECTS OF PESTICIDES ON WHOLE-BODY AND INDIVIDUAL TISSUE RESPIRATION

In probably the first study of this sort on fish exposed to an insecticide, Ferguson et al. (1966) observed that when mosquitofish (*Gambusia affinis*) were exposed to endrin (a chlorinated hydrocarbon) at a concentration of 20 ppb, there was no effect on metabolism for a period of days. Then suddenly hyperactivity and spasms would occur which caused the oxygen consumption to rise dramatically just before death. Similar results have been obtained on white suckers exposed to the organochlorine insecticide, methoxychlor (Waiwood and Johansen, 1974). The response appeared to be an all-or-nothing phenomenon in that fish which survived the exposure exhibited no change in oxygen consumption, while in those that did not survive, a big increase in metabolism took place shortly before death.

Bansal et al. (1979) found that chlordane also stimulates oxygen consumption but in their fish it was only in a transient way. When they exposed carp to this insecticide at concentrations ranging from the LC50 to 1/12 the LC50, they found that at concentrations below the LC50, there was an initial phase, which they called the sensitization phase, lasting 6 h irregardless of the concentration. Then there followed a period of hyperactivity and elevated oxygen consumption (two- to fourfold increase). By the end of 24 h of continued exposure, the level of activity and oxygen consumption of the fish had returned to normal, as if the receptors that were stimulating the activity had adapted to the insecticide. (Those fish exposed to the LC50 showed a steady decline in metabolic rate until death at 10 h of exposure.)

Using a system to quantify spontaneous swimming activity (described earlier in this chapter) Ellgaard et al. (1977) found a concentration-dependent increase in activity of bluegill exposed to sublethal concentrations of DDT. The maximum hyperactivty was generally achieved in about 8 days and continued for at least 16 days of continuous exposure. It is important to further note that the effects of DDT on activity persisted for at least 2 weeks in fish transferred to non-contaminated water.

In general then, it appears that organochlorine insecticides cause hyperactivity in fish. The character of the response depends on the species of fish. Hyperactivity and spasms are probably, at least in part, due to increases in the sensitivity of peripheral receptors to stimuli that would not normally be especially excitatory (Bahr and Ball, 1971). Of course there are probably also direct effects of the pesticide on the central nervous system.

The herbicide triclopyr seems to act on fish in a manner similar to organochlorine insecticides. Johansen and Green (1990) quantified both muscular activity and oxygen consumption in coho salmon exposed for 96 h to trichlopyr. At a concentration <0.10 mg/L it caused hypersensitivity to external stimuli (such as lights being turned on) and elevated activity and oxygen consumption. At somewhat higher concentrations the fish became lethargic and oxygen consumption declined.

Organophosphate insecticides appear to inhibit metabolic rate both in whole-body measurements and in tissues removed from exposed fish, although a transient stimulation in oxygen consumption may occur (Bansal et al., 1979; Rao et al., 1985).

Rath and Misra (1979) found a concentration-dependent inhibition that varied with the size of the fish. The amount of respiratory inhibition was size dependent in that small ones showed a greater inhibition after 15 days of exposure, but, they recovered more rapidly when returned to uncontaminated water. These observations would seem to suggest a faster uptake and rate of detoxification of the insecticide in smaller fish which fits with their overall faster rate of metabolism.

In a follow-up study, Rath and Misra (1980) measured *in vitro* respiration in brain, gill, and muscle tissues from fish exposed to diclorovros, as was done in their whole-animal investigation. Again there was an inverse relationship between body size and metabolic rate, and the percent inhibition was greater in tissues from smaller fish. Periods of exposure were for up to 28 days and, throughout, gill tissue exhibited the greatest percent of inhibition which might relate to its proximity to impact from the dissolved pesticide.

Methyl parathion (an organophosphate insecticide) also produces a marked decrease in metabolic rate of both whole-animal and specific tissue respiration (Murty et al., 1984;

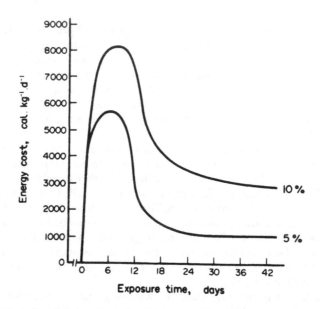

Figure 5 Energy cost of stress for rainbow trout due to stress of permethrin detoxification and tissue repair in relation to exposure time in 5 and 10% of the 96-h LC50. In calculating the energy cost, 3.25 cal/mg of oxygen consumed was used. (From Kunaraguru, A. K. and Beamish, W., *Comp. Biochem. Physiol.*, 75A, 247, 1983. With permission.)

Rao et al., 1985; Bala and Mohideen, 1992). The pyrethroid insecticide fenvalerate also produced a severe decrease in tissue and whole-animal respiration in 48-h exposures (Reddy and Philip, 1992). One possible problem with the latter two studies cited is that the investigators apparently measured control respiratory rate only before the pesticide exposures were commenced. With continued confinement in a respirometer, a fish may show a decline in respiration due to acclimation to the chamber and to starvation (Fry, 1971), thus a decrease could be normal rather than due to pesticide exposure.

Pyrethroid insecticides appear to stimulate respiration. In an excellent study on rainbow trout by Kumaraguru and Beamish (1983), permethrin caused the standard metabolic rate to rise over 7 days of exposure, but then under further exposure it gradually declined toward the level of the controls at a rate depending on concentration (Figure 5). Standard metabolism was measured in this study with swimming tunnels which permitted measurements of oxygen consumption at several rates of swimming. Then, by extrapolation to zero activity, the calculated standard rate is obtained (they refer to it as "basal"). Thus, the stimulation in metabolic activity caused by the pyrethroid was due at least in part to factors other than increased spontaneous activity of the fish (which was not measured). Kumaraguru and Beamish (1983) attribute the increased metabolic cost to detoxication and tissue repair. Figure 5 illustrates how they used the data obtained to obtain an actual energy cost of this increased maintenance which could be incorporated into an energy budget. The figure also illustrates the phenomenon of physiological acclimation to the pesticide as the curves return toward zero with time.

The implications of an elevated maintenance cost from pesticide presence in the water was also shown in work on carbaryl, which acts similarly to organophosphorus insecticides. Arunachalam et al. (1980) found that over a 27-day exposure to carbaryl, there was a decreased growth rate and conversion efficiency. Conversion efficiency refers to the percentage of food converted into growth so if this goes down, it implies either a reduction in assimilation or a greater cost of maintenance. Because assimilation was unchanged, the maintenance cost must have been higher even though respiration rate was not measured in this study.

Rotenone is another widely used insecticide. It is also used as a fish poison. Its mechanism of action is to block the flow of electrons from NAD to the cytochrome system and thus causes an inhibition of cellular respiration (Skadsen et al., 1980). Rainbow trout were forced to swim in a water tunnel at various speeds while being exposed to rotenone for only a few hours. Exposure to a concentration equivalent to the 96-h LC50 caused no effect on the standard metabolic rate (i.e., zero relative performance). It did, however, reduce the maximum metabolic rate so at that concentration it was acting as a limiting factor (Fry, 1971). At one half of the LC50, the standard metabolic rate was elevated (a masking response) but the maximum rate was not significantly reduced. Then at a still lower concentration of rotenone, an even further elevation in the standard rate of metabolism was found. The authors do not attempt to explain the basis of this inverse concentration response. Evidently, the stimulatory effect is only manifested at very low concentrations of rotenone. Perhaps, higher concentrations impair gas exchange at the gill surface to such an extent that even though the energy demand is up, the consumption of oxygen becomes inhibited. It would be interesting to see what chronic exposures to these low concentrations would do to the swimming capacity and resting rates of metabolism.

It has been repeatedly shown herein how difficult it is to determine the relative importance of spontaneous activity and intrinsic cellular respiration in explaining increases or decreases in whole-body metabolism during exposure to some chemical. Thus, the study by Peer et al. (1983) is of interest because they measured oxygen consumption and spontaneous activity simultaneously in mullet (*Rhinomugil corsula*) exposed to pentachlorophenol (PCP). By extrapolating to zero activity one can get a good estimate of the standard rate of metabolism. They observed that PCP is acting as a masking factor by increasing the maintenance cost. Because the slopes of activity vs. metabolism were not significantly different, the effect of the chemical is on cellular metabolism instead of on spontaneous activity.

Pentachlorophenol is classed as a metabolic stimulant which acts by uncoupling oxidative phosphorylation by the electron transport system in the mitochondria (Webb and Brett, 1973). Thus, oxygen is consumed but without the concomitant formation of ATP. Webb and Brett (1973) observed that growth of sockeye salmon was depressed in a concentration-dependent fashion by PCP. All fish received the same ration and assimilation was unaffected by the PCP. Therefore, the depressed growth is caused by the maintenance costs competing for the available energy.

The relationship between energy taken in, growth, and maintenance costs was emphasized in the equation and diagram at the beginning of this chapter. Using these concepts, one can prepare an energy budget for fish exposed to some pollutant. This was done for the cichlid, *Cichlasoma bimaculatum,* exposed to 0.2 ppm pentachlorophenate (Krueger et al., 1968). The heat of combustion of a sample of fish was determined on day one and again on the tenth day of exposure. In addition, the heat of combustion of the food eaten was obtained. Basal metabolic rate, nitrogen loss, etc., were obtained from other studies in that lab. It appears that PCP stimulated food intake but depressed growth in the cichlids. Webb and Brett (1973) found no effect of PCP on the appetite of salmon.

The cichlids exposed to PCP clearly had a higher "cost of living", or to quote the authors: "The pentachlorophenol did not destroy or damage the growth machinery; it merely made it more costly to operate." (Kruger et al., 1968, p. 124). When body composition was determined, it was found that the exposed fish stored less fat than controls. Unfortunately, the calories attributed to carbohydrate and protein were combined in their report, so it is not possible to determine changes in these individual stores of energy. Because fish exposed to PCP tend to mobilize liver glycogen into blood sugar (Thomas et al., 1981; Hanke et al., 1983), there were probably some large decreases in carbohydrate relative to protein.

Samis et al. (1993) report that bluegill exposed to PCP at a concentration of 173 µg/L for 22 days exhibited a depression in both growth and feeding during the final 10 days of exposure. However, when the PCP exposure concentration was at 48 µg/L, growth declined but food consumption was unaffected, thus appetite can be affected by PCP but the concentration must be fairly high.

The effluent from kraft pulpmills contains several potentially harmful substances including some chlorinated organics that resemble PCP in molecular structure (Davis, 1976). Exposure of salmon to a concentration of the effluent equivalent to one third of the 96 h LC50 causes an elevated oxygen consumption even though the arterial oxygen content is reduced (Davis, 1973). Thus, the increased energy demand is, at least in part, due to hyperventilation brought on by the internal hypoxia.

Based on an increased utilization of energy for maintenance, one would predict an inhibition of growth rate in fish receiving pulpmill effluent. Surprisingly, this does not necessarily occur. Indeed, several workers have noted stimulation of growth, providing the fish are fed an excess ration (see McLeay and Brown, 1979 for review). So the exposed fish apparently ate more food, if it was available, and this more than compensated for the increased energy demand.

Cyanide is considered a metabolic depressant as it binds in place of oxygen to the heme of cytochrome oxidase, the terminal enzyme of the electron transport system in cells. Because cyanide blocks the utilization of oxygen by the cells, oxygen consumption rate must decrease. This seems to have not been systematically investigated in fish, although a study by Dixon and Leduc (1981) on rainbow trout reveals an interesting phenomenon. Trout were exposed to two different concentrations of cyanide for 18 days after which time they were placed in respirometers with no cyanide present. The oxygen consumption was then determined over a period of 144 h. The cyanide-exposed fish initially respired at a lower rate than the controls and this was dose dependent, so the cyanide was acting as a depressant. Then, those that had previously been exposed to cyanide exhibited a greatly elevated rate which reached a peak at 72–96 h. Respiration then returned to a lower rate which was still above the controls. So, several days after the initial exposure, cyanide acted as a metabolic stimulant even though it was no longer present, and this was prolonged for at least a week.

An elevated rate of oxygen consumption after some sort of stress might be due to oxygen debt. However, the oxygen debt acquired after exhaustive exercise in trout is "paid back" in about 6 h (Scarbabello et al., 1991) so that cannot be the explanation for the prolonged high rate of respiration. Leduc (1984) notes that mammals chronically exposed to cyanide develop goiter due to failure of the thyroid to take up iodine. This causes an elevated amount of thyroid-stimulating hormone from the anterior pituitary. He further speculates that if the same thing occurs in fish, there would probably be an excess amount of thyroxine produced from the enlarged thyroid gland after the cyanide was no longer present. This might stimulate oxygen consumption and growth rate, both of which take place following exposure to low levels of cyanide.

The mudskipper (Boleophthalmus boddaerti) is a goby that inhabits burrows on mudflats that are frequently anoxic. Because it has high resistance to lack of oxygen, it was thought that it might also show good tolerance of cyanide. Chew and Ip (1992) found that indeed it is quite tolerant of cyanide but not because of its high anerobic capacity, nor because it slows metabolism down. When exposed to cyanide at concentrations that would normally kill other fish, this species shows no slowing of oxygen consumption even though cytochrome oxidase is inhibited by 50%. Evidently, it has a considerable excess of this enzyme present, although there is no obvious reason why such an adaptation evolved.

VII. EFFECTS OF PETROLEUM HYDROCARBONS ON ENERGY METABOLISM

Petroleum hydrocarbons may stimulate or inhibit oxygen uptake of juvenile or adult fish depending on the exposure concentrations. At concentrations approaching those that are lethal, the effect is apparently to inhibit oxygen uptake due to damage to the gill epithelium (Hawkes, 1977; Prasad, 1987). However, there may be an initial stimulatory effect for a few hours due to the increased breathing of the fish which has a considerable energy demand from the respiratory muscles (Thomas and Rice, 1979), but this is followed by a rapidly falling oxygen consumption.

Most studies on the effects of oil on fish have been done on embryos, larvae, or juveniles. At these stages, petroleum at chronic levels of exposure probably increases oxygen consumption, however, the evidence is mostly indirect. Vignier et al. (1992) exposed Atlantic salmon parr for up to 40 days. They observed a reduction in growth rate of oiled fish, even in those that were feeding normally. Higher levels of oil can inhibit feeding rate and thus growth rate (Fletcher et al., 1981; Kiceniuk and Khan, 1987). The results of Vignier et al. suggest that there is a reduction in food conversion efficiency which may be due to increased metabolic demand. It could also be due to direct effects of oil on the process of food absorption by the intestine, but no evidence is available on this.

Hose and Puffer (1984) observed an increased oxygen consumption by grunion embryos exposed to a very low concentration of benzo[a]pyrene. However, at higher concentrations, there was an inhibition of oxygen consumption. This can be contrasted with Ostrander et al. (1989) who were unable to detect any metabolic changes in coho salmon eggs or larvae exposed to benzo[a]pyrene. They did, however, note serious effects on behavior of the larvae as they attempted to emerge from the gravel after hatching so the petroleum hydrocarbon was not without effect.

Cod eggs exposed to North sea crude oil exhibited no changes in oxygen consumption, but the activity of the larvae was suppressed (Serigstad and Adoff, 1985). This is in contrast to the findings of Eldridge et al. (1977) in which Pacific herring eggs were more sensitive to benzene than were the larvae, although their data on the egg oxygen consumption is difficult to interpret. The higher dose caused increased respiration while the lower dose caused decreased respiration compared to controls. Clearly, the information currently available does not permit generalizations regarding the effect of petroleum on energetics. Indeed, there have even been examples of where it seemed to stimulate growth (Eldridge et al., 1977; Laughlin et al., 1981), possibly due to a stimulation of feeding.

VIII. METHODS APPLICABLE TO MEASUREMENT OF ENERGY EXPENDITURE IN THE FIELD

As the various studies of the effects of pollutants on oxygen consumption show, changes in bodily activity are frequently the reason for at least part of the changes in metabolic demand. In all these studies, however, there is the problem that the fish are confined in a chamber where "normal" movement is greatly restricted. A fish living in the wild will expend energy for feeding, escaping predation, and exploratory behavior. None of this is recorded in the present methods of measuring oxygen consumption, and yet if one is to prepare a complete energy budget, such data are required. Two techniques that have been used successfully with terrestrial animals to obtain measurements of energy expenditure in the wild are the telemetry of heart rate and the double-labeled water method.

Because blood flow is proportional to oxygen consumption, it follows that as the latter changes, so will the cardiac output. With the aid of biotelemetry, the heart rate can be recorded in a free-ranging animal on a continuous basis. This method has been used with considerable success in both birds and mammals (Gessamen, 1980). In fish, however, the relationship between heart rate and cardiac output is not always consistent because large changes in stroke volume may occur with little frequency change. With considerable effort, Priede and Tytler (1977) were able to calibrate individual fish and obtain some estimates of metabolism in free-ranging trout and cod by telemetry of the heart rate. More recently, Sureau and Lagardere (1991) were able to relate distance moved (and presumably energy expended) with telemetered heart rate in soles (*Solea solea*), but the technique did not work with sea bass (*Dicentrarchus*).

Rogers and Weatherly (1983) were able to detect the electromyogram from the opercular muscles and transmit this to a recorder. The voltages produced by the action potentials of the muscles were integrated and compared with the oxygen consumption of fish either swimming spontaneously in a chamber, or swimming at fixed speeds in a water tunnel. They obtained a good correlation between the average integrated voltage and oxygen consumption, although the curves describing these values were different if the fish was swimming spontaneously as opposed to being forced to swim in the tunnel. This was explained as being at least in part due to the use of different muscles for the two different types of swimming. In any case, this method offers a range of possibilities for continuous recording of expenditure of energy by fish in ponds, rivers, lakes, estuaries, and in laboratory settings as well. Kasello et al. (1992) have further refined this technique for measuring activity in free-ranging fish for months at a time.

The double-labeled water method is an ingenious way to measure energy expenditure over a period of time in free-ranging terrestrial animals. However, it depends upon several critical assumptions, and one of these precludes its use with aquatic animals. Namely, there is the requirement that little or no unlabeled water enter the body by way of the respiratory or skin surfaces (Nagy, 1980). With fish, there is a considerable flux of water across these surfaces so this method is not applicable to them.

The oxygen consumption rate of an animal is in reality a measure of the end result of many enzyme reactions in the cells. Biochemical research has revealed that certain enzymes are rate limiting in that they catalyze non-equilibrium reactions in certain metabolic pathways and their substrate is at concentrations approaching or exceeding saturation. In essence then, these key reactions set the pace for the other reactions in the metabolic chain (Hochachka and Somero, 1984; Newsholme and Crabtree, 1986). The activity of these enzymes can be assayed using fairly standard procedures. Selection of appropriate enzymes is based on two criteria: (1) enzyme has been shown to reflect whole-animal oxygen consumption rate or growth and (2) is stable when stored frozen for days or weeks. Enzymes that meet these criteria for fish include cytochrome oxidase and citrate synthase.

The relationship between oxygen consumption or growth and enzyme activity has been investigated to only a limited extent, but in all cases a good correlation exists (Goolish and Adelman, 1987; Kaupp and Somero, 1988; Koch et al., 1992). Others (Simon and Robin, 1971; Smith and Chong, 1982) have found the rate of enzyme activity changes in response to body size and/or starvation and season in exactly the same manner as oxygen consumption rate, although they did not measure the latter. In some of these studies, skeletal muscle was the test tissue; in others, it was liver or heart muscle. The various tissues gave similar trends. A possible weakness in the potential use of enzyme activities for assessing pollution effects is that they may not respond to increased swimming activity (Johnston and Moon, 1980; Goolish and Adelman, 1987). Furthermore, cytochrome oxidase in some species may be present in considerable excess, so it would not be a rate-limiting step (Chew and Ip, 1992).

Cytochrome oxidase has the largest database, but Somero (personal communication) reports that in marine organisms it is sometimes rather unstable when frozen compared to citrate synthase. Goolish and Adelman (1987), however, found good stability for this enzyme in frozen largemouth bass (a freshwater species).

Some environmental conditions including the presence of pollutants may stimulate increased amounts of anaerobic metabolism. A good enzyme for assessing changes in this type of metabolism is lactate dehydrogenase (Kaupp and Somero, 1988).

Assessing metabolic activity in free-ranging fish by biotelemetry of muscle electrical activity or by the measuring of tissue enzyme activities involves two different time frames. In the first method, the dynamics are important whereas the latter method will presumably give a sort of average of recent metabolic activity. Clearly, they are not mutually exclusive procedures, but instead they may offer opportunities to pursue different questions in the overall effort to understand how pollutants affect the physiological energetics of fish.

IX. EFFECTS OF POLLUTANTS ON LARVAL AND JUVENILE GROWTH

Growth is essentially the laying down of new flesh through the process of protein synthesis, but this is the endproduct where there are several physiological processes that precede and parallel it. These include feeding behavior, availability of food, type of food eaten, digestion, absorption, assimilation, excretion, energy expenditure for maintenance and movement, and changes in certain hormonal levels. All of these physiological phenomena are directly or indirectly affected by pollutants, some of which have been discussed at some length in this chapter. Thus, any change in growth produced by some exogenous chemical will usually have multiple causes. Therefore, growth is a sort of integrator of a variety of physiological and environmental factors (Weatherly, 1990).

Some stages of the life cycle are often more sensitive to chemical insult than is growth and thus lend themselves better to determinations of maximum acceptable toxicant concentrations (MATC) based on early life-stage studies (McKim, 1977; Woltering, 1984). However, growth is obviously a critical factor governing recruitment into the population and therefore deserves considerable attention.

It will come as no surprise for many people to learn that sublethal levels of a very wide variety of substances have been found to slow the growth of larvae or juveniles (reviewed in Woltering, 1984). In most cases, the mechanisms involved have not been investigated. In his comprehensive review of the influence of environmental factors (excluding pollution) on growth of fish, Brett (1978) emphasizes the interdependence of energy and growth. Metabolic expenditure by fish for maintenance and muscular contraction "steals" energy that could, at least in theory, be used for making more fish. This relationship between energy intake, energy expenditure, and growth is diagramed in Figure 1. In a sense, growth gets the energy that is left over after all the other needs are met. A pollutant may cause an increase in maintenance cost which then requires a greater food intake if the energy balance is to be maintained. This is nicely illustrated in a study wherein perch were exposed to copper at five concentrations (1–89 µg/L) for 30 days (Collvin, 1985). At concentrations above 22 µg/L there was a slowing of growth although no effect on food consumption. Instead, the slowing in growth was attributed to an increased metabolic maintenance cost due to detoxication. It might be recalled that Waiwood and Beamish (1978) have shown increased standard metabolic rate in rainbow trout exposed to copper at a concentration of 40 µg/L for 5 days. Prolonged exposures (40 days) of salmon parr to crude oil have been shown to depress food conversion and therefore growth, again by apparently increasing the metabolic cost of maintenance (Vignier et al., 1992). This has also been observed in fish exposed to cadmium and 2,4,dichlorophenol, two very different types of water pollutants (Borgmann and Ralph, 1986).

Feeding causes a surge in oxygen consumption (as mentioned earlier in this chapter). Feeding also causes a concomitant rise in protein synthesis but the time course of the change in protein turnover varies with different tissues (Houlihan, 1991). The reason for the stimulation in protein synthesis with feeding is unclear as it increases in liver before the blood levels of absorbed amino acids rise following assimilation of protein, so amino acids must not be the stimulus.

In a variety of animals it appears that the primary stimulus for muscle growth is dietary protein rather than energy (Millward, 1989). Of course, fish utilize protein for energy more than many other types of animals so the stimulus may be difficult to differentiate in them.

If food intake or assimilation is reduced then there will be less energy and protein available for growth so the rate will decline. For example, Farmanfarmian and Socci (1984) reported that the absorption of leucine (an essential amino acid) by the intestine was inhibited 20–80% by mercury concentrations that might be expected in the gut of *Fundulus* inhabiting waters contaminated with mercury. Methylmercury was somewhat less potent in this respect.

It is also common to observe that fish exposed to sublethal concentrations of some chemical may exhibit a reduced appetite for reasons that are unknown. Wilson et al. (1994) exposed rainbow trout to aluminum in acid water at a concentration of 162 μg/L. Even though the fish were fed to satiation, for the first 10 days of exposure they decreased their feeding rate which caused reduced growth that was only partially recovered during the subsequent 40-day exposures. The authors noted that the trout in the tanks contaminated with aluminum displayed reduced spontaneous activity which would lower the routine metabolic rate. This actually caused a rise in gross food conversion efficiency (38% compared with 31% for controls), but the combination of a higher standard metabolic rate and depressed appetite caused slowed growth. Acid water (pH 5.2) without added dissolved aluminum caused no effect on growth.

Growth hormone from the pituitary enhances appetite and also improves gross conversion efficiency [(Growth/Ration) × 100]. Donaldson et al. (1979) discuss some possible mechanisms by which it might do this. These include: (1) stimulation of fat mobilization and oxidation which would free amino acids for synthesis rather than utilization for energy; (2) increased protein synthesis; and (3) stimulation of insulin synthesis and release. Recent studies reviewed by Houlihan (1991) confirm increased protein synthesis occurring from growth hormone stimulation in fish.

In mammals (and possibly also fish) muscle growth is accomplished with nuclear replication which is considered to be under control of growth hormone. Cytoplasmic growth is under control of insulin, among other things. This latter hormone is influenced considerably by pollutant exposure, but not always in the same direction, as is discussed in the section in this chapter on carbohydrate changes. For example, cadmium causes damage to the insulin-secreting cells in the pancreas so the level of this hormone may fall, whereas many other pollutants stimulate insulin production indirectly by bringing about a hyperglycemia due to rises in "stress hormones". The high blood glucose then stimulates the beta cells to secrete more insulin. While this might be interpreted as stimulating growth, one of the primary stress hormones, cortisol, suppresses protein synthesis and stimulates protein breakdown (Van der Boon et al., 1991).

In general a physiological response to a stressor is amplified when that stressor is applied suddenly, as opposed to slowly. This was exemplified in a growth study by Seim et al. (1984). They measured growth and survival in steelhead trout exposed to several copper concentrations during development from 11–85 days post-fertilization. The unique part of the study was the comparison of intermittent with continuous exposure. The intermittent exposures were for 4.5 h daily and the concentration in the water followed a sort of square wave (concentration/time). The area under the concentration curve was used to calculate the average concentration as if it were given continuously. Growth was inhibited more with the intermittent exposures and they accumulated more copper in their tissues. If this phenomenon holds for other toxicants, it could have a marked impact on the estimates of MATC, for the natural environment, which we attempt to simulate in the laboratory, is rarely constant.

There have been reports of growth being stimulated in fish by exposure to a toxic pollutant. McLeay and Brown (1974) and Mason and Davis (1976) both reported elevated growth of alevins and juvenile salmon (*Oncorhynchus kisutch*) when exposed to kraft pulpmill effluent. It has also been seen more recently with European perch (*Perca fluviatilis*; Sandstrom et al., 1988), although decreases in growth were seen in lake trout (*Salmo trutta*) exposed to the same effluent (Oikari et al., 1988). In the last study cited, there was the presence of a fair amount of chlorophenolic compounds which would have caused an elevated maintenance metabolism.

A variety of hypotheses have been proposed to explain a stimulated growth rate. These include hormone analogs in the wood extractives that simulated growth hormone, stimulation of appetite, and inhibition of territorial behavior by the dark color. No direct evidence is available regarding any of these. In some cases, it may be a form of hormesis.

The term "hormesis" refers to an overcompensation to some inhibitory challenge. It was seen in crab zoeae which grew faster after being exposed for 5 days to petroleum hydrocarbons (Laughlin et al., 1981). Note that the exposure was for only 5 days, which simulates an oil spill. Compensatory growth has also been seen in larval or juvenile fish subjected to starvation or reduced ration and then given food *ad lib* (Weatherley and Gill, 1981; Dabrowski et al., 1986; Miglavs and Jobling, 1989). In one study (Miglavs and Jobling, 1989), compensatory growth was associated with improved food conversion efficiency.

Growth was also stimulated in minnows (*Phoxinus phoxinus*) that were fed a diet contaminated with PCBs (Bengtsson, 1979). Other work (Mayer et al., 1977) has shown that these chemicals stimulate thyroid activity in fish which provides a possible mechanism for the growth enhancement, although the influence of thyroid on growth in fish is very dependent on day length and/or season (Donaldson et al., 1979).

Finally, growth stimulation has been observed in larval striped bass exposed to the herbicide molinate (Heath et al., 1993). In this study, larvae were exposed to molinate for 4 days at a concentration of molinate less than one half the 96-h LC50. Then they were reared for 10 days in non-contaminated water. The exposed fish exhibited both an increased dry weight and RNA/DNA ratio.

Extensive studies utilizing isotopically labeled amino acids have begun to reveal the complex interactions between growth, protein synthesis, and protein degradation (i.e., protein turnover). The concepts emerging have been extensively reviewed by Houlihan (1991). Protein synthesis is obviously required for growth to occur, but degradation of proteins is also constantly occurring. It is now known that protein synthesis rates and degradation are intimately linked, but differ considerably in different tissues. For example, they are exceptionally high in gill tissue, but efficiency of protein retention is low there (around 4%) compared with white muscle where it is approximately 75%. In other words, there is a lot of protein turnover in gill tissue, perhaps because so much of it is epithelial tissue with a high degree of sloughing off.

There is a linear relationship between protein synthesis and growth rates as there is between growth and degradation rates. However, as growth rate increases, the rate of protein degradation does not increase as fast as synthesis (Houlihan, 1991). In other words, the slopes are different. Thirty to forty percent of the protein synthesis in a fish occurs in white skeletal muscle (Fauconneau, 1985), and this tissue is most responsive to the nutritional state of the fish (Lied et al., 1985). Houlihan (1991) argues that the fractional protein synthesis rate of white muscle can be used as a good indication of general growth rate of the whole organism.

Most growth studies of fish require the maintenance of the animals for weeks or months in the laboratory under well-defined conditions. Such long durations increase the likelihood of mechanical failures (e.g., floods) or disease ruining an experiment. Bulow (1970) was probably the first to report that the ratio RNA/DNA concentrations in the tissues of fish correlated with feeding intensity and, more importantly from our standpoint here, the recent growth rate of the animal. This ratio also changes with the season in wild fish and decreases when the fish are subjected to the stress of low dissolved oxygen (Bulow et al., 1981; Peterson and Peterson, 1992). Buckley (1984) further refined the technique for the assessment of growth of larval fish in the sea.

Barron and Adelman (1984) measured these nucleic acids and protein content in larval fathead minnows exposed for 96 h to hexavalent chromium, cyanide, p-cresol, ethyl acetate, or benzophenone. By using several doses and also measuring the growth for 28 days, they were able to relate the changes in biochemistry with the actual growth of the fish. Figure 6 shows the sort of data obtained in this study. In contrast to Bulow's work (which they do not cite), the absolute concentrations of the nucleic acids were more sensitive to the toxicant effect and more indicative of growth changes than were the ratios. Generally, the RNA content gave the best correlation with growth. In that regard, it has been reported (Venugopalan, 1967) that when growth hormone is injected into intact fish, there occurs a marked stimulation of liver RNA content, but a very variable effect on that of DNA. One cannot necessarily conclude from this that the pollutants were causing an inhibition of growth hormone because numerous other factors could also play a role, but it is an interesting relationship.

Evidently, tissue RNA can change quite rapidly when the organism is subjected to a toxicant. Barron and Adelman (1985) have found that within 24 h of exposure to a

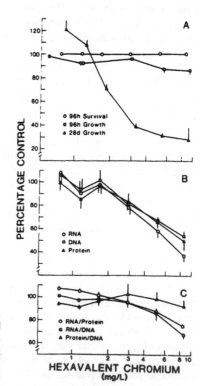

Figure 6 Effect of hexavalent chromium on growth, nucleic acid content, and protein of larval fathead minnows (*P. promelas*). Data are normalized as percentage of the mean control. All points are the means ±S.E. (From Barron, M. G. and Adelman, I. R., *Can. J. Fish. Aquat. Sci.*, 41, 141, 1984. With permission.)

sublethal concentration of cyanide, there is a significant difference in total body RNA between exposed and control fathead minnows.

The original procedure used by Bulow (1970) for measuring RNA and DNA was extremely tedious. Now several newer techniques (Bulow, 1987) and even more recent modifications make it possible to do over 50 samples per day (Heath et al., 1993).

Another biochemical methodology for assessing growth in larval and juvenile fish is the measurement of the enzymes citrate synthase, cytochrome oxidase, and lactate dehydrogenase. The activity of these enzymes, two aerobic and one glycolytic, correlated well with recent growth rate in the saithe, a marine teleost (Mathers, 1992). As was discussed earlier in this chapter, these enzymes also correlate well with overall metabolic rate so they, along with nucleic acid analyses, may offer considerable potential for assessing the impact of chronic exposures to various pollutants in the field and laboratory.

X. SWIMMING PERFORMANCE

In the foregoing discussions of the effect of various pollutants on energy metabolism, it was noted that some may reduce the maximum (active) rate of oxygen consumption. More accurately stated, aerobic scope was reduced. In this section, the effects of pollutants on swimming performance will be dealt with, realizing that aerobic scope and swimming performance are inextricably related.

Beamish (1978) provides an extensive review of the methodologies used in studying swimming capacity in fish. He classifies swimming performance into three major categories: sustained, prolonged, and burst. "Sustained swimming" includes all levels of activity from spontaneous movements to cruising for periods in excess of 200 min. In essence, sustained swimming can be maintained indefinitely. "Prolonged swimming speed" is of shorter duration and results in fatigue. The term "critical swimming speed" (under the category of prolonged swimming) is frequently used to designate a speed that is a maximum that can be maintained for a measured period of time. To determine this, a fish is placed in some sort of water tunnel and the speed is then increased stepwise with intervals of 30 min to 2 h at each step until the fish fails to maintain position for the full interval at a given step. Brett (1964) has devised a precise method of calculating the critical swimming speed which includes consideration of the duration of time a fish was able to breast the maximum speed. Most of the studies up to now involving pollutants have measured critical swimming speed (Ucrit).

The highest speed that a fish can attain is called the "burst speed", one which can be maintained for probably less than a minute, and in some species is probably reached for only a few seconds. It is powered primarily by anaerobic metabolic processes (Driedzic and Hochachka, 1979; Beamish, 1978). For all three swimming types, the time frames mentioned here are based largely on work using salmonids. Somewhat different times may be more appropriate for other fish groups and some species are incapable of maintaining a sustained speed at all.

Randall and Brauner (1991) recently reviewed the effects of temperature, environmental pH, and salinity on swimming performance of fish. They make the point that aerobic swimming speed is affected by muscle contractility and/or the rate of gas transfer by the gills and the circulatory system. An environmental factor such as temperature may affect swimming by altering muscle contractility whereas a change in pH affects the oxygen-carrying capacity of the blood. Although they do not discuss pollutants, it might be further pointed out that the nervous system can be an important target for many chemical contaminants and this could affect swimming ability.

This review is limited to studies done in the laboratory. While there have been attempts to measure swimming performance in the field, the problem of control and replicability make the results obtained rather anecdotal (see Beamish, 1978). Spontaneous locomotor activity and the avoidance of pollutants by fishes is taken up in Chapter 12.

In order to measure swimming performance in a controlled manner, the fish must be induced to swim at some specified speed. For this purpose a variety of swimming chambers have been developed ranging from a simple trough with a paddlewheel to move the water, to complex water tunnels, some quite large, where water is pumped through a swimming chamber in which the fish are confined (Fry, 1971; Beamish, 1978; Graham et al., 1990). In most investigations of the effects of pollutants on swimming performance, the fish were exposed to the test substance for some interval of time before being placed in the swimming chamber.

Blocking the transfer of oxygen across the gill should result in a reduction of aerobic swimming. This was shown with coniferous fibers from pulpwood which tend to clog the gill lamellae and thus interfere with normal water flow through them, and this probably reduces oxygen uptake. Exposure to these fibers in the water caused a dose-dependent reduction in endurance of fathead minnows forced to swim at a slow speed. Support for the hypothesis that oxygen transfer was impaired was the observation that lowering the oxygen in the water, or raising the temperature, accentuated the effect of the fiber on swimming (McLeod and Smith, 1966).

In the above-mentioned study, stamina at a fixed slow speed was measured. The critical swimming speed (Ucrit) of fish is also expected to be especially sensitive to impairment of oxygen transfer in gills because this is one of the primary determinants of aerobic scope. Kraft pulpmill effluent has been shown to impair the gas exchange process resulting in depressed levels of arterial oxygen (Davis, 1973). In a rather thorough study, Howard (1975) measured critical swimming speeds in coho salmon (*O. kisutch*) following exposure of the fish to pulpmill effluent at several different concentrations and lengths of time. There was a significant reduction in critical swimming speed at all concentrations above 1/10 the 96 h LC50 and the effect was concentration dependent. A plateau in the extent of swimming inhibition occurred after some 18–48 h of exposure. This may suggest that damage to the gills is not cumulative. Support for this hypothesis was found when coho were exposed to these same concentrations for 90 days, there was no impairment of critical swimming speed, so physiological compensation, or repair of the damaged tissue must have taken place. McLeay and Brown (1979; p. 1056) make the cogent point that "...these compensations should not be considered as acclimation responses unless it can be demonstrated that they occur without an increased energy cost or without a debilitation of other physiological zones of tolerance required for survival in the natural environment."

The ability of brook trout fry to swim in the presence of aluminum (300 μg/L) was found to be very dependent upon the pH of the water (Cleveland et al., 1986). The aluminum had no effect on swimming of the fry at pH 7.2 or 4.5. However, at the intermediate pH of 5.5, they were essentially unable to swim, but after the fish reached 37–67 days of age, the aluminum had no effect on Ucrit at any of the pHs. Aluminum in low pH waters causes problems for both ionoregulatory processes and oxygen transfer at the gills of adult trout (Wood, 1989). The effect, especially on oxygen transfer, seems to be greatest at a pH of around 5–6 which would fit with the larvae data in the Cleveland et al. (1986) work. Still, why there was no effect on Ucrit of juveniles is peculiar and no attempt is made by the authors to explain it.

Acclimation to aluminum has been explored using rainbow trout juveniles (Wilson and Wood, 1992). The fish were exposed for a period of 22 days to water of pH 5.2 in the presence of only 30 μg/L aluminum and other fish were kept without the metal. Figure 7 shows how the aluminum caused a rapid reduction in Ucrit and then a gradual restoration over time as the fish acclimated to the aluminum. Although complete Ucrit recovery did not occur in 22 days, the authors reported that complete restoration of ionoregulatory function occurred at that time.

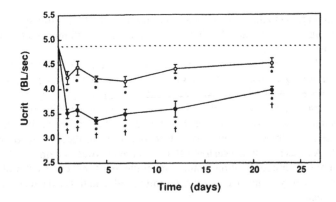

Figure 7 Critical swimming velocities (Ucrit) of rainbow trout expressed as body lengths per second, for fish exposed for 22 days to pH 5.2 (open circles) or pH 5.2 plus 30 µg Al/L (filled circles). The dashed line represents the combined mean value for control fish (maintained and swam at pH 6.5, zero aluminum) on days 0 and 24 (Ucrit = 4.82 ± 0.11 BL/s, n = 30). Asterisks indicate means different ($p < 0.05$) from the 6.5/0 control group, and daggers indicate means significantly different from the 5.2/0 group (From Wilson, R. W. and Wood, C. W., *Fish Physiol. Biochem.*, 10, 149, 1992. With permission.)

The reduction in swim speed from aluminum exposure is associated with histopathological changes in the gill epithelium (Mueller et al., 1991). These changes increase the diffusion distance for oxygen across the gill epithelium and thus are probably responsible for reducing the aerobic scope. Nikl and Farrel (1993) quantified both the changes in lamellar structure and Ucrit in juvenile salmonids exposed to a wood preservative at several concentrations and observed a strong relationship between reduction in Ucrit and extent of gill damage (Figure 4, Chapter 2).

Ucrit can also be reduced by pollutants that do not necessarily affect the transfer of oxygen across the gill. After only 5 h of exposure of trout to copper at 0.02 ppm (in soft water), there was a drop in critical swimming speed to about 55% of controls, but after 10 h of continuous exposure, swimming speed had returned to approximately 73% of normal where it remained for up to 30 days of exposure (Figure 8). The interaction with water hardness is noteworthy in this study. As the water gets harder, the effect on swimming is less. Waiwood and Beamish (1978) conclude that critical swimming performance showed impairment at copper concentrations as low as 0.25 of the 96-h LC50

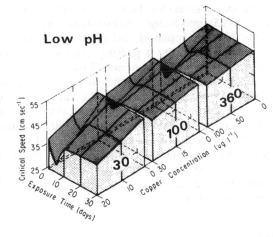

Figure 8 Effect of copper, water hardness, and the duration of exposure on the critical swimming speed of rainbow trout (pH 6, 12°C). Harness (mg/L CaCO₃) is shown by the larger numbers. (The general pattern at pH 7.5 was similar except the critical swimming speeds were higher.) (From Waiwood, K. G. and Beamish, W., *Water Res.*, 12, 611, 1978. With permission.)

which are comparable to those thresholds affecting reproduction in this species. However, under conditions of high water hardness and high pH, swimming performance is not nearly as sensitive to copper.

The physiological basis for the impairment of swimming speed by copper is complex and not clear. Oxygen transfer by the gills at these concentrations is probably not a problem as Sellers et al. (1975) exposed trout to copper concentrations considerably higher than these and observed no change in arterial oxygen in resting fish. The increased energy demand induced by the copper that was discussed earlier in this chapter would reduce the aerobic scope, and this was indeed observed in the Waiwood and Beamish (1978) work. As was mentioned earlier, effects on the nervous system which might cause some loss in muscular coordination would contribute to less efficient swimming resulting in an increased energy demand at any swimming speed, and thus a decrease in the critical swimming speed.

Lead seems to have relatively little effect on swimming speed. Freadman et al. (1985) tested striped bass at several concentrations of lead for 60 days and observed inhibition only at a concentration of 200 µg/L which is fairly high.

It was mentioned earlier in this chapter that petroleum hydrocarbons often stimulate resting metabolism in fish. It would therefore be predicted that a reduction in aerobic scope would follow. This seems to have been confirmed in work on the maximum swimming speed of coho salmon exposed for 48 h to a range of concentrations (25–75% of the LC50) of crude oil (Thomas and Rice, 1987). The effect on swimming performance was not, however, very large and a concentration at 75% of the LC50 was required to produce an effect on swimming performance. Rather large increases in plasma cortisol were seen so the fish were under significant stress from the crude oil.

Respiratory poisons exert their effects at the cellular level through more than one mechanism. For example, cyanide is known to be a blocker of the enzyme cytochrome oxidase so the cell is unable to utilize oxygen as a terminal electron acceptor. In a sense, then, it simulates hypoxia. Pentachlorophenol, on the other hand, blocks oxidative phosphorylation so that oxygen is used, but the generation of ATP aerobically is reduced. Cyanide has been found to be a potent inhibitor of swimming speed (reviewed in Leduc, 1984). Various workers have used exposures of up to 36 days and seen inhibitions in swimming speeds of 50% or more. Pentachlorophenol at chronic levels has been shown by two independent studies to have little effect on this measure of performance in trout or a cichlid (Webb and Brett, 1973; Peer et al., 1983). Thus, it cannot be predicted that metabolic poisons will necessarily alter the Ucrit.

Fish taken from polluted waterways have also been tested for swimming performance. Striped bass from the Hudson River exhibited reduced Ucrit when compared with the same species from other East Coast rivers or hatcheries (Freadman et al., 1985; Buckley et al., 1985). Interestingly, the two studies cited here were conducted approximately at the same time but presumably independent of each other. The explanation for the reduced performance capacity is probably complex as is the mix of pollutants in the Hudson. Freadman et al. (1985) measured PCBs in the parts per million range in the Hudson River fish. Sustained swimming (which they measured) is dependent almost exclusively on red muscle. PCBs are sequestered in the lipids, and, coincidentally the red muscles utilize lipids predominantly for energy so the authors suggest that these organochlorine compounds may alter lipid mobilization during swimming.

Buckley et al. (1985) found that their Hudson River striped bass had heavy parasitic infections in the muscles. Because several studies (reviewed in Beamish, 1978) have shown that parasitic infections can adversely affect swimming performance, the suggestion was made that this may be at least one mechanism for the poor performance of the contaminated fish.

Substances that act on the nervous system can have a marked effect on not only spontaneous activity, as was discussed above in the section on whole-body energy metabolism, but also on the critical swimming speed. Brook trout exposed to the organophosphate insecticide fenitrothion showed a dose-dependent inhibition of critical speed (Peterson, 1974). These insecticides are well-known acetylcholinesterase inhibitors. Post (1969) noted a 50% reduction in coho salmon "stamina" following exposure to malathion, another organophosphate insecticide, and this corresponded to an 86% reduction in acetylcholinesterase activity in the nervous system. When he tested rainbow trout there was a somewhat greater reduction in stamina but less inhibition in the enzyme activity. With the use of several concentrations of two organophosphates, Cripe et al. (1984) observed a marked dose-dependent inhibition of acetylcholinesterase but a relatively high concentration was required before critical swimming speed in sheepshead minnows was reduced. Whether there is an actual relationship between cholinesterase and swimming performance in fish is not clear from these studies. When one starts altering the functions of the nervous system, all sorts of indirect effects may take place, such as "motivational" disturbances.

Larval fish have varying abilities to swim, depending on species and stage of development. Heath et al. (1993) tested swimming performance of newly hatched striped bass larvae exposed for four days to the insecticides methyl parathion, carbofuran, or an herbicide, molinate. Two concentrations were tested, one at one half the 96-h LC50 and one more than an order of magnitude lower. Swimming performance was assessed in a simple fashion using a petri dish with a smaller dish in the center forming a circular "race track". Radial lines were marked under the petri dish. Single larvae were chased with a glass rod around the track for 1 min and the number of lines that were crossed were counted. This arrangement probably measures a combination of burst and prolonged swimming. Few of the larvae became exhausted during the 1-min duration. Significant decreases were seen in swimming performance of larvae in the methyl parathion at both concentrations and in the molinate at the one half LC50 concentration. Interestingly, in both the methyl parathion and molinate groups at the higher concentration, there was depressed swimming ability even after 10 days recovery in non-contaminated water. Acetylcholinesterase was inhibited in the exposed fish but this could not be unequivocally associated with the impaired swimming performance as this enzyme was also inhibited in the carbofuran-exposed fish as well, but no effect of the pesticide on their swimming was detected.

While several authors have noted an inhibition of swimming capacity by pesticides, Little et al. (1990) found no effect of methyl parathion, chlordane, or PCP on this capacity in small (0.5–1.0 g) rainbow trout. They used exposures of 96 h and concentrations as high as one half the 96-h LC50. They also tested the herbicides DEF (an organophosphate) and 2,4-DMA (a phenoxy compound) and found an enhanced swimming ability following exposure to low concentrations but an inhibition at high (but still sublethal) concentrations. The stimulation of swimming ability may be an example of hormesis wherein a modest stress causes enhanced ability to grow or perform other functions (Stebbing, 1987). Of course, there may be other explanations as well.

It should be emphasized here that swimming performance tests of Ucrit probably have validity only with those species that are migratory or normally swim for long periods (e.g., many salmonids). For, as Beitinger and McCauley (1990) have emphasized, swimming stamina is probably not important in many inactive fish species who rely more on burst swimming for the capture of food and evasion of predation. We currently know little about how pollutants affect this type of swimming, or if there is any effect at all. It seems most likely that those chemicals that affect the nervous system will have the most impact on burst swimming, since aerobic capacity is unimportant for that style of movement.

It is evident that a variety of mechanisms may be responsible for reductions in sustained swimming capacity. Under acute exposure to different substances, respiratory gas exchange is often inhibited due to gill damage, and this will reduce the aerobic scope and thereby the critical swimming speed. However, this is probably not as important during chronic exposure where gill damage is less, if there is any damage at all. Then, if swimming effects are seen, they may be due primarily to a greater metabolic demand for detoxification and repair which would reduce the aerobic scope.

Chronic exposure can also produce alterations in nervous function which reduces neuromuscular coordination and/or motivation. One of the most potent inhibitors of swimming performance is cyanide which reduces the ability of muscles to utilize oxygen. A final factor that may be important in limiting swimming stamina is the loss of energy stores (largely carbohydrate) that frequently occurs when fish are exposed to pollutants (see next section). While these are not particularly important for the resting fish, a fish swimming at maximum speed utilizes them rapidly and exhaustion takes place when they are expended (Driedzic and Hochachka, 1979).

XI. CHANGES IN CARBOHYDRATE, LIPID AND PROTEIN ENERGY STORES

A. INTRODUCTION

It was mentioned early in this chapter that fish utilize protein and lipids for energy to a greater extent than do mammals. However, when under acute stress from abiotic or biotic factors, fish are somewhat similar to other vertebrates in that they mobilize and use carbohydrates. Changes in the tissue concentration of carbohydrates have received a good deal of attention among students of the effects of pollution because of their relationship to the "traditional" stress responses of other vertebrate animals. First, some basic concepts will be reviewed so the changes induced by xenobiotics can be better understood. Figure 9 shows the major energy stores. The diagram emphasizes carbohydrates and their hormonal controls because these change rapidly during stress, but it should be emphasized that there is actually far more energy available to the fish in protein from muscle and fat in other parts of the body.

The greatest amount of carbohydrate is stored in the liver as glycogen. Skeletal muscle glycogen is also an important store, but the concentrations found in fish are generally at least an order of magnitude less than those in the liver (Black et al., 1962). Blood glucose is the form in which carbohydrate is transported and the concentration in the blood is controlled by cortisol, adrenaline, and insulin. In general, the homeostasis of blood glucose in fish is poor. The classical picture of insulin maintaining a tight control of blood glucose, as is typical of mammals, does not hold for fish. This hormone has relatively little effect on blood glucose; it instead is mostly involved in stimulating amino acid incorporation into protein and reducing the conversion of amino acids into glycogen (glyconeogenesis; Ablett et al., 1981). Thus, in fish insulin secretion rate by the beta cells in the pancreas is controlled more by amino acids than by glucose (Cowey et al., 1977).

The dynamics of intermediary metabolism are greatly influenced by any sort of stressor and these mechanisms are outlined in Figure 9. A stressor is detected as some sort of change that alters the homeostasis of the animal (including psychological, Schreck, 1981). Operating via the hypothalamus, the first effect is to activate the chromaffin cells which, in teleosts, are located in the walls of the cardinal veins and in some cases the head kidney (Mazeaud and Mazeaud, 1981). There is no well-defined adrenal medulla, as is found in mammals. Chromaffin cells release adrenaline and a small amount of noradrenaline which stimulate the conversion of liver glycogen into blood glucose and the utilization of glucose by muscle. These adrenergic effects may result in increases of blood sugar within minutes of the onset of the stress. Elevations in the concentration of blood lactic

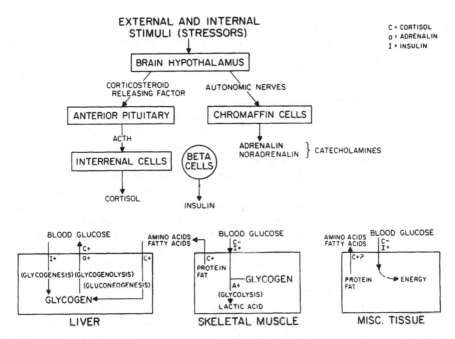

Figure 9 Generalized diagram of energy flows and their hormonal controls emphasizing carbohydrates and the generalized stress response. The letters designating specific hormones with a plus or minus beside them indicate the process is stimulated or inhibited by the hormone. (See text for further explanation.)

acid frequently occur coincident with the stress, but it is not clear whether this is due to adrenergic stimulation of glycolysis, or a direct effect of the stress (e.g., hypoxia) on the muscle tissues. It could, of course, involve both mechanisms.

If the stress persists, other mechanisms may also be involved in mobilization of energy stores. Classically, cortisol from the adrenal cortex (in mammals) acts to reduce glucose utilization by muscle and other tissues and helps form more glycogen in the liver. The net effect is to maintain an elevated blood glucose level. It has long been assumed that similar mechanisms occurred in fish since exogenous administration of cortisol (by injection) generally causes an elevated blood glucose (for review, see Van der Boon et al., 1991). However, this picture may need some revision, at least in trout. Andersen et al. (1991), in a rather sophisticated series of experiments, have cast doubt on the role of cortisol as a glucocorticoid. They implanted osmotic pumps into rainbow trout which provided an elevated level of cortisol without the usual attendant handling associated with injections. The fish were also fitted with dorsal aortic cannulae so serial blood samples could be taken, thus daily glucose, amino acid, and cortisol measurements were possible from the same fish. With this arrangement, they found upon elevation of cortisol that blood amino acids increased in concentration, but there was no elevation of blood glucose and no change in liver enzyme activities associated with glyconeogenesis. These findings suggest that prolonged blood glucose elevations in fish, as seen with chronic stress, must be maintained by mechanisms other than cortisol (adrenaline?). There may also be large species differences so generalizations from rainbow trout may not always be valid (Van der Boon et al., 1991).

While cortisol per se may or may not act directly to raise blood glucose levels, recent studies have revealed that it can alter the responsiveness of hepatocytes to adrenaline. Reid et al. (1992) have shown that chronically raising the blood cortisol level causes a big increase (fourfold) in number of adrenoreceptors on the hepatocyte membranes and this enhances glucose production in response to adrenaline. Thus, cortisol could contribute to hyperglycemia indirectly by this mechanism.

The dynamics of the stress response in mammals (and to a certain extent in fish) generally is characterized by an alarm phase during which the level of adrenaline rises rapidly causing a mobilization of liver glycogen into blood glucose. This is followed by a resistance phase in which the glucose levels remain elevated and liver glycogen may or may not be low, depending on the extent of hyperglycemia, diet, etc. Should the stressor continue, then either adaptation or exhaustion occurs.

Adaptation implies changes in several related physiological processes which permit the return of homeostasis. When blood sugar is involved, this means a return to near normal levels. Exhaustion may occur if the extent and duration of stress is sufficient and is characterized by a depletion of liver glycogen, decreasing levels of cortisol, depressed immune response, and a host of other changes that make the organism less able to survive, especially if other types of stressors are present (Selye, 1973; Pickering, 1981). There is also accumulating evidence that chronic high levels of cortisol in fish cause utilization of protein for energy so growth is inhibited (Van der Boon et al., 1991).

B. EFFECT OF PULPMILL EFFLUENT

Kraft pulpmill effluent has received considerable attention from fish physiologists, and investigation of the changes in energy stores is no exception. Exposure of coho salmon to a concentration of effluent equivalent to 0.8 of the 96-h LC50 produced an immediate hyperglycemia (80% increase in blood glucose). After 48 h of continuous exposure, blood glucose increased even further reaching levels almost four times those of controls. By this time, liver glycogen had decreased to nearly zero (McLeay and Brown, 1975). The authors feel these effects are primarily due to elevated adrenergic activity instead of cortisol because the latter stimulates synthesis of liver glycogen, and they observed a loss of glycogen. This interpretation may be correct, however, it seems just as likely that the extent of liver glycogen breakdown merely exceeded the somewhat slower process of glycogen formation.

The changes in concentrations of carbohydrate in response to pulpmill effluent were explained by McLeay and Brown (1975) as an indirect response to internal hypoxia. Work by Davis (1973) has shown that this effluent causes damage to the gills producing a lowering of arterial oxygen. Because environmental hypoxia (which also lowers arterial oxygen) has been shown to cause hyperglycemia and mobilization of liver glycogen (Heath and Pritchard, 1965), the hypothesis has support.

In a later study, McLeay and Brown (1979) exposed cohos to levels of pulpmill effluent ranging from 0.1–0.5 of the LC50 for periods of up to 200 days. They observed hyperglycemia and a lowering of liver glycogen, but the extent of the changes was not as great as with the acute exposure, and there were indications of recovery by 90 days in this chronic study. The authors mention a condition of "exhaustion" at 200 days, but the data obtained at that time were not much different than those obtained at 90 days. When Oikari and Nakari (1982) subjected trout to components of pulpmill effluent for 11 days, they did observe a clear-cut exhaustion in that the liver glycogen became almost completely depleted and the blood glucose decreased by 35%. They also noted elevated concentrations of ammonia in the blood. The ammonia apparently came from lateral muscle as its protein concentration decreased by 13%. These latter changes would be expected as cortisol stimulates the conversion of protein into free amino acids which are then deaminated by the liver for gluconeogenesis.

The dynamics of protein and lipid changes during exposure to pulpmill effluents are complex. Oikari et al. (1985) report elevated liver protein and muscle lipid after 30 days' exposure to a very low level of effluent. Muscle protein was depressed. They conclude this indicates a shift from protein to lipid metabolism. The high levels of liver protein at this time may be due to the induction of very large amounts of the biotransformation enzyme ethoxyresorufin-O-deethylase (EROD; Larsson et al., 1988). This also results usually in liver enlargement due to the fat and protein synthesis. In general, there are minimal changes in carbohydrates from these effluents if the exposures are realistically low (Andersson et al. 1987; Larsson et al., 1988).

C. METALS

As with pulpmill effluent, zinc at acute levels causes gill damage and consequently an internal hypoxia in fish exposed at acute concentrations (see Chapter 3). Such exposure results in elevated lactic acid and utilization of much of the muscle glycogen (Hodson, 1976). Such changes are in part, at least, a direct effect of hypoxia on the muscle tissue. Zinc also produces a hyperglycemia (presumably from adrenergic stimulation), which may be further potentiated by a depression of insulin secretion from damage to pancreatic beta cells (Wagner and McKeown, 1982) in much the same way that cadmium works (see below).

The dynamics of the response to some chemical insult can be as interesting as the final outcome. For example, Schreck and Lorz (1978) measured cortisol in coho salmon at several time intervals during exposure to different concentrations of copper. There was a dose-dependent rise in blood cortisol within 2 h of the start of exposure, but then this returned to normal at 24 h, only to again become elevated later. These changes in cortisol could be a physiological response to correct an altered blood electrolyte concentration (see Chapter 7), especially the delayed rise in hormone.

Copper causes hyperglycemia in fish, whether the exposure is for a short time or even up to a month. With a very high dose (5 mg/L), carp exhibited a marked hyperglycemia that, however, returned to control level at 6 days. With a much lower concentration (0.25 mg/L), the Indian catfish (*Heteropneustes fossilis*) exhibited a steadily rising blood glucose over a period of 30 days (Asztalos et al., 1990). Therefore, it seems probable that the return to control levels seen in the 1-week exposure of carp was due to exhaustion of liver glycogen, rather than any sort of adaptation to the presence of the metal.

A limited amount of work has shown that fish exposed to a stressor (e.g., a metal) for days or weeks will respond more when challenged by a subsequent but different stressor than do controls that have not had the previous exposure (reviewed in Schreck, 1990). Heath (1991) found that bluegill exposed to copper for a week exhibited more of a hyperglycemic response to environmental hypoxia than did controls. The copper also prolonged the time required for recovery from a hypoxic stress.

Elasmobranchs may respond to copper in a way considerably different than teleosts. At least, that seems to be evident in the study by Tort et al. (1987) in which they exposed dogfish sharks to five different concentrations of copper for 48 h. The highest concentration (16 mg/L) approximated the 24-h LC50 for this species of fish, which is very high compared to teleosts. There was a rather severe and dose-dependent decrease in blood glucose, which is the opposite of what would be expected in a teleost. They also found a marked increase in liver glycogen, but no changes in liver lipid or protein. They attribute the hypoglycemia to hemodilution (a decrease in hematocrit was seen) which is possible since most elasmobranchs are slightly hyperosmotic to seawater and copper has notable effects on osmoregulation (see Chapter 7). The rise in glycogen could be due to an inhibition by copper of glycogen conversion into blood sugar. An interesting hypothesis to test would be whether the enzymes responsible for mobilization of glycogen to blood sugar were being inhibited by the accumulation of copper in the liver.

The concentration of copper in the water has recently been shown to have a large qualitative effect on lipid metabolism. When exposed to a lethal concentration for three days, Sivaramakrishna et al. (1992) found that *Labeo* lost total lipids from nearly all tissues and there was an increase in free fatty acids and glycerol. On the other hand, exposures to sublethal concentrations for up to 30 days produced increases in total lipids and transitory increases in free fatty acids. The largest changes were seen in liver with lesser effects of the copper in gill, muscle, and brain, in that order. It seems probable that the rise in lipids could be due to conversion from protein although no direct evidence is available.

While metals, in general, may cause conversion of liver glycogen into blood glucose at acute and perhaps even at chronic concentrations, the mechanism by which this is done is certainly different for cadmium, depending on the fish species. In the Schreck and Lorz (1978) work on copper mentioned above, they also measured cortisol in cohos exposed to cadmium at a variety of concentrations including some that were clearly lethal. Much to their surprise, there were no significant changes in cortisol in the fish exposed to cadmium, even in those that were moribund, but a cadmium-induced hyperglycemia has been noted (Gill and Pant, 1983; Larsson and Haux, 1981; Das and Banerjee, 1980; Ghazaly, 1992; Pratap and Bonga, 1990) as well as hypoglycemia (Gill and Pant, 1983; Das and Banerjee, 1980), depending on the dose, and liver glycogen may or may not become depressed, depending on the species. For example, when rainbow trout and flounder (*P. flesus*) were exposed to cadmium at two different concentrations for 80 days, there was a marked reduction in liver glycogen in the trout, but an 80% elevation in the founder. Both exhibited a hyperglycemia, but the effect was greater in the trout (Larsson and Haux, 1981). The differences reflect the rather large difference in metabolic activity of these species and undoubtedly, there are some differences in their hormonal responses as well.

The cortisol response to cadmium may be quite species specific too. While Schreck and Lorz (1978) observed no change in coho salmon exposed to very high doses of cadmium, tilapia, on the other hand, showed a large increase in cortisol which persisted for 14 days' exposure. By 35 days of continuous cadmium exposure, the tilapia had returned to the control level suggesting an adaptation to the presence of the metal (Pratap and Bonga, 1990). Interestingly, blood glucose also went up, but returned to control level in only four days so the elevated cortisol did not correlate with the hyperglycemia. This perhaps fits in with the discussion at the beginning of this section relative to the possible lack of a glucocorticoid activity of cortisol in fish.

The effects of cadmium on carbohydrate metabolism are complicated by the fact that this element causes damage to the insulin-producing cells in the pancreas (Havu, 1969). In mammals cadmium has been shown to decrease the levels of circulating insulin (Ghafghazi and Mennear, 1973). A reduction in insulin secretion rate causes hyperglycemia, such as occurs during insulin diabetes. Thus, elevations seen in blood glucose during chronic exposure to cadmium might be in part due to this "diabetic" condition. There is probably also some adrenergic stimulation, especially if the doses are high, which could cause losses in liver glycogen, as occurred in the trout mentioned above. If chronic exposure results in hypoglycemia, it is probably due to a combination of liver glycogen depletion, reduced feeding, and even excretion of glucose by the kidney, as occurs in mammals poisoned by cadmium (Ghafghazi and Mennear, 1973).

In sharks, as was discussed above, copper produced a lowered blood sugar. In the same laboratory, it was found that cadmium caused just the opposite; a marked hyperglycemia and loss of liver glycogen and protein occurred during 96-h exposures (Hernandez-Pascual and Tort, 1989). In other words, cadmium caused a more "traditional" stress response in this elasmobranch.

A lowering of blood sugar was observed in trout exposed to lead for 16 weeks (Haux et al., 1981). The authors suggest this was due to damage of kidney tubules which they say occurs in lead-exposed mammals. Another possibility, which they did not mention, is the fact that lead depresses gluconeogenesis in the livers of rats (Cornell, 1974) and may do the same thing in fish. The situation regarding lead in fish is somewhat confusing in that the same laboratory observed an elevated blood glucose (60% increase) in fish from a lead-contaminated lake (Haux et al., 1985). Of course, in this field study, other factors besides lead could have been present to cause the elevated glucose levels.

Mercury (as HgCl) has also been shown to cause a lowering of blood glucose and liver glycogen in the freshwater snakehead (*Channa punctatus*) (Sastry and Rao, 1981). Sastry and Rao (1982) subsequently found that this element caused inhibition of the enzyme glucose-6-phosphatase in the liver of these fish. This is a key enzyme in the conversion of glycogen to glucose, so these findings provide a mechanism for the observed changes in the carbohydrates. These workers further found an elevated rate of activity for the enzymes glutamate dehydrogenase, amino acid oxidase, and xanthine oxidase which suggests a shift to more amino acid catabolism in the fish exposed to mercury. Such a change in metabolism could be viewed as a cellular-level compensation for the reduced energy available from carbohydrates.

We see from the foregoing review that the responses of fish to some metals follows a classical general adaptation syndrome of hyperglycemia, lowered liver glycogen, and elevated cortisol. Other metals such as mercury produce stress, but the response does not necessarily follow the classical adaptation syndrome.

D. PETROLEUM HYDROCARBONS AND SURFACTANTS

Petroleum and its individual constituents may be examples of substances that cause "stress", but the classical response of hyperglycemia is absent or minimal. McKeown and March (1978) used a 96-h exposure of rainbow trout to Bunker C oil and Zbanzszek and Smith (1984) a 1- to 72-h exposure of the same species to a model mixture of aromatic hydrocarbons. In the first study, a slight hypoglycemia was observed even though gill damage and some mortality took place. In the second investigation cited, depending on the concentration, significant changes occurred in hematological components but blood glucose remained the same, even in moribund fish. McKeown and March (1978) speculate that petroleum affects the kidney, causing loss of glucose through the urine.

Benzene, an important component of petroleum, was found to cause irritation of sensitive membranes such as gill epithelium and a considerable accumulation of benzene in tissues of the fish (MacFarlane and Benville, 1986). However, in spite of evident stress, striped bass adults exposed to this material for up to 21 days exhibited only transient elevations in cortisol, blood glucose, and lactate with apparent adaptation to the stress within 48 h.

Detergents have been shown to cause severe gill damage (see Chapter 3). Fingerling carp exposed to a sublethal concentration of linear alkyl benzene sulfonate for up to 96 h exhibited progressive loss of glycogen from gill, liver, and kidney but a gain in protein in these organs (Misra et al., 1991). Lactic acid increased considerably in all tissues so internal hypoxia must have occurred and this would explain the loss of glycogen due to enhanced anaerobic metabolism. The increased protein synthesis is difficult to explain but might conceivably reflect general repair of damaged tissue.

E. PESTICIDES

Pesticides seemingly cause a variety of changes in carbohydrate metabolism in fish, however, care must be taken in interpreting results. For example, high doses can cause an immediate hyperglycemia followed by hypoglycemia before death (Hanke et al., 1983). If the blood glucose were measured at only one time interval, one could conclude very different things regarding responses.

Exposure of eels (*Anguilla anguilla*) to 0.1 ppm of PCP for 8 days causes a marked hyperglycemia, but peculiarly, no changes in liver glycogen (Holmberg et al., 1972). High glucose levels persist even after 55 days of recovery in non-treated water. Holmberg and associates do not attempt to explain these findings other than to conclude the PCP caused a "hypermetabolic state". A high glucose level without a concomitant loss of some glycogen and the persistence of the hyperglycemia argues for some sort of alteration in the ability of tissues to metabolize glucose for energy, so it would tend to accumulate in the blood. Because PCP enhances tissue metabolism (see section on whole-animal metabolism), there would seem to be the possibility that the PCP caused damage to the insulin-producing beta cells in the pancreas. This mechanism has already been hypothesized to explain the effects of exposure to cadmium, but is only a tentative speculation at this time for PCP.

Organophosphate insecticides may suppress aerobic metabolism of tissues (see discussion in earlier part of this chapter); the action would therefore appear to be different from that of PCP. Hyperglycemia associated with liver glycogen mobilization has been observed in fish exposed to a variety of sublethal concentrations of organophosphate compounds (Koundinya and Ramamurthi, 1979; Sastry et al., 1982; Awasthi et al., 1984; Rani et al., 1990; Gill et al., 1990). Marked elevations in lactic acid, even after a month of exposure, suggest that the cause of the carbohydrate mobilization is probably due in part to adrenaline. A reduction in aerobic metabolism due to suppression of the aerobic metabolic machinery by the insecticides would also cause a stimulation of anaerobiosis and utilization of carbohydrate for energy.

Exposure of *C. punctatus* to quinalphos (another organophosphate insecticide) for 15 and 30 days produced slight but significant increases in liver glycogen and muscle glycogen and a mild hypoglycemia (Sastry et al., 1982). This is the opposite of the findings reviewed in the preceding paragraph; the authors feel their observations may in part be explained by elevations in insulin. As support for that hypothesis, they note that Eller (1971) found hyperplasia of pancreatic beta cells following chronic exposure to endrin (a chlorinated hydrocarbon insecticide). The mechanism of the hyperplasia induced by the pesticide is unknown, and we do not know whether an organophosphate insecticide can cause such an effect, nor is it known if the hyperplasia actually causes elevations in insulin, or is merely a compensatory response to inhibition of insulin synthesis or secretion by the pesticide.

A more definitive study of an organophosphate compound (dimethoate) may shed some light on the contradictory findings discussed above. Pant and Singh (1983) found that acute exposure produced a hyperglycemia and glycogenolysis, but when they exposed the fish to concentrations equivalent to 1/11 or 1/7 of the 96-h LC50 for 15 and 30 days they observed hypoglycemia even though liver glycogen was only partially depleted. The acute responses can easily be explained as a classical stress response mediated by adrenergic and possibly cortisol stimulations. They propose that hypoglycemia following chronic exposure may be due to the hyperplasia of beta cells already mentioned. Again, however, it is not known if there actually is an elevation in insulin.

It appears that chlorinated insecticides cause mobilization of liver glycogen into blood glucose in much the same fashion as seen with organophosphate compounds (Gill et al., 1991). For example, Verma et al. (1983) exposed three different species of freshwater fish to a chlorinated hydrocarbon and an organophosphate insecticide for 30 days at concentrations equivalent to 1/4, 1/8, and 1/16 of the 96-h LC50. With both compounds there was a dose-dependent hyperglycemia ranging up to almost a 100% increase and a dose-dependent loss in liver and muscle glycogen (up to 70%). Organochlorine compounds also cause big increases in lactic acid and the effect increases with duration of exposure (Gill et al., 1991).

Gill et al. (1990; 1991) noted a marked drop in brain glycogen from organophosphorus or organochlorine exposure. Because brain tissue is dependent upon carbohydrate for its metabolism (Hawkins, 1985), this depletion of that energy store could have serious implications and could provide a mechanism for neurological changes.

More recent studies have revealed changes in lipids and protein associated with pesticides. The effect depends on the tissue examined. The organochlorine endosulfan caused total lipids, free fatty acids, and proteins to increase in liver, whereas the total lipids decreased in skeletal muscles and ovary (Gill et al., 1991). A similar pattern of change was seen with an organophosphate insecticide (Gill et al., 1990).

Changes in lipids may have an impact on reproduction. Singh (1992) examined changes in female *Heteropneustes fossilis* (catfish) exposed to malathion (an organophosphate) for 4 weeks during various phases of the annual reproductive cycle. During the preparatory phase malathion caused a reduction in free fatty acids, monoglycerides, triglycerides, phospholipids, and free cholesterol in liver and gonad. Liver esterified cholesterol was increased. As spawning approached, the liver free fatty acids rose while all other lipids (except esterified cholesterol) remained low. At spawning, malathion caused the ovarian levels of triglycerides and esterified cholesterol to be elevated while other lipids were still low. Evidently, cholesterol biosynthesis is unaffected by the pesticide but the hydrolysis of esterified cholesterol to free cholesterol is inhibited. This could affect sex steroid biosynthesis. Also the mobilization of various hepatic lipids to the ovary may be inhibited by the malathion.

When carp were exposed to the synthetic pyrethroid cypermethrin for up to 96 h, they exhibited a steady decline in glycogen and lipid in liver, gill, and brain. Muscle, on the other hand, showed a steady increase in these energy stores over the same time (Piska et al., 1992). Interestingly, protein concentrations changed in exactly the opposite direction with increases in liver, gill, and brain, possibly due to synthesis of detoxification enzymes, while muscle protein exhibited a steady decline in concentration.

Lipid levels may have promise as one of several bioindicators of water contamination in field studies. Adams et al. (1992) measured serum and total body triglycerides in sunfish (*Lepomis*) taken from four sites along a stream receiving mixed contaminants. Declines in these measures were seen in the more contaminated sites with the serum triglycerides being the most sensitive of the two biochemical measures. The lower energy reserves were also reflected in lower growth rates and fecundity in the same populations of fishes.

F. MISCELLANEOUS POLLUTANTS

Ozone is a strong oxidizing agent which is an example of a toxicant that causes massive gill damage and subsequent internal hypoxia. Ozone is sometimes used as a substitute for chlorine in sewage disposal and in steam electric generating plants. Chronic exposure to 5 μg/L resulted in only mild gill damage and no changes in osmoregulation or blood glucose (Wedemeyer et al., 1979). However, only one day's exposure to 7 μg/L caused acute hyperglycemia, presumably mediated by adrenaline. This rather sharp threshold between a "no effect" level and a strong response has also been noted in simple toxicity tests on chlorine (Heath, 1977).

Lowering the the pH of the water to 5.2 caused a mild elevation of plasma glucose in rainbow trout (Brown et al., 1984). A further lowering to 4.7 resulted in a large hyperglycemic response which got steadily greater for up to 20 days of continuous exposure. Plasma cortisol was unaffected by lowering the pH to 5.2 but greatly elevated at pH 4.7, so the hyperglycemia may reflect some effect of cortisol as well as adrenergic action.

G. CONCLUDING COMMENT

From the foregoing discussion it is clear that levels of energy stores are sensitive to pollution stress. Thus, they may lend themselves to utilization as a biomarker in field monitoring programs, but if we are to understand the mechanisms involved in the changes observed, future studies should incorporate measures of hormones, especially catecholamines, insulin, and cortisol. Glucagon and growth hormone may also be important as well.

REFERENCES

Ablett, R. F., Sinnhuber, R. O. and Selivonchick, D. P., The effect of bovine insulin on glucose and leucine incorporation in fed and fasted rainbow trout (*Salmo gairdneri*), *Gen. Comp. Endocrinol.*, 44, 418, 1981.

Adams, S. M. and Breck, J. E., Bioenergetics, in, *Methods For Fish Biology*, Schreck, C. B. and Moyle, P. B., Eds., American Fisheries Society, Bethesda, MD, 1990, chap. 12.

Adams, S. M., Crumby, W., Greeley, M., Ryon, M. and Schilling, E., Relationship between physiological and fish population responses in a contaminated stream, *Environ. Toxicol. Chem.*, 11, 1549, 1992.

Andersen, D. E., Reid, S. D., Moon, T. W. and Perry, S. F., Metabolic effects associated with chronically elevated cortisol in rainbow trout (*Oncorhynchus mykiss*), *Can. J. Fish. Aquat. Sci.*, 48, 1811, 1991.

Andersson, T., Bengtsson, B., Forlin, L., Hardig, J. and Larsson, A., Long-term effects of bleached kraft mill effluents on carbohydrate metabolism and hepatic xenobiotic biotransformation enzymes in fish, *Ecotoxicol. Environ. Safety*, 13, 53, 1987.

Arillo, A., Margiocco, C., Melodia, F. and Mensi, P., Biochemical effects of long term exposure to Cr, Cd, Ni, on rainbow trout (*Salmo gairdneri*): influence of sex and season, *Chemosphere*, 11, 47, 1982.

Arunachalam, S., Jeyalakshmi, K. and Aboobucker, S., Toxic and sublethal effects of carbaryl on a freshwater catfish, *Mystus vittatus*, *Arch. Environ. Contam. Toxicol.*, 9, 307, 1980.

Asztalos, B., Nemcsok, J., Benedeczky, I., Gabriel, R., Szabo, A. and Refaie, O., The effects of pesticides on some biochemical parameters of carp (*Cyprinus carpio*), *Arch. Environ. Contam. Toxicol.*, 19, 275, 1990.

Awasthi, M., Shah, P., Dubale, M. and Gadhia, P., Metabolic changes induced by organophosphates in the piscine organs, *Environ. Res.*, 35, 320, 1984.

Bahr, T. G. and Ball, R. C., Action of DDT on evoked and spontaneous activity from the rainbow trout lateral line nerve, *Comp. Biochem. Physiol.*, 38A, 379, 1971.

Bala, S. T. and Mohideen, M. B., Oxygen consumption and opercular activity as good indicators of pesticidal (malathion and methyl parathion) stress in major carp, *Catla catla*, *Indian J. Comp. Anim. Physiol.*, 10, 20, 1992.

Bansal, S. K., Verma, S. R., Gupta, A. K., Rani, S. and Dalbla, R. C., Pesticide-induced alterations in the oxygen uptake rate of the freshwater major carp *Labeo rohita*, *Ecotoxicol. Environ. Safety*, 3, 374, 1979.

Barron, M. G. and Adelman, I. R., Nucleic acid, protein content and growth of larval fish sublethally exposed to various toxicants, *Can. J. Fish. Aquat. Sci.*, 41, 141, 1984.

Barron, M. G. and Adelman, I. R., Temporal characterization of growth of fathead minnow (*Pimephales promelas*) larvae during sublethal hydrogen cyanide exposure, *Comp. Biochem. Physiol.*, 81C, 341, 1985.

Bartholomew, G. A., Energy metabolism, in, *Animal Physiology*, 4th ed., Gordon, M. S., Ed., Macmillan, New York, 1982, chap. 3.

Bayne, B. L., Moore, M. N., Widdows, J., Livingstone, D. R. and Salkeld, P., Measurement of responses of individuals to environmental stress and pollution: studies with bivalve molluscs, *Philos. Trans. R. Soc. London Ser. B*, 286, 563, 1979.

Beitinger, T. L. and McCauley, R. W., Whole-animal physiological processes for the assessment of stress in fishes, *J. Great Lakes Res.*, 16, 542, 1990.

Bengtsson, B. E., Increased growth in minnows exposed to PCBs, *Ambio*, 8, 160, 1979.

Beamish, F. W. H., Influence of starvation on standard and routine oxygen consumption, *Trans. Am. Fish. Soc.*, 93, 103, 1964.

Beamish, F. W. H., Swimming capacity, in, *Fish Physiology*, Vol. 7, Hoar, W. S. and Randall, D. J., Eds., Academic Press, New York, 1978, chap. 2.

Beamish, F. W. H. and Mookherjii, P. S., Respiration of fishes with special emphasis on standard oxygen consumption, *Can. J. Zool.*, 42, 161, 1964.

Black, E. C., Connor, A. R., Lam, K. and Chin, W., Changes in glycogen, pyruvate and lactate in rainbow trout (*Salmo gairdneri*) during and following muscular activity, *J. Fish. Res. Bd. Can.*, 19, 409, 1962.

Borgmann, U. and Ralph, K., Effects of cadmium, 2,4-dichlorophenol and pentachlorophenol on feeding, growth, and particle-size-conversion efficency of white sucker larvae and young common shiners, *Arch. Environ. Contam. Toxicol.*, 15, 473, 1986.

Brafield, A. E. and Mathiessen, P., Oxygen consumption by sticklebacks exposed to zinc, *J. Fish Biol.*, 9, 359, 1976.

Brafield, A. E., Laboratory studies of energy budgets, in, *Fish Energetics, New Perspectives*, Tytler, P. and Calow, P., Eds., Johns Hopkins University Press, Baltimore, 1985, chap 10.

Brett, J. R., The respiratory metabolism and swimming performance of young sockeye salmon, *J. Fish. Res. Bd. Can.*, 21, 1183, 1964.

Brett, J. R. and Groves, T. D. D., Physiological energetics, in, *Fish Physiology*, Vol. 8, Hoar, W. S. and Randall, D. J., Eds., Academic Press, New York, 1979, chap. 6.

Brown, D. A. and Parsons, T. R., Relation between cytoplasmic distribution of mercury and toxic effects to zooplankton and chum salmon (*Oncorhynchus keta*) exposed to mercury in a controlled ecosystem, *J. Fish. Res. Bd. Can.*, 35, 880, 1978.

Brown, S. B., Eales, J. G., Evans, R. E. and Hara, T. J., Interrenal, thyroidal, and carbohydrate responses of rainbow trout (*Salmo gairdneri*) to environmental acidification, *Can. J. Fish Aquat. Sci.*, 41, 36, 1984.

Brown, P. B., Neill, W. H. and Robinson, E. H., Preliminary evaluation of whole body energy changes as a method of estimating maintenance energy needs of fish, *J. Fish Biol.*, 36, 107, 1990.

Buckley, L. J., RNA-DNA ratio: an index of larval fish growth in the sea, *Mar. Biol.*, 80, 291, 1984.

Buckley, L. J., Halavik, T., Laurence, G. C., Hamilton, S. J. and Yevich, P., Comparative swimming stamina, biochemical composition, backbone mechanical properties, and histopathology of juvenile striped bass from rivers and hatcheries of the Eastern United States, *Trans. Am. Fish. Soc.*, 114, 114, 1985.

Bulow, F., RNA-DNA ratios as indicators of recent growth rates of fish, *J. Fish. Res. Bd. Can.*, 27, 2343, 1970.

Bulow, F., RNA-DNA ratios as indicators of growth in fish: a review, in, *Age and Growth of Fish*, Summerfelt, R. C. and Hall, G. E., Eds., Iowa State University Press, Ames, 1987, 45.

Bulow, F., Zeman, M., Winningham, J. R. and Hudson, W. F., Seasonal variations in RNA-DNA ratios and in indicators of feeding, reproduction, energy storage and condition in a population of bluegill, *Lepomis macrochirus*, *J. Fish Biol.*, 18, 237, 1981.

Calabrese, A., Thurberg, F. P. and E. Gould, E., Effects of cadmium, mercury, and silver on marine animals, *Mar. Fish. Rev.*, 39, 5, 1977.

Calabrese, F. P., Thurberg, F. P., Dawson, M. A. and Wenzloff, D. R., Sublethal physiological stress induced by cadmium and mercury in the winter flounder, *Pseudopleuronectes americanus*, in, *Sublethal Effects of Toxic Chemicals on Aquatic Animals*, Koeman, J. H. and Strik, J. J. T. W. A., Eds., Elsevier, Amsterdam, 1975, 15.

Cech, J. J., Jr., Respirometry, in, *Methods for Fish Biology*, Schreck, C. B. and Moyle, P. B., Eds., American Fisheries Society, Bethesda, MD, 1990, chap. 10.

Chew, S. F. and Ip, Y. K., Cyanide detoxification in the mudskipper, *Boleophthalmus boddaerti*, *J. Exp. Zool.*, 261, 1, 1992.

Cleveland, L., Little, E. E., Hamilton, S. J., Buckler, D. R. and Hunn, J. B., Interactive toxicity of aluminum and acidity to early life stages of brook trout, *Trans. Am. Fish. Soc.*, 115, 610, 1986.

Colgan, P., Motivational analysis of fish feeding, *Behavior*, 45, 38, 1973.

Collvin, L., The effect of copper on growth, food consumption and food conversion of perch *Perca fluviatilis* L. offered maximal food rations, *Aquat. Toxicol.*, 6, 105, 1985.

Cornell, R., Depression of hepatic gluconeogenesis by acute lead poisoning in rats, *Physiologist*, 17, 199, 1974 .

Cowey, C. B., Higuera, M. and Adron, J., The effect of dietary composition and of insulin on gluconeogenesis in rainbow trout (*Salmo gairdneri*), *Br. J. Nutr.*, 38, 385, 1977.

Cripe, G. M., Goodman, L. R. and Hansen, D. J., Effect of chronic exposure to EPN and to guthion on the critical swimming speed and brain acetylcholinesterase activity of *Cypriodon variegatus*, *Aquat. Toxicol.*, 5, 255, 1984.

Dabrowski, K., Takashima, F. and Strussmann, C., Does recovery growth occur in larval fish?, *Bull. Jpn. Soc. Sci. Fish.*, 52, 1869, 1986.

Das, K. and Banerjee, K., Cadmium toxicity in fishes, *Hydrobiologia*, 75, 117, 1980.

Davis, J. C., Sublethal effects of bleached kraft pulpmill effluent on respiration and circulation in sockeye salmon (*Oncorhynchus nerka*), *J. Fish. Res. Bd. Can.*, 30, 369, 1973.

Davis, J. C., Progress in sublethal effect studies with kraft pulpmill effluent and salmonids, *J. Fish. Res. Bd. Can.*, 33, 2031, 1976.

210

Daxboeck, C., Davie, P. S., Perry, S. F. and Randall, D. J., Oxygen uptake in a spontaneously ventilating blood-perfused trout preparation, *J. Exp. Biol.,* 101, 35, 1982.

Diana, J. S., An experimental analysis of the metabolic rate and food utilization of northern pike, *Comp. Biochem. Physiol.,* 71A, 395, 1982.

Dixon, D. G. and Leduc, G., Chronic cyanide poisoning of rainbow trout and its effects on growth, respiration, and liver histopathology, *Arch. Environ, Contam. Toxicol.,* 10, 117, 1981.

Donaldson, E. M., The pituitary-interrenal axis as an indicator of stress in fish, in, *Stress and Fish,* Pickering, A. D., Ed., Academic Press, New York, 1981, chap. 2.

Donaldson, E. M., Fagerlund, U. H., Higgs, D. A. and McBride, J. R., Hormonal enhancement of growth, in, *Fish Physiology,* Vol. 8, Hoar, W. S., Randall, D. J. and Brett, J. R., Eds., Academic Press, New York, 1979, chap. 9.

Driedzic, W. R. and Hochachka, P. W., Metabolism in fish during exercise, in, *Fish Physiology,* Vol. 7, Hoar, W. S. and Randall, D. J., Eds., Academic Press, New York, 1978, chap. 8.

Eldridge, M. B., Echeverria, T. and Whipple, J., Energetics of Pacific herring (*Clupea harengus pallasi*) embryos and larvae exposed to low concentrations of benzene, a monaromatic component of crude oil, *Trans. Am. Fish. Soc.,* 106, 452, 1977.

Eller, L. L., Histopathological lesions in cutthroat trout (*Salmo clarki*) exposed chronically to the insecticide endrin, *Am. J. Pathol.,* 64, 321, 1971.

Ellgaard, E. G., Ochsnedr, J. C. and Cox, J. K., Locomotor hyperactivity induced in the bluegill sunfish, *Lepomis macrochirus,* by sublethal concentrations of DDT, *Can. J. Zool.,* 55, 1077, 1977.

Ellgaard, E. G., Tusa, J. E. and Malizia, A. A., Locomotor activity of the bluegill *Lepomis macrochirus:* hyperactivity induced by sublethal concentrations of cadmium, chromium, and zinc, *J. Fish Biol.,* 1, 19, 1978.

Ellgaard, E. G., Breaux, J. G. and Quina, D. B., Effects of South Louisiana crude oil on the bluegill sunfish, *Lepomis macrochirus, Proc. La. Acad. Sci.,* 42, 49, 1979.

Ellgaard, E. G. and Rudner, T. W., Lead acetate: toxicity without effects on the locomotor activity of the bluegill sunfish, *Lepomis macrochirus, J. Fish Biol.,* 21, 411, 1982.

Ellgaard, E. G. and Gilmore, J. Y., Effects of different acids on the bluegill sunfish, *Lepomis macrochirus, J. Fish Biol.,* 25, 133, 1984.

Ellgaard, E. and Guillot, J., Kinetic analysis of the swimming behavior of bluegill sunfish, *Lepomis macrochirus* Raf., exposed to copper: hypoactivity induced by sublethal concentrations, *J. Fish Biol.,* 33, 601, 1988.

Engel, D. W. and Fowler, B. A., Copper and cadmium-induced changes in the metabolism and structure of molluscan gill tissue, in, *Marine Pollution: Functional Responses,* Vernberg, W. B., Thurberg, F. P., Calabrese, A. and Vernberg, F. J., Eds., Academic Press, New York, 1979.

Evans, R. M., Purdie, F. C. and Hickman, C. P., The effect of temperature and photoperiod on the respiratory metabolism of rainbow trout, (*Salmo gairdneri*), *Can. J. Zool.,* 40, 107, 1962.

Farmanfarmian, A. and Socci, R., Inhibition of essential amino acid absorption in marine fishes by mercury, *Mar. Environ. Res.,* 14, 185, 1984.

Fauconneau, B., Protein synthesis and protein deposition in fish, in, *Nutrition and Feeding in Fish,* Cowey, C., Mackie, A. and Bell, J., Eds., Academic Press, New York, 1985.

Felts, P. A. and Heath, A. G., Interactions of temperature and sublethal environmental copper exposure on the energy metabolism of bluegill, *Lepomis macrochirus* Rafinesque, *J. Fish Biol.,* 25, 445, 1984.

Ferguson, D. E., Ludke, J. L. and Murphy, G. G., Dynamics of endrin uptake and release by resistant and susceptible strains of mosquitofish, *Trans. Am. Fish. Soc.,* 95, 335, 1966.

Fletcher, D. J., The physiological control of appetite in fish, *Comp. Biochem. Physiol.,* 78A, 617, 1984.

Fletcher, G. L., Kiceniuk, J. W. and Williams, U. P., Effects of oiled sediments on mortality, feeding and growth of winter flounder *Pseudopleuronectes americanus, Mar. Ecol. Prog. Ser.,* 4, 91, 1981.

Freadman, M. A., Thurberg, F. P. and Calabrese, A., Swimming/locomotor capacity of Hudson River striped bass, in, *Marine Pollution and Physiology: Recent Advances,* Vernberg, F. J., Ed., University of South Carolina Press, Columbia, 1985, 31.

Fry, F. E. J., The effect of environmental factors on the physiology of fish, in *Fish Physiology,* Vol. 6, Hoar, W. S. and Randall, D. J., Eds., Academic Press, New York, 1971, chap. 1.

Gessman, J. A., An evaluation of heart rate as an indirect measure of daily energy metabolism of the American kestrel, *Comp. Biochem. Physiol.,* 65A, 273, 1980.

Ghafghazi, T. and Mennear, J., Effects of acute and subacute cadmium administration on carbohydrate metabolism in mice, *Toxicol. Appl. Pharmacol.,* 26, 231, 1973.

Ghazaly, K. S., Hematological and physiological responses to sublethal concentrations of cadmium in a freshwater teleost, *Tilapia zilli, Water, Air, Soil Pollut.,* 64, 551, 1992.

Gill, T. S. and Pant, J. C., Cadmium toxicity: inducement of changes in blood and tissue metabolites in fish, *Toxicol. Lett.,* 18, 195, 1983.

Gill, T. S., Pande, J. and Tewari, H., Sublethal effects of an organophosphorus insecticide on certain metabolite levels in a freshwater fish, *Puntius conchonius, Pest. Biochem. Physiol.,* 36, 290, 1990.

Gill, T. S., Pande, J. and Tewari, H., Effects of endosulfan on the blood and organ chemistry of freshwater fish, *Barbus conchonius, Ecotoxicol. Environ. Safety,* 21, 80, 1991.

Gnaiger, E., Symbols and units: toward standardization, in, *Polarographic Oxygen Sensors, Aquatic and Physiological Applications,* Gnaiger, E. and Forstner, H., Eds., Springer-Verlag, New York, 1983.

Goolish, E. M. and Adelman, I. R., Tissue-specific cytochrome oxidase activity in largemouth bass: the metabolic costs of feeding and growth, *Physiol. Zool.,* 60, 454, 1989.

Graham, J. B., Dewar, H., Lai, N. C., Lowell, W. R. and Arce, S. M., Aspects of shark swimming performance determined using a large water tunnel, *J. Exp. Biol.,* 151, 175, 1990.

Hanke, W., Gluth, G., Bubel, H. and Muller, R., Physiological changes in carps induced by pollution, *Ecotoxicol. Environ. Safety,* 7, 229, 1983.

Haux, C., Sjobeck, M. and Larsson, A., Some toxic effects of lead on fish, in, *Stress and Fish,* Pickering, A. D., Ed., Academic Press, New York, 1981, 340

Haux, C., Larsson, A., Lithner, G. and Sjobeck, M., A field study of physiological effects on fish in lead-contaminated lakes, *Environ. Toxicol. Chem.,* 5, 283, 1985.

Havu, N., Sulfhydryl inhibitors and pancreatic islet tissue, *Acta Endocrinol.,* (Suppl.), 139, 1969.

Hawkes, J. W., The effects of petroleum hydrocarbon exposure on the structure of fish tissues, in, *Fate and Effects of Petroleum Hydrocarbons in Marine Ecosystems and Organisms,* Wolfe, D. A., Ed., Pergamon Press, New York, 1977.

Hawkins, R., Cerebral energy metabolism, in, *Cerebral Metabolism and Metabolic Encephalopathy,* McCandless, D. W., Ed., Plenum Press, New York, 1985, 3.

Heath, A. G., Toxicity of intermittent chlorination to freshwater fish: influence of temperature and chlorine form, *Hydrobiologica,* 56, 39, 1977.

Heath, A. G., Effect of water-borne copper on physiological responses of bluegill (*Lepomis macrochirus*) to acute hypoxic stress and subsequent recovery, *Comp. Biochem. Physiol.,* 100C, 559, 1991.

Heath, A. G. and Pritchard, A. W., Changes in the metabolic rate and blood lactic acid of bluegill sunfish, *Lepomis macrochirus,* Raf. following severe muscular activity, *Physiol. Zool.,* 35, 323, 1962.

Heath, A. G. and Pritchard, A. W., Effects of severe hypoxia on carbohydrate energy stores and metabolism in two species of freshwater fish, *Physiol. Zool.,* 38, 325, 1965.

Heath, A. G., Cech, J. J., Zinkl, J. G., Finlayson, B. and Fujimura, R., Sublethal effects of methyl parathion, carbofuran and molinate on larval striped bass, *Morone saxatilis, Am. Fish. Soc. Symp.,* 14, 17, 1993.

Hernandez-Pascual, M. and Tort, L., Metabolic effects after short-term sublethal cadmium exposure to dogfish (*Scyliorhinus canicula*), *Comp. Biochem. Physiol.,* 94C, 261, 1989.

Hewett, S. and Johnson, B., Fish Bioenergetics Model 2, University of Wisconsin-Madison, Sea Grant Institute, Communications, 1800 University Ave., Madison, WI, 1992.

Hochachka, P. W. and Somero, G., *Biochemical Adaptation,* Princeton University Press, Princeton, NJ, 1984.

Hodson, P. V., Temperature effects on lactate-glycogen metabolism in zinc-intoxicated rainbow trout (*Salmo gairdneri*), *J. Fish. Res. Bd. Can.,* 33, 1393, 1976.

Holmberg, B., Jensen, S., Larsson, A., Lewander, K. and Olsson, M., Metabolic effects of technical pentachlorophenol (PCP) on the eel *Anguilla anguilla* L., *Comp. Biochem. Physiol.,* 43B, 171, 1972.

Hose, J. E. and Puffer, H. W., Oxygen consumption rates of grunion (*Leuresthes tenuis*) embryos exposed to the petroleum hydrocarbon, Benzo[a]pyrene, *Environ. Res.,* 35, 413, 1984.

Houlihan, D. F., Protein turnover in ectotherms and its relationships to energetics, in, *Advances in Comparative and Environmental Physiology,* Vol. 7., Gilles, R., Ed., Springer-Verlag, Berlin, 1991, chap. 1.

Houlihan, D. F., Biochemical correlates of growth rate in fish, in, *Fish Ecophysiology,* Rankin, J. C. and Jensen, F. B., Eds., Chapman and Hall, London, 1993, chap. 2.

Howard, T. E., Swimming performance of juvenile coho salmon (*Oncorhynchus kisutch*) exposed to bleached kraft pulpmill effluent, *J. Fish. Res. Bd. Can.,* 32, 789, 1975.

Itazawa, Y. and Oikawa, S., Metabolic rates in excised tissues of carp, *Experentia*, 39, 160, 1983.

Jobling, M., Influence of feeding on the metabolic rate of fishes: a short review, *J. Fish Biol.*, 18, 385, 1981.

Johansen, J. A. and Green, G. H., Sublethal and acute toxicity of the ethylene glycol butyl ester formulation of triclopyr to juvenile coho salmon (*Oncorynchus kisutch*), *Arch. Environ. Contam. Toxicol.*, 19, 610, 1990.

Johansen, K. and Pettersson, K., Gill O_2 consumption in a teleost fish, *Gadus morhua*, *Respir. Physiol.*, 44, 277, 1981.

Johnston, I. A. and Moon, T. M., Endurance exercise training in the fast and slow muscles of a teleost fish (*Pollachius virens*), *J. Comp. Physiol.*, 135B, 147, 1980.

Kasello, P., Weatherley, A., Lotimer, J. and Farina, M., A biotelemetry system recording fish activity, *J. Fish Biol.*, 40, 165, 1992.

Kaupp, S. E. and Somero, G., Biochemical indices of metabolic and growth rates in fish, *Am. Zool.*, 28, 127, 1988.

Kiceniuk, J. W. and Kahn, R. A., Effect of petroleum hydrocarbons on Atlantic cod, *Gadus morhua*, following chronic exposure, *Can. J. Zool.*, 65, 490, 1987.

Koch, F., Wieser, W. and Niederstatter, H., Interactive effects of season and temperature on enzyme activities, tissue and whole animal respiration in roach, *Rutilus rutilus*, *Environ. Biol. Fishes*, 33, 73, 1992.

Koundinya, P. R. and Ramamurthi, R., Effect of organophosphate pesticide (Fenitrothion) on some aspects of carbohydrate metabolism in a freshwater fish, *Sarotherodon (Tilapia) mossambicus* (Peters), *Experentia (Basel)*, 35, 1632, 1979.

Krueger, H. M., Saddler, J. B., Chapman, G. A., Tinsley, L. J. and Lowry, R. R., Bioenergetics, exercise, and fatty acids of fish, *Am. Zool.*, 8, 119, 1968.

Kumaraguru, A. K. and Beamish, F. W. H., Bioenergetics of acclimation to permethrin (NRDC-143) by rainbow trout, *Comp. Biochem. Physiol.*, 75A, 247, 1983.

Larsson, A. and Haux, C., Effects of cadmium on carbohydrate metabolism in fish, in, *Stress and Fish*, Pickering, A. D., Ed., Academic Press, New York, 1981, 341.

Larsson, A., Andersson, T., Forlin, L. and Hardig, J., Physiological disturbances in fish exposed to bleached kraft mill effluents, *Water. Sci. Technol.*, 20, 67, 1988.

Laughlin, R. B., Ng, J. and Guard, H. E., Hormesis: a response to low environmental concentrations of petroleum hydrocarbons, *Science*, 211, 705, 1981.

Leduc, G., Cyanides in water: toxicological significance, in, *Aquatic Toxicology*, Vol. 2, Weber, L. J., Ed., Raven Press, New York, 1984, chap. 4.

LeFavre, M. E., Wyssbrod, H. R. and Brodsky, W. A., Problems in the measurement of tissue respiration with the oxygen electrode, *Bioscience*, 20, 761, 1970.

Lied, E., Lie, O. and Lambertsen, G., Nutritional evaluation in fish by measurement of *in vitro* protein synthesis in white trunk muscle tissue, in, *Nutrition and Feeding in Fish*, Cowey, C., Mackie, A. and Bell, J., Eds., Academic Press, New York, 1985.

Lett, P. F., Farmer, G. J. and Beamish, F. W. H., Effect of copper on some aspects of the bioenergetics of rainbow trout (*Salmo gairdneri*), *J. Fish. Res. Bd. Can.*, 33, 1335, 1976.

Little, E. E., Archeski, R. D. Flerov, B. and Kozlovskaya, V. I., Behavioral indicators of sublethal toxicity in rainbow trout, *Arch. Environ. Contam. Toxicol.*, 19, 380, 1990.

MacFarlane, R. and Benville, P., Primary and secondary stress response of striped bass (*Morone saxatilis*) exposed to benzene, *Mar. Biol.*, 92, 245, 1986.

MacLeod, J. C. and Pessah, E., Temperature effects on mercury accumulation, toxicity, and metabolic rate in rainbow trout, (*Salmo gairdneri*), *J. Fish. Res. Bd. Can.*, 30, 485, 1973.

Malte, H., Effects of aluminum in hard, acid water on metabolic rate, blood gas tensions and ionic status in the rainbow trout, *J. Fish Biol.*, 29, 187, 1986.

Mason, B. J. and Davis, J. C., Growth in underyearling coho salmon, *Oncorhynchus kisutch*, during chronic exposure to sublethal levels of neutralized bleached kraft pulp mill waste, *J. Fish Res. Bd. Can.*, Tech. Rep. 1976, cited in Davis (1976).

Mathers, E. M., Houlihan, D. F. and Cunningham, M. J., Nucleic acid concentrations and enzyme activities as correlates of growth rate of the saithe *Pollachius virens*: growth rate estimates of open-sea fish, *Mar. Biol.*, 112, 363, 1992.

Mayer, F. L., Mehrle, P. M. and Sanders, H. O., Residue dynamics and biological effects of polychlorinated biphenyls in aquatic organisms, *Arch. Environ. Contam. Toxicol.*, 5, 501, 1977.

Mazeaud, M. M. and Mazeaud, F., Adrenergic responses to stress in fish, in, *Stress and Fish*, Pickering, A. D., Ed., Academic Press, New York, 1981, chap. 3.

McKeown, B. and March, G. L., The acute effect of bunker C oil and an oil dispersant on: serum glucose, serum sodium, and gill morphology in both fresh-water and sea-water acclimated rainbow trout, *(Salmo gairdneri), Water Res.,* 12, 157, 1978.

McKim, J. M., Evaluation of tests with early life stages of fish for predicting long-term toxicity, *J. Fish. Res. Bd. Can.,* 34, 1148, 1977

McLeay, D. J. and Brown, D. A., Growth stimulation and biochemical changes in juvenile coho salmon *(Oncorhynchus kisutch)* exposed to bleached kraft pulpmill effluent for 200 days, *J. Fish. Res. Bd. Can.,* 31, 1043, 1974.

McLeay, D. J. and Brown, D. A., Effects of acute exposure to bleached kraft pulpmill effluent on carbohydrate metabolism of juvenile coho salmon *(Oncorhynchus kisutch)* during rest and exercise, *J. Fish. Res. Bd. Can.,* 32, 753, 1975.

McLeay, D. J. and Brown, D. A., Stress and chronic effects of untreated and treated kraft pulpmill effluent on the biochemistry and stamina of juvenile coho salmon *(Oncorhynchus kisutch), J. Fish. Res. Bd. Can.,* 36, 1049, 1979.

McLeod, J. C. and Smith, L. L., Jr., Effect of pulpmill fiber on oxygen consumption and swimming endurance of the fathead minnow, *Pimephales promelas, Trans. Am. Fish. Soc.,* 95, 71, 1966.

Miglavs, I. and Jobling, M., Effects of feeding regime on food consumption, growth rates and tissue nucleic acids in juvenile Arctic charr, *Salvelinus alpinus,* with particular respect to compensatory growth, *J. Fish Biol.,* 34, 947, 1989.

Millward, D., The nutritional regulation of muscle growth and protein turnover, Aquaculture, 79, 1, 1989.

Mishra, J. and Srivastava, A. K., Malathion induced hematological biochemical changes in the Indian catfish *Heteropneustes fossilis, Environ. Res.,* 30, 393, 1983.

Misra, V., Kumar, V., Pandey, S. and Viswanathan, P., Biochemical alterations in fish fingerlings *(Cyprinus carpio)* exposed to sublethal concentration of linear alkyl benzene sulphonate, *Arch. Environ. Contam. Toxicol.,* 21, 514, 1991.

Mommesen, T. P., Metabolism of the fish gill, in, *Fish Physiology,* Hoar, W. S. and Randall, D. J., Eds., Academic Press, New York, 1984, chap. 7.

Mueller, M., Sanchez, P., Bergman, H., McDonald, D., Rhem, R. and Wood, C., Nature and time course of acclimation to aluminum in juvenile brook trout *(Salvelinus fontinalis).* 2. Histology, *Can. J. Fish. Aquat. Sci.,* 48, 2016, 1991.

Murty, A. S., Ramami, A., Christopher, K. and Rajabhushanam, B., Toxicity of methyl parathion and fensulfothion to the fish *Mystus cavasius, Environ. Pollut. Ser. A.,* 34, 37, 1984.

Nagy, K. A., CO_2 production in animals: analysis of potential errors in the doubly labeled water method, *Am. J. Physiol.,* 238, R466, 1980.

Newsholme, E. A. and Crabtree, B., Maximum catalytic activity of some key enzymes in provision of physiological useful information about metabolic fluxes, *J. Exp. Zool.,* 239, 159, 1986.

Nikl, D. and Farrell, A. P., Reduced swimming performance and gill structural changes in juvenile salmonids exposed to 2-(thiocyanomethylthio)benzothiazole, *Aquat. Toxicol.,* 27, 245, 1993.

O'Hara, J., Alterations in oxygen consumption by bluegills exposed to sublethal treatment with copper, *Water Res.,* 5, 321, 1971.

O'Hara, J., Relationship of the metabolic rate of excised gill tissue to body size in two species of sunfish, *Can. J. Zool.,* 49, 373, 1971.

Oikawa, S. and Itazawa, Y., Examination of techniques for manometric determination of the rate of tissue respiration, *Bull. Jpn. Soc. Sci. Fish.,* 49, 23, 1983.

Oikari, A. and Nakari, T., Kraft pulp mill effluent components cause liver dysfunction in trout, *Bull. Environ. Contam. Toxicol.,* 28, 266, 1982.

Oikari, A., Holmbom, B., Anas, E., Miilunpalo, M., Kruzynski, G. and Castren, M., Ecotoxicological aspects of pulp and paper mill effluents discharged to an inland water system: distribution in water, and toxicant residues and physiological effects in caged fish *(Salmo gairdneri), Aquat. Toxicol.,* 6, 219, 1985.

Oikari, A., Lindstrom-Seppa, P. and Kukkonen, J., Subchronic metabolic effects and toxicity of a simulated pulp mill effluent on juvenile lake trout, *Salmo trutta* m. lacustris, *Ecotoxicol. Environ. Safety,* 16, 202, 1988.

Olson, K. R., Fromm, P. D. and Franz, W. L., Ultrastructural changes of rainbow trout gills exposed to methyl mercury or mercuric chloride, *Fed. Proc.,* 32, 261, 1973.

Olson, K. R., Squibb, K. S. and Cousins, R. J., Tissue uptake, sublethal distribution, and metabolism of $^{14}CH_3HgCl$ and $CH_3{}^{203}HgCl$ by rainbow trout, *Salmo gairdneri, J. Fish. Res. Bd. Can.,* 35, 381, 1978.

214

Ostrander, G. K., Landolt, M. L. and Kocan, R. M., Whole life history studies of coho salmon (*Oncorhynchus kisutch*) following embryonic exposure to benzo[a]pyrene, *Aquat. Toxicol.*, 15, 109, 1989.

Pamila, D., Subbaiyan, P. and Ramaswamy, M., Toxic effects of chromium and cobalt on *Sarotherodon mossambicus, Indian J. Environ. Health*, 33, 218, 1991.

Pandian, T. J., Fish, in, *Animal Energetics*, Vol. 2, Pandian, T. J. and Vernberg, F. J., Eds., Academic Press, New York, 1987, chap. 7.

Panigrahi, A. K. and Misra, B. N., Toxicological effects of mercury on a freshwater fish, *Anabasscandens,* and their ecological implications, *Environ. Pollut.*, 16, 31, 1978.

Pannevis, M. C. and Houlihan, D., The energetic cost of protein synthesis in isolated hepatocytes of rainbow trout (*Oncorhynchus mykiss*), *J. Comp. Physiol. B*, 162, 393, 1992.

Pant, J. and Singh, T., Inducement of metabolic dysfunction by carbamate and organophosphorus compounds in a fish, *Puntius conchonius, Pest. Biochem. Physiol.*, 20, 294, 1983.

Perry, S. and Walsh, P., Metabolism of isolated fish gill cells: contribution of epithelial chloride cells, *J. Exp. Biol.*, 144, 507, 1989.

Peer, M. M., Nirmala, J. and Kutty, M. N., Effects of pentachlorophenol (Na PCP) on survival, activity and metabolism in *Rhinomugil corsula* (Hamilton), *Cyprinus carpio* (Linnaeus) and *Tilapia mossambica* (Peters), *Hydrobiologia*, 107, 19, 1983.

Peterson, R. H., Influence of fenitrothion on swimming velocity of brook trout (*Salvelinus fontinalis*), *J. Fish. Res. Bd. Can.*, 31, 1757, 1974.

Peterson, M. S. and Peterson, N., Growth under stressed conditions in juvenile channel catfish *Ictalurus punctatus* as measured by nucleic acids, *Comp. Biochem. Physiol.*, 103A, 323, 1992.

Pickering, A. D., Ed., *Stress and Fish*, Academic Press, New York, 1981.

Piska, R. S., Waghray, S. and Devi, I., The effect of sublethal concentration of synthetic pyrethroid, cypermethrin to the common carp, *Cyprinus carpio*, fry, *J. Environ. Biol.*, 13, 89, 1992.

Post, G., Acetyl cholinesterase inhibition and stamina in salmonids, in, *Progress in Sport Fisheries Research,* Bureau of Sport Fishing and Wildlife, Division of Fishing Research, Resources Publication, 1969, 88.

Prasad, M., Toxicity of crude oil to the metabolism of freshwater minor carp, *Puntius sophore, Bull. Environ. Contam. Toxicol.*, 39, 188, 1987.

Pratap, H. and Bonga, S. E. W., Effects of water-borne cadmium on plasma cortisol and glucose in the cichlid fish *Oreochromis mossambicus, Comp. Biochem. Physiol.*, 95C, 313, 1990.

Precht, H., Concepts of the temperature adaptations of unchanging reaction systems of cold-blooded animals, in, *Physiological Adaptation,* Prosser, C. L., Ed., American Physiological Society, Washington, D.C., 1958, 50.

Priede, L. G. and Tytler, P., Heart rate as a measure of metabolic rate in teleost fishes; *Salmo gairdneri, Salmo trutta* and *Gadus morhua, J. Fish Biol.*, 10, 231, 1977.

Rani, V. J., Venkateshwarlu, P. and Janaiah, C., Impact of sublethal concentration of malathion on certain aspects of metabolism in freshwater fish, *Clarias batrachus, Comp. Physiol. Ecol.*, 15, 13, 1990.

Randall, D. and Brauner, C., Effects of environmental factors on exercise in fish, *J. Exp. Biol.*, 160, 113, 1991.

Rao, G. M. M., Oxygen consumption of rainbow trout (*Salmo gairdneri*) in relation to activity and salinity, *Can. J. Zool.*, 46, 781, 1968.

Rao, K. S. P., Sahib, L. K. and Rao, K. V., Methyl parathion (*O-O*-dimethyl *O*-4-nitrophenyl thiophosphate) effects on whole-body and tissue respiration in the teleost *Tilapia mossambica* (Peters), *Ecotoxicol. Environ. Safety*, 9, 339, 1985.

Rath, S. and Misra, B. N., Sub-lethal effects of Dichlorvos (DDVP) on respiratory metabolism of *Tilapia mossambica,* Peters of 3 age groups, *Exp. Gerontol.*, 14, 37, 1979.

Rath, S. and Misra, B. N., Age related changes in oxygen consumption by the gill, brain and muscle tissues of *Tilapia mossambica* Peters exposed to Dichlorvos (DDVP), *Environ. Pollut.*, 23A, 95, 1980.

Reddy, P. M. and Philip, G. H., Changes in the levels of respiration and ions in the tissues of freshwater fish, *Labeo rohita* under fenvalerate stress, *Chemosphere*, 25, 843, 1992.

Reid, S. D., Moon, T. W. and Perry, S. F., Rainbow trout hepatocyte β-adrenoceptors, catecholamine responsiveness, and the effects of cortisol, *Am. J. Physiol.*, 262 (*Regulatory Integrative Comp. Physiol.*, 31), R794, 1992.

Rogers, S. C. and Weatherly, A. H., The use of opercular muscle electromyograms as an indicator of the metabolic costs of fish activity in rainbow trout, *Salmo gairdneri* Richardson, as determined by radiotelemetry, *J. Fish. Biol.*, 23, 535, 1983.

Samis, A., Colgan, P. and Johansen, P., Pentachlorophenol and reduced food intake of bluegill, *Trans. Am. Fish. Soc.*, 122, 1156, 1993.

Sandstrom, O., Neuman, E. and Karas, P., Effects of bleached kraft mill effluent on growth and gonad function in Baltic coastal fish, *Water Sci. Technol.,* 20, 107, 1988.

Sandheinrich, M. and Atchison, G., Modeling bioenergetics effects of cadmium on bluegill growth, *Soc. Environ. Toxicol. Chem. Meeting Poster Presentation,* p. 134, 1993.

Sastry, K. V. and Rao, D. R., Enzymological and biochemical changes produced by mercuric chloride in a teleost fish, *Channa punctatus, Toxicol. Lett.,* 9, 321, 1981.

Sastry, K. V. and Rao, D. R., Chronic effects of mercuric chloride on the activities of some enzymes in certain tissues of the freshwater murrel, *Channa punctatus, Chemosphere,* 11, 1203, 1982.

Scarabello, M., Heigenhauser, G. J. F. and Wood, C. M., The oxygen debt hypothesis in juvenile rainbow trout after exhaustive exercise, *Respir. Physiol.,* 84, 245, 1991.

Schreck, C. B., Stress and compensation in teleostean fishes: response to social and physical factors, in, *Stress and Fish,* Pickering, A. D., Ed., Academic Press, New York, 1981, chap. 13.

Schreck, C. B., Physiological, behavioral and performance indicators of stress, in, *Biological Indicators of Stress in Fishes,* Adams, S. M., Ed., American Fisheries Society Symposium, 8, 1990, 29.

Schreck, C. B. and Lorz, H. W., Stress response of coho salmon (*Oncorhynchus kisutch*) elicited by cadmium and copper and potential use of cortisol as an indicator of stress, *J. Fish. Res. Bd. Can.,* 35, 1124, 1978.

Seim, W., Curtis, L., Glenn, S. and Chapman, G., Growth and survival of developing steelhead trout (*Salmo gairdneri*) continuously or intermittently exposed to copper, *Can. J. Fish. Aquat. Sci.,* 41, 433, 1984.

Sellers, C. M., Heath, A. G. and Bass, M. L., The effect of sublethal concentrations of copper and zinc on ventilatory activity, blood oxygen, and pH in rainbow trout, *Water Res.,* 9, 401, 1975.

Selye, H., The evolution of the stress concept, *Am. Sci.,* 61, 692, 1973.

Serigstad, B. and Adoff, G. R., Effects of oil exposure on oxygen consumption of cod eggs and larvae, *Mar. Environ. Res.,* 17, 266, 1985.

Simon, L. M. and Robin, E. D., Relationship of cytochrome oxidase activity to vertebrate total and organ oxygen consumption, *Int. J. Biochem.,* 2, 569, 1971.

Singh, P. B., Impact of malathion and Y-BHC on lipid metabolism in the freshwater female catfish, *Heteropneustes fossilis, Ecotoxicol. Environ. Safety,* 23, 22, 1992.

Singh, H. S. and Reddy, T., Effect of copper sulfate on hematology, blood chemistry, and hepato-somatic index of an Indian catfish, *Heteropneustes fossilis,* and its recovery, *Ecotoxicol. Environ. Safety,* 20, 30, 1990.

Sivaramakrishna, B., Suresh, A., Venkataramana, P. and Radhakrishnaiah, K., Copper influenced changes of lipid metabolism in the tissues of the freshwater teleost *Labeo rohita, Biochem. Intern.,* 26, 335, 1992.

Skadsen, J. M., Webb, P. W. and Kostecki, P. T., Measurement of sublethal metabolic stress in rainbow trout (*Salmo gairdneri*) using automated respirometry, *J. Environ. Sci. Health,* B, 15, 193, 1980.

Smith, J. C. and Chong, C. K., Body weight, activities of cytochrome oxidase and electron transport system in the liver of the American Plaice *Hippoglossoides platessoides.* Can these activities serve as indicators of metabolism?, *Mar. Ecol. Prog. Ser.,* 9, 171, 1982.

Somero, G., Yancy, P., Chow, T. and Snyder, C., Lead effects on tissue and whole organism respiration of the estuarine teleost fish, *Cillichthys mirabilis, Arch. Environ. Contam. Toxicol.,* 6, 349, 1977.

Soofiani, N. M. and Hawkins, A. D., Field studies of energy budgets, in, *Fish Energetics, New Perspectives,* Tytler, P. and Calow, P., Eds., Johns Hopkins University Press, Baltimore, 1985, chap. 11.

Stebbing, A. R. D., Growth hormesis: a by-product of control, *Health Phys.,* 52, 543, 1887.

Storozhuk, N. G. and Smirnov, B. F., Effect of mercury on the respiration rate of larval pink salmon, *Oncorhynchus gorbuscha, J. Ichthyol.,* 22, 168, 1982.

Sureau, D. and Lagardere, J., Coupling of heart rate and locomotor activity in sole, *Solea solea,* and bass, *Dicentrarchus labrax,* in their natural environment by using ultrasonic telemetry, *J. Fish. Biol.,* 38, 399, 1991.

Thomas, P., Carr, R. S. and Neff, J. M., Biochemical stress responses of mullet *Mugil cephalus* and polychaete worms *Neanthes virens* to pentachlorophenol, in *Biological Monitoring of Marine Pollutants,* Vernberg, J., Calabrese, A., Thurberg, F. and Vernberg, W., Eds., Academic Press, New York, 1981, 73.

Thomas, R. E. and Rice, S. D., The effect of exposure temperatures on oxygen consumption and opercular breathing rates of pink salmon fry exposed to toluene, naphthalene, and water-soluble fractions of cook inlet crude oil and No. 2 fuel oil, in, *Marine Pollution: Functional Responses,* Vernberg, W. B., Thurberg, F. P., Calabrese, A. and Vernberg, F. J., Eds., Academic Press, New York, 1979, 39.

Thomas, R. E. and Rice, S. D., Effect of water-soluble fraction of cook inlet crude oil on swimming performance and plasma cortisol in juvenile coho salmon (*Oncorhynchus kisutch*), *Comp. Biochem. Physiol.,* 87, 177, 1987.

Thurberg, F. P. and Dawson, M. A., Physiological response of the cunner, *Tautogolabrus adspersus,* to cadmium. III. Changes in osmoregulation and oxygen consumption. NOAA Tech. Rep. NMFSSSRF-681, 1974, 11.

Tort, L., Crespo, S. and Balasch, J., Oxygen consumption of the dogfish gill tissue following zinc treatment, *Comp. Biochem. Physiol.,* 72C, 145, 1982.

Tort, L., Torres, P. and Hidalgo, J., Short-term cadmium effects on gill tissue metabolism, *Mar. Pollut.,* 15, 448, 1984.

Tort, L. Torres, P. and Flos, R., Effects on dogfish haematology and liver composition after acute copper exposure, *Comp. Biochem. Physiol.,* 87C, 349, 1987.

Van den Thillart, G. and Verbeek, R., Anoxia-induced oxygen debt of goldfish (*Carassius auratus*), *Physiol. Zool.,* 64, 525, 1991.

Van der Boon, J., Van den Thillart, G. and Addink, A., The effects of cortisol administration on intermediary metabolism in teleost fish, *Comp. Biochem. Physiol.,* 100A, 47, 1991.

Venugopalan, V. K., Effect of growth hormone injection on the level of nucleic acids in the liver of intact fish (*Ophicephalus striatus*), *Gen. Comp. Endocrinol.,* 8, 332, 1967.

Verma, S. R., Rani, S., Tonk, L. P. and Dalela, R. C., Pesticide-induced dysfunction in carbohydrate metabolism in three freshwater fishes, *Environ. Res.,* 32, 127, 1983.

Vignier, V., Vandermeulen, J., and Fraser, A., Growth and food conversion by Atlantic salmon parr during 40 days exposure to crude oil, *Trans. Am. Fish. Soc.,* 121, 322, 1992.

Wagner, G. F. and McKeown, B. A., Changes in plasma insulin and carbohydrate metabolism of zinc-stressed rainbow trout (*Salmo gairdneri*), *Can. J. Zool.,* 60, 2079, 1982.

Waiwood, K. G. and Johansen, P. H., Oxygen consumption and activity of the white sucker (*Catostomus commersoni*), in lethal and nonlethal levels of the organochlorine insecticide, Methoxychlor, *Water Res.,* 8, 401, 1974.

Waiwood, K. G. and Beamish, F. W. H., Effects of copper, pH and hardness on the critical swimming performance of rainbow trout (*Salmo gairdneri* Richardson), *Water Res.,* 12, 611, 1978.

Weatherly, A. H., Approaches to understanding fish growth, *Trans. Am. Fish. Soc.,* 119, 662, 1990.

Weatherly, A. H. and Gill, H. S., Recovery growth following periods of restricted rations and starvation in rainbow trout *Salmo gairdneri, J. Fish Biol.,* 18, 195, 1981.

Webb, P. W. and Brett, J. R., Effects of sublethal concentrations of sodium pentachlorophenate on growth rate, food conversion efficiency, and swimming performance in underyearling sockeye salmon (*Oncorhynchus nerka*), *J. Fish. Res. Bd. Can.,* 30, 499, 1973.

Wedemeyer, G. A., Nelson, N. C. and Yasutake, W. T., Physiological and biochemical aspects of ozone toxicity to rainbow trout, (*Salmo gairdneri*), *J. Fish. Res. Bd. Can.,* 36, 605, 1979.

Wieser, W. and Medgyesy, N., Metabolic rate and cost of growth in juvenile pike (*Esox lucius* L.) and perch (*Perca fluviatilis* L.): the use of energy budgets as indicators of environmental change, *Oecologica,* 87, 500, 1991.

Wilson, R. W., Bergman, H. and Wood, C., Metabolic costs and physiological consequences of acclimation to aluminum in juvenile rainbow trout (*Oncorhynchus mykiss*). 1. Assimilation specificity, resting physiology, feeding and growth, *Can. J. Fish. Aquat. Sci.,* 51, 527, 1994.

Wilson, R. W. and Wood, C. M., Swimming performance, whole body ions, and gill Al accumulation during acclimation to sublethal aluminum in juvenile rainbow trout, *Fish Physiol. Biochem.,* 10, 149, 1992.

Woltering, D. M., The growth response in fish chronic and early life stage toxicity tests: a critical review, *Aquat. Toxicol.,* 5, 1, 1984.

Wood, C. M., The physiological problems of fish in acid waters, in, *Acid Toxicity and Aquatic Animals,* Morris, R., Taylor, E., Brown, D. and Brown, J., Eds., 1989, 125.

Wood, C. M. and McDonald, D. G., Physiological mechanisms of acid toxicity in fish, in, *Acid Rain/Fisheries,* Johnson, R. E., Ed., American Fisheries Society, Bethesda, MD, 1982, 197.

Zaba, B. N. and Harris, E. J., Accumulation and effects of trace metal ions in fish liver mitochondria, *Comp. Biochem. Physiol.,* 61C, 89, 1978.

Zbanzszek, R. and Smith, L. S., The effect of water-soluble aromatic hydrocarbons on some haemotological parameters of rainbow trout, *Salmo gairdneri,* during acute exposure, *J. Fish Biol.,* 24, 545, 1984.

Alterations in Cellular Enzyme Activity, Antioxidants, Adenylates, and Stress Proteins

I. INTRODUCTION

Essentially all chemical reactions in cells are catalyzed by enzymes, thus the action of a foreign chemical in the cell almost always involves disturbances in enzyme function. The subject of biochemical toxicology is immense and much is beyond the scope of this treatise as it involves subjects such as induction of neoplasms, etc. This chapter follows naturally from the one on whole-body energetics (Chapter 8), because in attempting to understand the mechanisms by which pollutants produce alterations in energetics, we are often led to the cellular level of biological complexity in order to gain a fuller understanding of the underlying mechanisms. Because many of the cellular enzyme systems in cells are associated with the generation of ATP, adenylates will be considered here. Antioxidants and stress proteins are included in this chapter, somewhat arbitrarily, because they are cellular constituents.

A recent book on aquatic animal biochemical toxicology (Malins and Ostrander, 1994) deals with xenobiotic transformation processes, cellular histological biomarkers, DNA adducts, neoplasia, etc. In the presentation here, coverage will be limited to the action of particular chemicals on intermediary metabolism enzymes, adenylates, and stress proteins and antioxidants. Enzymatic transformations of xenobiotic materials and changes in tissue glutathione are covered in Chapter 5. Alterations in tissue ascorbic acid are discussed in Chapter 6 and of acetylcholinesterase in Chapter 12.

II. SOME COMMENTS ABOUT ENZYME METHODOLOGY

It is often assumed that xenobiotic "A" affects only enzyme "B" (or perhaps C, D, and E); such specificity of action rarely occurs. Rather, there is usually a broader spectrum of effects in that many chemicals affect the same enzyme, and many enzymes are affected by the same chemical. Nevertheless, the direction of the change (i.e., increase or decrease in activity) and the relative extent of the effect make these sort of studies fruitful, providing one realizes that reports of an altered activity of some enzyme does not mean that it is the only one affected, and is therefore the "mode of action" of that contaminant. Other enzymes were probably also affected but were not assayed. Because cells have many hundreds of types of enzymes, the possibilities are considerable.

The measurement of an enzyme's activity is usually done using only partially purified systems. The test tissue (e.g., liver) is homogenized in order to break up the cells. Differential centrifugation may or may not then be used to separate constituents such as mitochondria from other components. The activity of the enzyme in question is then quantified by measuring the production of a product, or the change in a coenzyme (e.g., NAD) while the reaction mixture is provided with an excess of substrate specific for the enzyme in question. Assuming the pH and temperature are appropriate, the activity of the enzyme is then limited by the amount of enzyme present. The activity is expressed as micrograms of enzyme in the extract, as units of activity, or as the amount of some product formed per unit time. When foreign chemicals are present, there may be inhibition or acceleration of the catalyzed reaction rate. The mechanism of these effects can involve changes in enzyme quantity, or they may be due to a direct effect of the chemical on the enzyme molecules affecting affinity for substrate, etc.

Obviously, one measures enzyme actvity *in vitro*. There have been numerous studies where the test substance (e.g., a trace metal such as cadmium) was added directly to the reaction mixture and the activity of some enzyme is then measured. It is quite obvious to anyone who has studied biology that cells are far more than a mere bag of enzymes. Cells are incredibly complex in both structure and function. Enzymes are packaged, for the most part, in organelles, and the ultramicroscopic structure of these is altered by common water pollutants (e.g., Somasundaram et al., 1984). The penetration of foreign chemicals into these structures may be considerably different than into the cell interior as a whole. As was indicated in Chapter 5, cells often have mechanisms for sequestering and/or excreting potentially harmful substances (Fowler, 1987).

From all these considerations it is obvious that the effects on enzymes may be quite different if the material is added *in vitro* as opposed to exposing the animal to the test substance for a period of time and then measuring the enzyme activity in tissues removed from the exposed animals (i.e., an *in vivo* test). For example, Jackim et al. (1970) and Hilmy et al. (1985) found little correlation between the *in vitro* and *in vivo* effects of several trace metals on a variety of liver enzymes in fish. Gill et al. (1990) examined this question in several organs of the rosy barb following mercury exposure and found little correspondence between the effects seen. Changes produced by a chemical can be in opposite directions when comparing *in vitro* and *in vivo* exposures. This is strikingly seen in the work of Bostrom and Johansson (1972) on eels exposed to pentachlorophenol (PCP). Of the seven liver enzymes they measured, cytochrome oxidase was the most sensitive to *in vitro* inhibition by PCP. However, *in vivo* exposure of the eels to the same pesticide resulted in a considerable stimulation of cytochrome oxidase activity. Thus, there may be enzyme induction in the intact tissue but enzyme inhibition when exposed to the chemical *in vitro*.

Another problem with *in vitro* studies, and this is an especially important one, has to do with the metabolism of organic molecules, such as pesticides. Many, if not most, xenobiotic organic chemicals are rapidly converted to other metabolites in the liver (see Chapter 5). Therefore, the lesions produced in enzymatic activity observed from exposure of a fish to some organic chemical may be markedly different both quantitatively and qualitatively than that seen if the chemical is simply included in the homogenate reaction mixture. In DeBruin's (1976; p. 690) massive review of the largely mammalian literature, he concludes at one point: "Examples of additional inhibitory effects of poisons on enzyme systems may be multiplied almost indefinitely. However, many of the studies in this field apply to *in vitro* experiments and offer, in general, little value for elucidating basic mechanisms of toxic action *in vivo*."

In the review here, we will limit our consideration to those studies where the exposure was to intact fish, and the extensive mammalian literature will, for the most part, be omitted.

For the discussion that follows, a point that should be kept in mind is that enzymes of energy metabolism are sensitive to nutritional stress so that some of them decrease during starvation. These include but are probably not limited to several Kreb's cycle enzymes (Smith and Chong, 1982; Goolish and Adelman, 1987; Houlihan et al., 1993). It is common for fish to greatly decrease their food consumption when under pollutant stress, so some apparent enzyme inhibitions from long-term exposures may be due to this mechanism rather than some direct effect of the pollutant on the enzyme.

III. ALTERATIONS IN CELLULAR ENZYME ACTIVITY RESULTING FROM METAL EXPOSURE

Table 1 summarizes many of the studies of the action of metals on enzyme activity. For all of the work listed, exposures were to the waterborne element and tissues were then removed from the fish following the interval of time indicated. In one of the first studies on this topic, Jackim et al. (1970) exposed killifish (*Fundulus heteroclitus*) held in seawater to a variety of metals individually at a concentration equal to their 96-h LC50. Only surviving fish were used. The enzymes chosen were all metal-containing enzymes, so the presumption was that traces of other cations might displace or affect the metal moiety. From examination of Table 1 it is obvious that there is considerable qualitative variety in the effect of the five different metals on these enzymes. For example, lead was a stimulator of alkaline phosphatase but a strong inhibitor of both acid phosphatase and xanthine oxidase. Copper was the lone stimulator of xanthine oxidase, although it was a weak inhibitor at all concentrations when tested *in vitro*. This would seem to indicate a compensatory biosynthesis of the enzyme in order to "adapt" to the presence of copper, a phenomenon noted by Gould (1977) in flounders exposed to cadmium. Jackim et al. (1970) found cadmium was the weakest inhibitor of catalase and xanthine oxidase, but the strongest inhibitor of RNAase.

Cadmium, along with mercury, has received considerable attention because of its high toxicity and the fact that there is no known function in cells for these elements. Not surprisingly, high concentrations of cadmium can result in enzyme inhibition, (Gould and Karolus 1974). However, chronic exposures, which are more relevant to the real world, cause some far more interesting biochemical effects. When the the same species (the marine cunner) was exposed to quite low levels (50–100 ppb) of cadmium for 30 days, liver aspartate aminotransferase was inhibited by about 20%. In the same fish, the enzyme glucose-6-phosphate dehydrogenase increased in activity in a dose-dependent manner (MacInnes et al., 1977). The authors attribute this to increased pentose shunt activity in order to provide metabolites for increased biosynthesis (for repair of cadmium damaged tissues?). These enzyme changes occurred even though there were no detectable increases of cadmium in the liver itself.

There was a clear-cut increase in carbonic anhydrase activity in flounders emersed in water with cadmium (Gould, 1977). This is probably the enzyme associated with erythrocytes, rather than the kidney tubules, as there is little renal carbonic anhydrase present in marine teleosts (Gould, 1977). Christensen and Tucker (1976) found a strong inhibition by heavy metals, including cadmium, of carbonic anhydrase from fish erythrocytes when measured *in vitro*. Gould (1977) further notes that this enzyme contains zinc, and cadmium is known to displace metal ions from enzymes and thereby reduce their activity.

Table 1 Changes in Enzyme Activity Following Exposure of Fish to Test Metal

Chemical	Enzyme(s)	Effect	Species	Duration of exposure	Concentration	Organ(s)	Ref.
Pb, Hg	Alkaline phosphatase	∧	*Fundulus heteroclitus*	96 h	96 h LC50	Liver	1
Cu, Cd	Alkaline phosphatase	ns	*F. heteroclitus*	96 h	96 h LC50	Liver	1
Ag	Alkaline phosphatase	∨	*F. heteroclitus*	96 h	96 h LC50	Liver	1
Pb, Cd, Cu, Hg	Acid phosphatase	∨	*F. heteroclitus*	96 h	96 h LC50	Liver	1
Pb, Hg, Ag, Cd	Xanthine oxidase	∨	*F. heteroclitus*	96 h	96 h LC50	Liver	1
Cu	Xanthine oxidase	∧	*F. heteroclitus*	96 h	96 h LC50	Liver	1
Ag, Pb, Cu, Hg	Catalase	∨	*F. heteroclitus*	96 h	96 h LC50	Liver	1
Cd	Alkaline phosphatase	ns	*Barbus conchionius*	48 h	96 h LC50	Liver, gills	2
Cd	Alkaline phosphatase	∧	*B. conchionius*	48 h	96 h LC50	Kidney, ovary	2
Cd	Alkaline phosphatase	∨	*B. conchionius*	48 h	96 h LC50	Gut	2
Cd	Acid phosphatase	∧	*B. conchionius*	48 h	96 h LC50	Gut, ovary	2
Cd	Glutamate and pyruvate transaminase	∨	*B. conchionius*	48 h	96 h LC50	Liver, muscle	2
Cd	Lactic dehydrogenase	∧	*B. conchionius*	48 h	96 h LC50	Heart	2
Cd	Cu-Zn-Superoxide dismut.	∨	*Scorpaena guttata*	30 d	1/3 96 h LC50	Intestine	3
Cd	Aspartate aminotransferase	∨	*Tautogolabrus*	96 h	3 and 24 ppm	Liver	4
Cd	Aspartate aminotransferase	∨	*Tautogolabrus*	30 d	50 and 100 ppb	Liver	5
Cd	Glucose-6-PO₄ dehydrogenase	∧	*Tautogolabrus*	30 d	50 and 100 ppb	Liver	5
Cd	Leucine aminopeptidase	∨	*Pseudopleuronectes*	60 d	5–10 ppb	Kidney	6
Cd	Carbonic anhydrase	∧	*Pseudopleuronectes*	60 d	5–10 ppb	Kidney	6
Cd	Aspartate aminotransferase	∧	*Mugil cephalus*	1–4 d	96 h LC50	Heart, gill	7
Cd	Alanine aminotransferase	∧	*M. cephalus*	1–4 d	96 h LC50	Heart, gill	7
Cd	Alanine aminotransferase	∨	*M. cephalus*	1–4 d	96 h LC50	Liver	7
Cd	Acid phosphatase	∨	*M. cephalus*	1–4 d	96 h LC50	Heart, gill, liver	7
Cd	Lactic dehydrogenase	∨	*M. cephalus*	1–4 d	96 h LC50	Heart, gill, liver	7
Cd	Several enzymes	∨	Herring eggs		10 ppm		15
Hg	Glycolytic enzymes	∨	*Carassius auratus*	3–6 wk	0.25 ppb	Muscle	8
Hg	Alkaline phosphatase	∧	*C. auratus*	3–6 wk	0.25 ppb	Liver	8
Hg	Protein synthetase enzymes	∨	*C. auratus*	3–6 wk	0.25 ppb	Liver, muscle	8
Hg	Acid phosphatase	∨	*Puntius conchonius*	48 h	96 h LC50	Liver, gills, kidney	9
Hg	Alkaline phosphatase	∧	*P. conchonius*	48 h	96 h LC50	Liver, gills, kidney	9
Hg	Two amino acid transferase	∨	*P. conchonius*	48 h	96 h LC50	Liver, gills, kidney	9

Metal	Enzyme/parameter	Effect	Species	Concentration	Duration	Tissue	Ref.
Hg	Glycolysis	<	*Gambusia holbrooki*	0.86 ppm	28 h	Muscle	10
Hg	Kreb's cycle activity	>	*G. holbrooki*	0.86 ppm	28 h	Muscle	10
Hg	Na,K, Mg ATPases	<	*Notopterus notopterus*	17–88 ppb	30 d	Brain, gill, kidney	17
Hg	Na,K, ATPase	<	*Channa punctatus*	0.3 ppm	30 d	Brain	18
Hg	Acid phosphatase	<	*C. punctatus*	0.3 ppm	30 d	Brain	18
Hg	Alkaline phosphatase	>	*C. punctatus*	0.3 ppm	30 d	Brain	18
Hg	Glucose-6-phosphatase, Lactic and pyruvate dehydrase	<	*C. punctatus*	0.3 ppm	30 d	Brain	18
Hg, Pb	Succinic, malic, lactic dehydrogenases	<	3 Freshwater teleosts	"Sublethal"	48 h	Liver, muscle, brain	11
Pb, Hg	Delta-aminolevulinic A. dehydrase	<	*Fundulus heteroclitus*	0.8–10 ppm	4–14 d	Liver, kidney	19
Ag, Zn	Delta-aminolevulinic A. dehydrase	>	*F. heteroclitus*	0.8–10 ppm	4–14 d	Liver	19
Pb	Alkaline, acid and glucose phosphatase	<	*Heteropneustes*	3.8 ppm	4–30 d	kidney, ovary	23
Cu	Alkaline phosphatase	<	*C. punctatus*	7.5 ppm	24 h	Liver	20
Cu	Acid and alkaline phosphatase	>	*Puntius conchonius*	96 h LC50	48 h	Gut, kidney	12
Cu	Acid phosphatase	<	*P. conchonius*	96 h LC50	48 h	Liver	12
Cu	Alanine aminotransferase	<	*P. conchonius*	96 h LC50	48 h	Gill, kidney, muscle	12
Cu	Lactate dehyrogenase	<	*P. conchonius*	96 h LC50	48 h	Liver, gill	12
Cu	Alkaline phosphatase	>	*C. punctatus*	7.5 ppm	24 h	Kidney	20
Cu	Acid phosphatase	>	*C. punctatus*	7.5 ppm	24 h	Liver, kidney	20
Cu	Glycogen phosphorylase, glucose-6-phosphorylase, lactate dehydrogenase	>	*Labeo rohita*	0.2 ppm	1–30 d	Liver, muscle	21
Cu	Succinic dehydrogenase	<	*Labeo rohita*	0.2 ppm	1–30 d	Liver, muscle	21
Cu	Succinic dehydrogenase	<	*Tilapia mossambica*	1.5 ppm	1–14 d	Muscle, liver	22
Cu	Succinic dehydrogenase	>	*T. mossambica*	1.5 ppm	1–14 d	Liver, muscle	22
Cu, Zn	Succinic dehydrogenase	<	*Oreochromis*	96 h LC50	96 h	Liver, brain, muscle	13
Cu, Zn	Glyceraldehyde dehydrogenase	>	*Oreochromis*	96 h LC50	96 h	Liver, brain, muscle	13
Cr	Three oxidases	ns	*Salmo gairdneri*	2.5 ppm	48 h	Several	14
Cr	Na,K ATPase	<	*S. gairdneri*	2.5 ppm	48 h	Kidney, intestine	15
Cr	Na,K ATPase	ns	*S. gairdneri*	2.5 ppm	48 h	Gill, liver	15

Note: Where more than one chemical or tissue is listed in sequence, arranged in order of decreasing effect; ns = no significant change. References are as follows: 1. Jackim et al., 1970; 2. Gill et al., 1991; 3. Bay et al., 1990; 4. Gould and Karolus, 1974; 5. MacInnes et al., 1977; 6. Gould, 1977; 7. Hilmy et al., 1985; 8. Nicholls et al., 1989; 9. Gill et al., 1990; 10. Kramer et al., 1992; 11. Shaffi, 1993; 12. Gill et al., 1992; 13. James et al., 1992; 14. Buhler et al., 1977; 15. Kuhnert et al., 1976; 16. Mounid et al., 1975; 17. Verma et al., 1983; 18. Sastry and Sharma, 1980; 19. Jackim, 1973; 20. Srivastava and Pandley, 1982; 21. Radhakrishnaiah et al., 1992; 22. Balavenkatasubbaiah et al., 1984. 23. Sastry and Agrawal, 1979.

Therefore, the induction of more enzyme in chronically exposed fish is interpreted as an adaptation at the molecular level to the partial enzyme inhibition caused by the poison. This is a nice example of the greater value of *in vivo* studies compared to those done *in vitro*.

The enzyme glucose-6-phosphate dehydrogenase is a key control point in the pentose phosphate shunt. It is not a metalloenzyme but its activity is normally modified by the concentration of magnesium. Cadmium caused a loss of sensitivity to this magnesium activiation (Gould, 1977). Metabolic flexibility is necessary for adapting to rapidly changing temperatures (Hochachka and Somero, 1984) so this could have subtle implications for the flounders in their estuarine habitat where temperatures are notoriously variable.

Enzymes from different tissues may be affected very differently by cadmium. An example of this is seen in the study by Hilmy et al. (1985) in which juvenile mullet were exposed to cadmium at a fairly high concentration. Enzyme activities were measured daily for 4 days. In contrast to the inhibition of aspartate aminotransferase in the liver of cunner, which was mentioned above, they observed a strong stimulation of this enzyme in the heart and gill by cadmium, and this increased over the 4-day period. Enzyme from liver exhibited an initial stimulation followed by inhibition, which suggests that a low tissue concentration of cadmium stimulated activity, but as the concentration built up over time, inhibition of enzyme activity occurred. Another transferase, alanine aminotransferase, was also induced by cadmium in the two non-hepatic tissues of the mullet, but inhibited in the liver.

Tissue-specific effects of cadmium were also observed in the freshwater rosy barb. Alkaline phosphatase was unaffected in liver and gills, stimulated in kidney and ovary, and inhibited in the gut (Gill et al., 1991). It is difficult to understand stimulation of alkaline phosphatase in kidney as cadmium tends to accumulate there (Chapter 5). Acid phosphatase may be stimulated or inhibited depending on species of fish and tissue examined (Gill et al., 1991).

From the studies discussed here, it seems that cadmium can cause either induction or inhibition of a variety of cellular enzymes during acute exposures. However, more chronic exposures may yield a different picture. When rainbow and brown trout were exposed to high but still sublethal concentrations of cadmium for either 2 weeks or 3 months, no changes were noted in the activities of some 10 different enzymes in a variety of tissues (Roberts et al., 1979). This is in spite of the fact that significant amounts of cadmium accumulated in the liver, kidney, and gills of the test fish. It is noteworthy that two of the enzymes (glucose-6-phosphate dehydrogenase; aspartate aminotransferase) examined in the trout are the same ones that showed changes in marine fish exposed acutely to cadmium. Perhaps, with more prolonged exposures, adaptation at the molecular level takes place. Time-course studies are needed to confirm this.

In developing herring eggs, the enzyme phosphoenol-pyruvate carboxykinase increases in activity by about two orders of magnitude during the time between the early blastodisk and just before hatching. When herring eggs were exposed to 10 ppm cadmium during this time, a decreased activity was observed (Mounid et al., 1975). Somewhat less extensive decreases in enzyme activity were also noted for the NAD- and NADP-malic enzymes and propionyl-CoA-carboxylase. Rosenthal and Alderdice (1976; p. 2055) point out that: "Since the enzymes studied are involved in biosynthetic processes, it is speculated that the depressive effect of cadmium on activity of the enzymes explains why cadmium exposure results in smaller, inactive larvae at hatching".

Long-term exposures of scorpionfish (a marine species) to cadmium resulted in no changes in alkaline phosphatase, succinic dehydrogenase, or glyceraldehyde phosphate dehydrogenase in several tissues (Bay et al., 1990). In other words, they obtained results similar to those of Roberts et al. (1979) on trout. However, they did detect a severe (eightfold) inhibition of Cu-Zn-superoxide dismutase in intestine with a trend (although not significant) toward less of this enzyme in the other tissues. It might be speculated that the high sensitivty of the intestine might be due to the fact that this is a marine species which must drink water to correct for osmotic loss. Freshwater fish would not have that mode of cadmium entry into the body.

The dismutase enzyme helps protect cells from superoxide anion radicals produced during oxidative metabolism so lipid peroxidation of membranes might follow cadmium exposure. Bay et al. (1990) further present evidence that cadmium displaced copper and zinc from the enzyme pool of the cells and thereby disrupted copper and zinc homeostasis. It also should be noted that the same laboratory (Brown et al., 1989) has found that the subcellular distribution of cadmium differs greatly depending on whether exposures are acute or chronic.

While cadmium seemingly has a relatively small effect on oxidative enzymes, mercury has a considerable effect on these enzymes. Lactate, pyruvate, and succinic dehydrogenase enzymes are all key control points in the flow of carbons through glycolysis and the citric acid cycle. Mercury is apparently a strong inhibitor of these enzymes as exposure of the freshwater teleost, commonly called the murrel, for 60 days to 3 ppb of mercury caused a greater than 85% inhibition of lactate and pyruvate dehydrogenases in muscle and of succinic dehydrogenase in gill (Sastry and Rao, 1981). Such a level of inhibition would certainly slow the ability to produce ATP, especially when the fish is under some sort of stress or when forced to swim (see MacLeod and Pessah, 1973).

In a follow-up study, Sastry and Rao (1982) measured the activities of several enzymes in six different tissues of the murrell following 15, 30, and 60 days of exposure to 3 ppb of mercury. Most of the changes seen were in the 60-day samples and are summarized in Table 2 . The percent change from control fish varies between different tissues; brain and muscle seem to be the least affected, while kidney and liver show the greatest changes. This is most likely due to differential rates of uptake and retention of mercury in these tissues (see Chapter 5). This illustrates how it would be useful in future investigations of this sort to have simultaneous measures of the tissue concentrations of xenobiotic chemicals.

Inhibition of glucose-6-phosphatase and hexokinase (Table 2) would cause a reduction in both glycogenolysis and gluconeogenesis (i.e., the breakdown of glycogen and the synthesis of glucose from non-carbohydrate molecules, respectively). Thus, we have a possible mechanism for the decreases in blood glucose of fish exposed to mercury (Sastry and Rao, 1981), instead of the hyperglycemia which is generally observed in fish under stress (see section on carbohydrate metabolism in Chapter 8). Such a reduction in the availability of carbohydrate for energy was partially compensated for by increases in the activity of glutamate dehydrogenase and L-amino acid oxidase, both of which are central to the utilization of amino acids for energy. In a different study on the same species,

Table 2 Percent Change in Enzyme Activity Following 60 Days Exposure of the Freshwater Murrel, *Channa punctatus,* to 3 ppb of Mercuric Chloride

	Brain	Gill	Intestine	Kidney	Liver	Muscle
Glucose-6-phosphatase	ns	39	48	40	37	
Hexokinase	ns	ns	ns	52	49	ns
Malate dehydrogenase	27	29	18	27	34	33
Glutamate dehydrogenase	ns	ns	41	30	51	ns
L-Amino acid oxidase	23	43	ns	36	33	ns
Xanthine oxidase	ns	ns	52	ns	38	ns

Note: ns indicates no significant change.

Sastry and Sharma (1980) observed inhibition (10–43%) of several key enzymes involved in energy metabolism in the brain.

Recently, Shaffi (1993) reported on activities of lactate dehydrogenase, succinic dehydrogenase, and malate dehydrogenase in three freshwater species following 24- or 48-h exposures to mercury. For all three enzymes the greatest inhibition occurred in liver with progressively less inhibition in muscle, brain, kidney, and gill, in that order. Unfortunately, the actual concentration of mercury used is not specified. Because there was a marked inhibition of enzyme activity in 24 h, it must have been fairly high, even though the author refers to it as "sublethal".

According to the findings of Verma et al. (1983), Na,K-ATPase is inhibited by mercury more in brain tissue than gill, liver, or kidney. Thus, even though the brain does not accumulate much mercury compared to other tissues (Massaro, 1974), at least one of the enzymes there is quite sensitive to the metal. In spite of this seeming anomaly, there is indirect evidence that accumulation of mercury may be an important variable in several of the other tissues. In nearly all the above-mentioned studies, in addition to the long-term exposures, observations were also made at shorter time intervals, such as 7 or 15 days. In most cases, there were non-significant changes seen in enzyme activity at those shorter exposure periods which could reflect the time required for the metal to accumulate in the tissues (see Chapter 5).

It is well known that mercury in natural waterways is frequently methylated by microbes located in sediments (Bryan, 1976). More relevant to the laboratory studies discussed in this review is the observation that the liver of fish may also methylate the element, although demethylation can also occur (Bryan, 1976). The reason for raising this point is that there is some evidence that methylmercury and mercury may not always act in the same manner in cells. For example, Fox et al. (1974) found inhibition of Kreb's cycle activity in brain slices of guinea pig when subjected to methylmercury but not when exposed to the inorganic form of mercury at the same concentration. Thus there is a possibility that the inhibition of Kreb's cycle enzymes found in these fish studies may be due to methylmercury rather than inorganic mercury. Furthermore, Southward et al. (1974) claim that inorganic mercury did not affect ATPase activity in their *in vitro* mammalian cell extract studies, so the inhibition of ATPase could also be due to methylmercury. However, the two studies mentioned were done *in vitro* using mammalian tissues, so their relevance to fish *in vivo* activity is uncertain.

The enzyme delta-aminolevulinic acid dehydrase catalyzes the first step in heme synthesis and the ultimate formation of hemoglobin. A depression in the activity of this enzyme is used in human medicine as a diagnostic indicator of chronic or acute lead poisoning (DeBruin, 1976). Not surprisingly, when fish are exposed to lead, this enzyme is inhibited (Jackim, 1973; Johansson-Sjobeck and Larsson, 1979). For the most part, this effect is limited to lead although mercury may also cause a slight inhibition (Jackim, 1973). Silver and zinc actually stimulated the liver to produce more enzyme (Jackim, 1973) so delta-aminolevulinate dehydrase inhibition seems to be relatively specific for lead poisoning in both fish and humans. This is one of the only examples of a specific action of a contaminant. Johansson-Sjobeck and Larsson (1979) observed that 30-day exposures of trout to 75 ppb lead caused a 74% lowering of the activity of this enzyme but the hemoglobin, hematocrit, and red blood cell count remained unchanged. So delta-aminolevulinate dehydrase must have a large "safety factor" in its activity and/or the rate of turnover of erythrocytes in fish may be relatively slow. Because they are nucleated, their lifespan is probably greater than those of mammals who have an average lifespan of 3 months.

Moving on to copper (Table 1), a 24-h exposure of the green snakehead (*Ophiocephalus punctatus*) to this element at a concentration approximating 1/10 the 48-h LC50 concentration produced a 24 and 37% inhibition of acid phosphatase in the liver and kidney, respectively (Srivastava and Pandley, 1982). The effect on alkaline phosphatase was, on the other hand, rather different. There was a 10% inhibition in the liver, but a 20% stimulation in the kidney. Gill et al. (1992) also observed a stimulation of alkaline phosphatase in kidney (and intestine). Srivastava and Pandley (1982) speculate that since this enzyme is involved in glucose reabsorption in the kidney tubules, there may be an enhanced activity of it because of the high glucose levels that occur in the blood during exposure to copper. This kind of response is, however, not limited to the kidney or gut, or to this element. Recall that Sastry and Sharma (1980) observed a stimulation of this same enzyme by mercury in the brain.

An increase in acid phosphatase was seen after copper exposure in some tissues (Srivastava and Pandley, 1982; Gill et al., 1992). This enzyme is associated with lysosomal activity and Gill et al. (1992) speculate that its elevation reflects proliferation of lysosomes in an attempt to sequester the toxic copper ions. They note that Weiss et al. (1986) have found large secondary and tertiary lysosomes in fish following exposure to copper and hepatic copper tends to be sedimented in this fraction of homogenate.

Copper exposure causes mobilization of liver glycogen into blood glucose (Chapter 8). At the enzyme level, this is seen in the stimulation of glycogen phosphorylase (Radhakrishnaiah et al., 1992). Copper also seems to inhibit oxidative metabolism at both the whole animal level (Chapter 8) and at the enzyme level and this is compensated for by a stimulation of glycolysis which is reflected in an enhanced lactate dehydrogenase and glyceraldehyde dehydrogenase activity (James et al., 1992; Radhakrishnaiah et al., 1992; Balavenkatasubbaiah et al., 1984).

Limited evidence suggests that chromium may have little effect on enzymes of energy metabolism (Buhler et al., 1977). This element did inhibit the enzyme Na,K ATPase in kidney and intestine, but not in gill or liver. Na,K ATPase is critical in the function of the so-called "sodium pump" located in cellular membranes. From these results, one could predict that chromium would cause loss of sodium in the urine of freshwater fish and a reduction in the uptake of sodium from the water, thereby producing an altered electrolyte homeostasis (Chapter 7).

IV. ENZYME EFFECTS FROM ORGANIC CHEMICALS

Most of the information on the action of organic chemicals is, as is to be expected, on pesticides (Table 3). Before considering them, however, mention should be made of the petroleum hydrocarbons. There has been a great deal of physiological work done on their effects in fish and invertebrates (see Anderson, 1979). Enzymatic studies have mostly been directed at understanding the biotransformations and detoxification reactions (see Chapter 5). Little attention has been paid to their possible impact on enzymes associated with energy metabolism. Heitz et al. (1974) exposed mullet to a crude oil for 4 days and then assayed 13 enzymes in several tissues. They found essentially no significant changes. With crude oil it is difficult to determine what the LC50 is, but the authors note that the concentration they used is well below the lethal level. Perhaps it was below the threshold for any enzyme effects, or as they concluded, crude oil has little effect on the enzymes of energy metabolism in fish.

Such a conclusion may depend on the species of fish and the doses of petroleum used in experiments. Elevation (Davison, 1992) and suppression (Prasad, 1987) of whole-animal metabolic rates have been reported upon exposure to sublethal concentrations of petroleum components. Prasad (1987) also reported a severe dose-dependent suppression of *in vitro* tissue respiration rate gill, liver, kidney, and muscle excised from oil-exposed carp. Thus, there is a high probability that petroleum hydrocarbons would suppress enzyme activity.

Pentachlorophenol is a powerful uncoupler of oxidative phosphorylation in mitochondria (see Rao, 1978 for review). When eels were exposed for 4 days to this pesticide,

Table 3 Changes in Enzyme Activity Following Exposure of Fish to Test Chemical

Chemical	Enzyme(s)	Effect	Species	Exposure	Concentration	Organ	Ref.
Crude oil	13 enzymes	ns	Mugil cephalus	4 d	ca. 4 ppm	Muscle, brain, liver and gill	1
Pentachlorophenol	Pyruvate kinase and lactic dehydrogenase	∨	Anguilla anguilla	4 d	0.1 ppm	Liver	2
Pentachlorophenol	Hexokinase, G-6-PO4, dehydrogenase, fumerase cyto. oxidase, succinic and pyruvate dehydrogenase	∧	A. anguilla	4 d	0.1 ppm	Liver	2
Pentachlorophenol	Succinic and pyruvate dehydrogenase	∨	Notropterus	30 d	2.8–8.3 ppb	Brain, liver, gill	3
Pentachlorophenol	Lactic dehydrogenase	∧	Notropterus	30 d	2.8–8.3 ppb	Brain, liver, gill	3
Parathion	Aldolase and phosphorylase "a"	∧	Tilapia mossambica	48 h	1/3 LC50	Gill, liver, brain	4
Parathion	Phosphorylase "a"	∨	T. mossambica	48 h	1/3 LC50	Muscle	4
Parathion	Phosphorylase "b"	∨	T. mossambica	48 h	1/3 LC50	Gill, liver, brain	4
Metasystox	Succinic dehydrogenase	∨	Channa striatus	30 d	1/3 LC50	Gill, brain, muscle, liver, kidney	5
Metasystox	Lactic dehydrogenase	∧	C. striatus	30 d	1/3 LC50	Muscle, brain, gill, liver, kidney	5
Quinalphos	Hexokinase, lactate, pyruvate and succinic dehydrogenases	∨	C. punctatus	30 d	0.025 ppm	Liver, gill, brain	6
Malathion	Acid and alkaline phosphatase	∨	Branchydanio	7 d	0.5–1.1 ppm	Liver	8
Phosphamidon	Alkaline phosphatase	∧	Puntius conchonius	48 h	96 h LC50	Gill	7
Phosphamidon	Acid phosphatase	∨	P. conchonius	48 h	96 h LC50	Liver	7
Phosphamidon	Two transaminases	∧	P. conchonius	48 h	96 h LC50	Heart, muscle	7
Phosphamidon	Lactic dehydrogenase	∨	P. conchonius	48 h	96 h LC50	Heart, liver, muscle, gill	7

Pesticide	Enzyme		Species	Duration	Dose	Tissue	Ref
Aldicarb	Alkaline phosphatase	>	*P. conchonius*	48 h	96 h LC50	Kidney, gill	7
Aldicarb	Two transaminases	>	*P. conchonius*	48 h	96 h LC50	Gill, heart	7
Aldicarb	Lactic dehydrogenase	◇	*P. conchonius*	48 h	96 h LC50	>Gill, <heart	7
Carbofuran	Acid and alk. phosphatase	>	*Channa punctatus*	6 m	4.5 ppm	Liver	9
Endosulfan	Acid and alk. phosphatase	>	*P. conchonius*	48 h	96 h LC50	Kidney, liver, ovary	7
Endosulfan	Lactic dehydrogenase	<	*P. conchonius*	48 h	96 h LC50	Heart, muscle, liver	7
Endosulfan	Lactic and malic dehydrogenase	<	*Clarias batrachus*	7 d	0.001 ppb	Liver, muscle	10
Endosulfan	Acid and alk. phosphatase	<	*Channa gachua*	15–30 d	2.25–5.6 ppb	Liver, kidney, muscle	11
Dieldrin	Glutamate dehydrogenase	<	*Salmo gairdneri*	240 d	Dietary	Brain	12
Dieldrin	Glutamate dehydrogenase	>	*S. gairdneri*	240 d	Dietary	Liver	12
	Glutamine synthetase						
Dieldrin	Glutamate synthetase	>	*S. gairdneri*	240 d	Dietary	Brain	13
Paraquat	Glycogen phosphorylase	>	*S. gairdneri*	1–7 d	0.5–10 ppm	Liver	13
	G-6-Phosphatase						

Note: Reference numbers are as follows: 1. Heitz et al., 1974; 2. Bostrom and Johansson, 1972; 3. Verma et al., 1982; 4. Rao and Rao, 1983; 5. Natarajam, 1984; 6. Sastry et al., 1982; 7. Gill et al., 1990; 8. Kumar and Ansari, 1986; 9. Ram and Singh, 1988; 10. Shukla and Tripathi, 1991; 11. Sharma, 1990; 12. Mehrle and Bloomfield, 1974; 13. Simon et al., 1983.

Bostrom and Johansson (1972) observed increases in several enzymes of the citric acid cycle, the respiratory chain, and the pentose shunt but decreases in the enzymes of glycolysis (Table 3). This is to be expected as there would be a need for increased respiration if the efficiency of ATP synthesis was reduced because of the uncoupling mentioned above.

The opposite results were obtained by Verma et al. (1982) on *Notopterus* in which it was found that glycolysis was stimulated and aerobic respiratory enzymes were inhibited by PCP. They make no attempt at reconciling these differences with previous work (they are not even mentioned). Rather their findings are interpreted as indicating that the PCP induced a hypoxic condition in the fish by damaging the gill tissue. An unpublished Ph.D. dissertation is cited in support of this hypothesis. The work on eels mentioned above involved even a more acute type of exposure so one would think that such a treatment would have caused hypoxia there too. Krueger et al. (1966) obtained results on a cichlid that support the Bostrom and Johansson (1972) work. Other than a species difference, there does not appear to be an obvious explanation for the contradictory results.

Organophosphorus (OP) insecticides are well-known inhibitors of acetylcholinesterase in a variety of organisms. This aspect of organophosphorus action is discussed in Chapter 12. There is accumulating evidence that they also cause some interesting effects on cellular enzymes, especially those involved in energy metabolism in fish (Table 3).

Parathion has been found to cause an increase in active phosphorylase "a" in a variety of tissues except muscle, where the enzyme activity went down after parathion exposure (Rao and Rao, 1983). The inactive phosphorylase "b" enzyme decreased in the different tissues, but again muscle was the "non-conformist" in that it remained unchanged there. These phosphorylase changes are required for the rapid mobilization of glycogen and this phenomenon was observed in the same study. It seems paradoxical that muscle enzymes did not respond enzymatically in the same way as the other tissues, since there was a 50% loss in muscle glycogen (Rao and Rao, 1983). Parathion also caused an increase in aldolase, and since this is a key enzyme in glycolysis, there must have been an increased utilization of glucose via that metabolic pathway in the liver.

Other work (Natarajan, 1984), also on an OP insecticide (metasystox), has shown an inhibition of succinic dehydrogenase, an enzyme of Kreb's cycle, but a stimulation of the glycolytic enzyme lactate dehydrogenase. Under these circumstances, an increase in lactate should occur, however, this metabolite was not measured. Natarajan (1984) cites work done in India which showed that Sevin® (a carbamate insecticide) also decreased the activities of Kreb's cycle enzymes. And finally, the OP pesticide sumithion has been shown to suppress these oxidative enzymes while stimulating lactic dehydrogenase (Koundinya and Ramamurthi, 1979).

Increased glycolysis and reduced oxidative metabolism can be interpreted in at least two different ways. Natarajan (1984) attributes these alterations to gill damage by the insecticide resulting in "histotoxic anoxia". No data on gill histology are presented, but an increased red blood cell count and hemoglobin content of the blood was noted, responses frequently observed in hypoxic fish (see Chapter 2). However, when one measures enzyme activity, the maximum rate of activity is actually measured and there seems little reason for oxidative enzyme activity to decrease as a result of a hypoxic condition. More to the point, *in vitro* studies of OP insecticides suggest a direct inhibition of the pesticide on the enzymes involved in aerobic respiration (Hiltibran, 1974), and *Channa striatus* has the ability to breath air so gill damage should not be as much of a problem for it as it might be for some other teleost.

Sastry et al. (1982) reported inhibition of both aerobic and glycolytic enzymes in most of the tissues assayed from snakehead fish which had been exposed chronically to quinalphos. Unfortunately, they make no attempt at reconciling their seemingly contradictory findings. It is noteworthy that they also found decreases in blood glucose, lactic acid, and hemoglobin along with increases in glycogen. Thus, the response they report to this insecticide is almost exactly opposite to that seen with other organophosphates!

A large dose-dependent inhibition of alkaline and acid phosphatase was found in livers of zebra danios exposed to malathion (Kumar and Ansari, 1986). The investigators attribute the decrease in acid phosphatase to leakage of lysosomal enzyme into the plasma, an interesting hypothesis supported by other work where elevations were seen in this enzyme in the plasma of chickens exposed to malathion. They go on to point out that alkaline phosphatase inhibitions have several possible explanations including the fact that they have serine residues at active sites. Organophosphorus insecticides inhibit enzymes, such as acetylcholinesterase, which possess this characteristic.

The OP insecticide phosphamidon also caused a decrease in acid phosphatase activity, but an increase in alkaline phosphatase (Gill et al., 1990). They attribute the stimulation of the latter enzyme to glucocorticoid influence but the evidence is tenuous at best. The transaminase enzyme stimulation could be associated with gluconeogenesis which is stimulated by glucocorticoids so it is surprising that liver transaminase activity was not affected by phosphamidon (Gill et al., 1990), for that organ is where traditionally most gluconeogenesis occurs.

Phosphamidon caused a strong inhibition of lactate dehydrogenase from all tissues examined by Gill et al. (1990). Such a response is probably due to leakage from the cells into the plasma. It is perhaps noteworthy that the same authors found a marked stimulation of this same enzyme from heart when the pesticide was added directly *in vitro*.

The carbamate insecticides seem to cause a consistent stimulation of phosphatase activity in a variety of tissues (Ram and Singh, 1988; Gill et al., 1990). Aldicarb also stimulated transaminase activity in several tissues but inhibited lactate dehydrogenase in the same tissues (Gill et al., 1990).

The organochlorine (OC) insecticides are no longer used much in industrialized countries because of their persistence in the environment, but they still get heavy usage in many so-called "third world" countries. Nearly all the studies of enzyme effects in fish have been conducted *in vitro*, with all the weaknesses associated with that methodology. The general conclusion from that work is that these insecticides reduce metabolic energy production by cells (see Hiltibran, 1982 for review).

With *in vivo* exposure to the OC insecticide endosulfan, Gill et al. (1990) found marked inhibition of lactate dehydrogenase in several tissues. This was confirmed in a different species by Schukla and Tripathi (1991) who also found inhibition of malate dehydrogenase, a Kreb's cycle enzyme. The latter workers found that the fish synthesized new enzyme over a period of 28 days if transferred to non-contaminated water, thus the loss of activity was not permanent. They point out that low doses of OC pesticides are antithyroid so the effect on the enzymes of cellular energy metabolism may in part be mediated by decreases in these thyroid hormones. Support for this hypothesis was obtained when they treated their fish with the thyroid hormones T3 and T4 and got normal enzyme activity even after endosulfan exposures.

The evolution of insecticide resistance by insect populations has been a continuing problem that even gets into the popular press. Less well known is the resistance which is developed by some mosquitofish (*Gambusia*) populations exposed for many generations in nature to a particular pesticide. Ferguson (1967) concluded this was genetically based. A biochemical mechanism for it was proposed by Yarbrough and Wells (1971) who were able to show that the membrane of mitochondria in livers and brains from resistant mosquitofish apparently excluded the OC endrin from contact with the dehydrogenase enzymes therein. When they disrupted the mitochondrial membranes by freezing and thawing, they got inhibition of succinic dehydrogenase by endrin in mitochondria from both susceptible and resistant fish, but when the mitochondria were intact, there was no inhibition of this enzyme if the mitochondria were obtained from endrin-resistant fish.

Overall, the limited evidence suggests that OC compounds inhibit various enzymes associated with the release of energy in cells. The action on alkaline and acid phosphatase enzymes appears to depend on duration of exposure. Short term, the effect was to stimulate activity (Gill et al., 1990) whereas exposures of 15 or 30 days (Sharma, 1990) caused a consistent strong inhibition that was dose-dependent (as much as 66%). For OC compounds which have very long persistence in the environment, such long-term exposures are probably more realistic than those short term. However, for OP insecticides with their limited persistence, exposures for only a few days may be quite realistic.

Dieldrin given in the diet at several doses to rainbow trout caused a decreased brain glutamate dehydrogenase activity, however, the activity of this enzyme increased in the liver (Mehrle and Bloomfield, 1974). The higher doses caused the concentration of ammonia in the brain to rise, which is to be expected when the ammonia detoxifying enzyme is suppressed. The authors suggest that part of the mechanism of dieldrin intoxication is due to ammonia acting on the brain. Ammonia is certainly toxic to the brain as it seems to inhibit ATP production of nerve cells (see Tomasso et al., 1980 for review).

A variety of ATPases have been found to be sensitive to OC insecticides when tested *in vitro* (Hiltibran, 1982). In some cases even stimulation is observed (Hiltibran, 1982). However, Davis et al. (1972) made the important observation that *in vivo* exposure of trout to oral doses of DDT that produced death in 36 h caused no significant changes in Na,K ATPase from gills, brain, and kidney. Thus, the *in vitro* effects were not confirmed in the intact animal.

Paraquat is a widely used herbicide. When carp were exposed to doses ranging from 0.5–10 ppm, there was a doubling of the liver phosphorylase activity in fish from the higher concentrations. Glucose-6-phosphatase activity in the liver also was elevated some 40–50%. Using a dose of 0.5 ppm for several days produced a gradual rise in these enzyme activities reaching a fourfold rise at 7 days for the phosphorylase (Simon et al., 1983). No information was given on the LC50 for this chemical, but the fish would appear to have been under considerable stress as these enzyme activations would mobilize carbohydrate. The authors also cite unpublished work that showed that paraquat caused increased lactate dehydrogenase activity. All these changes mean a mobilization of energy from glycogen and an enhanced rate of glycolysis which were probably mediated by increased epinephrine and cortisol.

V. CONCLUDING COMMENTS ON ENZYME EFFECTS

From the above discussion, we see that a variety of enzymes can be affected by a given substance, and many substances affect the same enzyme. Stimulation of an enzyme may occur in one tissue while inhibition is seen in another, within the same species of fish. Thus, with a few exceptions, the idea of there being a single mode of action at the enzyme level is generally not a viable concept for a wide variety of pollutants. Of course cells have a multitude of different enzymes so further work may yield some examples of specific metabolic lesions induced by certain chemicals such as the inhibition by lead of delta-aminolevulinic acid dehydrase. In order to make meaningful interpretations, there is a need for measures of the concentration of xenobiotic chemicals in the same tissues in which enzyme activity is assayed. Furthermore, partitioning of the chemical into specific organelles should not be overlooked.

The effects seen in the *in vivo* studies reviewed herein probably involve a combination of endocrine influences (especially the stress hormones cortisol and epinephrine), direct action of the chemical on enzyme molecules, and/or enzyme induction by inhibition of negative feedback. While *in vitro* methods may indicate relative sensitivities of different enzymes to the direct action of a particular chemical, they probably have little relevance to the "real world" because they may or may not simulate the concentration that the enzyme is exposed to *in vivo*, and of course, the other two potential mechanisms mentioned above are eliminated in an *in vitro* preparation.

Finally, at the risk of sounding unintentionally provincial, it should be noted that the majority of the *in vivo* work has been done in India, using local species of fish, many of which are air breathers. Whether similar findings will be obtained on other freshwater and marine teleosts remains to be seen.

VI. ANTIOXIDANTS

Oxygen free radicals are one, two, or three electron reduction products of molecular oxygen. They include the superoxide anion radical (O_2^-), hydrogen peroxide, and the hydroxyl radical ($\cdot OH$). They are produced as by-products of many cellular reactions including both cytoplasmic (e.g., xanthine oxidase, mixed function oxidase) and mitochondrial (e.g., electron transport chain) enzymes. These oxyradicals can cause considerable cellular damage including oxidations of membrane lipids, proteins, and nucleic acids. The polyunsaturated fatty acids of membranes are especially susceptible to peroxidation and the techniques for the detection and measurement of this have been reviewed by Gutteridge and Halliwell (1990). The oxidative lesions in cells are also related to induction of tumors in fish (Malins and Haimanot, 1991).

Normally, the accumulation of oxyradicals is controlled by antioxidant enzymes and low molecular weight molecules such as glutathione and ascorbate. Metallothionein, which is recognized as a metal detoxification agent, has also been suggested as a possible antioxidant (Shi, 1990).

A wide variety of organic and inorganic toxic contaminants cause an increased production of oxyradicals in cells (Stegeman et al., 1992). This in turn serves to induce production of more of the antioxidant enzymes superoxide dismutases, catalases, peroxidases, glutathione reductase, and glutathione. Ascorbate is also involved in antioxidant activities and is discussed in this book in Chapter 6 while glutathione is taken up in Chapter 5. Here we limit our discussion primarily to the antioxidant enzymes.

Basal antioxidant enzyme activities tend to be higher in tissues which exhibit high oxidative metabolism. These include red muscle, the gas gland in the swim bladder, heart, and blood (Filho et al., 1993). Enzyme activity levels also differ between fish species; those that are more physically active tend to have the higher activities (Filho et al., 1993).

When trout were exposed to cadmium for 180 days at concentrations 13 and 25% of the LC50, elevations in liver superoxide dismutase were seen. This induction of the antioxidant enzyme could be interpreted as a response to an increase in oxyradicals. On the other hand, glutathione reductase went down and catalase was unchanged (Palace et al., 1993). Ascorbate also showed a marked decline with cadmium exposure presumably due to its utilization as an antioxidant. In the same study, they observed that a diet deficient in ascorbate caused elevations in superoxide dismutase in the absence of cadmium, which the authors suggest may be due to a compensatory antioxidant enzyme induction.

When dab (a flatfish) were exposed to sediments contaminated with polynuclear aromatic hydrocarbons (PAHs) and polychlorobiphenyls (PCBs) for 25 to 140 days, the activity of all three antioxidant enzymes declined in both contaminated and non-contaminated (reference) fish. However, superoxide dismutase, catalase, and lipid peroxidation were higher in contaminated fish compared to those from the reference sediments after 80 days, but not at 140 days. Glutathione peroxidase did not change (Livingston et al., 1993). The authors conclude that the PAHs and PCBs caused transient oxidative stress.

Fish sampled from areas along the coast of Spain contaminated with iron, copper, and several aromatic hydrocarbons exhibited elevations in antioxidant enzymes when compared with a reference site (Rodriquez-Ariza et al., 1993). Enzymes which increased included superoxide dismutase, glutathione peroxidase, catalase, and glutathione reductase. The largest increase was in glutathione peroxidase, especially the Se-dependent isoenzyme. This enzyme is particularly important in protection against lipid peroxidation and seems to have been successful in these fish as little lipid peroxidation was detected. These observations exemplify an important concept in that toxic chemicals may induce various types of biochemical changes that are an adaptation to the chemical stressors. These include detoxification enzymes such as those associated with cytochrome P450, metallothionein, and the antioxidant enzymes described here. It is useful, as was done in the study just mentioned, to determine whether the adaptive changes are successful in reducing or eliminating toxicity.

VII. ADENYLATES

The release of energy from organic foodstuffs in cells by the various metabolic processes occurs in a stepwise fashion. The energy so obtained is "stored" temporarily in high energy phosphate bonds. The term adenylates refers to the nucleotides adenosine triphosphate (ATP), adenosine diphosphate (ADP), and adenosine monophosphate (AMP). Creatine phosphate, while being a high-energy phosphate molecule, is not a nucleotide, although it is often measured along with the adenylates. Clearly, ATP has the most energy stored in it so the relative concentrations of these adenylates reflects the energy immediately available to the cell for muscular contraction, maintenance, ionic pumping, biosynthesis etc. They also act as modulators of enzyme activity in glycolysis and the Kreb's cycle (Atkinson, 1977). The adenylate energy charge (AEC) is the ratio of ATP + 0.5 (ADP) to total adenylate concentration in the tissue. This value can vary from 0–1. In invertebrates and microorganisms, unstressed individuals show an AEC above about 0.8, and decreases in the AEC have been useful in assessing environmental stressors in these forms (Atkinson, 1977; Ivanovici, 1980; Livingston, 1985).

ATP is the most variable component of the adenylate pool. In fish it decreases during muscular exercise (Schulte and Hochachka, 1990) and hypoxia (see Chapter 2), and different temperatures can result in significant changes (Walesby and Johnston, 1980; Kindle and Whitmore, 1986). For the most part, ADP and AMP change comparatively little with these kinds of alterations in activity or environment. With decreases in ATP, one might expect a rise in AMP, but this does not necessarily occur because it can be converted to inosine monophosphate (IMP).

Seasonal changes in adenylates can be quite large (Dehn, 1992). The redear sunfish (a centrarchid) showed the lowest muscle AEC (0.5–0.75) in the winter but liver dropped to below 0.5 in April during the spawning season. Dehn also noted that liver exhibited a lower AEC than muscle at all times of the year.

MacFarlane (1981) tested the effect of water pH on brain, muscle, gill, and liver tissue adenylate concentrations of gulf killifish (*F. grandis*). The test organisms were exposed for periods of up to 96 h at four different pHs ranging downward from the control level of 7.8. The greatest decreases in ATP, total adenylates, and AEC were observed in brain and gill. There were relatively small changes in total adenylate and energy charge compared to ATP which decreased as much as 72%. The effect of hydrogen ions in the water on brain ATP appeared to be dose dependent and this depression increased with time of exposure, as well.

McFarlane (1981) explained the declines in ATP and total adenylates in the brain as being due to decreased sodium levels in the blood induced by the acid. He cites a study on fish electric organs to the effect that neural activity causes release of ATP into the extracellular fluid which is then degraded extracellularly to adenosine before being taken back into the cells. When the sodium concentration is reduced, this uptake of adenosine is inhibited. Another factor (discussed by MacFarlane) which probably contributes to lowered ATP in tissues other than brain is an elevated energy demand in the gills from pumping hydrogen ions out of the fish and in the muscles from the considerable muscular activity caused by the acid.

Copper at high sublethal levels has been shown to cause a lowering of ATP in liver and muscle (Heath, 1984). Bluegill (*Lepomis macrochirus*) were exposed to copper at a concentration equivalent to the LC20 and tissues sampled at intervals of up to 96 h of exposure. Significant decreases in ATP occurred first at 48 h with the greatest changes in liver. Brain exhibited no change in ATP. Measurements of tissue lactic acid established that tissue hypoxia did not occur, therefore other hypotheses to explain the lowered level of ATP must be proposed. Heath (1984) noted increases in tissue water content concomitant with the drop in ATP. It was calculated that this dilution of the cell contents accounted for 25–30% of the decrease in nucleotide concentration. The remainder of the decrease is attributable to one or more of the following causes: (1) bluegill exposed to sublethal copper show elevated liver tissue energy demand (Felts and Heath, 1984), presumably due to detoxification processes and increased demand for energy by cells generally pulls ATP down (Walesby and Johnston, 1980); (2) when fish liver mitochondria are exposed to copper (*in vitro*), there is a large increase in permeability and oxygen consumption (Zaba and Harris, 1978), so a direct effect of copper on mitochondrial respiration is probable; (3) inhibition of some enzyme systems probably occurred (Table 1); and (4) copper may alter cellular concentrations of glutathione. Stacey and Klaassen (1981) noted a lowering of glutathione in rat hepatocytes exposed to copper (*in vitro*). Schole (1982) has proposed a central role for glutathione in the regulation of energy metabolism in cells so changes in the concentration of this compound could have large effects on these chemical reactions. This latter mechanism is, of course, highly speculative.

There seems to have been little work on the effects of pesticides on adenylates. The carbamate insecticide carbofuran and the synthetic pyrethroid fenvalerate had only slight effects on these parameters during 10-day exposures to doses a bit less than one half the 96-h LC50 (Hohreiter et al., 1991). This relatively small effect may in part be due to the failure of these substances to be taken up by the fish during short-term exposures. Because some insecticides certainly affect energy metabolism at the molecular level (Table 2) and whole-animal metabolism (Chapter 8), it seems likely that adenylates would also be affected as well. The kinetics of the response may be as important to reveal as the overall effect.

There has been some interest in the use of adenylates as biomarkers in field and laboratory studies (partially reviewed in Mayer et al., 1992). Thus far, mollusks seem to yield useful results better than fish. In part, this is probably because of the effect of struggling during capture on muscle adenylates. This can be considerable in fish and occurs within seconds so accurate baseline values are extremely difficult to obtain (Schulte and Hochachka, 1990). Another problem is the large effect of season on these parameters which was discussed earlier. In general, it seems that substances that accumulate in the liver, such as copper, or some of the lipophilic organics, might have by far the greatest effect on adenylates there. Also, liver adenylates are not affected by handling stress so variability should be less (assuming spawning season is avoided).

VIII. STRESS PROTEINS

The stress proteins are a group of cellular proteins whose synthesis increases when the cell is subjected to any sort of stress. They were first discovered in *Drosophila* that had experienced a fairly rapid increase in temperature and were therefore called "heat shock proteins" (hsp; Welch, 1993). It is now known that the induction of these molecules is much broader than just from heat so the name "stress proteins" is becoming more accepted. There are four major groups of stress proteins based on their size and these are designated hsp90, 70, 60, and 16–24. The newer terminology is "stress 90", "stress 70", and "chaperonin", for the 60-kDa family and low molecular weight (LMW) for the 16- to 24-kDa group (Sanders, 1993).

Stress proteins are not limited to stressed cells. They are found in nearly all cells and their function is to regulate folding, assembly, and aggregation of other proteins. They are often collectively called chaperones for this reason. A primary mechanism of toxicity involves protein denaturation which results in molecular aggregation and misfolding (Hightower, 1991). This causes binding of a protein called heat shock factor (HSF) to controlling genes (promoter) which then activate transcription of the stress protein genes. Synthesis of stress proteins then occurs to facilitate the repair of the denatured proteins. As the proteins are repaired, stimulation of the HSF declines slowing the induction process so the feedback loop is completed.

The specific stress proteins induced vary with different tissues, probably in part due to different tissue sensitivities to particular toxicants. Within a tissue, the stress protein induction process appears to be non-specific (i.e., all stressors cause essentially the same response although the kinetics may differ). Also, work with a number of groups other than fish indicates that stress 70, chaperonin, and the LMW stress proteins have different isoforms and their induction may differ with the specific stressor. Future studies may reveal more stressor-specific proteins involved in other types of cellular repair as well (Sanders, 1993).

Part of the process of acclimation to a toxicant may involve stress proteins. In a variety of animal cells, it has been found that previous treatment with a heat shock or toxicant stress makes the cells more tolerant of a subsequent stress (Welch, 1993). This is, incidentally, separate from the induction of detoxifying enzymes such as those associated with cytochrome P450 or metallothionein which are discussed in Chapter 5.

Thus far, research on stress protein induction relative to water pollution has largely been limited to work on mollusks and grass shrimp (reviewed in Sanders, 1993). The situation will undoubtedly change rapidly in the future, although the techniques for measuring induction are currently difficult for those not familiar with molecular biology methodologies. The potential for future utilization of the stress protein response is great because it should integrate damage due to multiple stressors while at the same time possibly yielding diagnostic information as to the cause of the damage (Sanders, 1993).

REFERENCES

Anderson, J. W., An assessment of knowledge concerning the fate and effects of petroleum hydrocarbons in the marine environment, in, *Marine Pollution: Functional Responses,* Vernberg, W., Thurberg, F., Calabrese, A. and Vernberg, F., Eds., Academic Press, New York, 1979, 3.

Atkinson, D. E., *Cellular Energy Metabolism and its Regulation,* Academic Press, New York, 1977.

Balavenkatasubbaiah, M., Rani, A. U., Geethanjali, K., Purushotham, K. and Ramamurthi, R., Effect of cupric chloride on oxidative metabolism in the freshwater teleost, *Tilapia mossambica, Ecotoxicol. Environ. Safety,* 8, 289, 1984.

Bay, S., Greenstein, D., Szalay, P. and Brown, D., Exposure of scorpionfish (*Scorpaena guttata*) to cadmium: biochemical effects of chronic exposure, *Aquat. Toxicol.,* 16, 311, 1990.

Bhatnagar, R. S., Ed., *Molecular Basis of Environmental Toxicity,* Ann Arbor Science, Ann Arbor, MI, 1980.

Bostrom, S. L. and Johansson, R. G., Effects of pentachlorophenol on enzymes involved in energy metabolism in the liver of the eel, *Comp. Biochem. Physiol.,* 41B, 359, 1972.

Brown, D. A., Bay, S., Alfafara, J., Hershelman, G. and Rosenthal, K., Exposure of scorpionfish (*Scorpaena guttata*) to cadmium: effects of acute and chronic exposures on the subcellular distribution of cadmium, copper and zinc, *Aquat. Toxicol.*, 16, 295, 1989.

Bryan, G. W., Some aspects of heavy metal tolerance in aquatic organisms, in, *Effects of Pollutants on Aquatic Organisms*, Lockwood, A. P. M., Ed., Cambridge University Press, Cambridge, 1976, 7.

Buhler, D. R., Stokes, R. M. and Caldwell, R. S., Tissue accumulation and enzymatic effects of hexavalent chromium in rainbow trout (*Salmo gairdneri*), *J. Fish. Res. Bd. Can.*, 34, 9, 1977.

Christensen, G. M. and Tucker, J. H., Effects of selected water toxicants on the *in vitro* activity of fish carbonic anhydrase, *Chem. Biol. Interact.*, 13, 181, 1976.

Davis, P. W., Friedhoff, J. M. and Wedemeyer, G. A., Organochlorine insecticide, herbicide and polychlorinated biphenyl (PCB) inhibition of NaK-ATPase in rainbow trout, *Bull. Environ. Contam. Toxicol.*, 8, 69, 1972.

Davison, W., Franklin, C. E., McKenzie, J. and Dougan, M. C. R., The effects of acute exposure to the water soluble fraction of diesel fuel oil on survival and metabolic rate of an Antarctic fish (*Pagothenia borchgrevinki*), *Comp. Biochem. Physiol.*, 102C, 185, 1992.

DeBruin, A., *Biochemical Toxicology of Environmental Agents*, Elsevier, New York, 1976.

Dehn, P., Seasonal changes in adenylate energy metabolism in the muscle and liver of the redear sunfish, *Lepomis microlophus*, *Aquat. Living Res.*, 5, 197, 1992.

Felts, P. and Heath, A. G., Interactions of temperature and sublethal environmental copper exposure on the energy metabolism of bluegill, *Lepomis macrochirus* Rafinesque, *J. Fish Biol.*, 25, 445, 1984.

Ferguson, D., The ecological consequences of pesticide resistance in fishes, *Trans. 32nd N. Am. Wildl. Nat. Resour. Conf.*, Wildlife Management Institute, Washington, D.C., 1967, 103.

Filho, D., Giulivi, C. and Boveris, A., Antioxidant defences in marine fish. I. Teleosts, *Comp. Biochem. Physiol.*, 106C, 409, 1993.

Fowler, B. A., Intracellular compartmentation of metals in aquatic orgranisms: roles in mechanisms of cell injury, *Environ. Health Perspect.*, 71, 121, 1987.

Fox, J. H., Patel-Mandlik, K. and Cohen, M. M., Comparative effects of organic and inorganic mercury on brain slice respiration and metabolism, *J. Neurochem.*, 24, 757, 1974.

Gill, T., Tewari, H. and Pande, J., Use of the fish enzyme system in monitoring water quality: effects of mercury on tissue enzymes, *Comp. Biochem. Physiol.*, 97C, 287, 1990.

Gill, T., Pande, J. and Tewari, H., Enzyme modulation by sublethal concentrations of aldicarb, phosphamidon and endosulfan in fish tissues, *Pest. Biochem. Physiol.*, 38, 231, 1990.

Gill, T., Tewari, H. and Pande, J., *In vivo* and *in vitro* effects of cadmium on selected enzymes in different organs of the fish *Barbus conchonius* Ham. (Rosy barb)., *Comp. Biochem. Physiol.*, 100C, 501, 1991.

Gill, T. S., Tewari, H. and Pande, J., Short- and long-term effects of copper on the rosy barb (*Puntius conchonius*), *Ecotoxicol. Environ. Safety*, 23, 294, 1992.

Gould, E., Alterations of enzymes in winter flounder, *Pseudopleuronectes americanus*, exposed to sublethal amounts of cadmium chloride, in, *Physiological Responses of Marine Biota to Pollutants*, Vernberg, F. J., Calabrese, A., Thurberg, F. P. and Vernberg, W. B., Eds., Academic Press, New York, 1977, 209.

Gould, E. and Karolus, J., Physiological response of the cunner, *Tautogloabrus adspersus*, to cadmium. V. Observations on the biochemistry, U.S. Department of Commerce, Washington, D.C., NOAA Technical Report NMFS SSRF-681, 21, 1974.

Goolish, E. M. and Adelman, I. R., Tissue-specific cytochrome oxidase activity in largemouth bass: the metabolic costs of feeding and growth, *Physiol. Zool.*, 60, 454, 1987.

Gupta, P. K. and Sastry, K. V., Alterations in the activities of three dehydrogenases in the digestive system of two teleost fishes exposed to mercuric chloride, *Environ. Res.*, 24, 15, 1981.

Gutteridge, H. and Halliwell, B., The measurement and mechanism of lipid peroxidation in biological systems, *Trends Biochem. Sci.*, 15, 129, 1990.

Heath, A. G., Changes in tissue adenylates and water content of bluegill, *Lepomis macrochirus*, exposed to copper, *J. Fish Biol.*, 24, 299, 1984.

Heitz, J. R., Lewis, L., Chambers, J. and Yarbrough, J., The acute effects of empire mix crude oil on enzymes in oysters, shrimp, and mullet, in, *Pollution and Physiology of Marine Organisms*, Vernberg, F. and Vernberg, W., Eds., Academic Press, New York, 1974, 311.

Hightower, L. E., Heat shock, stress proteins, chaperones, and proteotoxicity (meeting review), *Cell,* 66, 1, 1991.

Hilmy, A. M., Shabana, M. B. and Daabees, A. Y., Effects of cadmium toxicity upon the *in vivo* and *in vitro* activity of proteins and five enzymes in blood serum and tissue homogenates of *Mugil cephalus, Comp. Biochem. Physiol.,* 81C, 145, 1985.

Hiltibran, R. C., Oxygen and PO uptake by bluegill liver mitochondria in the presence of some insecticides, *Trans. Il. Acad. Sci.,* 74, 228, 1974.

Hiltibran, R. C., Effects of insecticides on the metal-activated hydrolysis of adenosine triphosphate by bluegill liver mitochondria, *Arch. Environ. Contam. Toxicol.,* 11, 709, 1982.

Hochachka, P. W. and Somero, G. N., *Biochemical Adaptation,* Princeton University Press, Princeton, NJ, 1984.

Hohreiter, D., Reinert, R. and Bush, P., Effects of insecticides carbofuran and fenvalerate on adenylate parameters in bluegill sunfish (*Lepomis macrochirus), Arch. Environ. Contam. Toxicol.,* 21, 325, 1991.

Ivanovici, A. M., Application of adenylate energy charge to problems of environmental assessment in aquatic organisms, *Helgol. Wiss. Meeresunters.,* 33, 556, 1980.

Jackim, E., Hamlin, J. and Sonis, S., Effects of metal poisoning on five liver enzymes in the killifish (*Fundulus heteroclitus), J. Fish. Res. Bd. Can.,* 27, 383, 1970.

Jackim, E., Influence of lead and other metals on fish delta-aminolevulinate dehydrase activity, *J. Fish. Res. Bd. Can.,* 30, 560, 1973.

James, P., Sampath, K. and Ponmani, P., Effect of metal mixtures on activity of two respiratory enzymes and their recovery in *Oreochromis mossambicus, Indian J. Exp. Biol.,* 30, 496, 1992.

Johansson-Sjobeck, M. and Larsson, A., Effects of inorganic lead on delta-aminolevulinic acid dehydratase activity and hematological variables in the rainbow trout, *Salmo gairdneri, Arch. Environ. Contam. Toxicol.,* 8, 419, 1979.

Kindle, K. R. and Whitmore, D. H., Biochemical indicators of thermal stress in *Tilapia aurea* (Steindachner), *J. Fish Biol.,* 29, 243, 1986.

Koundinya, P. R. and Ramamurthi, R., Effect of organophosphate pesticide sumithion (Fenitrothion) on some aspects of carbohydrate metabolism in a freshwater fish, *Sarotherodon (Tilapia) mossambicus* (Peters), *Experientia (Basel),* 35, 1632, 1979.

Kramer, V. J., Newman, M., Mulvey, M. and Ultsch, G., Glycolysis and krebs cycle metabolites in mosquitofish, *Gambusia holbrooki,* exposed to mercuric chloride: allozyme genotype effects, *Environ. Toxicol. Chem.,* 11, 357, 1992.

Krueger, H., Lu, S. D., Chapman, G. and Cheng, J. T., Effects of pentachlorophenol on the fish, *Cichlasoma bimaculatum,* Abstracts from the 3rd Int. Pharmacological Congr. Sao Paulo, Brazil, 1966, 24, (cited in Bostrom and Johansson, 1972).

Kuhnert, P., Kuhnert, B. and Stokes, B., The effect of *in vivo* chromium exposure on Na/K- and Mg-ATPase activity in several tissues of the rainbow trout (*Salmo gairdneri), Bull. Environ. Contam. Toxicol.,* 15, 383, 1976.

Kumar, K. and Ansari, B., Malathion toxicity: effect on the liver of the fish *Branchydanio rerio* (Cyprinidae), *Ecotoxicol. Environ. Safety,* 12, 199, 1986.

Livingston, D. R., Biochemical measurements, in, *The Effects of Stress and Pollution on Marine Animals,* Praeger Publishers, New York, 1985, chap. 4.

Livingston, D. R., Lemaire, P., Matthews, A., Peters, L. Bucke, D. and Law, R., Pro-oxidant, antioxidant and 7-(ethoxyresorufin)-Deethylase (EROD) activity responses in liver of Dab (*Limanada limanda*) exposed to sediment contaminated with hydrocarbons and other chemicals, *Mar. Pollut. Bull.,* 26, 602, 1993.

MacFarlane, R. B., Alterations in adenine nucleotide metabolism in the gulf killifish (*Fundulus grandis*) induced by low pH water, *Comp. Biochem. Physiol.,* 68B, 193, 1981.

MacInnes, J., Thurberg, F., Greig, R. and Gould, E., Long-term cadmium stress in the cunner, *Tautogolabrus adspersus, Fish. Bull. U.S.,* 75, 199, 1977.

MacLeod, J. and Pessah, E., Temperature effects on mercury accumulation, toxicity and metabolic rate in rainbow trout (*Salmo gairdneri), J. Fish. Res. Bd. Can.,* 30, 485, 1973.

Malins, D. C. and Haimanot, R., The etiology of cancer: hydroxyl radical-induction DNA lesions in histologically normal livers of fish from a population with liver tumours, *Aquat. Toxicol.,* 20, 123, 1991.

Malins, D. C. and Ostrander, G. K., Eds., *Aquatic Toxicology: Molecular, Biochemical, and Cellular Perspectives*, Lewis Publishers, Boca Raton, FL, 1994.

Massaro, E. J., Pharmacokinetics of toxic elements in rainbow trout, *U.S.E.P.A. Research Report Series* EPA-660/3-74-027, U.S. Environmental Protection Agency, Washington, D.C., 1974.

Mayer, F. L., Versteeg, D., McKee, M., Folmar, L., Graney, R., McCume, D. and Rattner, B., Physiological and nonspecific biomarkers, in, *Biomarkers: Biochemical, Physiological, and Histological Markers of Anthropogenic Stress*, Huggett, R., Kimerle, R., Mehrle, P. and Bergman, H., Eds., Lewis Publishers, Boca Raton, FL, 1992, chap. 1.

Mehrle, P. M. and Bloomfield, R. A., Ammonia detoxifying mechanisms of rainbow trout altered by dietary dieldrin, *Toxicol. Appl. Pharmacol.*, 27, 355, 1974.

Meunier, F. M. and Morel, N., Adenosine uptake by cholinergic synaptosomes from *Torpedo* electric organ, *J. Neurochem.*, 31, 845, 1978.

Mounid, M. S., Rosenthal, H. and Eisan, J. S., Some effects of cadmium on the metabolism of developing eggs of Pacific herring, *Int. Counc. Exp. Sea. Counc. Meet.*, 1975/E, 20, 1, 1975, (cited in Rosenthal, H. and Alderdice, D. F.).

Natarajan, G. M., Effect of sublethal concentration of metasystox on selected oxidative enzymes, tissue respiration, and hematology of the freshwater air-breathing fish, *Channa striatus (Bleeker)*, *Pest. Biochem. Physiol.*, 21, 194, 1984.

Nicholls, D., Teichert-Kuliszewska, K. and Girgis, G., Effect of chronic mercuric chloride exposure on liver and muscle enzymes in fish, *Comp. Biochem. Physiol.*, 94C, 265, 1989.

Packer, R. K., Acid-base balance and gas exchange in brook trout (*Salvelinus fontinalis*) exposed to acidic environments, *J. Exp. Biol.*, 79, 127, 1979.

Palace, V. P., Majeski, H. S. and Klaverkamp, J. F., Interactions among antioxidant defenses in liver of rainbow trout (*Oncorhynchus mykiss*) exposed to cadmium, *Can. J. Fish. Aquat. Sci.*, 50, 156, 1993.

Prasad, M., Toxicity of crude oil to the metabolism of freshwater minor carp, *Puntius sophore*, *Bull. Environ. Contam. Toxicol.*, 39, 188, 1987.

Radhakrishnaiah, K., Venkataramana, P., Suresh, A. and Sivaramakrishna, B., Effects of lethal and sublethal concentrations of copper on glycolysis in liver and muscle of the freshwater teleost, *Labeo rohita*, *J. Environ. Biol.*, 13, 63, 1992.

Ram, R. and Singh, S., Carbofuran-induced histopathological and biochemical changes in liver of the teleost fish, (*Channa punctatus*), *Ecotoxicol. Environ. Safety*, 16, 194, 1988.

Rao, K. R., *Pentachlorophenol, Chemistry, Pharmacology, and Environmental Toxicology*, Plenum Press, New York, 1978.

Rao, K. S. P. and Rao, K. V. P., Regulation of phosphorylases and aldolases in tissues of the teleost (*Tilapia mossambica*) under methyl parathion impact, *Bull. Environ. Contam. Toxicol.*, 31, 474, 1983.

Roberts, K. S., Cryer, A., Kay, J., Solbe, J. F., Wharfe, J. R. and Simpson, W. R., The effects of exposure to sublethal concentrations of cadmium on enzyme activities and accumulation of the metal in tissues and organs of rainbow and brown trout (*Salmo gairdneri* and *Salmo trutta*), *Comp. Biochem. Physiol.*, 62C, 135, 1979.

Rodriquez-Ariza, A., Peinado, J., Pueyo, C. and Lopez-Barea, J., Biochemical indicators of oxidative stress in fish from polluted littoral areas, *Can. J. Fish. Aquat. Sci.*, 50, 2568, 1993.

Rosenthal, H. and Alderdice, D. F., Sublethal effects of environmental stressors, natural and pollutional, on marine fish eggs and larvae, *J. Fish. Res. Bd. Can.*, 33, 2047, 1976.

Sanders, B. M., Stress proteins in aquatic organisms: an environmental perspective, *Crit. Rev. Toxicol.*, 23, 49, 1993.

Sastry, K. V. and Agrawal, M. K., Effects of lead nitrate on the activities of a few enzymes in the kidney and ovary of *Heteropneustes fossilis* (sic), *Bull. Environ. Contam. Toxicol.*, 22, 55, 1979.

Sastry, K. V. and Rao, D. R., Enzymological and biochemical changes produced by mercuric chloride in a teleost fish, *Channa punctatus*, *Toxicol. Lett.*, 9, 321, 1981.

Sastry, K. V. and Rao, D. R., Chronic effects of mercuric chloride on the activities of some enzymes in certain tissues of the fresh water murrel, *Channa punctatus*, *Chemosphere*, II, 1203, 1982.

Sastry, K. V. and Sharma, K. S., Effects of mercuric chloride on the activities of brain enzymes in a freshwater teleost, *Opiocephalus (Channa punctatus)*, *Arch. Environ. Contam. Toxicol.*, 9, 425, 1980.

Sastry, K. V., Siddiqui, A. A. and Singh, S. K., Alteration in some biochemical and enzymological parameters in the snake head fish *Channa punctatus*, exposed chronically to quinalphos, *Chemosphere*, 11, 1211, 1982.

Schole, J., Theory of metabolic regulation including hormonal effects on the molecular level, *J. Theor. Biol.*, 96, 579, 1982.

Schulte, P. M. and Hochachka, P. W., Energy metabolism during recovery from exhaustive exercise in rainbow trout white muscle, *Physiologist*, 33, A-88, 1990.

Shaffi, S., Comparison of the sublethal effect of mercury and lead on visceral dehydrogenase system in three inland teleosts, *Physiol. Res.*, 42, 7, 1993.

Sharma, R. M., Effect of endosulfan on acid and alkaline phosphatase activity in liver, kidney and muscles of *Channa gachua, Bull. Environ. Contam. Toxicol.*, 44, 443, 1990.

Shi, C., Metallothionein as a scavenger of free radicals, *Trace Elem. Med.*, 7, 48, 1990.

Shukla, S. and Tripathi, G., Endosulfan toxicity on metabolic enzymes and protective role of triiodothyroine in the catfish *Clarias batrachus, Pestic. Sci.*, 32, 363, 1991.

Simon, L. M., Nemcsok, J. and Boross, L., Studies on the effect of paraquat on glycogen mobilization in liver of common carp (*Cyprinus carpio* L), *Comp. Biochem. Physiol.*, 75C, 167, 1983.

Smith, J. C. and Chong, C. K., Body weight, activities of cytochrome oxidase and electron transport system in the liver of the American plaice *Hippoglossoides platessoides*. Can these activties serve as indicators of metabolism?, *Mar. Ecol. Prog. Ser.*, 9, 171, 1982.

Somasundaram, B., King, P. E. and Shackley, S., The effects of zinc on the ultrastructure of the brain cells of the larva of *Clupea harengus, Aquat. Toxicol.*, 5, 323, 1984.

Southward, J., Nitisewojo, P. and Green, D. E., Mercurial toxicity and the perturbation of the mitochondrial control system, *Fed. Am. Soc. Exp. Biol.*, 33, 2147, 1974.

Srivastava, D. and Pandley, K., Effect of copper on tissue acid and alkaline phosphatases in the green snakehead, *Ophicephalus punctatus* (Bloch), *Toxicol. Lett.*, II, 237, 1982.

Stacey, N. H. and Klaassen, C. D., Copper toxicity in isolated rat hepatocytes, *Toxicol. Appl. Pharmacol.*, 58, 211, 1981.

Stegeman, J. J., Brouwer, M., Ki Giulio, R., Forlin, L, Fowler, B., Sanders, B. and Van Veld, P., Molecular responses to environmental contamination: enzyme and protein systems as indicators of chemical exposure and effect, in, *Biomarkers: Biochemical, Physiological and Histological Markers of Anthropogenic Stress,* Huggett, R., Kimerle, R., Mehrle, P. and Bergman, H., Lewis Publishers, Boca Raton, FL, 1992, chap. 6.

Tomasso, J. R., Goudie, C. A., Simco, B. A. and Davis, K. B., Effects of environmental pH and calcium on ammonia toxicity in channel catfish, *Trans. Am. Fish. Soc.*, 109, 229, 1980.

Ultsch, G. R., Ott, M. E. and Heisler, N., Acid-base and electrolyte status in carp (*Cyprinus carpio*) exposed to low environmental pH, *J. Exp. Biol.*, 93, 65, 1981.

Welch, W. J., How cells respond to stress, *Sci. Am.,* May, 56, 1993.

Weis, P., Bogden, J. and Enslee, E., Mercury and copper induced hepatocellular changes in the mummichog *Fundulus heteroclitus, Environ. Health Perspect.*, 65, 167, 1986.

Verma, S. R., Madjur, J. and Tonk, L. P., *In vivo* effect of mercuric chloride on tissue ATPases of *Notopterus notopterus, Toxicol. Lett.*, 16, 305, 1983.

Verma, S. R., Rani, S. and Dalela, R. C., Effects of sodium pentachlorophenate on enzymes of energy metabolism in tissues of *Notopterus notopterus, Toxicol. Lett.*, 10, 297, 1982.

Walesby, N. J. and Johnston, I. A., Temperature acclimation in brook trout muscle: adenine nucleotide concentrations, phosphorylation state and adenylate energy charge, *J. Comp. Physiol.*, 139, 127, 1980.

Yarbrough, J. D. and Wells, M. R., Vertebrate insecticide resistance: the in vitro endrin effect on succinic dehydrogenase activity on endrin-resistant and susceptible mosquitofish, *Bull. Environ. Contam. Toxicol.*, 6, 171, 1971.

Zaba, B. N. and Harris, E. J., Accumulation and effects of trace metal ions in fish liver mitochondria, *Comp. Biochem. Physiol.*, 61C, 89, 1978.

Chapter 10

Acid Pollution

I. INTRODUCTION

Acid precipitation has become a widespread and increasing problem in several countries in Europe and North America. Acid falls as rain and snow and even dry gases and small particles. Some people may be surprised to learn that the latter two account for approximately two thirds of the total acid precipitation in the U.K. (Mason, 1989). The major acids of concern are sulfuric and nitric which are formed in the atmosphere from emissions of coal-fired power plants and automobiles.

The pH of uncontaminated rainwater is not neutral, instead due to equilibrium with carbon dioxide, it is around 5.6. Rain and snow even in remote areas are almost always more acidic than this and in areas where air pollution is acute, the pH can easily fall below 5. The effect on streams and lakes depends greatly on the hydrogeology (Henriksen, 1982; Norton, 1982). Those bodies of water surrounded with soils or rock rich in base minerals (e.g., Ca/Mg carbonates and bicarbonates) are relatively immune from effects of acid precipitation. In large areas of North America and Scandinavia, the buffering capacity of the fresh surface waters is extremely limited so over the past decades there has occurred a rapid (in the historic sense) drop in their pH.

Atmospheric pollutants are deposited episodically such that there is usually an autumn and a spring decline in pH. The autumn decline coincides with the spawning and early embryonic development of species like lake trout (*Salvelinus namaycush*), lake white fish (*Coregonus clupeaformis*), and lake herring (*C. artedii*) (Peterson et al., 1982). Very rapid spring reductions of pH can occur due to snow melt and this may coincide with the late embryonic and/or larval development of fall-spawned eggs. As will be seen later, the larval (alevin in salmonids) stages are often the most sensitive developmental stage to acute acid stress. Furthermore, the spring spawning run of many species such as the rainbow trout may occur at a time of very low ambient pH.

The reductions in pH of surface waters mobilize metals, especially aluminum (Baker, 1982). The increases in aluminum from acid precipitation are generally from natural input due to the ubiquitous distribution of this element. Because fish are more often than not receiving the impact of both acid water and aluminum, in recent years there has been a tendency for laboratory studies to combine these. This chapter will, however, emphasize pure acid toxicity with some reference to action of aluminum where appropriate. Aluminum is also discussed in some detail in Chapter 7.

The sensitivity of different fish species to low pH differs, even between different salmonids. For example, cutthroat trout (*Oncorhynchus clarki*), and golden trout (*O. aguabonita*) are probably the most sensitive of the family while rainbow trout are intermediate followed by brook (*S. fontinalis*) and brown trout (*Salmo truta*) (Delonay et al., 1993). As a general rule, warmwater species are more tolerant of acid conditions than are salmonids.

The literature on acid rain is immense and certain aspects of the physiological effects of acid toxicity to fish have been reviewed by Fromm (1980), Peterson et al. (1982), Wood and McDonald (1982), Howells (1984), and Wood (1989). Throughout this book, individual pollutants have been discussed within the context of how they affect particular organs or physiological processes. Here we devote a whole chapter to the one pollutant because of the extensive information available. In addition, it gives an opportunity to attempt to integrate several physiological processes in the context of one pollutant. The organization of this chapter will therefore be different from others in this book. Instead of the usual normal physiology followed by a consideration of the effects of pollutants on that physiology, we will instead use a life-cycle approach. We will start with spawning and then consider the effects of acid (and often aluminum) on the different life stages. For those relatively unfamiliar with basic fish physiology, reference to the early parts of the chapters on reproduction, respiration, and osmoregulation may be helpful.

II. SPAWNING

Failure to lay eggs when the mature adults are subjected to chronic low pH has been noted in many studies involving both salmonids and non-salmonid species (e.g., Weiner et al., 1986; Peterson et al., 1982; Tam and Payson, 1986). Some reduction in numbers of eggs laid has been seen at nearly any pH below control levels, thus there seems to be no overall threshold pH that can be applied to all fish species, or even to all salmonids. Moreover, the effects of a lowered pH on egg production is also very dependent on environmental calcium levels; the lower the calcium, the greater the effect (Parker and McKeown, 1987). It has been suggested that the primary physiological mechanism for decreased numbers of eggs is an inhibition in the production of yolk proteins (Peterson et al., 1982).

Many female fish exhibit elevations in plasma calcium during ovarian maturation. This elevation may be important in the process of vitellogenesis. Vitellogenin, a phosphoprotein synthesized in the liver, is transported to the ovaries as a calcium-vitellogenin complex for use in yolk synthesis. Female white suckers from acidified lakes were reported to have abnormally low blood calcium levels (Beamish et al., 1975). This often cited reference has been used to explain failure of reproduction in fish residing in acidified waters. However, Munkittrick (1991) has pointed out that the calcium levels in the Beamish et al. (1975) female fish did not differ significantly from controls. Instead, Beamish et al. based their conclusion on the ratio of female to male calcium level, and it turns out that the four male control fish had abnormally low concentrations of calcium in their blood. Indeed, Weiner et al. (1986) found no effect on plasma calcium of low pH per se in the laboratory on rainbow trout females, although it did cause a reduction in calcium in the males. This finding can be contrasted with that of Parker and McKeown, (1987) in which a low pH caused a failure of female rainbow trout to exhibit the normal surge in plasma calcium during oogenesis. They also noted that low calcium levels in the water reduced the plasma concentration of vitellogenin and lowered the gonadosomatic index (an indication of reduced egg production). It was concluded that the depressed calcium availability inhibited vitellogenin formation by the liver. More recently, Roy et al. (1990) utilized a homologous radioimmunoassay to measure vitellogenin in rainbow trout at three pHs (7.6, 5.6, 4.5) during spawning. There was a surge in vitellogenin during the 20-day exposure period in the pH 7.6 and 5.6 groups, but the fish in pH 4.5 failed to exhibit the surge. Thus, low pH can suppress vitellogenin formation, but whether calcium is involved may depend on the species, hardness of water, and pH.

Assuming there is a reduction in blood calcium associated with low pH water, the question may be asked as to what is the mechanism of this reduction? Limited data suggest that sublethal low pH alone does not affect the activity of Ca^{2+} ATPase in the gills so calcium uptake from the environment may not be a critical problem, rather it appears that increased calcium loss from the body may be more important (Parker et al., 1985; Parker and McKeown, 1987). A further mechanism that could be involved is the effect of acid on sex hormone levels in the female fish. Estradiol induces hypercalcemia in maturing female fish (Whitehead et al., 1978), but Weiner et al. (1986) reported no effect of pH 4.5, 5.0, or 5.5 on female sex hormones during the final 6 weeks of reproductive maturation in rainbow trout. Interestingly, in contrast to just about everybody else, they also saw no effect of pH on plasma calcium levels in female fish.

The above discussion suggests that egg production per se is quite sensitive to reduced pH. However, a recent study casts some doubt on that hypothesis. Mount (1988) worked with brook and rainbow trout in soft acid water and observed reduced fecundity, but this was due to decreased growth in the female. Egg production per unit body weight of the female was actually normal in those exposed to acidified water, but those fish experiencing stress, as reflected in loss of plasma sodium and decreased feeding, showed a severely reduced body size. Mount also made the important observation that progeny from brook trout that had been in very low calcium were more sensitive to acid than were those spawned in water with higher calcium levels. He concludes that recruitment failure in soft acid waters is not due to failure of egg production but is instead due to the progeny being excessively sensitive to subsequent acid.

This hypothesis gets some support from the work of Weiner et al. (1986) who found that exposure of rainbow trout females or males to acid waters (pH 4.5–5.5) during reproductive maturation causes the subsequent eggs and larvae to be more sensitive to low pH. This occurs even if only one of the parents was exposed to acidity (Weiner et al., 1986). It is tempting to speculate that this increased sensitivity of progeny might be due to reduced vitellogenin which could result in a decreased yolk deposition. However, since exposure of males only also produced increased sensitivity (although not as severe), other mechanisms are additionally possible.

Androgen levels in male salmon have been shown to be sensitive to low pH. Normally, these hormones undergo a considerable surge just before spawning. Examination of fish collected from two rivers (pH 5.6 and 4.7) revealed a reduction in this surge in those from the low pH river (Freeman and Sangalang, 1985). The cause of this could be due to a generalized stress response and reduced feeding in the acid-exposed fish, although direct effects on either gonadotropin release or sex hormone synthesis are not excluded (Freeman and Sangalang, 1985; Wendelaar Bonga and Balm, 1989).

Sperm motility and fertilization has also been shown to be sensitive to low pH. Daye and Glebe (1984) reported that duration of movement by Atlantic salmon sperm declined in a linear fashion with pH until about pH 4.5. Below this point, motility rapidly declined to zero. Fertilization success was unaffected by a pH above 5 but at pH of 4 it was zero. Even a very short exposure duration can have a marked effect on subsequent development. An acute exposure to pH 5 at the time of fertilization caused high mortality in early development in both salmonids and cyprinids (Gillet and Rombard, 1986).

III. EMBRYONIC DEVELOPMENT AND HATCHING

As embryonic development of fish eggs proceeds, sensitivity to acid changes. The most sensitive stage is early cleavage before organ development occurs (Peterson et al., 1982). At this time, the zona radiata (chorion) acts as a reservoir for hydrogen ions rather than a barrier allowing diffusion of these ions into the perivitelline fluid. Apparently, this thick membrane decreases in permeability to hydrogen ions later so as the embryo develops, resistance to acid waters increases by about 0.5 pH units (Daye and Garside, 1979). However, sublethal morphological effects on the developing embryos of Atlantic salmon occur at a pH as high as 5. These effects include abnormal development of various organs such as gills, brain, and heart (Peterson et al., 1982). At hatching there is a big increase in sensitivity to acidity which emphasizes the protective effect of the chorion (Siddens et al., 1984).

Newly shed salmon eggs take up water until they become turgid (water hardening). This process is inhibited when the pH is 4.5 or lower (Peterson and Martin-Robichaud, 1982). The developing embryo is surrounded by perivitelline fluid which is encased in the chorion. This fluid accounts for about 7% of the egg and contains macromolecules which facilitate water hardening. It also has some buffering capacity so it, along with the chorion, help protect the developing embryo from hydrogen ions (Eddy and Talbot, 1985).

When the embryo reaches the eyed stage it begins to accumulate sodium from the environment. This process is severely inhibited at a pH of 4 (higher pHs were not tested). Eddy and Talbot (1985) showed that uptake only is affected and there is no effect of the acid on sodium efflux from the egg. The perivitelline fluid acts as an "ion trap" binding sodium and other ions thus reducing efflux (Eddy and Talbot, 1985).

Studies of the direct effect of low pH on hatching are made more difficult by the fact that pH adversely affects development of the embryo so a delay or failure in hatching could be due to that. A key step in the hatching process is the softening of the egg capsule which is catalyzed by a hatching enzyme. This enzyme is apparently a protease and like all enzymes exhibits an optimum pH. Several studies (cited in Peterson et al., 1982) have shown this optimum to be above pH 7, yet the pH of the perivitelline fluid (in spite of its buffering capacity) is generally only 0.5 pH units above that of the ambient water (Peterson et al., 1980). Inhibition of chorionase activity at pH 4.5 delayed or prevented hatching (Waiwood and Haya, 1983). Problems with hatching at low pH may, in addition to chorionase enzyme inhibition, be due to reduced embryo activity because of the effects on development mentioned above (Peterson et al., 1980).

IV. LARVAE FROM HATCHING THROUGH SWIM-UP

The physiological process most affected in larvae by acid (or aluminum) is probably ionoregulation. During larval development in a wide variety of species, there normally occurs a marked uptake of sodium, potassium, and calcium from the water while at the same time body magnesium declines (Peterson et al., 1982; Wood et al., 1990a,b). The ionic uptake apparently occurs via chloride cells in the yolk-sac epithelium, gills, and skin (Hwang and Hirano, 1985). Wood et al. (1990a) demonstrated that brook trout fry exposed continuously from fertilization to swim-up to either acid water (pH 4–5.2) or acid water plus aluminum failed to take up ions at the normal rate. However, perhaps a more important finding of this study was that the environmental calcium concentration influenced the net electrolyte accumulation much more than either pH or aluminum. Decreased uptake of electrolytes could have been due to altered ion exchange rates or to reduced rate of development of the larvae, or both. Of course, as Wood et al. (1990a) point out, the rate of development might be limited by the net electrolyte accumulation. They conclude that both development and ion exchange rates are important with the latter dominating in the yolk-sac stage.

The importance of environmental calcium for electrolyte uptake in larvae is striking, and at pH 4.8, a calcium level of 8 mg/L provides almost complete protection from the acid (Wood et al., 1990a). This level of calcium is at the upper end of what is usually considered the soft water range which, unfortunately, is not generally found in areas where acid stress occurs.

Spawning brook trout have been shown to actively avoid low pH by selecting areas of alkaline upwellings from groundwater in which to build their redds (Johnson and Webster, 1977). Thus, the young hatchlings avoid the most severe acid conditions until emerging from the gravel. Because this emergence may occur at the time of snow melt, there has been some interest in the effect of acid and/or aluminum exposures at that time. When brook trout fry at the yolk-sac stage were exposed for 21 days to several different combinations of acid, aluminum, and calcium, Wood et al. (1990b) found that water pH was the primary determinant of whole-body ion status, rather than calcium as in the above-mentioned study. In part this is due to the fact that in the previous study, eggs and early larvae did not survive the lowest pHs so no electrolyte data were obtained on them. Nevertheless, to quote Wood et al. (1990b, p. 1612): "Fry surviving a continuous 91-day exposure from fertilization will maintain an ionic status most dependent upon the water Ca during exposure, while alevins surviving a 21-day challenge in the yolk-sac or swim-up phase will exhibit a status most dependent upon the pH of the challenge." They go on to emphasize that the latter is probably the more environmentally realistic as calcium is usually low in waters that are impacted by acid. Because growth was not influenced by pH, aluminum, or calcium in the yolk-sack and swim stages, the authors conclude that the mechanism of ionoregulatory failure is depressed active uptake and changes in permeability, rather than a reduced development.

A surprising finding of the Wood et al. (1990a,b) studies was the protective effect of aluminum at intermediate concentrations. Ionoregulatory failure was reduced at aluminum concentrations of 37 and 111 μg/L. Indeed, in the exposures at the later larval stages, it seemed to stimulate electrolyte uptake so that aluminum exposed fish had higher ion concentrations than controls. At aluminum levels higher and lower than these, the metal had a harmful effect on ionoregulation. The same laboratory had earlier shown a protective effect of aluminum in adults from acute acid exposure. The physiological basis of this phenomenon is unclear; long-term exposures of adult salmonids to aluminum causes proliferation of chloride cells in the gill epithelium (Tietge et al., 1988). If this occurred in the larvae, it would facilitate uptake of electrolytes, but it is not known if a similar adaptation to aluminum occurs in larvae, and the bimodal nature of this response to aluminum is perplexing.

Exposures to acid or aluminum do not necessarily need to be prolonged to result in significant effects on subsequent development in some species. When lake trout embryos were exposed for 5 days to aluminum (100, 200 μg/L) in water of pH 5, there were no mortalities, but alevins examined after 21 or 32 days in water lacking aluminum were smaller, had less calcified skeletons, and whole-body calcium and potassium concentrations were down by 18–22% (Gunn and Noakes, 1987). Evidently, the short exposures caused permanent damage to ionoregulatory mechanisms as well as to feeding ability and yolk utilization.

A number of studies (see De Lonay et al, 1993 for references) have documented acid effects on various aspects of larval behavior. The most consistent finding in both salmonids and centrarchids is a reduction of spontaneous swimming activity. This would probably contribute to reduced feeding and therefore growth. Furthermore, the ability to maintain position in a current was reduced in brook trout exposed to acid waters (Cleveland et al., 1986).

V. JUVENILE AND ADULT: BLOOD ACID-BASE BALANCE AND ELECTROLYTE CHANGES FROM ACUTE EXPOSURES

The arterial pH can be affected by ambient water pH, but the calcium concentration in the water very much controls this effect. At a low calcium level, lowering the water pH around rainbow trout to 4.3 (equal to the 7-day LC50 for this species) has no effect on arterial pH. On the other hand, a considerable acidosis occurs at higher calcium levels (see Figure 1). Even with a decrease in blood pH as occurs in high calcium acid water, there is little effect on intracellular pH of brain, heart, or skeletal muscle. Thus, the mechanism by which acid causes death at realistic pHs in trout is not acidosis (Wood, 1989).

Acid may act on carp (*Cyprinus carpio*) a bit differently than it does on trout. At a pH of 4 and an intermediate level of calcium in the water carp experienced a drop in arterial pH concomitant with electrolyte alterations. Ultsch et al. (1980) attributed this blood pH drop largely to a loss of base through the gills. It appears then that the impact of acidified water on blood pH differs with the species and may be more severe in warmwater species.

Numerous studies (beginning with Packer and Dunson, 1970) of fish exposed acutely to acid have shown a considerable drop in blood or whole-body electrolytes (see reviews by Wood, 1989; McDonald et al., 1989). Both influxes and effluxes of sodium and chloride through the gills are affected; but of the two, however, influx may be the more sensitive process. Influx is inhibited and efflux stimulated so the net effect is a massive loss of ions from the body. This becomes progressively worse at lower pHs so that at a pH of approximately 4 in trout influx is completely stopped and efflux is increased almost threefold over that of controls at pH 7. Wood (1989) concludes that Na^+ influx is inhibited by acid due to the competition of H^+ with Na^+ for transporter at the gill. The inhibition of chloride uptake is more obscure because it is not clear to what extent, if any, this process is linked to sodium transport. Normally, chloride in the water is exchanged for bicarbonate and hydroxyl ions in the blood so there may be a depletion of these in the gill cells of fish in acid conditions, or the chloride carrier may be damaged by the acidified water (Wood, 1989).

The large increases in outward electrolyte diffusion are believed due to a massive increase in ion permeability produced by the leaching of calcium away from the paracellular channels in the gill epithelium (McDonald et al., 1989). The tight junctions between pavement cells and chloride cells in the gills of rainbow trout are very sensitive to changes in pH (Freda et al., 1991). These changes can occur very rapidly (i.e., within minutes of exposure) and may involve only minor changes in permeability of the gills to water (Jackson and Fromm, 1980).

As with other physiological process affected by acid, water hardness (expressed as $CaCO_3$) is an extremely important variable. Figure 2 illustrates the effect that water hardness has on body sodium of white suckers (*Catostomus commersoni*) exposed to acid. Note that the exposure duration for the fish in pH 4 soft water was for only 1.5 days compared to 6–19 days for the fish in hard water. This was because in soft water at pH 4, there was 100% mortality within 48 h, but only 16% in hard water at the same pH after 19 days. The pumpkinseed (*Lepomis gibbosus*) is a more acid-tolerant species than is the white sucker. When it was exposed to pH 4 water, (Fraser and Harvey, 1984) it eventually lost the same amount of sodium as did the sucker, but the latter lost sodium at an 11-fold faster rate. These observations support the suggestion (McDonald, 1983) that lethality from acid water is determined more by the rate of sodium loss than the total amount.

Milligan and Wood (1982) have proposed that the etiology of death in acid water can be traced to the ionoregulatory failure discussed above, although the final cause is actually circulatory, rather than ionic. In this proposed sequence of events, (Figure 3), if the initial branchial ion loss causes a rapid drop in plasma Na^+ and/or Cl^- by more than

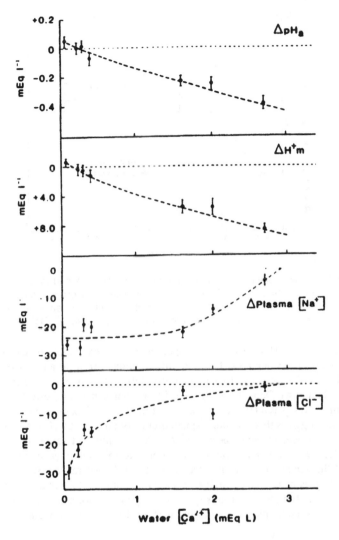

Figure 1 Relationship between concentration of calcium in the water and the extent of acid-base and ionic disturbances in the arterial blood of adult rainbow trout exposed for 3 days to pH 4.3. (From Wood, C., in, *Acid Toxicity and Aquatic Animals*, Morris, R., Taylor, E., Brown, D. and Brown, J., Eds., Cambridge University Press, New York, 1989, 125. With permission.)

30%, it triggers a reduction in plasma volume due to water loss to the environment and tissues. The increased hematocrit and plasma protein produce a rise in blood viscosity which in turn causes arterial blood pressure to rise. Hematocrits as high as 70% and massive increases in blood viscosity have been observed in fish exposed to acid (Wood and McDonald, 1982). Wood (1989) has suggested that catecholamines may facilitate these various changes, a point supported by the findings of Ye et al. (1991) in which they measured increases in catecholamines during acid exposure in rainbow trout. Wood (1989) further points out that the blood acidosis seen when fish in hard water are exposed to a low pH is probably secondary to the electrolyte loss as a cause of death. This mechanism seems to apply to all fish species and even amphibians (Freda and McDonald, 1988).

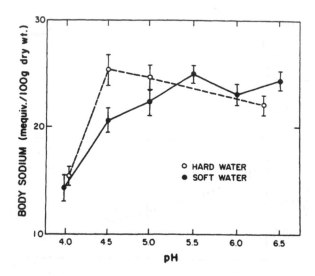

Figure 2 Whole-body sodium concentration in white sucker exposed to acidified soft water (Ca = 0.207 mE/L) and hard water (Ca = 2.11 mE/L. Exposures were 6–9 days except for pH 4 in soft water which was for 36 h. Points are means of 15–19 fish except for pH 4 in soft water where N = 4. (Redrawn from Fraser, G. A. and Harvey, H. H., *Can. J. Zool.*, 62, 249, 1984.)

Tolerance to low pH varies between species and at least some of the mechanisms involved have been revealed that facilitate this tolerance. In trout and shiners (*Notropis cornutus*), both rather sensitive to acid, calcium definitely moderates the rate of sodium loss, but Freda and McDonald (1988) have found that it has little effect on the loss of sodium from yellow perch (*Perca flavescens*), a rather acid-insensitive species. This and other evidence suggests that calcium displacement from the gill by H+ ions does not occur in the perch as it does in other species which are less able to tolerate acid waters.

The sunfish (*Enneacanthus obesus*) is rather acid tolerant, even more so than other members of the genus (Gonzalez and Dunson, 1989). Its tolerance rests in part with the fact that it has gills which have a high affinity for calcium so it is not displaced as readily by H+ ions. In addition, however, this species compensates for sodium losses by shifting this element from bone into the extracellular fluid. In addition, low ambient pH causes

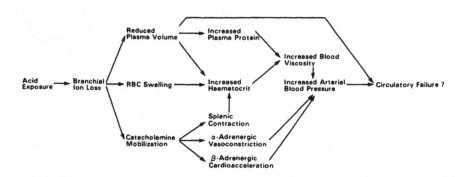

Figure 3 Proposed sequence of events through which environmental acid exposure may cause death of a fish. See text for discussion. (From Wood, C. M. and McDonald, D. G., *Acid Rain/Fisheries*, Johnson, R. E., Ed., American Fisheries Society, Bethesda, MD, 1982, 197. With permission.)

decreases in body water and potassium concentration which helps to reduce the net movement of water from the extracellular fluid compartment into the intracellular fluid.

Electrolyte losses via the kidney are small in soft acid water, probably because of fluid shifts into the muscle intracellular compartment and a subsequent rise in plasma oncotic pressure which causes urine volume to fall (Wood, 1989). In hard water, however, it should be recalled that a rapid influx of H^+ through the gills occurs. Under these circumstances, the kidney plays an important role in excreting the excess hydrogen ions. Thus, the response of fish to acid in hard water is one of the few examples in which the teleost kidney plays a significant roll in acid-base regulation (Heisler, 1989).

The above discussion regarding responses to acute exposures simulates conditions such as a rapid snow melt or heavy precipitation of low pH rain. A more gradually induced acidification may shift threshold pHs to a lower level. For example, Van Dijk et al. (1993) exposed carp to pH 4 water over 4 h and then monitored them for 48 h. Contrary to the findings of Ultsch et al. (1981) where severe losses of plasma ions and a decline in plasma pH occurred in carp with a rapid drop in water pH, they saw no changes, nor were there any changes in stress hormones. A somewhat similar finding was reported by Balm and Pottinger (1993) with rainbow trout which were gradually exposed to pH 4 waters. Thus, the kinetics of the acid exposure may have a big effect on the ultimate response. This may be a nice example of the need to consider the "psychological" aspects of a stress (Schreck, 1981). The sudden imposition of any sort of change in a fish's environment may induce a stress response out of proportion to the severity of the stressor. This is a point of which we physiologists need to be more cognizant.

VI. JUVENILE AND ADULT: BLOOD CHANGES FROM CHRONIC EXPOSURE

Many fish experience a chronic type of exposure that may last for many days or even months and years. Basically, there are two types of response that might occur with this prolonged exposure: (1) there may be a chronic depression of plasma electrolytes that persists more or less indefinitely or (2) over time in the acidified water there may be a partial or complete compensation indicating the fish were successful in adjusting to the acid water. Both types of response have been seen, but the character of the response depends on species and on the presence of other ions such as calcium or aluminum. In addition, a third possibility must be considered in that populations may become genetically selected for better tolerance of acid waters.

Juvenile Atlantic salmon were kept in soft water of pH 4.5–5.2 for one or three months and it was found that ionoregulation was impaired throughout the whole time (Lacroix et al., 1985). Moreover, the fish exhibited reduced growth and branchial ATPase activity and failed to undergo smoltification (Johnston et al., 1984). When adult rainbow trout were exposed to pH 4.8 in soft water for three months there was an initial net loss of sodium and chloride (Audet et al., 1988). By 30–50 days of continuous exposure to the low pH, a new equilibrium was reached in which influx and efflux rates were both lower than in controls and plasma levels of these ions were stabilized at an abnormally low, but not lethal level. Plasma glucose rose steadily during the exposure reaching a maximum of 20 mmol/L which is approximately seven times the control level. No acid-base disturbance was seen but ammonia excretion rate increased with time, which suggests an increased protein catabolism. The extreme hyperglycemia along with the increased ammonia excretion rate suggests the fish were under a considerable degree of stress (Audet et al., 1988). In a subsequent study using the same experimental regime (Audet and Wood, 1993), an elevated cortisol level was found throughout the exposure time but catecholamines remained unchanged. Prolonged sublethal acid stress apparently weakens the fish for, if they are subsequently exposed to a lethal pH level, they lose electrolytes more rapidly and are less tolerant of the acid than those that had not received a pre-exposure. Thus, rainbow trout do not acclimate to low pH alone (Audet and Wood, 1988).

Plasma potassium and calcium remained unchanged in the Audet et al. (1988) study. The lack of an effect on calcium is interesting in light of the debate mentioned in the beginning of this chapter concerning the relationship between a presumed decline in blood calcium and reproductive failure in female salmonids from acid lakes.

There has been recently published an apparent contradiction to the conclusion that rainbow trout cannot acclimate to acid. Balm and Pottinger (1993) exposed trout to pH 4 for 14 days in soft water and found an increased turnover rate of choride cells, little change in plasma electrolytes, no increased sensitivity to confinement stress, and no elevation in cortisol. In other words, their fish seemed to adjust to the acid fairly well, although they did not test their tolerance to a subsequent more acute acid stress. They were using two European strains of rainbow trout whereas the Audet and Wood work was done with Canadian fish. This may or may not explain the discrepancy in the results. Another difference is that Balm and Pottinger subjected their fish to a gradual lowering of the pH whereas Audet and Wood's fish experienced a more rapid decline.

The ability to acclimate to low pH is quite species dependent. African *Tilapia* showed fairly good recovery of osmolarity when exposed to pH 3.5 by day 15 of exposure but goldfish did not (Wendelaar Bonga et al., 1984). There are also populations within a trout species that are more resistant to acid waters (McWilliams, 1982), but since acclimation to acid by trout may not occur, this is probably due to genetic selection (Swaarts et al., 1978).

Adaptation to acid by salmonid fishes can certainly occur in the presence of hard water (see Audet et al., 1988 for references), a condition which is not common in areas receiving acid precipitation. Interestingly, aluminum is frequently associated with acid pollution and it seems to actually aid adaptation by fish to these conditions. Wood et al. (1988a) and Wilson and Wood (1992) have found that if sublethal concentrations of aluminum are combined with a chronic acid stress, the fish show an initial ionoregulatory dysfunction which may be greater than with acid alone, but then as time goes on, those with the combination of aluminum and low pH show partial or complete restoration of normal plasma sodium and chloride. Also, fish acclimated to acid + aluminum are able to better tolerate a subsequent challenge with a lower pH and high concentrations of aluminum (Wood et al., 1988b). Thus, trout in soft water can acclimate to aluminum but not acid alone. Indeed, Wood et al. (1988b) propose that low levels of aluminum may be beneficial to fish under chronic acid stress.

The mechanism of this beneficial effect of chronic aluminum exposure seems to rest with a proliferation of chloride cells (McDonald and Milligan, 1988) and reduced permeability of the branchial epithelium. This latter phenomenon may be due to epithelial swelling (Tietge et al., 1988) which increases the blood to water diffusion distance. Such a change would probably reduce the transfer capacity for oxygen which would limit exercise ability, a point taken up later.

Yellow perch (*P. flavescens*) is a species with high acid tolerance, but even within this species, there are populations that have diverged physiologically (Nelson and Mitchell, 1992). Environmental acidity may act synergistically with swimming to upset blood acid-base. Thus, it is interesting that fish from populations in acid lakes were able to maintain acid-base balance in their blood better than those from circumneutral lakes when exercised in acid waters (Nelson and Mitchell, 1992).

Exhaustive exercise increases the toxicity of acid waters to trout (i.e., increases the lethal pH) (Graham and Wood, 1981). This is not surprising as severe exercise depresses blood pH, cellular pH, and causes losses in blood ions via the gills; it can even cause death of fish in circumneutral pH waters (Wood et al., 1983). In addition, a fish actively swimming in acid waters brings much more water in contact with its gills than when at rest. When carp were exposed gradually to pH 4 water, they exhibited no alterations in blood electrolytes, even after 24 h. However, if they were forced to swim constantly at two body lengths per second, while being gradually subjected to the acid water, they exhibited a steady decline in plasma pH, Na^+, and Cl^-, so the increased gill water flow exacerbated the effect of acid (Van Dijk et al., 1993).

Warmwater species of fish frequently experience a greater seasonal temperature change than do salmonids. This can have a serious implication if they are also exposed to low pH conditions. When young largemouth bass (*Micropterus salmoides*) were exposed to acid waters along with a temperature simulating overwintering (3.8°C), they lost blood osmolality more than did those exposed to the acid at a warmer temperature. McCormick and Jensen (1992) propose that the first winter for juveniles may be the most critical stage in the life cycle in northern populations which experience both severe winter and acid conditions.

Before leaving this section, brief consideration should be given to what happens to fish when exposed to alkaline conditions. The critical alkaline pH for salmonids appears to be between 9 and 10. Acute exposure to these conditions causes an inhibition of ammonia excretion by the gills and respiratory alkalosis (i.e., elevation of plasma pH). The inhibition of ammonia excretion is "...probably due to a reduction of the diffusion gradient for NH_3 across the gills when the environmental pH approaches the negative logarithm of the dissociation constant for ammonia" (Wilkie and Wood, 1991, p. 1070). When rainbow trout were exposed to pH 9.5 for 72 h, they initially showed alkalosis but blood pH was eventually stabilized around 8, in part by accumulation of lactate even though blood oxygen was not limiting. Perhaps more interesting was the observation that nitrogen excretion shifted toward formation of urea, but even so, ultimately, plasma ammonia stabilized at a level about six times that of controls (Wilkie and Wood, 1991). Changes in plasma sodium and chloride were comparatively minor with alkaline exposure. Wilkie and Wood (1991) point out that while the rainbow trout seems to be able to adapt to pH 9.5, it is not without cost so the fish may become more susceptible to death from other causes.

There is a strain of cutthroat trout (*O. clarki henshawi*) that thrives in Pyramid Lake, NV, at pH 9.4. Apparently these fish have a lower rate of ammonia excretion, a higher proportion of nitrogen excretion via urea, a higher rate of renal excretion, a lack of coupling between ammonia excretion and sodium influx, and a high plasma ammonia level which facilitates diffusive excretion of ammonia across the gills (Wright et al., 1993). When these fish were challenged with water of pH 10, they exhibited a doubling of plasma ammonia concentration and urea excretion and a decreased level of plasma sodium chloride. Survival was poor during 72-h exposures and those that did not survive showed highly elevated plasma ammonia and severely depressed sodium and chloride (Wilke et al., 1993). Thus, these fish which are living at pH 9.4 are existing at near their limit which may be exceeded in the future as Pyramid Lake becomes more desiccated. There is an African species of *Tilapia* living in a pH of around 10 and is able to do this in part because it excretes its nitrogen primarily as urea (Randall and Wright, 1989).

VII. HORMONAL RESPONSES

Cortisol secretion rate increases rapidly with acid exposure (Wendelaar Bonga and Balm, 1989), as it does with most environmental stressors. In *Tilapia*, the cortisol level returns to normal in a few days even though the fish is still in the acid water. This is not due to a decrease in secretion rate of the interrenal cells, however. Instead, there is an elevation in peripheral cortisol clearance rate which could be due to metabolism of the hormone in the liver or excretion by the kidney. Such an enhanced rate of peripheral clearance of cortisol was not found in brook trout exposed to high acidity. Instead, Wood et al. (1988b) found high cortisol levels even after 10 weeks of continuous exposure in this species.

The elevated cortisol levels seen in acid-exposed fish are probably in response to the ionoregulatory dysfunction since cortisol stimulates chloride cell proliferation, and Na,K ATPase activity in the gills (Wendelaar Bonga and Balm, 1989). Furthermore, the cortisol elevates blood glucose which reduces the osmoregulatory load produced by the loss of sodium and chloride ions.

Two hormones from the anterior pituitary have been shown to be sensitive to acid stress: prolactin and arginine vasotocin. Prolactin rises in fish experiencing acid exposure and it can remain elevated for months (Wendelaar Bonga and Balm, 1989). This hormone is important in freshwater fish where it reduces ion and water permeability of epidermis and gill epithelium (Bern, 1983).

Arginine vasotocin may or may not be involved in osmoregulation (Evans, 1993). The concentration of this hormone is generally measured in the pituitary rather than the blood because of extremely low concentrations in blood. Hontela et al. (1991) showed that this hormone rose in concentration in brook trout whenever acid exposure was sufficient to lower plasma sodium levels. The same laboratory (Hontela et al., 1993) went on to test this hormone as a biomarker for acid stress in feral fish. It appears to have some promise but they had to develop ANCOVA models to adjust for body size and season, which have a considerable effect on the pituitary levels of this hormone.

Hormones from the caudal neurosecretory system are also stimulated by acid waters (Hontela et al., 1989). These hormones along with cortisol, prolactin, and possibly vasotocin are all involved in regulating blood ion concentrations so their increase in fish experiencing osmoregulatory dysfunction is probably an attempt to restore homeostasis.

Thyroid activity has also been shown to be sensitive to acid exposure. In general, a decrease in the blood level of thyroid hormones is seen and they may or may not show recovery when the fish is transferred to circumneutral pH waters (Brown et al., 1990). The mechanism by which acid waters depress thyroid activity is currently unknown.

Mention has already been made at several points about catecholamines. Changes in the level of these hormones are probably very much influenced by the degree of stress the fish is psychologically perceiving. Insofar as the regulation of plasma electrolytes is concerned, catecholamines may not necessarily be considered a homeostatic group of hormones because they increase gill blood perfusion and increase the area of the gill lamellae in direct contact with the water (Wendelaar Bonga, 1993). Thus, catecholamines could actually do more harm than good for fish in acid waters, although this hypothesis needs direct confirmation.

VIII. VENTILATION AND BLOOD GASES

Measuring ventilation in fish, especially salmonids, requires an assessment of both breathing frequency and depth (i.e., stroke volume), for the latter may change while the former does not. Better still is to measure the actual ventilation volume but this requires a two- (or three-) compartment chamber in which the fish can be restrained with a membrane around the head separating the mouth from the postopercular waters. For example, Janssen and Randall (1975) found that rainbow trout in low pH increased their ventilation volume by increases in ventilatory stroke volume with little change in breathing frequency. The response to the low pH was not immediate, but instead developed gradually which may have been due to a production of mucus on the gills which reduced oxygen transfer. Neville (1979) found that ventilation increased with mild acid exposure in this species only if there was a rise in external PCO_2 and Janssen and Randall (1975) found that this gas was a more powerful stimulant to ventilation than was acid alone.

The observation that CO_2 is such a strong stimulant is important in that addition of H^+ to water can cause the PCO_2 to rise due to the reaction $H^+ + HCO3^- > H_2CO_3 > CO_2 + H_2O$. Thus the increases in ventilation often observed with elevated acidity may in fact be due to the elevated CO_2 (Giles, 1984; Neville, 1979; Walker et al., 1988).

Arterial Po_2 seems to be relatively unaffected by low but sublethal pH levels in soft water. Indeed, it may actually increase with acid exposure due to the carbon dioxide-induced hyperventilation just discussed (Walker, 1988). If the water pH is low enough, however, mucus formation on the gill surface may restrict oxygen diffusion and cause a decrease in arterial Po_2 (see Chapter 3). The threshold for this in carp is between pH 3.5 and 4.0 (Ultsch et al., 1980). It is not clear where that threshold is for salmonids, but in soft water it must be above pH 4.8 (Walker et al., 1988). In hard water, it is probably higher than in soft water as alterations in gas exchange seem to be greater in hard water at a given pH (Wood et al., 1988b). Also, the addition of aluminum at 300 µg/L causes a marked drop in arterial Po_2 of brook trout in pH 4.8 (Wood et al., 1988b).

Increases in arterial PCO_2 cause a drop in HBO_2 due to the well-known Bohr effect. The pH of red blood cells does not necessarily fall as much as that of the plasma, however. This is due to compensatory Na/H exchanges through the erythrocyte membrane which are facilitated by catecholamines (Nikinmaa, 1983). In spite of this, in acid-exposed rainbow trout, the Bohr effect is not completely eliminated (Ye et al., 1991). Thus, a drop in actual concentration of oxygen in the blood occurs and this, rather than changes in Po_2 or pH, may be the major stimulus for increased ventilation in salmonid fishes during acid exposure (Randall, 1982).

IX. OXYGEN CONSUMPTION, SWIMMING PERFORMANCE, AND SWIM BLADDER INFLATION

The rate of standard oxygen consumption is generally not influenced much by changes in pH of the water, except at very low levels (pH <4.0) where it begins to decline (Ultsch, 1978; Ultsch et al., 1980; Ye et al., 1991; Van Dijk et al., 1993). However, the critical oxygen tension (i.e., that level of oxygen in the water below which oxygen consumption becomes dependent on oxygen tension) is affected by water pH in both carp and trout. Ultsch et al. (1980) found that at lower pHs, the critical oxygen tension increased, thus the fish were less able to deal physiologically with environmental hypoxia if the pH was also low.

Critical swimming speed of trout appears to not be influenced much by pH over a range of 6 to 8 (Waiwood and Beamish, 1978). However, 7-day exposures of rainbow trout to various levels of pH revealed a threshold of 4.4–4.6 below which critical swimming speed declined linearly by about 4% per 0.1 pH unit (Graham and Wood, 1981). A somewhat similar finding was reported for arctic charr where the effect of low pH was somewhat greater but the threshold was the same (Hunter and Scherer, 1988). The above studies involved fish that had been exposed for a week so some adaptation to the acid may have occurred. When trout were exposed for only 24 h to water of pH 4 and 5, a 45 and 33% decline, respectively, in critical swimming velocity was observed (Ye and Randall, 1991). It is interesting that in the same study, exposure to a pH of 10, produced a 39% impairment in swimming capacity. The effects on swimming could be due to a reduction in oxygen content of the blood caused by the plasma pH decline and resulting Bohr shift, as swimming speed is very dependent on blood oxygen content (Jones, 1971). This hypothesis sounds attractive, but arterial pH is not very sensitive to the water pHs used in these studies. Also, as was mentioned above, catecholamines help maintain intracellular pH of the erythrocytes, so altered oxygen content of the blood is probably not the primary mechanism of decreased swimming capacity, except perhaps at a very low pH.

While oxygen content of the blood may not change too much with minor changes in pH, oxygen transport could. This is because the blood may become more viscous due to hemoconcentration brought on by the acid and the blood circulation to the muscles could then be impaired (Butler et al., 1992). A further mechanism for impaired swimming capacity has been recently revealed by Butler and Day (1994). Using electromyography and biochemical measurements of lactate and glycogen, they found that brown trout exposed to pH 4 failed to recruit white muscle fibers as swimming speed increased. Because most of the muscle bulk of a fish is white muscle, this could have a considerable effect on maximum swimming speed.

Acid water can also affect recovery from exhaustion. Fish that had been fatigued were able to restore blood acid-base balance within 2 h of rest if they were in neutral or alkaline waters, but not if the water was acidic (Ye and Randall, 1991).

Swim bladder function in fish may be especially sensitive to depressions in pH, however, only limited information is available on this. When fathead minnows (*Pimephales promelas*) experienced a 4-day exposure to pH 5.3 water they had difficulty in adjusting buoyancy with the swim bladder (Jansen and Gee, 1988). This dysfunction persisted for at least 32 days and the authors suggest it could contribute to the elimination of fathead minnows (and other species?) from acidified environments. Overall, this would appear to be an interesting subject to pursue in other species, and if confirmed, to examine possible mechanisms.

X. BEHAVIOR

Laboratory experiments using pH gradients have shown that a wide variety of fish species can detect and then avoid water of a particular level of acidity (assuming that less acid water is available). Peterson et al. (1989) report that the threshold for avoidance ranges from pH 4.1 for yellow perch, through 4.7 for brook charr, to 5.9 for rainbow trout. The threshold avoidances correspond fairly well to the limiting pH suggested by field surveys. Of course, as the authors point out, the presence of aluminum in field situations may be more important than the pH per se. Avoidance of alkaline conditions seems to be slight although waters above pH 10 were not tested.

When fish experience acid stress, they are usually less active and tend to lose their appetite (Jones et al., 1987). At an intermediate pH of 5, Arctic charr were found to look normal as far as activity was concerned but were less responsive to an extract of food (Jones et al., 1985). A pH of 4.5 caused a considerable effect on behavior (decreased) and an increase in thigmotaxis (Jones et al., 1987). When returned to control conditions following 16 days in the acid treatment, behavior returned to normal within 2 weeks. Some may have recovered sooner, so effects were not permanent.

Chemoreception is an important process in feeding behavior for many fish species. Lemly and Smith (1987) devised a behavior assay to test the effect of low ambient pH on chemoreception by fathead minnows. They observed a sharp threshold of pH 6. At that pH and lower, there was a complete cessation of response to a pulsed liquid food stimulus. The olfactory sensory organs were examined with scanning electron microscopy but no lesions were observed. It is interesting that this seems to be a rather high pH for blockage of feeding response but it corresponds to the level where fathead minnows become eliminated from natural waters.

Finally, mild acid exposure may even affect the ability of fish to learn a maze. Goldfish were exposed to pH 5.2 for 11 days and then, using operant conditioning, taught a maze.

Those exposed to the low pH were significantly slower to learn the maze than were controls (Garg and Garg, 1992).

XI. CONCLUDING COMMENT

The amount of information we now have on the effects of acid on fishes exceeds that for any other single pollutant. Acid stress clearly affects a multitude of functions in fishes, but it is interesting that many, if not most, of these dysfunctions can be traced to an initial effect on ionoregulation at the gills wherein the fish loses sodium, chloride, and to a lesser extent calcium more rapidly than it can be recovered. This same etiology appears to hold for both acute as well as more chronic types of exposure and for different stages of development. The major difference being that with chronic exposures, the fish may achieve a different equilibrium level, while under a severe acute exposure they die of circulatory failure.

REFERENCES

Audet, C. and Wood, C. M., Do rainbow trout (*Salmo gairdneri*) acclimate to low pH?, *Can. J. Fish. Aquat. Sci.*, 45, 1399, 1988.

Audet, C. and Wood, C. M., Branchial morphological and endocrine responses of rainbow trout (*Oncorhynchus mykiss*) to a long-term sublethal acid exposure in which acclimation did not occur, *Can. J. Aquat. Sci.*, 50, 198, 1993.

Audet, C., Munger, R. and Wood, C., Long-term sublethal acid exposure in rainbow trout (*Salmo gairdneri*) in soft water: effects on ion exchanges and blood chemistry, *Can. J. Fish. Aquat. Sci.*, 45, 1387, 1988.

Baker, J. P., Effects on fish of metals associated with acidification, in, *Acid Rain/Fisheries*, Johnson, R. E., Ed., American Fisheries Society, Bethesda, MD, 1982, 165.

Balm, P. H. M. and Pottinger, T. G., Acclimation of rainbow trout (*Oncorhynchus mykiss*) to low environmental pH does not involve an activation of the pituitary-interrenal axis, but evokes adjustments in branchial ultrastructure, *Can. J. Fish. Aquat. Sci.*, 50, 2532, 1993.

Beamish, R. J., Lockhart, W. L., Van Loon, J. C. and Harvey, H. H., Long-term acidification of a lake and resulting effects on fishes, *Ambio*, 4, 98, 1975.

Bern, H., Functional evolution of prolactin and growth hormone in lower vertebrates, *Am. Zool.*, 23, 663, 1983.

Brown, S., Evans, R., Majewski, H., Sangalang, G. and Klaverkamp, J., Responses of plasma electrolytes, thyroid hormones, and gill histology in Atlantic salmon (*Salmo salar*) to acid and limed river waters, *Can. J. Fish. Aquat. Sci.*, 47, 2431, 1990.

Butler, P. J. and Day, N., Acid water and white muscle recruitment in the brown trout (*Salmo trutta*), in, *Proceedings of High Performance Fish Symposium*, Mackinlay, D. D., Ed., Fish Physiology Association, Vancouver, 1994, 312.

Butler, P. J., Day, N. and Namba, K., Interactive effects of seasonal temperature and low pH on resting oxygen uptake and swimming performance of adult brown trout, *Salmo trutta*, *J. Exp. Biol.*, 165, 195, 1992.

Cleveland, L., Little, E. E., Hamilton, S., Buckler, D. R. and Hunn, J. B., Interactive toxicity of aluminum and acidity to early life stages of brook trout, *Trans. Am. Fish. Soc.*, 115, 610, 1986.

Daye, P. G. and Garside, E. T., Development and survival of embryos and alevins of the Atlantic salmon, *Salmo salar*, continuously exposed to acidic levels of pH from fertilization, *Can. J. Zool.*, 57, 1713, 1979.

Daye, P. G. and Glebe, P. D., Fertilization success and sperm motility of Atlantic salmon (*Salmo salar*) in acidified water, *Aquaculture*, 43, 307, 1984.

Delonay, A., Little, E., Woodward, D., Brumbaugh, W., Farag, A. and Rabeni, C., Sensitivity of early life stage golden trout to low pH and elevated aluminum, *Environ. Toxicol. Chem.*, 12, 1223, 1993.

Eddy, F. B. and Talbot, C., Sodium balance in eggs and dechorionated embryos of the Atlantic salmon *Salmo salar*, exposed to zinc, aluminum and acid waters, *Comp. Biochem. Physiol.*, 81C, 259, 1985.

Evans, D. H., Osmotic and ionic regulation, in, *The Physiology of Fishes,* Evans, D. H., Ed., CRC Press, Boca Raton, FL, 1993, chap. 11.

Fraser, G. A. and Harvey, H. H., Effects of environmental pH on the ionic composition of the white sucker (*Catastomus commersoni*) and pumpkinseed (*Lepomis gibbosus*), *Can. J. Zool.,* 62, 249, 1984.

Freda, J. and McDonald, D., Physiological correlates of interspecific variation in acid tolerance in fish, *J. Exp. Biol.,* 136, 243, 1988.

Freda, J., Sanchez, D. and Bergman, H., Shortening of branchial tight junctions in acid-exposed rainbow trout, *Oncorhynchus mykiss, Can. J. Fish. Aquat. Sci.,* 48, 2028, 1991.

Freeman, H. C. and Sangalang, G. B., The effects of an acidic river, caused by acidic rain, on weight gain, steroidogenesis, and reproduction in the Atlantic salmon (*Salmo salar*), in, *Aquatic Toxicology and Hazard Assessment: Eighth Symposium,* ASTM STP 891, Bahner, R. C. and Jansen, D. J., Eds., American Society for Testing and Materials, Philadelphia, 1985, 339.

Fromm, P. O., A review of some physiological and toxicological responses of freshwater fish to acid stress, *Environ. Biol. Fish,* 5, 79, 1980.

Garg, R. and Garg, A., Operant learning of goldfish exposed to pH depression in water, *J. Environ. Biol.,* 13, 1. 1992.

Giles, M., Majewski, H. and Hobden, B., Osmoregulatory and hematological responses of rainbow trout (*Salmo gairdneri*) to extended environmental acidification, *Can. J. Fish. Aquat. Sci.,* 41, 1686, 1984.

Gillet, C. and Rombard, P., Prehatching embryo survival of nine freshwater fish eggs after a pH shock during fertilization or early stages of development, *Reproduct., Nutr., Develop.,* 26, 1319, 1986.

Gonzalez, R. and Dunson, W., Mechanisms for tolerance of sodium loss during exposure to low pH of the acid-tolerant sunfish *Enneacanthus obesus, Physiol. Zool.,* 62, 1219, 1989.

Graham, M. S. and Wood, C. M., Toxicity of environmental acid to the rainbow trout: interactions of water harness, acid type, and exercise, *Can. J. Zool.,* 59, 1518, 1981.

Gunn, J. and Noakes, D., Latent effects of pulse exposure to aluminum and low pH on size, ionic composition and feeding efficiency of lake trout (*Savlelinus namaycush*) alevins, *Can. J. Fish. Aquat. Sci.,* 44, 1418, 1987.

Heisler, N., Acid-base regulation in fishes: mechanisms, in, *Acid Toxicity and Aquatic Animals,* Morris, R., Taylor, E. W., Brown, D. J. A., and Brown, J. A., Eds., Cambridge University Press, New York, 1989, 85.

Henriksen, A., Susceptibility of surface waters to acidification, in, *Acid Rain/Fisheries,* Johnson, R. E., Ed., American Fisheries Society, Bethesda, MD, 1982, 103.

Hontela, A., Roy, Y., VanCoillie, R., Lederis, K. and Chevalier, G., Differential effects of low pH and aluminum on the caudal neurosecretory system of the brook trout, *Salvelinus fontinalis, J. Fish Biol.,* 35, 265, 1989.

Hontela, A., Rasmussen, J., Ko, D., Lederis, K. and Chevalier, G., Arginine vasotocin, an osmoregulatory hormone, as a potential indicator of acid stress in fish, *Can. J. Fish. Aquat. Sci.,* 48, 238, 1991.

Hontela, A., Rasmussen, J. B., Lederis, K., Tra, H. and Chevalier, G., Elevated levels of arginine vasotocin in the brain of brook trout (*Salvelinus fontinalis*) from acid lakes: a field test of a potential biomarker for acid stress, *Can. J. Fish. Aquat. Sci.,* 50, 1717, 1993.

Howells, G. D., Fishery decline: mechanisms and predictions, *Philos. Trans. R. Soc. London,* B, 305, 529, 1984.

Hunter, L. A. and Scherer, E., Impaired swimming performance of acid-exposed Arctic charr, *Salvelinus alpinus, Water Pollut. Res. J. Can.,* 23, 301, 1988.

Hwang, P. P. and Hirano, P., Effects of environmental salinity on intercellular organization and junctional structure of chloride cells in early stages of teleost development, *J. Exp. Zool.,* 236, 115, 1985.

Jackson, W. F. and Fromm, P. O., Effect of acute acid stress on isolated perfused gills of rainbow trout, *Comp. Biochem. Physiol.,* 67C, 141, 1980.

Jansen, W. A. and Gee, J. H., Effects of water acidity on swimbladder function and swimming in the fathead minnow *Pimephales promelas, Can. J. Fish. Aquat. Sci.,* 45, 65, 1988.

Janssen, R. G. and Randall, D. J., The effect of changes in pH and PCO_2 in blood and water on breathing in rainbow trout, *Salmo gairdneri, Respir. Physiol.,* 25, 235, 1977.

Johnson, D. W. and Webster, D. A., Avoidance of low pH in selection of spawning sites by brook trout (*Salvelinus fontinalis*), *J. Fish. Res. Bd. Can.,* 34, 2215, 1977.

Johnston, C., Saunders, R., Henderson, E., Harman, P. and Davidson, K., Chronic effects of low pH on some physiological aspects of smoltification in Atlantic salmon (*Salmo salar*), *Can. Tech. Rep. Fish. Aquat. Sci.,* 1294, 77, 1984. (as cited in Wood 1989).

Jones, D., The effects of hypoxia and anaemia on the swimming performance of rainbow trout (*Salmo gairdneri*), *J. Exp. Biol.,* 55, 541, 1971.

Jones, K. A., Hara, T. J. and Scherer, E., Behavioral modifications in arctic char (*Salvelinus alpinus*) chronically exposed to sublethal pH, *Physiol. Zool.,* 58, 400, 1985.

Jones, K. A., Brown, S. B. and Hara, T. J., Behavioral and biochemical studies of onset and recovery from acid stress in arctic char (*Salvelinus alpinus*), *Can. J. Fish. Aquat. Sci.,* 44, 373, 1987.

Lacroix, G., Gordon, D. and Johnston, D., Effects of low environmental pH on the survival, growth, and ionic composition of postemergent Atlantic salmon (*Salmo salar*), *Can. J. Fish. Aquat. Sci.,* 42, 768, 1985.

Lemly, A. D. and Smith, R. J. F., Effects of chronic exposure to acidified water on chemoreception of feeding stimuli in fathead minnows (*Pimephales promelas*): mechanisms and ecological implications, *Environ. Toxicol. Chem.,* 6, 225, 1987.

Lockhart, W. L. and Lutz, A., Preliminary biochemical observations on fishes inhabiting an acidified lake in Ontario, Canada, *Water, Air, Soil Pollut.,* 7, 317, 1977.

Mason, J., Introduction, the causes and consequences of surface water acidification, in, *Acid Toxicity and Aquatic Animals,* Morris, R., Taylor, E., Brown, D. and Brown, J., Eds., Cambridge University Press, New York, 1989, 1.

McDonald, D. G., The interaction of environmental calcium and low pH on the physiology of the rainbow trout, *Salmo gairdneri.* I. Branchial and renal net ion and H^+ fluxes, *J. Exp. Biol.,* 102, 123, 1983.

McDonald, D. G. and Milligan, C. L., Sodium transport in the brook trout, *Salvelinus fontinalis:* effects of prolonged exposure to low pH in the presence and absence of aluminum, *Can. J. Fish. Aquat. Sci.,* 45, 1606, 1988.

McDonald, D. G., Reader, J. P. and Dalziel, T., The combined effects of pH and trace metals on fish ionoregulation, in, *Acid Toxicity and Aquatic Animals,* Morris, R., Taylor, E., Brown, D. and Brown, J., Eds., Cambridge University Press, New York, 1989, 221.

McCormick, J. H. and Jensen, K., Osmoregulatory failure and death of first-year largemouth bass (*Micropterus salmoides*) exposed to low pH and elevated aluminum, at low temperature in soft water, *Can. J. Fish. Aquat. Sci.,* 49, 1189, 1992.

McWilliams, P. G., A comparison of physiological characteristics in normal and acid exposed populations of the brown trout, *Salmo trutta, Comp. Biochem. Physiol.,* 72A, 515, 1982.

Milligan, C. L. and Wood, C. M., Disturbances in haematology, fluid volume distribution and circulatory function associated with low environmental pH in the rainbow trout, *Salmo gairdneri, J. Exp. Biol.,* 99, 397, 1982.

Mount, D. R., Physiological and toxicological effects of long-term exposure to acid, aluminum, and low calcium on adult brook trout (Salvelinus fontinalis) and rainbow trout (*Salmo gairdneri*), Ph.D. dissertation, University of Wyoming, p. 184, 1988.

Munkittrick, K. R., Calcium-associated reproductive problems of fish in acidified environments: evolution from hypothesis to scientific fact, *Environ. Toxicol. Chem.,* 10, 977, 1991.

Nelson, J. and Mitchell, G., Blood chemistry response to acid exposure in yellow perch (*Perca flavescens*): comparison of populations from naturally acidic and neutral environments, *Physiol. Zool.,* 65, 493, 1992.

Neville, C. M., Ventilatory response of rainbow trout (*Salmo gairdneri*) to increased H^+ ion concentration in blood and water, *Comp. Biochem. Physiol.,* 63A, 373, 1979.

Nikinmaa, M., Adrenergic regulation of haemoglobin oxygen affinity in rainbow trout red cells, *J. Comp. Physiol.,* 152, 67, 1983.

Norton, S. A., The effects of acidification on the chemistry of ground and surface waters, in *Acid Rain/ Fisheries,* Johnson, R. E., Ed., American Fisheries Society, Bethesda, MD, 1982, 93.

Packer, R. K. and Dunson, W. A., Effects of low environmental pH on blood pH and sodium balance of brook trout, *J. Exp. Zool.,* 174, 65, 1970.

Parker, D. B., McKeown, B. A. and MacDonald, J. S., The effect of pH and/or calcium-enriched freshwater on gill Ca-ATPase activity and osmotic water inflow in rainbow trout (*Salmo gairdneri*), *Comp. Biochem. Physiol.,* 81A, 149, 1985.

Parker, D. B. and McKeown, B. A., Effects of pH and/or calcium-enriched freshwater on plasma levels of vitellogenin and Ca^{2+} and on bone calcium content during exogenous vitellogenesis in rainbow trout (*Salmo gairdneri*), *Comp. Biochem. Physiol.,* 87A, 267, 1987.

Peterson, R. H., Daye, P. G. and Metcalf, J. L., Inhibition of Atlantic salmon hatching at low pH, *Can. J. Fish. Aquat. Sci.,* 37, 770, 1980.

Peterson , R. H., Daye, P. G, Lacroix, G. and Garside, E., Reproduction in fish experiencing acid and metal stress, in *Acid Rain/Fisheries,* Johnson, R. E., Ed., American Fisheries Society, Bethesda, MD, 1982, 176.

Peterson, R. H. and Martin-Robichaud, D. J., Water uptake by Atlantic salmon ova as affected by low pH, *Trans. Am. Fish. Soc.,* 111, 772, 1982.

Peterson, R. H., Coombs, K., Power, J. and Paim, U., Responses of several fish species to pH gradients, *Can. J. Zool.,* 67, 1566, 1989.

Randall, D., The control of respiration and circulation in fish during exercise and hypoxia, *J. Exp. Biol.,* 100, 275, 1982.

Randall, D. and Wright, P., The interaction between carbon dioxide and ammonia excretion and water pH in fish, *Can. J. Zool.,* 67, 2936, 1989.

Roy, R. L., Ruby, S., Idler, D. and So, Y., Plasma vitellogenin levels in pre-spawning rainbow trout, *Oncorhynchus mykiss,* during acid exposure, *Arch. Environ. Contam. Toxicol.,* 19, 803, 1990.

Schreck, C., Stress and compensation in teleostean fishes: response to social and physical factors, in, *Stress and Fish,* Pickering, A. D., Ed., Academic Press, New York, 1981, chap. 13.

Siddens, L. K., Siem, W. K., Curtis, L. R. and Chapman, G. A., Toxicity of environmental acidity on various life stages of brook trout, *Proc. West. Pharmacol. Soc.,* 27, 265, 1984.

Swaarts, F. A., Dunson, W. A. and Wright, J., Genetic and environmental factors involved in increased resistance of brook trout to sulfuric acid solutions and mine acid polluted waters, *Trans. Am. Fish. Soc.,* 107, 651, 1978.

Tam, W. H. and Payson, P. D., Effects of chronic exposure to sublethal pH on growth, egg production, and ovulation in brook trout, *Salvelinus fontinalis, Can. J. Fish. Aquat. Sci.,* 43, 275, 1986.

Tietge, J., Johnson, R. D. and Bergman, H., Morphometric changes in gill secondary lamellae of brook trout (*Salvelinus fontinalis*) after long-term exposure to acid and aluminum, *Can. J. Fish. Aquat. Sci.,* 45, 1643, 1988.

Ultsch, G., Oxygen consumption as a function of pH in three species of freshwater fishes, *Copeia,* 1978, 272, 1978.

Ultsch, G., Ott, M. and Heisler, N., Acid-base and electrolyte status in carp (*Cyprinus carpio*) exposed to low environmental pH, *J. Exp. Biol.,* 93, 65, 1980.

Ultsch, G., Ott, M. and Heisler, N., Standard metabolic rate, critical oxygen tension, and aerobic scope for spontaneous activity of trout (*Salmo gairdneri*) and carp (*Cyprinus carpio*) in acidified water, *Comp. Biochem. Physiol.,* 67A, 329, 1981.

Van Dijk, P., van den Thillart, G., Balm, P. and Wendelaar Bonga, S., The influence of gradual water acidification on the acid/base status and plasma hormone levels in carp, *J. Fish Biol.,* 42, 661, 1993.

Van Dijk, P., van den Thillart, G. and Wendelaar Bonga, S., Is there a synergistic effect between steady-state exercise and water acidification in carp?, *J. Fish Biol.,* 42, 673, 1993.

Waiwood, B. A. and Haya, K., Levels of chorionase activity during embryonic development of *Salmo salar* under acid conditions, *Bull. Environ. Contam. Toxicol.,* 30, 511, 1983.

Waiwood, K. and Beamish, F. W. H., Effects of copper, pH and hardness on the critical swimming performance of rainbow trout (*Salmo gairdneri*), *Water Res.,* 12, 285, 1978.

Walker, R. L., Wood, C. M. and Bergman, H. L., Effects of low pH and aluminum on ventilation in the brook trout (*Salvelinus fontinalis*), *Can. J. Fish. Aquat. Sci.,* 45, 1614, 1988.

Weiner, G. S., Schreck, C. B. and Li, H. W., Effects of low pH on reproduction of rainbow trout, *Trans. Am. Fish. Soc.,* 115, 75, 1986.

Wendelaar Bonga, S. E., Van der Meij, J. C. A., Van der Krabben, W. A. and Flik, G., The effect of water acidification on prolactin cells and pars intermedia PAS positive cells in the teleost fish *Orechromis* (formerly *Sarotherodon*) *mossambicus and Carassius auratus, Cell Tissue Res.,* 238, 601, 1984.

Wendelaar Bonga, S. E. and Balm, P. L. M., Endocrine responses to acid stress in fish, in, *Acid Toxicity and Aquatic Animals,* Morris, R., Taylor, E., Brown, D. and Brown, J., Eds., Cambridge University Press, New York, 1989, 243.

Wendelaar Bonga, S. E., Endocrinology, in *The Physiology of Fishes,* Evans, D. H., Ed., CRC Press, Boca Raton, FL, 1993, chap. 15.

Whitehead, C., Bromage, N. and Forster, J., Seasonal changes in reproductive function of the rainbow trout (*Salmo gairdneri*), *J. Fish Biol.,* 12, 601, 1978.

Wilkie, M., Wright, P. A., Iwama, G. K. and Wood, C. M., The physiological responses of the Lahontan cutthroat trout (*Oncorhynchus clarki henshawi*), a resident of highly alkaline Pyramid Lake (pH 9.4), to challenge at pH 10, *J. Exp. Biol.,* 175, 173, 1993.

Wilkie, M. P. and Wood, C., Nitrogenous waste excretion, acid-base regulaton, and ionoregulation in rainbow trout (*Oncorhynchus mykiss*) exposed to extremely alkaline water, *Physiol. Zool.,* 64, 1069, 1991.

Wilson, R. and Wood, C., Swimming performance, whole-body ions, and gill Al accumulation during acclimation to sublethal aluminum in juvenile rainbow trout (*Oncorhynchus mykiss*), *Fish Physiol. Biochem.,* 10, 149, 1992.

Wood, C. M., Turner, J. D. and Graham, M. S., Why do fish die after severe exercise?, *J. Fish Biol.,* 22, 189, 1983.

Wood, C., McDonald, D., Booth, C., Simons, B., Ingersoll, C. and Bergman, H., Physiological evidence of acclimation to acid/aluminum stress in adult brook trout (*Salvelinus fontinalis*). I. Blood composition and net sodium fluxes, *Can. J. Fish. Aquat. Sci.,* 45, 1587, 1988a.

Wood, C., Simons, B., Mount, D. and Bergman, H., Physiological evidence of acclimation to acid/aluminum stress in adult brook trout (*Salvelinus fontinalis*). 2. Blood parameters by cannulation, *Can. J. Fish. Aquat. Sci.,* 45, 1597, 1988b.

Wood, C. M., The physiological problems of fish in acid waters, in, *Acid Toxicity and Aquatic Animals,* Morris, R., Taylor, E., Brown, D. and Brown, J. Eds., Cambridge University Press., New York, 1989, 125.

Wood, C. M. and McDonald, D. G., Physiological mechanisms of acid toxicity to fish, in, *Acid Rain/ Fisheries,* Johnson, R. E., Ed., American Fisheries Society, Bethesda, MD, 1982, 197.

Wood, C. M., McDonald, D. G., Ingersoll, C. G., Mount, D. R., Johannsson, O. E., Landsberger, S. and Bergman, H., Effects of water acidity, calcium and aluminum on whole-body ions of brook trout (*Salvelinus fontinalis*) continuously exposed from fertilization to swim-up: a study by instrumental neutron activation analysis, *Can. J. Fish. Aquat. Sci.,* 47, 1593, 1990a.

Wood, C. M., McDonald, D. G., Ingersoll, C. G., Mount, D. R., Johannsson, O. E., Landsberger, S. and Bergman, H., Whole body ions of brook trout (*Salvelinus fontinalis*) alevins: responses of yolk-sac and swim-up stages to water acidity, calcium, and aluminum, and recovery effects, *Can. J. Fish. Aquat. Sci.,* 47, 1604, 1990b.

Wright, P., Iwama, G. and Wood, C. M., Ammonia and urea excretion in Lahontan cutthroat trout (*Oncorhynchus clarki henshawi*) adapted to the highly alkaline Pyramid Lake (pH 9.4), *J. Exp. Biol.,* 175, 153, 1993.

Ye, X. and Randall, D., The effect of water pH on swimming performance in rainbow trout (*Salmo gairdneri,* Richardson), *Fish Physiol. Biochem.,* 9, 15, 1991.

Ye, X., Randall, D. and Xiqin, H., The effect of acid water on oxygen consumption, circulating catecholamines and blood ionic and acid-base status in rainbow trout (*Salmo gairdneri*), *Fish Physiol. Biochem.,* 9, 23, 1991.

Chapter 11

The Immune System

I. OVERVIEW OF FISH IMMUNOLOGY

In common with other vertebrates, the fish immune system is used to defend against harmful bacteria, viruses, fungi, and parasites. The fish immune system has many similarities with that of the mammalian system, but there are also some important differences. For example, temperature and life style have a considerable influence on immune function in fish. Species living a solitary existence in cold waters have rather poorly developed immune systems compared to social species in warm waters where pathogens are more easily passed between fish and the pathogens have shorter generation times due to the higher temperatures (Post, 1983). Within a species, temperatures at the low end of a species thermal range may reduce immunological competency. Zeeman (1986) emphasizes how daily and seasonal cycles can greatly modify the immune response in fish.

Our current understanding of fish immunology is far behind that of mammals even though many of the techniques and concepts worked out on the latter are now being applied to fish (Stolen, 1993, 1994). Reviews of fish immunology include those by Post (1983), Satchell (1991), Weeks et al. (1992), and the book on the subject edited by Manning and Tatner (1985).

The first line of defense in a fish includes the physical barriers of scales and skin and the chemical barrier, mucus. The latter contains lysozymes which are bacteriolytic enzymes (Lie et al., 1989). A wide variety of pollutants and low pH as well as the stress from handling stimulate increased secretion by mucus glands in the gills and skin. Whether the composition of the mucus changes as a result of such a hypersecretion is apparently unknown, but it does serve to potentially trap harmful organisms and then slough them off (Anderson, 1990). Another initial defense is the acidity of the stomach which, at least in mammals and presumably also in fish, kills many bacteria.

If the initial barriers are penetrated by a pathogen, then the other components of the immune system come into play. At first, certain immune cells are attracted to the site of entry. These include tissue macrophages and monocytes in the blood which phagocytize the microbes. Macrophages may actually be monocytes (a type of leukocyte) that have entered the tissues and differentiated. Macrophages are considered to be of greater immunological importance in fish than in the other vertebrates (Weeks et al., 1992), so their activity is frequently assayed as a test for immune function.

Neutrophils (another leukocyte) are attracted to the site of invasion, where they release enzymes, including lysozyme, that destroy harmful organisms. Neutrophils and thrombocytes (which are primarily involved in blood clotting) have also been reported to be phagocytic (MacArthur and Fletcher, 1985). All the reactions thus far described are collectively termed the "non-specific response" because they are essentially the same irregardless of the pathogen involved. If these defense mechanisms are overcome, the microorganisms may rapidly multiply and kill the fish unless the specific immune response is induced in time.

The specific immune response must be induced in reaction to individual antigens. These are specific molecules, usually protein, that are on the surface of foreign organisms such as bacteria or viruses, or are toxins secreted by those organisms. Antigens are basic to the recognition of non-self by the various components of the immune system. An antigen is transported by the monocytes to melanomacrophage "centers" of the anterior kidney and spleen where the antigen is presented to lymphocytes that are responsible for producing the antibodies, and to T cells which are specific for that particular antigen. The transporting and presenting of an antigen by monocytes is referred to as antigen processing.

The classic picture in mammals of B lymphocytes formed in the bone marrow and residing in the lymph nodes and responsible for humoral immunity must be modified in fish. Fish have no bone marrow or lymph nodes and the production of lymphocytes is apparently in the head kidney, gut-associated tissue, and the spleen. Some B lymphocytes become activated to divide (thus becoming plasma cells) by an antigen and start secreting antibodies toward the specific antigen. Other B lymphocytes differentiate into memory cells that serve to "remember" what the antigen was so that upon subsequent exposure to the same antigen, the antibody response (i.e., the secondary response) is much more rapid and larger.

The antibodies are proteins which are referred to as immunoglobulins and, in mammals, are grouped into five classes designated with the prefix "Ig". In fish, however, only the IgM class has been found (Jurd, 1985). Antibodies do not destroy antigen-bearing invaders. They instead inactivate antigens and mark them for destruction by macrophages and complement, another protein component of the immune armament. They also may neutralize viruses or toxins and agglutinate them, which makes them more accessible for phagocytosis by macrophages.

The other major arm of the specific immune response is designated "cell mediated". It depends on the function of T lymphocytes, which, in mammals, are formed in the very young by the thymus from cells that originated in the bone marrow. The thymus then largely degenerates as the animal matures. The presence of T lymphocytes in fish was for some time controversial but now they are generally considered to be present (Anderson, 1990; Satchell, 1991). Mammals, and presumably fish, have subpopulations of these cells which aid in antibody production and others that suppress it. The thymus in fish involutes at sexual maturity in some species but retains its normal size in others (Satchell, 1991).

There are other parts of the immune system of fish that are not well understood. These include natural killer (NK) cells, which destroy tumors; interferon, which interferes with replication of viruses; transferrin in the blood, which deprives microorganisms of iron; and an assortment of other molecules which agglutinate pathogens and aid phagocytosis (Alexander, 1985; Satchell, 1991).

II. EFFECTS OF POLLUTANTS ON IMMUNE FUNCTION

The field of immunotoxicology, even in mammals, is relatively young (National Research Council, 1992). It has been presumed for decades that environmental pollutants or other stressors (e.g., handling) can affect one or more of the immunological functions in a fish, for it is almost common knowledge that fish frequently then become more susceptible to various diseases (Snieskzko, 1974). The field of fish immunotoxicology (or immunomodulation as it is sometimes called) has, however, only begun to really grow since the mid-1980s. Earlier observations were usually peripheral to other aims of a toxicological or immunological study and have been extensively reviewed by Zeeman and Brindley (1981). In many cases, the major observation was the occurrence of histological lesions in kidney or spleen following exposures to some pollutant at often unrealistically high environmental concentrations.

There are a wide variety of tests that have been developed to assess how well a particular component of the immune system is functioning. Most of these are modified from similar ones used in mammalian clinical laboratories and vary considerably in how difficult they are to perform. They fall into five broad groups: (1) susceptibility of fish to infection or tumor formation; (2) changes in leukocyte numbers or differential count; (3) alterations in phagocytic response or melanomacrophage centers; (4) measures of antibody production; and (5) scale or fin graft rejection times (as a measure of cell-mediated immune function). Anderson (1990), Stolen (1993, 1994), and Weeks et al. (1992) provide useful guidance for those wishing to begin investigations in this rapidly growing discipline.

A. METALS
1. Cadmium
This metal causes both immunological inhibition and stimulation in mammals, depending on a variety of factors, especially the particular immune function measured (Koller, 1981). For example, T-lymphocyte activities are usually suppressed by cadmium whereas the effects on B lymphocytes are more varied. The same sort of variability may also prevail for fish. Thuvander (1989) tested rainbow trout after 12 weeks in very low concentrations of cadmium (0.7 or 3.6 µg Cd/L) and observed suppression of T-lymphocyte function whereas the B-antibody response to a bacterial challenge was enhanced in the fish exposed to cadmium. The metal exposure caused no other clinical effects in the fish. One cannot, however, conclude that cadmium stimulates antibody capability in fish. Tests in the same species at comparable cadmium concentrations, but using a response by the intact fish to the injection of human blood cells, found a modest inhibition of antibody production (Viale and Calamari, 1984).

The data presented by Robohm (1986) points up how it is hazardous to conclude much about effects of cadmium on immune responses in fish if it is based on only one species. He exposed cunners and striped bass to cadmium for 96 h and tested the antibody response of intact fish to a bacterium. Cadmium caused inhibition of serum antibody titers in cunners, but it caused a sixfold stimulation in striped bass. Exposures were at the same concentration in both fish species (10–12 µg/ml) a cadmium concentration that is about half the LC50. There does not appear to be a clear explanation for this large species difference.

Cadmium can have a marked effect on differential leukocyte counts in fish. Newman and MacLean (1974) reported a dose-dependent increase (threefold) in neutrophils and a dose-dependent decrease (also nearly threefold) in lymphocytes in the cunner, a marine teleost. Three-week exposures of goldfish to cadmium yielded rather similar results except the effect was not as large or was it dose dependent (Murad and Houston, 1988). Murad and Houston speculate the lack of dose dependency may reflect differences in cadmium sensitivity of lymphocyte subpopulations so that as a threshold is reached, essentially all of those from one population are eliminated. Then, unless the threshold of the other population(s) is achieved, there is no further change in lymphocyte number with rising cadmium concentration. The primary reason for the decrease in lymphocytes is evidently due to a reduction in their rate of production. Evidence for this is that the blast cells, which are lymphoid progenitors, also decreased and other investigators (Stromberg et al., 1983) report lesions in the hematopoietic areas following cadmium exposure.

Stimulation of the numbers of neutrophils does not currently have an explanation. Because it has now been observed following exposure to cadmium in two very different sorts of fish, it may be a fairly widespread phenomenon. While the depression of lymphocytes by cadmium probably compromises immune function, it is not clear what effect stimulation of neutrophils may have. It might conceivably compensate for the lack of lymphocytes. A suggestion of this is seen in the work of MacFarlane et al. (1986) who exposed juvenile striped bass to cadmium at 12 or 30 µg/L for 5 days and then challenged them with the bacterium which causes columnaris disease. The members exposed to cadmium were slightly less susceptible to the disease than were controls. It would be premature, however, to jump to any broader conclusions from this single study.

A variable that could have a considerable effect on the modulation of immune response in fish exposed to cadmium is their lack of a cortisol response to this metal. Most fish exhibit an elevated plasma cortisol level in response to nearly any stressor, however, cadmium appears to be an exception to this generalization (Schreck and Lorz, 1978; Thomas and Neff, 1985). As will be discussed later in this chapter, elevated cortisol may be a primary mechanism for immune system suppression in fish exposed to a variety of pollutants. It is not clear why cadmium alone among the metals fails to induce this hormonal change, but it may help explain the sometimes unexpected results when fish are exposed to this metal.

2. Copper

Copper sulfate is frequently used to control external columnaris infections of pond fishes. Thus, it is not surprising that exposure of juvenile striped bass for 5 days to this metal improved their resistance to this disease bacterium (MacFarlane et al., 1986). However, this metal caused immunosuppression of antibody-producing cells in rainbow trout when tested *in vitro* (Anderson et al., 1989), and air-breathing catfish (*Saccobranchus fossilis*) exposed to copper at several very low concentrations for 28 days exhibited a dose-dependent depression of antibody production, phagocytic activity of spleen and kidney macrophages, and a prolongation of eye allograft rejection time (Khangarot and Tripathi, 1991). The latter indicates a suppression of T-cell activity.

Hetrick et al. (1982) exposed striped bass to copper at 7 and 10 µgL^{-1} for 96 h prior to challenge with two naturally occurring bacterial species (*Vibrio anguillarum* and *Pasteurella piscicida*). Copper exposure increased susceptibility of the fish to the disease organisms. So, while copper may be protective for some external diseases, defense against internal infections can be compromised by prolonged exposure to the metal.

When zebrafish were exposed to copper or zinc at several different concentrations for 7 days, a dose-dependent suppression of kidney lymphocyte number and natural cytotoxic cells was observed (Rougier et al., 1994). Copper was more potent in this regard than was zinc. The effect on macrophage activity was, however, more complex. Copper caused a marked decrease in macrophage activity both *in vitro* and *in vivo*, but zinc caused a modest increase in macrophage activity under the same conditions. The authors do not attempt to explain this stimulation of activity by zinc, but the fact that it occurs with both *in vitro* and *in vivo* experiments suggests it is not an artifact.

3. Miscellaneous Metals

In a rather old study, exposure of brown bullhead catfish to lead for up to 183 days was reported to produce a reduction in spleen size but an increase in leukocyte number and an especially large increase in the number of thrombocytes (Dawson, 1935). Crandall and Goodnight (1963) obtained somewhat similar findings in guppies exposed to lead for at least 60 days. They also noted that lymphoid tissue in the head kidney was greatly reduced in those exposed to lead. The same workers observed similar changes in guppies chronically exposed to zinc.

Chronic exposure of trout and carp to nickel, zinc, copper, or chromium was found to suppress to a variable extent the primary humoral response to an intraperitoneal inoculation of MS2 bacteriophage (O'Neill, 1981). O'Neill (1981; p. 329) notes that metals may "...block the active sites of antibody molecules and disturb the metabolism, ionic balance and cellular division of immunocompetent cells".

Among the metals, manganese has relatively low toxicity to fish (Jones, 1964). It is an essential trace element for various functions and stimulates NK cell activity in mammals (Keen et al., 1985). Thus, it is interesting to see that it also stimulated NK cell activity in carp based on both an *in vitro* and an *in vivo* test (Ghanmi et al., 1990). The authors cite work on mammals which suggest that the stimulatory effect is due to factors such as lymphokines or interferon in those species. This metal was also found to have a "strong enhancing effect on phagocytosis" by carp macrophages when added to the test medium in an *in vitro* test (Cossarini-Dunier, 1987), somewhat similar to what was reported above for zinc.

When phagocytes undergo phagocytosis they show a burst of oxygen consumption and chemiluminescence which can be measured using luminol as an amplifier. The chemiluminescence response is therefore a measure of phagocytosis of a group of cells. Using an *in vitro* preparation, Wishkovsky et al. (1989) found inhibition of this response in macrophages from three species of estuarine fish when in the presence of tributyltin in the test medium.

On the basis of one study (Roales and Perlmutter, 1977), it appears that methylmercury can severely inhibit at least some components of the immune system. Blue gourami (*Trichogaster*, a tropical freshwater species) were exposed to methylmercury for 4 or 5 weeks and then tested for viral neutralizing or bacterial agglutination titers. Some inhibition of the neutralization occurred and bacterial agglutination was non-detectable in those exposed to mercury.

There has been some interest in the use of *in vitro* systems for testing various toxicants because it requires fewer animals and does not require extensive facilities for exposing the animals to the test chemical (Anderson et al., 1989). Elsasser et al. (1986) used this approach with trout phagocytes and the chemiluminescent response mentioned above. They incubated the phagocytes at a concentration of copper, aluminum, or cadmium equivalent to 1/10 of the lowest concentration that showed toxicity for a cultured indicator cell line. Copper caused a strong inhibition of the phagocytic response, aluminum somewhat less so. Cadmium, however, caused an initial stimulation followed by a variable decrease. These findings are interesting but need validating in intact systems. A real problem with *in vitro* studies is simulating the concentrations of test chemicals that might prevail in the intact animal. Another potential confounding factor is the duration of the exposure which is usually for only a few hours when *in vitro*, but when *in vivo*, it may be for days or weeks. However, the *in vitro* approach could be useful for screening a wide variety of potential immunomodulators which could then be tested in the intact fish.

B. ORGANIC POLLUTANTS

A major class of organic contaminants that fish may get exposed to is, of course, pesticides. Given the extreme variety of pesticides used, it is somewhat surprising so little is known about how they affect the immune systems of fish (Plumb and Areechon, 1990). Zeeman and Brindley (1981) reviewed the earlier work, nearly all of which suggested decreased disease resistance in fish exposed to various pesticides. They note, however, that this may often be explained as a generalized stress effect (also see below), especially given that the test concentrations used were usually quite large.

Atrazine, a triazine herbicide, and lindane, an organochlorine insecticide, were tested for their direct effect on carp macrophage phagocytosis *in vitro*. No effect was found at concentrations up to the limit of their solubility in water (Cossarini-Dunier, 1987). Cossarini-Dunier cites other work in the same laboratory which showed no effect on antibody production by these pesticides when given in the food to carp for 2.5 months. There was no effect even though the lymphoid organs were highly contaminated, nor did the pesticides have any effect on graft rejection times. Another organochlorine insecticide (Mirex) was tested in trout via inclusion in the diet for a year at 50 ppm (Cleland and Sonstegard, 1987; Cleland et al., 1988). This dietary exposure caused no effect on humoral immune expression or on NK cell activity. Similar negative results were found when humoral immune expression was measured in carp given food contaminated with lindane (up to 1000 ppm) for a year (Cossarini-Dunier et al., 1987). When trout were given a daily body dose (1 mg/kg) of lindane in their diet for 30 days, a decrease in chemiluminescent response was observed but there was no effect on lymphocytic proliferation and on the number of circulating B lymphocytes (Dunier et al., 1994). Taken together, these findings suggest that organochlorine insecticides may have relatively little effect on the fish immune system, at least when exposures are at low levels, even for long times, but, if exposures produce a generalized stress response reflected in an elevated serum cortisol level, then immunosuppression may occur as was seen when trout were exposed to endrin for 30–60 days (Bennett and Wolke, 1987a,b).

From the very limited work that has been done, it appears that organophosphorus insecticides may also exhibit relatively little immunotoxicity. Dunier et al. (1991) tested the effect of trichlorofon and dichlorvos on lymphocytic proliferation and phagocytosis *in vitro* using carp cells. The doses were high and caused suppression of these two functions. This may have relevance where these pesticides are used for ectoparasite treatments of fish in aquaculture. However, when dichlorvos was administered in the ambient water (trichlorofon is rapidly converted to this molecule in water), no effect was detected on the humoral response to *Yersinia ruckeri*.

Channel catfish exposed to malathion for 30 days exhibited suppression of antibody agglutination. Two doses were used (0.5 and 1.75 mg/L), only the higher one caused suppression (Plumb and Areechon, 1990). Because this organophosphorus insecticide breaks down rapidly in the environment, it seems doubtful that concentrations this high would persist for long.

The responses of fish to sediment contaminated by diesel fuel are complex and difficult to interpret. Tahir et al. (1993) report that exposure of dab to this sediment for 4 weeks caused an increase in leukocyte number with low doses but a suppression with high doses. Serum lysozyme activity and kidney phagocytic respiratory burst activity were depressed at all doses whereas the number of antibody-secreting cells in the kidney, serum bacteriocidal activity, and anti-protease activities tended to increase with exposure. The authors speculate that the stimulatory effect on some immune functions may be a compensatory response to suppression of others by immunotoxins in the sediment.

C. MISCELLANEOUS POLLUTANTS

Lymphocyte proliferation and antibody production were both suppressed in channel catfish exposed to acidified waters (pH 4) for 2 weeks. The secondary antibody response was also suppressed. The author (Morra, 1993) presents limited evidence that the mechanism may have been the suppression of the ability of B cells to process the antigen.

The enzyme lysozyme is found in the lysosomes of neutrophils and macrophages and is secreted into the blood by these cells. Mock and Peters (1990) found that fairly high doses of ammonia can cause a depression in its activity in the blood. This may be a generalized stress response because the same investigators found that the stress of handling and transport also caused suppression of lysozyme activity.

The finding of a suppression of lysozyme activity may only occur with acute stressors. Secombes et al. (1991) found no effect on lysozyme activity of chronic exposures of fish to sewage sludge, but they did get decreases in thrombocyte numbers and increases in melanomacrophage centers. In a subsequent study Secombes et al. (1992) found that the sludge caused a decrease in leukocyte bactericidal activity and in antibody-secreting cells but no change in specific Ig or phagocytosis. In the same study, it was noted that immunization overcame the inhibitory effect of sewage sludge exposure on leukocyte bactericidal activity.

D. IMMUNE EFFECTS IN FISH FROM CONTAMINATED NATURAL WATERS

The potential for using measurements of immune function in field monitoring programs is probably large, however, they have not been used much (Wester et al., 1994). This may be largely due to the shortage of easily applied procedures that would lend themselves to routine monitoring. The usual approach has been to look for diseases in sampled fish (Sindermann, 1979). For this purpose, O'Connor et al. (1987) developed an index of disease conditions (primarily external) that may be induced by pollution in marine teleosts and shellfish. Their publication also has a section to help non-professionals interpret the findings and evaluate their importance.

A rather extensive disease study was carried out by Malins et al. (1988) in which they sampled English sole in Puget Sound over a 5-year period in order to investigate the etiological relationships between prevalences of disease (primarily hepatic neoplasms) and concentrations of aromatic hydrocarbons in the sediment and their metabolites in the fish bile. They obtained good correlations between concentrations of chemicals in sediment and tissues and prevalence of disease. However, Malins et al. (1988; p. 61) wisely point out that "...such relationships cannot be interpreted as *de facto* evidence of specific cause and effect." This is because not all potential harmful chemicals can be identified in the samples and even the ones measured may act through synergistic or antagonistic interactions with others. Even with these caveats, it seems that disease surveys may be a useful biomarker (among others) for environmental health (McCarthy and Shugart, 1990).

Increased occurrence of neoplasms in fish exposed to contaminants suggests suppression of a primary defense that vertebrates have against the spread of transformed or malignant cells. Key players in this are the cytotoxic leukocytes often called NK cells. Thus, it is interesting that Faisal et al. (1991) report a depression in NK activity in *Fundulus* taken from a river in Virginia which passes through a heavily industrialized urban area and the water and sediments contain a wide variety of chemicals including polynuclear aromatic hydrocarbons and creosote. In another study, the chemotactic response of fish macrophages was tested in two other species of fish from the same river and these were compared with fish from another river that was relatively clean (Weeks et al., 1986). A 30 to 40% decrease in macrophage chemotactic activity was found in those fish from the contaminated river. When those fish were held in clean water for 3 weeks the chemotactic activity returned to normal, showing that the effect of the polluted water was reversible. This raises the question (not asked by the authors) whether the return to normal immune function was due to depuration of the toxic chemicals from the tissues, or was it due to changes in hormonal status?

In the same laboratory that found suppression of macrophage chemotactic activity in fish from a contaminated river, a stimulation of macrophage chemoluminescence was observed in *Fundulus* from the same river (Kelley-Reay and Weeks-Perkins, 1994). This apparent stimulation by the contaminants was further confirmed by observing chemoluminescence stimulation of macrophages in fish transferred from the control river to waters from the contaminated one. The mechanism for this remains obscure but may involve increased lymphokines or a direct effect of some chemical constituent of the waters.

Following a chemical spill (including chlorinated and heterocyclic hydrocarbons, aromatic nitro, and heavy metal compounds) eels were found to have considerable histopathological damage in the spleen (Spazier et al., 1992). In addition, melanomacrophage centers were lacking in eels suffering from the aftermath of the spill.

Melanomacrophage centers are fairly easy to assay using standard histological methods. Wester et al. (1994) suggest they may be a useful biomarker for immunotoxicology in part because they are considered as the primitive analog of the mammalian lymph follicle. Their density may decrease in fish from contaminated waters or along a pollution gradient, although as was noted above, some chemicals may cause increased density of the centers. Wester et al. (1994) suggest that the latter response may reflect accumulation of cytotoxic waste.

III. HORMONAL MODULATION OF IMMUNE RESPONSE

The general adaptation response (GAS) of fish to any sort of stress is characterized by elevations in cortisol and catecholamines (Pickering, 1981). While these hormones are involved in the maintenance of homeostasis in the face of the stress, they can also have a marked effect on numerous immunological functions.

The evidence is fairly strong now that cortisol suppresses several of the immune functions in fish, even at concentrations that are only slightly higher than those found in "unstressed" animals (Thomas and Lewis, 1987; Pickering and Pottinger, 1989; Ainsworth et al., 1991). These investigations were performed using fish in which cortisol was given either in the food or via intraperitoneal injection, thus the confounding effects of contaminant chemicals or other stressors were eliminated. Moreover, in studies (Tripp et al., 1987) carried out *in vitro*, physiological concentrations of cortisol suppressed the mitogenic response of coho salmon lymphocytes so at least one molecular mechanism has been revealed. The suppression of immune function by exogenous cortisol has also been shown to make trout more susceptible (in a cortisol dose-dependent manner) to common bacterial and fungal diseases (Pickering and Pottinger, 1989).

The interrelations of cortisol and immune function may be more complex than a mere suppression of the latter by the former. Schreck and Bradford (1990) have found that fish lymphokines can suppress the secretion rate of cortisol from interrenal cells, thus exhibiting a form of negative feedback. This is, incidentally, the reverse of what occurs in mammals (Sapolsky et al., 1987).

Although cortisol alone can have a considerable effect on a variety of immune system activities, it is premature to conclude that all effects of pollution (or other stressors) are due to this hormone (Maule and Schreck, 1990). In mammals, catecholamines which become elevated during stress, mobilize blood granulocytes and macrophages (Weeks et al., 1992) thereby causing their numbers to rise in the blood. This could be a form of compensation for the effect of some contaminant on immune function. Clearly, we are only beginning to touch the surface of this field of pollution physiology.

REFERENCES

Ainsworth, A. J., Dexiang, C. and Waterstrat, P. R., Changes in peripheral blood leukocyte percentages and function of neutrophils in stressed channel catfish, *J. Aquat. Anim. Health*, 3, 41, 1991.

Alexander, J. B., Non-immunoglobulin humoral defence mechanisms in fish, in, *Fish Immunology*, Manning, M. J. and Tatner, M. F., Eds., Academic Press, New York, 1985, 133.

Anderson, D. P., Immunological indicators: effects of environmental stress on immune protection and disease outbreaks, *Am. Fish. Soc. Symp.*, 8, 38, 1990.

Anderson, D. P., Dixon, O. W., Bodammer, J. E. and Lizzio, E. F., Suppression of antibody-producing cells in rainbow trout spleen sections exposed to copper *in vitro*, *J. Aquat. Anim. Health*, 1, 57, 1989.

Bennett, R. O. and Wolke, R. E., The effect of sublethal endrin exposure on rainbow trout, *Salmo gairdneri* Richardson. I. Evaluation of serum cortisol concentrations and immune responsiveness, *J. Fish Biol.*, 31, 375, 1987a.

Bennett, R. O. and Wolke, R. E., The effect of sublethal endrin exposure on rainbow trout, *Salmo gairdneri* Richardson. II. The effect of altering serum cortisol concentrations on the immune response, *J. Fish Biol.*, 31, 387, 1987b.

Cleland, G. B. and Sonstegard, R. A., Natural killer cell activity in rainbow trout (*Salmo gairdneri*): effect of dietary exposure to Aroclor 1254 and/or mirex, *Can. J. Fish. Aquat. Sci.*, 44, 636, 1987.

Cleland, G. B., McElroy, P. J. and Sonstegard, R. A., The effect of dietary exposure to Aroclor 1254 and/or mirex on humoral immune expression of rainbow trout (*Salmo gairdneri*), *Aquat. Toxicol.*, 2, 141, 1988.

Cossarini-Dunier, M., Effects of the pesticides atrazine and lindane and of manganese ions on cellular immunity of carp, *Cyprinus carpio, Fish Biol.*, 31 (Supplement A), 67, 1987.

Cossarini-Dunier, M., Monod, G., Demael, A. and Lepot, D., Effect of y-hexachlorocyclohexane (Lindane) on carp (*Cyprinus carpio*). I. Effect of chronic intoxication on humoral immunity in relation to tissue pollutant levels, *Ecotoxicol. Environ. Safety*, 13, 339, 1987.

Crandall, C. A. and Goodnight, G. J., The effects of sublethal concentrations of several toxicants to the common guppy, *Lebistes reticulatus, Trans. Am. Microsc. Soc.*, 82, 59, 1963.

Dawson, A. B., The hemopoietic response in the catfish, *Ameiurus nebulosus*, to chronic lead poisoning, *Biol. Bull.*, 68, 335, 1935 (As cited in Zeeman and Brindley, 1981).

Dunier, M., Siwicki, A. K. and Demael, A., Effects of organophosphorus insecticides: effects of trichlorofon and dichlorvos on the immune response of carp (*Cyprinus carpio*). III. *In vitro* effects on lymphocyte proliferation and phagocytosis and *in vivo* effects on humoral response, *Ecotoxicol. Environ. Safety*, 22, 79, 1991.

Dunier, M., Siwicki, K., Scholtens, J., Dal Molin, S., Vergnet, C. and Studnicka, M., Effects of lindane exposure on rainbow trout (*Oncorhynchus mykiss*) immunity. III. Effect on nonspecific immunity and B lymphocyte functions, *Ecotoxicol. Environ. Safety*, 27, 324, 1994.

Elsasser, M. S., Robertson, B. S. and Hetrick, F. M., Effects of metals on the chemiluminescent response of rainbow trout (*Salmo gairdneri*) phagocytes, *Vet. Immun. Immunopathol.*, 12, 243, 1986.

Faisal, M., Weeks, B. A., Vogelbein, W. K. and Huggett, R. J., Evidence of aberration of the natural cytotoxic cell activity in *Fundulus heteroclitus* (Pices: Cyprinodontidae) from the Elizabeth River, Virginia, *Vet. Immunol. Immunopathol.*, 29, 339, 1991.

Ghanmi, Z., Bouabhaia, M., Alifuddin, M., Troutaud, D. and Deschaux, P., Modulatory effect of metal ions on the immune response of fish: *in vivo* and *in vitro* influence of MnCl2 on NK activity of carp pronephros cells, *Ecotoxicol. Environ. Safety*, 20, 241, 1990.

Hetrick, F. M., Robertson, B. S. and Tsai, C. F., Effect of heavy metals on the susceptibility and immune response of striped bass to bacterial pathogens, *National Oceanographic Atmospheric Administration Publication*, 82112603, p. 32, 1982.

Jones, J. R. E., *Fish and River Pollution*, Butterworths, London, 1964.

Jurd, R. D., Specialisation in the teleost and anuran immune response, in, *Fish Immunology*, Manning, M. J. and Tatner, M. F., Eds., Academic Press, New York, 1985, 9.

Keen, C. L., Lonnerdal, B. and Hurley, L. S., Manganese, in, *Biochemistry of the Essential Ultratrace Elements*, Freiden, E., Ed., Plenum Press, New York, 1985, 89.

Kelley-Reay, K. and Weeks-Perkins, B., Determination of macrophage chemiluminescent response in *Fundulus heteroclitus* as a function of pollution stress, *Fish Shellfish Immunol.*, 4, 95, 1994.

Kjangarot, B. S. and Tripathi, D. M., Changes in humoral and cell-mediated immune responses and in skin and respiratory surfaces of catfish, *Saccobranchus fossilis*, following copper exposure, *Ecotoxicol. Environ. Safety*, 22, 291, 1991.

Koller, L. D., Immunological effects of cadmium, in, *Cadmium in the Environment*, Nriagu, J., Ed., Wiley and Sons, New York, 1981, 719.

Lie, O., Evensen, O., Sorensen, A. and Froysadal, E., Study on lysosome activity in some fish species, *Dis. Aquat. Organisms*, 6, 105, 1989.

MacArthur, J. I. and Fletcher, T. C., Phagocytosis in fish, in, *Fish Immunology,* Manning, M. J. and Tatner, M. F., Eds., Academic Press, New York, 1985, 29.

MacFarlane, R. D., Bullock, G. L. and McLaughlin, J. J. A., Effects of five metals on susceptibility of striped bass to *Flexibacter columnaris, Trans. Am. Fish. Soc.,* 115, 227, 1986.

Malins, D. C., McCain, B. B., Landahl, J. T., Meyers, M. S., Krahn, M. M., Brosn, D. W., Chan, S. L. and Roubal, W. T., Neoplastic and other diseases in fish in relation to toxic chemicals: an overview, *Aquat. Toxicol.,* 11, 43, 1988.

Manning, M. J. and Tatner, M. F., Eds., *Fish Immunology,* Academic Press, New York, 1985.

Maule, A. G. and Schreck, C. B., Changes in numbers of leukocytes in immune organs of juvenile coho salmon after acute stress or cortisol treatment, *J. Aquat. Anim. Health,* 2, 298, 1990.

McCarthy, J. F. and Shugart, L. R., Eds., *Biomarkers of Environmental Contamination,* Lewis Publishers, Boca Raton, FL, 1990.

Mock, A. and Peters, G., Lysozyme activity in rainbow trout, *Oncorhynchus mykiss* stressed by handling, transport and water pollution, *J. Fish Biol.,* 37, 873, 1990.

Morra, D. S., Effects of acidic water on immune responses in the channel catfish, *Ictalurus punctatus, J. Immunol.,* 150, Part 2, 16A, 1993.

Murad, A. and Houston, A. H., Leucocytes and leucopoietic capacity in goldfish, *Carassius auratus,* exposed to sublethal levels of cadmium, *Aquat. Toxicol.,* 13, 141, 1988.

National Research Council, *Biologic Markers in Immunotoxicology,* National Academy Press, Washington, D.C., 1992.

Newman, M. W. and MacLean, S. A., Physiological response of the cunner, *Tautogolabrus adspersus,* to cadmium. VI. Histopathology, *National Oceanic and Atmospheric Administration Technical Report,* NMFS SSRF-681, 27, 1974.

O'Connor, J. S., Ziskowski, J. J. and Murchelano, R. A., Index of pollutant-induced fish and shellfish disease, *NOAA (National Oceanic and Atmorspheric Administration) Special Report,* Washington, D.C., 1987.

O'Neill, J. G., Heavy metals and the humoral immune response of freshwater teleosts, in, *Stress and Fish,* Pickering, A. D., Ed., Academic Press, New York, 1981, 328.

Pickering, A. D., Ed., *Stress and Fish,* Academic Press, New York, 1981.

Pickering, A. D. and Pottinger, T., Stress responses and disease resistance in salmonid fish: effect of chronic elevation of plasma cortisol, *Fish Physiol. Biochem.,* 7, 253, 1989.

Plumb, J. A. and Areechon, N., Effect of malathion on humoral immune response of channel catfish, *Dev. Comp. Immunol., 14,* 355, 1990.

Post, G. W., *Textbook of Fish Health,* TFH Publishing, Neptune City, NJ, 1983, chap. 8.

Roales, R. R. and Perimutter, A., The effects of sub-lethal doses of methylmercury and copper, applied singly and jointly, on the immune response of the blue gourami (*Trichogaster trichopoterus*) to viral and bacterial antigens, *Arch. Environ. Contam. Toxicol.,* 5, 325, 1977.

Robohm, R. A., Paradoxical effects of cadmium exposure on antibacterial antibody responses in two fish species: inhibition in cunners (*Tautogolabrus adspersus*) and enhancement in striped bass *(Morone saxatilis), Vet. Immunol. Immunopathol.,* 12, 251, 1986.

Rougier, F., Troutaud, D., Ndoye, A. and Deschaux, P., Non-specific immune response of zebrafish, *Branchyodanio rerio* (Hamilton-Buchanan) following copper and zinc exposure, *Fish Shellfish Immunol.,* 4, 115, 1994.

Sapolsky, R., Rivier, C., Yamamoto, G., Plotsky, P. and Vale, W., Interleukin-1 stimulates the secretion of hypothalamic corticotropin releasing factor, *Science,* 238, 522, 1987.

Satchell, G. H., *Physiology and Form of Fish Circulation,* Cambridge University Press, Cambridge, 1991, chap. 5.

Schreck, C. B. and Lorz, H. W., Stress response of coho salmon (*Oncorohynchus kisutch*) elicited by cadmium and copper and potential use of cortisol as an indicator of stress, *J. Fish. Res. Bd. Can.,* 35, 1124, 1978.

Schreck, C. and Bradford, C., Interrenal corticosteroid production: potential regulation by the immune system in the salmonid, in, *Progress in Comparative Endocrinology,* Epple, A., Scanes, C. and Stetson, M., Eds., Wiley-Liss, New York, 1990, 480.

Secombes, C. J., Fletcher, T. C. , Oflynn, J. A., Costello, M., Stagg, R. and Houlihan, D., Immunocompetence as a measure of the biological effects of sewage sludge pollution in fish, *Comp. Biochem. Physiol.,* 100C, 133, 1991.

Secombes, C. J., Fletcher, T. C., White, A., Costello, M., Stagg, R. and Houlihan, D., Effects of sewage sludge on immune responses in the dab, *Limanda mimanda, Aquat. Toxicol.,* 23, 217, 1992.

Sindermann, C. J., Pollution-associated diseases and abnormalities of fish and shellfish: a review, *Fish. Bull.,* 76, 717, 1979.

Snieskzko, S. F., The effects of environmental stress on outbreaks of infectious diseases of fishes, *J. Fish Biol.,* 6, 197, 1974.

Spazier, E., Storch, V. and Braunbeck, T., Cytopathology of spleen in eel *Anguilla anguilla* exposed to a chemical spill in the Rhine river, *Dis. Aquat. Organisms,* 14, 1, 1992.

Stolen, J. S., Fletcher, T. C., Anderson, D. P., Robertson, B. S. and van Muiswinkel, W. B., Eds., *Techniques in Fish Immunology,* Vols. 1–4, SOS Publications, Fair Haven, NJ, 1993–1994.

Stromberg, P. C., Ferrante, J. G. and Carter, S., Pathology of lethal and sublethal exposure of fathead minnows, *Pimephales promelas,* to cadmium: a model for aquatic toxicity assessment, *J. Toxicol. Environ. Health,* 11, 247, 1983.

Tahir, A., Fletcher, T., Houlihan, D. and Secombes, C., Effect of short-term exposure to oil-contaminated sediments on the immune response of dab, *Limanda limanda, Aquat. Toxicol.,* 27, 71, 1993.

Thomas, P. and Neff, J., Plasma corticosteroid and glucose responses to pollutants in striped mullet: different effects of naphthalene, benzo[a]pyrene and cadmium exposure, in, *Marine Pollution and Physiology: Recent Advances,* Vernberg, F. J., Thurburg, F. P., Calabrese, A. and Vernberg, W., Eds., University of South Carolina Press, 1985, 63.

Thomas, P. and Lewis, D. H., Effects of cortisol on immunity in red drum, *Sciaenops ocellatus, J. Fish Biol.,* 31 (Suppl. A), 123, 1987.

Thuvander, A., Cadmium exposure of rainbow trout, *Salmo gairdneri* Richardson: effects on immune functions, *J. Fish Biol.,* 35, 521, 1989.

Tripp, R. A., Maule, A. G., Schreck, C. B. and Kaattari, S. I., Cortisol mediated suppression of salmonid lymphocyte responses *in vitro, Dev. Comp. Immunol.,* 11, 565, 1987.

Viale, G. and Calamari, D., Immune response in rainbow trout *Salmo gairdneri* after long-term treatment with low levels of Cr, Cd, and Cu, *Environ. Pollut.,* 35A, 247, 1984.

Weeks, B. A., Anderson, D. P., DuFour, A. P., Fairbrother, A., Goven, A. J., Lahvis, G. P. and Peters, G., Immunological biomarkers to assess environmental stress, in, *Biomarkers: Biochemical, Physiological, and Histological Markers of Anthropogenic Stress,* Huggett, R. J., Kimerle, R. A, Mehrle, P. M. and Bergman, H. L., Eds., Lewis Publishers, Boca Raton, FL, 1992, chap. 5.

Weeks, B. A., Warinner, J. E., Mason, P. L. and McGinnis, D. S., Influence of toxic chemicals on the chemotactic response of fish macrophages, *J. Fish Biol.,* 28, 653, 1986.

Wester, P. W., Vethaak, A. D. and van Muiswinkel, W. B., Fish as biomarkers in immunotoxicology, *Toxicology,* 86, 213, 1994.

Wishkovsky, A., Mathews, E. S. and Weeks, B. A., Effect of tributyltin on the chemiluminescent response of phagocytes from three species of estuarine fish, *Arch. Environ. Contam. Toxicol.,* 18, 826, 1989.

Zeeman, M. G. and Brindley, W. A., Effects of toxic agents upon fish immune systems: a review, in, *Immunologic Considerations in Toxicology,* Vol. II, Sharma, R. P., Ed., CRC Press, Boca Raton, FL, 1981, chap. 1.

Zeeman, M., Modulation of the immune response in fish, *Vet. Immunol. Immunopathol.,* 12, 235, 1986.

Chapter 12

Behavior and Nervous System Function

I. INTRODUCTION

In Chapter 1, where the levels of integration (complexity) spectrum were discussed, it was pointed out that behavior is toward the end indicating a high level of integration. Thus, behaviors are influenced by a multitude of factors around and within the animal; however, it is safe to say that all of these operate through the nervous system. The sensory receptors are constantly sampling the environment and feeding this information to the central nervous system which integrates this with other external and internal information before making a behavioral response. In addition, changes in hormonal levels and blood chemistry, whether caused by natural factors or pollutants, may alter activities of particular areas of the brain, thereby influencing behavior.

Categories of behavior in fish (called "ethological units" by Smith, 1984) include those behaviors associated with schooling, feeding, migration, aggression, fear, learning, rheotropism, and attraction to or avoidance of a chemical or temperature. Even as broad a classification as this has some overlap, but it does give a rough framework. (Some workers include breathing frequency as a behavior category. In this book, breathing is covered in Chapter 3, swimming capacity is reviewed in Chapter 8, and reproductive behaviors are taken up in Chapter 13.)

For a good overview of "normal" fish behavior, the interested reader should see Pitcher (1993). The effects of pollutants on fish behavior have received increasing attention over the past decade or so, and several recent reviews have appeared (Atchison et al., 1987; Beitinger, 1990; Henry and Atchison, 1991; Blaxter and Hallers-Tjabbes, 1992; Scherer, 1992; Birge et al., 1993; Little et al., 1993). The physiology of fish sensory organs including vision, olfaction, audition, electrical, and taste are extensively reviewed in Evans (1993). In an earlier review, Smith (1984) made a good attempt to relate altered behavior to changes in function of specific brain regions in the fish, but it is clear that this aspect of neurotoxicology of fishes remains in its infancy.

First, a few general comments about the brain of teleost fishes. These organisms have the same major brain structures as the higher vertebrates, but the relative sizes of these parts are quite different. Thus, among the structures that can be easily observed, fish have relatively large cerebellums and optic and olfactory lobes, but usually a very small cerebrum. (The latter is called the telencephalon in many references.) There are marked differences in brain structure between different fish species based largely on their life style (Gutherie, 1983). For example, those species which detect food primarily by vision have large optic lobes whereas those that use olfactory sense for feeding have large olfactory bulbs. While learning in higher vertebrates relies heavily on the cerebrum, fish appear to have considerable capacity for learning in the cerebellum, as well.

In the vertebrates, muscular activity and equilibrium is largely coordinated by the cerebellum. The integration of sensory information and the initiation of an appropriate locomotor activity, to a large degree, depend on the reticulomotor (reticular) system. This large group of neurons is located in the anterior end of the medulla oblongata and projects into the mesencephalon. In fish, the portion of the reticular system called the ventrolateral peduncular neuropil is the key to nearly all motor activity (Smith, 1984). For a more extensive discussion of fish brain functional anatomy see Gutherie (1983), Northcutt and Davis (1983), and Davis and Northcutt (1983). Smith (1984) provides a more general summary of the brain regions of fish and the neurotransmitters and behaviors associated with those regions. One of the generalizations to be drawn from these reviews is that there is a good deal of functional overlap between the major portions of the brain of the fish, which compounds the problems of physiological analysis.

II. LOCOMOTOR ACTIVITY

In a sense, nearly all behaviors of fish involve locomotor activity. Indeed, it is the predictable body movements of an animal that the behavioral scientist observes and uses to draw conclusions as to purpose, causality, etc. However, here we are referring to seemingly random movements. Because the extent of muscular activity considerably influences the energetic cost of existence for a fish, this topic is also taken up from a different perspective in Chapter 8.

A. METHODS FOR MEASURING LOCOMOTOR ACTIVITY

Observations of changes in locomotor activity in response to the presence of a pollutant tended in earlier studies to be anecdotal and often associated with other experimental purposes. The fish were described as becoming "restless", "excitable", "lethargic", "dashing wildly", etc. These subjective observations can be of some qualitative use, but it is much better to quantify the extent of bodily activity. For this purpose, various devices have been developed, some of considerable complexity and/or ingenuity.

As a fish moves through the water, it will cause small water currents that have the potential of being detected if the fish is in a sufficiently small chamber. One of the earlier schemes to measure fish movements by this means was that of Spoor (1946) who suspended small paddles in the water which, when moved, caused an electrical contact to close and record on a strip-chart recorder. A more recent development along this line is that of Fisher et al. (1983) who used a movement transducer attached to the paddle to detect its oscillations. Small water currents can also be detected by changes in the rate of heat loss from a small thermoregulated chamber in the aquarium (Beamish and Mookherjii, 1964) or a thermistor in the water (LaBarbera and Vogel, 1976).

The small voltage changes produced by fish moving in the water can be detected with electrodes in the water of a small aquarium (Spoor et al., 1971), but this method is not especially quantitative. It mostly indicates whether the fish is moving around in the test tank or sitting still. It detects the breathing movements quite well (Heath, 1972), however, so it can be useful in measuring that physiological variable. These various methods offer a way of quantifying fish locomotor activity while measuring some other physiological activity such as oxygen consumption or heart rate, where the fish must be rather severely confined. However, when only the measurement of locomotor activity is desired, other techniques discussed below are probably more suitable.

Perhaps the most elaborate of the devices for quantifying locomotor activity of fish is that of Kleerekoper (1977). Briefly, this is a round steel tank either 200 or 549 cm in diameter. The central open section (100- or 250-cm diameter) is surrounded by radial dividers which separate the outer area into 16 equal chambers. These have photoelectric "gates" which are used to detect when a fish enters a chamber and into which one it goes. As a fish moves from one chamber to another, its movements are quantified by computer. This gives an indication of the speed of movement, how much exploratory activity is present, and the orientation taken when a fish exits a chamber. An improved version of this system has been developed by Scarfe et al. (1985).

A video camera coupled to a minicomputer has been used to measure movement patterns and chemical avoidance by bluegill (Lubinski et al., 1980). The fish were in a tank 50 cm square with the camera viewing from above. The position of the fish was recorded as a pair of coordinates every 4 s, and every 10 min behavioral parameters, such as direction of turn and angular size of the turn, were calculated. This system is limited to small fish in a relatively small tank. An array (1900) of photoconductive cells embedded in the floor of a larger tank permits monitoring of fish in a more natural setting or the use of larger fish (Kleerekoper, 1969). Fish locomotor patterns have also been monitored with a set of infrared emitters and photoelectric sensors dividing the tank into eight equal areas. Movement from one area to another activates a counter (Morgan, 1979).

Unfortunately, the above schemes are expensive and complicated to operate. A relatively simple way of quantifying locomotor activity has been developed by Ellgaard et al. (1977). Groups of fish are confined to a 10-gal aquarium. To run a test, they are moved to one half of the tank with a partition and a separate partition containing four holes of appropriate size is inserted to divide the aquarium in half. Then the original partition is removed and the number of fish in the empty half is counted at intervals. The data are then treated as an opposed first-order reaction to determine rate constants.

Another fairly simple although labor intensive technique is described by Little et al. (1990). They used single small trout in a circular cylinder viewed from above with a video camera. A fish was watched for a period of 2 min and with the aid of a stopwatch, the duration of time in which the fish was moving was determined. This would appear to work very well with trout, which are a relatively active species. It remains to be seen how it would work with species that are not so active.

B. METALS

The effect of copper on locomotor activity appears to vary with the species. At low sublethal concentrations copper stimulated locomotor activity in brook trout (*Salvelinus fontinalis*) (Drummond et al., 1973), however, the external electrode technique which was used for this work failed to detect a dose-dependent response. Instead of a stimulation of activity as seen in the trout, bluegill exhibited a decreased level of activity which was dependent both upon concentration and duration of exposure (Ellgaard and Guillot, 1988). Four marine teleost species were tested in the Kleerekoper apparatus and a considerable species variation was noted in their response to copper (Scarfe et al., 1982). Immediately upon being exposed to the metal (0.1 mg/L), the sea catfish (*Arius felis*) and sheepshead (*Archosargus probatocephalus*) increased their activity while the pinfish (*Lagodon rhomboides*) and 33% of the Atlantic croaker (*Micropogon undulatus*) became hypoactive. Species differences also occurred in the data on angular orientation which, according to the authors, is probably controlled by different neurophysiological mechanisms. From a toxicological standpoint (i.e., LC50), the sea catfish was found to be the most sensitive of the four species tested. Subsequent work with the sea catfish has shown the response to be dose dependent with a reasonably well-defined threshold of 0.1 mg/L (Steele, 1983).

Although copper produced hypoactivity in bluegill, the metals chromium, zinc, and cadmium have been shown in the same laboratory to cause a dose-dependent increase in locomotor activity in the same fish species (Ellgaard et al., 1978; See Table 1, Chapter 8). The authors attributed the stimulatory effect of the metals to a direct effect on cellular energy metabolism. However, this appears to me to be a confusion of cause and effect. Metals stimulate some enzymes and inhibit others (see Chapter 9) and a primary determinant of whole-animal energy metabolism is muscular activity, rather than the other way around (see Chapter 8). Moreover, there is no particular reason for alterations in basal energy expenditure, as might be produced by pollutants affecting the enzymes, to affect spontaneous muscular activity.

Henry and Atchison (1979a,b, 1986) used a variety of body movements of bluegill as indicators of sublethal concentrations of zinc, cadmium, or copper. Coughs, yawns, jerk swimming, fin flickering, agonistic behaviors, and chafing against objects were counted visually and found to increase in a dose-dependent manner. The social rank of a fish influenced the response, especially in the copper where the most subordinate and most dominant individuals were affected by the metal to the greatest extent. Agonistic behaviors by dominant individuals were increased by exposure to the metals (but see section below on aggression). Concentrations as low as 21 µg Cd/L, 99 µg Zn/L, and 34 µg Cu/L were detectable with this method. These concentrations are roughly an order of magnitude below the 96-h LC50, so the behavioral measures proved quite sensitive to the three metals tested.

In summary, many metals apparently stimulate activity in fishes which is then reflected in an elevated metabolic rate, a topic discussed at some length in Chapter 8. The mechanism(s) for metal stimulation of activity in fish is largely open to conjecture. Smith (1984) suggests they may act somewhat similar to organochlorine insecticides which are known to enhance sensory and motor nerve activity. Increased exploratory activity associated with avoidance behavior may be a key process, especially upon the initial exposure to a metal (Scarfe et al., 1982). In other words, the metals probably act as a physical irritant to a potentially wide assortment of external tissues of the fish. Where suppressions of activity are seen (e.g., copper and bluegill), inhibition of some peripheral sensory function is possible and, of course, effects on the central nervous system areas cannot be ruled out.

C. MISCELLANEOUS CHEMICALS

High levels of petroleum hydrocarbons have been reported to cause initial hyperactivity which then progresses to a general lethargy. If exposure continues, loss of equilibrium is seen (Anderson, 1975). Lower concentrations may result in a dose-dependent depression in muscular activity (Ellgaard et al., 1979). Overall, these changes have been likened to those observed with general anesthesia in fish or acute alcohol intoxication in mammals. The implication is that such behavioral alterations are due to disruption of nervous activity in the reticular formation in the brain stem.

Organochlorine insecticides tend to cause hyperactivity in fish, at least upon initial exposure to the chemical, but the quantitative measurement of this effect has generally been indirect via determination of changes in the rate of oxygen consumption (Waiwood and Johansen, 1974; Bansal et al., 1979). DDT (and possibly some other organochlorine compounds as well) causes altered function of the lateral line organs (Anderson, 1971; Bahr and Ball, 1971). Electrical recording from the lateral line nerves showed that a low frequency pressure wave in the water normally caused a short burst of nerve impulses.

Exposure to sublethal levels of DDT increased the duration of the burst, and the DDT effect is especially dramatic at lower temperatures. This latter observation is interesting in that DDT has been shown to be more toxic at lower temperatures (Cope, 1965) and fish exposed to DDT actively avoid colder waters when given a choice (Ogilvie and Anderson, 1965). We should hasten to note here that the temperature effect on organochlorine pesticide toxicity is not the same for all of them. For some of these pesticides, increased temperature increases toxicity, for others it is decreased (see Cairns et al., 1975 for review).

The hyperactivity induced by organochlorine insecticides would appear to be related to an increased sensitivity to external stimuli, a phenomenon noted in mammals exposed to pesticides of this group (Murphy, 1975). These substances could also cause altered neurochemical activity. Fingerman and Russel (1980) did a combined study of the action of the polychlorinated biphenyl Aroclor 1242 on locomotor activity and brain neurotransmitters in killifish (*Fundulus grandis*). Activity was measured visually with a grid chamber and Aroclor caused a roughly tenfold increase in activity on the first day which gradually reduced to a fivefold elevation over controls by day three of exposure. Because the solution was not changed, this decrease may reflect a declining Aroclor concentration. The catecholamines norepinephrine and dopamine in the brains decreased after 24 h of exposure. The interpretation of these findings is unclear because the function of these neurotransmitters in the brains of fishes has received little experimental investigation. Furthermore, discrete levels of the transmitter in specific brain regions were not measured. These may be more critical than the overall brain concentration.

While a number of investigators have seen stimulation of activity with organochlorines (discussed in preceding paragraph), more prolonged exposure may produce an inhibition. Little et al. (1990) observed a dose-dependent decrease in activity of juvenile trout after 96-h exposures to chlordane. At a concentration one half the 96-h LC50 (the highest dose tested), activity in the trout was reduced by approximately 50%. Pentachlorophenol caused a similar effect. Part of this discrepancy in results between the various studies may be due to the kinetics of the response. Ellgaard et al. (1977) tested bluegill at several very low concentrations of DDT over a period of up to 16 days. There was a dose-dependent stimulation in spontaneous activity, however, at concentrations in excess of 10 µg/L the stimulation was transient and an inhibition of activity was revealed after several days. The higher the dose, the less the duration of hyperactivity before the suppression.

The effect of organophosphorus insecticides on activity is contradictory; stimulation and inhibition have both been observed (Bansal et al., 1979; Rath and Misra, 1979; Johnson, 1978; Henry and Atchison, 1984; Little et al., 1990) (also see Chapter 8). The particular preparation, fish species, duration of exposure, and dose used all may contribute to the disparity of responses. It is well known that these chemicals, along with the carbamate insecticides, affect the central nervous system by inhibition of acetylcholinesterase, the enzyme that prevents the buildup of acetylcholine (Zinkl et al., 1991). As Smith (1984) points out, they inhibit acetylcholinesterase in peripheral tissues as well, which could also affect locomotor activity.

Residual chlorine is a strong oxidizing agent that irritates gill tissue (Bass and Heath, 1977). In spite of this irritation, sublethal exposures of brook charr for up to 6 days caused lethargy and an increase in thigmotaxis (Jones and Hara, 1988). In the same study it was further noted that the aberrant behavior persisted in some individuals for at least 46 days after being returned to non-chlorinated water. This long persistence is interesting as Bass and Heath (1977) observed rapid recovery in gill function following fairly high but intermittent (i.e., for less than 2 h) chlorine exposures. Thus, normal behavior recovered more slowly than some other physiological functions.

III. AVOIDANCE OF OR ATTRACTANCE TO WATERBORNE CHEMICALS

Many of the changes observed in locomotor activity of fish when they are exposed to chemical pollutants probably reflect a simple attempt to get away from the irritant, especially when test concentrations are high. In order to avoid low concentrations, the fish must first be able to detect the chemical in the water, and then go down a gradient toward "clean" water. Detection is probably the more critical factor here, and there have been several ingenious devices built to quantify this ability in fishes. They fall roughly into two types: the steep gradient chambers and the shallow gradient chambers (see Figure 1). In the former, the fish is given a rather distinct choice while in the latter, it is exposed to a range of concentrations over some defined distance. The usual procedure is to place one or more fish into the test chamber and, before adding the chemical, allow them to adjust to the new environment. Then, the test chemical is added to the water inflow on the dosed side and observations are begun. Generally, the concentration is then raised in a stepwise progression over a period of several hours. The data gathered may be the time spent in the dosed water, the number of entries into that water, or distribution of the fish after some time interval. In most instances, visual observation (often with the aid of a video camera) is used, but photocells coupled to a computer have also been successfully adapted to this sort of study (Kleerekoper et al., 1973).

There have been several reviews of fish avoidance behavior studies (Cherry and Cairns, 1982; Giattina and Garton, 1983; Atchison et al., 1987; Beitinger, 1990). Of considerable interest have been the attempts to define a threshold, or that concentration of chemical which is required to cause a significant avoidance (or preference). Table 1 gives some values for metals that have been observed. Water hardness is included in the table as this is known to greatly influence metal toxicity to fish (Sprague, 1985). With the

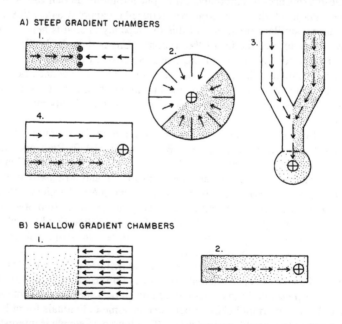

Figure 1 Schematic diagram of several of the systems used to test for avoidance or attractance of fishes to aquatic contaminants. (A) 1. Countercurrent flow chamber; 2. "rosette apparatus"; 3. "Y" trough; 4. modified fluvarium; (B) 1. fluvarium; 2. linear gradient. (From Giattina, J. D. and Garton, R. R., *Residue Rev.*, 87, 43, 1983. With permission.)

Table 1 Minimum Concentration of Waterborne Metal Producing an Avoidance Response in the Laboratory

Chemical	Concentration (mg/ℓ)	Hardness (mg/ℓ)	Species
Copper	0.1	89.5	Rainbow trout
	2.4	20	Atlantic salmon
	5.0	5.4	Goldfish
	6.4	25	Rainbow trout
	70	112	Rainbow trout
Zinc	5.6	14	Rainbow trout
	50	112	Rainbow trout
	54	20	Atlantic salmon
Cadmium	50	112	Rainbow trout
Nickel	23.9	26	Rainbow trout
Lead	26	28	Rainbow trout
Mercury (attractance)	0.2	112	Rainbow trout

References and data in Giattina, J. D. and Garton, R. R., *Residue Rev.*, 87, 43, 1983.

exception of cadmium, fish exhibited an avoidance (attractance for mercury) at concentrations well below the 96-h LC50. Indeed, it may be as low as the LOEC determined by standard chronic tests (Atchison et al., 1987). Thus, they possess the ability to sense the presence of some metals at quite low concentrations, and in some instances at a concentration below that which causes reductions in reproduction. Unfortunately, it is not possible to generalize this high ability to avoid all metals. Cadmium concentrations may actually become lethal before fish avoid it in the water (Black and Birge, 1980; Chapman, 1978), or with some species, it is not detected at concentrations manyfold higher than lethal, (Hartwell et al., 1989). Recently, McNicol and Scherer (1991) found that lake whitefish exhibited a dichotomous response to cadmium. Some of the fish were attracted to cadmium at nearly all test concentrations, others were repelled by the metal. If the direction was ignored, it was also found that lake whitefish reacted to very low and very high concentrations of cadmium, but virtually ignored the intermediate ones.

Selenium also appears to not be avoided by fish (Hartwell et al., 1989). Test conditions contribute markedly both to the qualitative and quantitative data obtained in avoidance studies. For example, goldfish in a shallow gradient actually were attracted to copper, but the same concentrations in a steep gradient resulted in avoidance (Kleerekoper et al., 1972; Westlake et al., 1974). Furthermore, high copper concentrations may attract rainbow trout, while low ones cause an avoidance behavior (Giattina et al., 1982).

Long-term acclimation to metals can affect tolerance (Anadu et al., 1989) and avoidance behavior. Hartwell et al. (1987) tested fathead minnows with a blend of metals which simulate the effluent from a fly ash settling pond of an electric power plant. Acclimation to this blend composed of copper, arsenic, chromium, and selenium at a total concentration of 48 µg/L for 3 months caused the fish to lose avoidance behavior to a concentration (245 µg/L) that was clearly avoided by controls. Surprisingly, acclimation to twice the previous level for 3 months resulted in an actual preference for elevated metals by the fish. Acclimated fish mildly avoided concentrations 5 times the holding concentration after 6 months, but by the end of 9 months of acclimation to the metals, they were not responsive to concentrations as high as 10 times the exposure level. Thus acclimation to a blend of metals caused the fish to largely lose the ability to avoid them. It seemingly is not known whether acclimation to single metals would cause a similar loss of behavioral sensitivity.

Just why under certain circumstances some metals are attractive while they are avoided or ignored under other conditions is not at all clear. Currently, there is no knowledge of the sensory modality involved in the detection of metals by fish (but see next section on sensory receptors). Receptor organs, such as those around the eyes and the gills, may be important in detection of metals in the water but no direct evidence appears to be available.

Information on avoidance of DDT is somewhat variable, depending on species and method of testing (Giattina and Garton, 1983). One of the more interesting observations is that DDT-susceptible mosquitofish (*Gambusia affinis*) avoid DDT, but those resistant do not (Kynard, 1974). Thus resistance which has been acquired to a pesticide, which may make them able to tolerate massive tissue levels (Ferguson, 1967), may also cause a loss in sensory sensitivity to that chemical. On the other hand, lest one think that the situation is that straightforward, Kynards (1974) also tested different populations of mosquitofish avoidance to parathion (an organophosphorus insecticide) and found the susceptible ones exhibited an avoidance threshold of 200 µg/L and the resistant ones 1000 g/L. This sounds like the same general trend as with the DDT; however, the 24-h LC50s were found to be 20 and 2000 µg/L, respectively. So the resistant fish avoided acutely toxic concentrations while those susceptible did not, thus selection may have favored the ones able to avoid harmful concentrations of the parathion.

A variety of other chemicals has been shown to cause avoidance responses in fish. These include pulp and paper mill effluents, the herbicide 2,4-D, and residual chlorine (see Giatinna and Garton, 1983 for references). Chlorine has probably received more attention in this regard than any other chemical. In freshwater, residual chlorine is composed of two major constituents, "free" chlorine made up of hypochlorous acid and hypochlorite ions; and combined chlorine made up mostly of monochloramine. Free chlorine is more toxic than the combined form (Heath, 1977) and fish avoid it at lower concentrations than they do the combined form (Cherry et al., 1979). They also avoid sulfur dioxide, which is used as a dechlorination chemical (Hall et al., 1984). In general, a wide variety of species, both marine and freshwater, have been found to avoid chlorine at concentrations well below the lethal level, but temperature, salinity, body size, and time of exposure all influence the results (Cherry et al., 1982; Hall et al., 1983). Temperature is an especially critical factor as an elevated temperature can attract fish into concentrations of chlorine that would otherwise be avoided. This, incidentally, is an aspect that could be a complicating factor for a variety of substances in addition to chlorine, because many effluents containing toxicants also have waste heat.

Not all chemicals are avoided by fish at concentrations below the lethal levels. These include the already mentioned metals cadmium and mercury. Other chemicals that appear to not be avoided at all or require acutely lethal concentrations to induce avoidance are PCBs, phenol, parathion, the herbicides Garlon and Vision (Giattina and Garton, 1983; Beitinger, 1990; Morgan et al., 1991), and probably some others that have not been tested. It would seem reasonable to hypothesize that this apparent failure to respond is due to an inability to detect the presence of the substance at low concentrations. However, using classical conditioning methods, Hasler and Wisby (1950) showed that fish can detect phenol at a concentration as low as 0.0005 mg/L, which is far below the lethal level. Thus, failure to detect the chemical cannot be the only reason for this lack of response. From a practical standpoint, it means that the predictive value of using this behavior for estimating chronic toxicity is severely limited. Further, it cannot be assumed that fish will avoid lethal levels of all chemicals in the environment and thereby provide some sort of safety factor.

There are examples of contaminants acting as an attractant rather than a repellant. Hara and Thompson (1978) found that whitefish (*Coregonus clupeaformis*) were attracted to the detergent sodium lauryl sulfate (SLS) at all concentrations they tested except the highest (10 mg/L), which is above the 96-h LC50. Mercuric chloride may also be an attractant for rainbow trout at a sublethal level (0.2 mg/L) (Black and Birge, 1980), and mention was made above of how copper under some circumstances can act as an attractant. However, these examples of attraction are clearly the exception.

Avoidance of chemicals can have considerable ecological importance. A habitat may be effectively eliminated by this mechanism, and thus it could explain the absence of some fish in polluted areas. For migratory species, such as salmon and eels, a polluted stretch of river could block either upstream or downstream movement. Saunders and Sprague (1967) reported that adult Atlantic salmon on their upstream migration actively avoided areas contaminated with copper and zinc at concentrations that were clearly sublethal. Based on both their field and laboratory data, they concluded that concentrations above 38 μg Cu/L and 480 μg Zn/L would effectively block migration.

The combination of a chemical that is avoided with one that is attracted can produce some interesting problems for the fish, as well as the experimenters. Alanine is a ubiquitous constituent of prey odor so many fish are attracted to it. Steele et al. (1990) arranged experiments to test the effect of copper on that attractance behavior in zebrafish. A concentration of copper of only 10 μg/L was avoided by itself, but in the presence of alanine it was not avoided. Perhaps more importantly, a copper concentration of only 1 μg/L suppressed attraction to alanine! This concentration is well below the USEPA criterion for copper (USEPA, 1986).

Determining a threshold concentration for avoidance behavior is important for pollution biologists, but extrapolations from laboratory determinations to the field may be misleading as a host of conditions such as hardness, temperature, acidity, previous exposure to the pollutant, and the sharpness of the gradients can have a large effect on the response of the fish.

IV. SENSORY RECEPTORS

The sense organs are the way the central nervous system samples the environment. Most of the locomotor and avoidance behaviors which have been discussed herein are actually the result of some sensory input. In this section, the concern will be with the action of pollutant chemicals on receptor function per se. Because of the technical difficulties of investigating these physiological functions, the database is severely limited. Research on the effects of pollutants has generally fallen into three areas: (1) histological lesions in specific sensory receptor organs, (2) alterations in nerve impulse frequency as recorded from sensory nerves during or following exposure to a test chemical, and (3) recording of summed electrical activity directly from the olfactory organ or eye while stimulating the organ with a chemical or light. The resulting recording is referred to as an electro-olfactogram (EOG) and electroretinogram (ERG), respectively.

In addition to the usual senses that most people think of such as sight, touch, temperature, etc., fish are especially well endowed with chemical senses (Hara, 1993). They have olfactory receptors, generally located in the paired nostrils, which sense material dissolved in the water and the gustatory sensors which are located in the roof of the mouth, on the barbels (if present), gill rakers, and in some species, over the entire body. These chemoreceptors serve for feeding, kin recognition, alarm communication, reproduction, and navigation.

The olfactory receptors have received the bulk of attention from fish researchers interested in pollution effects on sensory function. For example, the olfactory epithelium has been examined with scanning electron microscopy before and after exposure to benzene, a common petroleum hydrocarbon (Babcock, 1985). It was found that the epithelium experiences physical damage from exposure to sublethal levels of this hydrocarbon in the water.

The lake whitefish is normally attracted to a food extract and it has been shown by cauterizing the nares that they are attracted to the odor. This olfactory function for detecting food is significantly reduced by chronic exposure to 50 µg HgCl/L (Kamchen and Hara, 1980). The well-known homing migrations of adult salmon have been shown to depend on the olfactory sense (Stabell, 1992). Salmon follow one or more odors up the stream until they find the home where they spawn. This olfactory homing is extremely sensitive to "foreign" substances in the water as is exemplified by the observation that rinsing one's hands in the stream will cause the migrating adults to reverse direction and retreat downstream (Smith, 1982).

When the olfactory bulb of a salmon or trout is perfused with a solution containing any of several different amino acids (alanine is a popular one to use), a characteristic burst of nerve impulses can be recorded from the olfactory nerves, or a distinct EOG can also be detected. However, if the fish are first exposed to inorganic mercury (>0.10 mg/L) or copper (>0.008 mg/L) there is a depression of the response within 2 h. These metal concentrations are 3–5 times lower than the lethal levels. The degree of neurological depression increased with time of exposure and concentration (Hara et al., 1976). Increasing the calcium level in the water reduces the inhibitory effect of copper on the receptor function (Bjerselius et al., 1993). Copper at concentrations as low as 0.044 mg/L have been shown to change the attractiveness of homestream water to migrating salmonids (Sutterlin and Gray, 1973), and this is well above the concentration causing inhibition of olfaction sensitivity to amino acids. (Also see the above discussion on avoidance of metals in the field and laboratory.)

The mechanism by which metals alter olfactory response of fish has received some attention. In high doses of copper (0.5–5.0 mg/L) estuarine fishes exhibit histological lesions of the olfactory mucosa (Gardner and LaRoche, 1973). At lower doses, Hara et al. (1976) hypothesized the effect to be primarily due to the metals tying up sulfhydryl and amino groups in the membranes of the receptor cells. Because they observed a rather rapid response and quick recovery, it was suggested the effect may be limited to the membrane, at least at the lower levels they used. Winberg et al. (1992) present evidence that low concentrations of copper do not inhibit binding of odorants to the receptor membrane. Instead, they propose that the copper inhibits the transduction mechanism of the receptor cells (i.e., the conversion of the presence of odorant into nerve impulses).

Further work on the mechanisms of inhibition has been done using methylmercury and inorganic mercury. Baatrup et al. (1990) found that methylmercury accumulates mostly in the olfactory receptor cells of the sensory epithelium whereas inorganic mercury accumulates mainly along the borders of the receptor and in the supporting cells of the epithelium. What was especially striking was how the methylmercury caused a more permanent damage to the olfactory function in that rinsing with freshwater failed to restore normal activity. Baatrup et al. (1990) suggest that the mercury ion acts on the calcium gate of the olfactory receptors and inhibits the binding of amino acids to the membrane. Methylmercury, on the other hand, has little effect on calcium channels for diffusion but does affect active transport of calcium. Methylmercury further accumulates in the cell more than the inorganic form thereby potentially causing effects on enzyme function there (Baatrup et al., 1990).

In addition to the receptors for smell, the function of taste receptors is also blocked by mercury and copper (Hikada, 1970; Sutterlin and Sutterlin, 1970) so these metals appear to have a rather broad inhibitory effect on chemoreceptors. Recall that copper is avoided at very low concentrations by fish whereas high concentrations or steep gradients may actually cause attraction. This raises the interesting question of what is being detected in the avoidance/preference responses. One could speculate that since copper and mercury are such effective chemosensory inhibitors, perhaps it is the inhibition of detection of one or more substances (rather than the metal, per se) that is the effective stimulus when fish are exposed to either of these elements.

As mentioned earlier, the detergent SLS is an attractant for whitefish (Hara and Thompson, 1978). The ABS ("hard") detergents have also been found to inhibit feeding behavior in catfish (Bardach et al., 1965) and in the flagfish (*Jordanella floridae*) (Foster et al., 1966). Hara and Thompson (1978) have shown neurologically how SLS may also affect feeding. They recorded nervous activity from the olfactory bulb of fish exposed to this detergent and a standard food stimulus. The response to the food stimulus was reduced almost immediately upon exposure to the detergent and the response was dose dependent over a range of 0.1–10 mg/L. This is, of course, similar to the effects seen with mercury and copper, however, the mechanism is undoubtedly different. With a detergent, the observed depression of sensory function may involve removal of the thin mucous layer over the receptor membranes. This layer is a source for the inorganic ions required for the sensory transduction process (Hara, 1993).

The petroleum hydrocarbon naphthalene has been found to induce histological lesions in the olfactory, gustatory, and lateral line receptors at concentrations that cause no effect on other organs except gill epithelium (DiMichelle and Taylor, 1978), thus these sensory receptors may be one of the first organs to suffer from this chemical insult. Other components of petroleum appear to cause specific alterations in the olfactory organ epithelium. According to Gardner (1978), whole crude oil causes hyperplasia, water soluble components produce epithelial metaplasia, and water insoluble components induce dilation and congestion of submucosal vasculature. Scanning electron microscopic examination of the olfactory epithelium of sole following an 8-day exposure to the water-soluble fraction of Prudhoe Bay crude oil "revealed degenerative changes in the chemosensory cilia and a loss of the microridges that circumscribe the perimeter of the epithelial cells surrounding the olfactory organs in six of eight fish" (Hawkes 1980; p. 3230).

Rather specific lesions in olfactory tissue may also be characteristic of the action of copper, mercury, and silver, at least in some species (Gardner, 1975). However, cadmium does not seemingly affect olfactory tissue (Gardner, 1978). This last observation may also relate to the finding that cadmium is not avoided by fish except at high concentrations (see section on avoidance).

The lesions in the olfactory organs caused by metals such as copper and zinc apparently are not permanent. Moran et al. (1987) noted that fish that had experienced nearly complete loss of receptor epithelium from having been accidentally exposed to metals in the water showed good regeneration within 8 days of being placed in non-contaminated waters.

Lead and zinc may, in addition to affecting olfactory organs, cause histological effects to the lateral line organs if the metal concentrations are fairly high (Haider, 1979). The lateral line organs of fish serve primarily to detect water movements and transient pressure changes and hydrodynamic interactions in the immediate vicinity of a fish (Popper and Platt, 1993). Temperature has a marked effect on the neurogenic activity of these receptor organs (Peters and Weber, 1977).

In trout the impulse frequency in nerves from the lateral line exhibits a rather steady spontaneous rate which is directly proportional to temperature. Peters and Weber (1977) investigated the action of DDT on the output from the lateral line organs and found that the insecticide caused the impulse frequency from the receptors to take on a bursting character at temperatures below 12°C. This altered pattern of activity may help explain the changes in orientation and schooling activity observed in goldfish exposed to DDT at sublethal levels (Weis and Weis, 1974).

It has frequently been observed that fish exposed to DDT exhibit hyperexcitability to mechanical stimuli, such as those produced by tapping on the side of an aquarium. This is sensed by the lateral line so Peters and Weber (1977) tested the threshold of the lateral line organs by providing them with a controlled mechanical stimulus and found they were not altered by DDT, even though Anderson (1968) had earlier reported that DDT caused the lateral line organs in trout to produce a more prolonged burst of activity following a single mechanical stimulus. Thus, the DDT effect on reflex responses to external stimuli (Anderson, 1970; Peters and Weber, 1977) must involve central nervous system changes, perhaps in part brought about by the abnormal input from the bursting activity seen at lower temperatures and more direct effects of DDT on neuronal function within the nervous system (Narahashi and Haas, 1968).

There has been little work on the effects of pollutants on the eyes and associated structures of fish. Copper has been reported to cause damage to the cornea of larval striped bass (Bodammer, 1985). The visual system is quite sensitive to mercury poisoning in mammals (Evans et al., 1975) and it may also affect that of fish. Panigrahi and Misra (1978) reported blindness and exophthalmia (protrusion of the eyeball) in Indian catfish exposed to mercury for periods of 4 weeks. They attribute the blindness to effects of the mercury on the brain and optic nerves, although no direct evidence is provided. It is perhaps noteworthy that methylmercury tends to concentrate in the mammalian brain preferentially in the visual and motor areas (Shaw et al., 1975). Using a psychophysical technique, Hawryshyn et al. (1982) found that rainbow trout injected with methylmercury 15 days previous to testing exhibited a loss of both rod and cone sensitivity. This is in contrast to the finding for mammals that only the rod reception was affected by mercury (Fox, 1979).

V. FEEDING AND PREDATOR–PREY BEHAVIOR

A wide variety of pollutants at low sublethal concentrations have been shown to depress feeding rate. These include residual chlorine (Jones and Hara, 1988), copper (Drummond et al., 1973; Waiwood and Beamish, 1978), petroleum components (Woodward et al., 1987), detergents (Foster et al., 1966), dioxins (Mehrle et al., 1988), organophosphorus and organochlorine insecticides (Little et al., 1990), and low pH water (Jones et al., 1985). The mechanism for this suppression of feeding is largely unknown. Feeding and the behaviors associated with it, such as predation, represent a high level of integration of input from gustatory, lateral line, visual, olfactory, and internal visceral receptors. A pollutant may act on any or all of these receptors as well as influence the central nervous system directly by causing lesions in certain areas or changing the rate of neurotransmitter production or breakdown. Regarding the latter, in a very interesting study, Pavlov et al. (1993) noted reduced feeding in bream contaminated with an organophosphorus insecticide and the suppression of food consumption correlated with reduced acetylcholinesterase in the brain. They then followed this observation up with intraperitoneal injection of cholinergic drugs and one which restores acetylcholinesterase. These treatments caused a return to normal feeding thus providing strong evidence for the involvement of the neurotransmitter acetylcholine in the control of feeding. The control of appetite by the brain of fish (Fletcher, 1984) (as well as other animals) is poorly understood and undoubtedly involves several areas with the ventrolateral peduncular neuropil probably being the primary one.

The depression of feeding by acute exposures to toxicants may reflect some deficit in peripheral sensory function rather than brain activity, particularly where the effect occurs rapidly upon pollutant exposure (see preceding section on sensory effects). Elevations in the hormone cortisol in response to stress could also suppress appetite because, at least in some species, it causes a rise in blood glucose (see Chapter 8). Little et al. (1990)

conclude from their work with juvenile rainbow trout and several agricultural pesticides that inhibited motivation for feeding occurred at higher test concentrations; but at lower concentrations, reduced feeding efficiency and fewer strike frequencies (at prey) predominated.

The behaviors associated with feeding have received considerable attention and there has been some interest in the effect previous exposure to a toxicant has on the predator–prey interaction. Essentially two experimental approaches have been used, which Giddings (1981) refers to as the "mechanistic approach" and the "population approach". In the former, various aspects of the specific behaviors, such as reactive distance, handling time, or capture success are measured. The population approach is done by enclosing prey in the same chamber with the predator and counting survivors of the prey after some interval of time. Refuge areas are often provided for prey and hiding spots for predators are available in some systems. If the chemical causes altered behavior, such as jerks or hyperactivity in the prey, this presumably will be reflected in a greater rate of predation. The methodology involved in these studies has been discussed at some length by Giddings (1981) and Little et al. (1985).

One of the first to use the predator–prey interaction in a toxicological test was Goodyear (1972) who irradiated mosquitofish with ionizing radiation and then tested their survival rate in groups with a single largemouth bass. Irradiation caused a much more rapid loss of fish by predation due to these fish tending to wander out of the refuge which was provided in the test chamber. Organophosphorus insecticides (Hatfield and Anderson, 1972), mercury (Kania and Ohara, 1974), cadmium (Sullivan et al., 1978), hydrazine (Fisher et al., 1980), pentachlorophenol (Brown et al., 1985), and methyl parathion (Farr, 1978), all at low sublethal levels, have been shown to increase predation rate on prey, or in some way alter the interaction. In a carefully conducted study, Sullivan et al. (1978) showed that exposure to 50 µg/L cadmium for 48 h actually decreased prey vulnerability of fathead minnows. They attribute this to an example of "sufficient challenge" in which an organism apparently benefits from a small dose of some substance that is harmful if prolonged or at a higher dose (Smyth, 1967). However, when the minnows were exposed for 21 days to the same or lower concentration of cadmium, predation was increased significantly. The actual behavioral differences observed in the prey were rather subtle. This is a schooling species and the researchers noted that the cadmium-treated fish often would briefly change direction in the school and/or orient at right angles to the others for several seconds (they were branded so as to be recognizable among non-exposed members). Holcombe et al. (1980) also observed a decline in schooling activity in fathead minnows exposed to the herbicide 2,4,-D. The lateral line organs are important in schooling behavior (Popper and Platt, 1993) so these findings suggest, but do not prove, an effect of the test chemical on that sensory receptor. Because schooling is considered to be of survival value to avoid predation (Taylor, 1976), it is easy to see how this altered behavior of individuals would make them more vulnerable to predators.

Apparently the action of metals on schooling behavior differs with the metal, or perhaps fish species. Koltes (1985) used a computerized video technique to examine the effect of copper on Atlantic silverside (*Medidia*) schools. She observed that low concentrations of copper caused the schools to increase activity, individual fish decreased their nearest neighbor distances and swam more in parallel orientation. In a sense, the schools became more uniform although swimming at a higher speed. It is not clear whether this would affect predation.

Most studies of predator–prey behavior use fish that have been raised in hatcheries or kept in the laboratory for a considerable period of time. Weis and Khan (1991), on the other hand, obtained mummichogs (*F. heteroclitus*) from a highly polluted estuary and conspecifics from a pristine habitat. Those from the polluted habitat exhibited a reduced ability to capture juvenile guppies.

A rather novel type of predator–prey interaction was investigated by Fisher et al. (1980) using the dorsal light response of bluegill (*Lepomis macrochirus*). In this case, the predator was the test animal. Most fish will tilt the dorsal surface of the body toward a strong light source when it is first turned on, and the degree of tilt increases if the fish sees desirable prey. Indeed, the more "desirable" the prey, the greater the tilt and this can be quantified using an appropriate chamber fixed with a protractor (Vinyard and O'Brian, 1975). In this study, the effect of hydrazine (a widely used reagent in jet fuels, herbicides, pharmaceutical drugs, etc.) on the predator was measured by exposing the fish for only 10–15 min before testing. An artificial prey was used for which the bluegills were to respond. A significant reduction in tilt occurred at a concentration of hydrazine approximately 1/10 the 96-h LC50 for this species. The authors wisely avoid drawing any ecological implications from this study.

Exposing only the prey, or the predator, to the test chemical is actually a bit of an artificial situation since, in nature, presumably both will be inhabiting the same water. Thus, the work of Woltering et al. (1978) is of interest because they exposed both the predator (largemouth bass) and prey (mosquitofish) to ammonia. At concentrations above 0.34 mg/L there was actually a lower rate of predation. When prey densities were raised, the ammonia effect was greater and the prey even harassed the predator!

One of the most thorough predator–prey studies is that of Sandheinrich and Atchison (1989). They used a mechanistic approach which yielded considerably more information than merely the feeding rate. Bluegill were exposed for 4 days to 4 concentrations of copper ranging from 5–1700 µg/L. Prey used were two species of *Daphnia*, an amphipod, and two sizes of zygoptera. Both copper-treated and untreated prey were tested. Prey handling time was the most consistent parameter to be altered. It increased significantly with increasing copper concentrations and this effect occurred with both treated and untreated prey. This caused overall consumption rate of prey to go down in a dose-dependent fashion, which, of course can be predicted to depress growth rate. Prey handling time refers to the interval between when a prey item is taken in and the fish begins to hunt another item. The mechanism by which copper, or anything else, changes prey handling time is apparently unknown.

In summary, we see that the behaviors associated with predator-prey interactions are extremely sensitive to a variety of chemicals in the water. Increased vulnerability to being eaten could have population consequences for the prey, but if the same water is causing the predator to feed less, for whatever reason, then the two effects may even cancel each other out, or the predator may experience more of an effect than the prey. Sandheinrich and Atchison (1990) make a strong plea for a more mechanistic approach to these sorts of studies so the findings can be indicative of effects that might occur in natural communities. They show how with the use of foraging theory and bioenergetics modeling that predictions of effects on growth rate can be made based on things like alterations in food consumption and increased cost of food handling by predators. These predictions can then be field verified. The possibilities for future research on this seem promising.

VI. AGGRESSION

Here we are referring to intraspecific aggressive behaviors associated with territoriality and social hierarchies. A particular species often shows characteristic aggressive displays which can be quantified, although for the inexperienced observer, there would be a large degree of subjectivity. Smith (1984) has reviewed the literature on the neurobiological basis of aggression in fish. In short, it involves mostly the telencephalon portion of the forebrain (which incidentally also serves for olfaction). Evidently, the telencephalon of the fish brain is analogous to the limbic system of higher vertebrates where most of the emotional responses are integrated. There are lower brain nuclei in fish where the behaviors may initiate, but they are then facilitated in the telencephalon.

Bluegill are prone to develop social hierarchies, at least in captivity. Henry and Atchison (1979a) exposed this species for 15 days to several concentrations of a mixture of cadmium and zinc. Aggression, as measured by number of nips, increased at the low concentrations while it decreased at the higher ones, so the effect was dependent on dose but not in a linear fashion. Henry and Atchison (1986) later observed that copper at only 34 µg/L for 96 h caused dominant bluegill to increase their number of threats and nips.

Exposure of bluegill groups to methyl parathion at concentrations less than 1/10 the 96-h LC50 caused a reduction in aggressive behaviors with the greatest effect seen in the dominant fish during the first 24 h (Henry and Atchison, 1984). This same organophosphorus insecticide caused male Siamese fighting fish to ignore each other after being exposed for 5 days to a fairly high (1.0 mg/L) dose (Welsh and Hanselka, 1972).

Aggressive interactions can cause the "loser" to become more susceptible to a toxicant. Sparks et al. (1972) found that the submissive bluegill of a pair was less able to resist a lethal level of zinc than was the dominant one. However, if a shelter (overturned flowerpot) was provided, the aggressive interactions were reduced and the differential sensitivity to zinc was eliminated. The authors note that these sorts of behavioral hierarchies may contribute to variability in toxicity bioassays.

Because toxicants usually alter the frequency of aggressive encounters, rather than their character, it has been suggested that the mode of action may be on the telencephalon, because the same thing occurs when certain neural pathways there are severed (Smith, 1984). In general, it appears that aggressive behaviors may not be as sensitive to pollutant exposures as other behaviors such as general locomotor activity and prey–predator interactions.

VII. LEARNING

Laboratory studies of learning in fish are largely limited to conditioned responses. The literature on conditioning in fishes has been reviewed by Marcucella and Abramson (1978) in a chapter entitled, "Behavioral Toxicology and Teleost Fish". Unfortunately, that title is rather misleading as there is nothing on the action of environmental pollutants included. There is a good discussion of the action of various drugs on conditioned responses which can be useful in interpreting the neurogenic basis of a particular response.

In classical conditioning studies, a stimulus, generally a light, is administered to the fish at the same time it receives an electrical shock through the water. Of course, the fish exhibits an immediate escape maneuver. The same procedure is repeated until the fish "learns" to associate the light with the unconditioned stimulus and responds before the shock is given. It is then deemed trained. Weir and Hine (1970) trained goldfish in this manner and then exposed them to one of a variety of metals for either 24 or 48 h before testing again. Therefore, this was an investigation of the effect of the metals on retention, rather than learning. They observed a marked dose- and time-dependent impairment of the conditioned response. The metals tested were arsenic, lead, mercury, and selenium, and in all cases, significant impairment was observed at concentrations below the estimated LC 1, so the behavioral criterion was quite sensitive to the metals in the water.

There has been a fair amount of interest in the effect of insecticides, especially DDT, on fish learning, but the results have been unclear. When a mild stimulus is applied to the gular region of a fish, it will twist its tail in a propeller-like manner. Anderson and Prins

(1970) found that this reflex could be conditioned using light as the conditioned stimulus and a mild electric shock applied to the gular area as the unconditioned stimulus. A 24-h exposure to a sublethal dose of DDT caused a greatly reduced ability to "learn" this reflex suggesting that DDT had some effect on the central nervous system to inhibit learning. Using a more conventional type of conditioning setup, they also showed a reduction in learning from DDT exposure (Anderson and Peterson, 1969), but when this was reevaluated in the same laboratory with a modified procedure, no effect of the DDT was noted (Jackson et al., 1970). The previous findings may have been due to an impairment of performance, rather than learning, an important point to consider when doing "learning" studies.

Several different experiments were performed by Hatfield and Johansen (1972) to evaluate the action insecticides of different types have on both learning and retention. A 24-h exposure to the 96-h LC50 concentration caused a slight enhancement of learning with DDT, no effect from methoxychlor, and a mild to severe retardation of learning with the organophosphorus compounds Abate and sumithion. Retention was relatively unaffected by the insecticides and the alterations in learning were apparently not permanent as 7 days of recovery in non-contaminated water returned the fish to "normal". Treatment at 1/10 the 96-h LC50 with each of the chemicals caused no effect on learning, but whether a chronic exposure would result in changes is unknown from these data.

The organophosphate parathion clearly inhibits both learning and retention in goldfish. Sun and Taylor (1983) found that 24-h exposure prior to training had a dose-dependent effect on learning, and it is noteworthy that the lowest concentration found effective (100 µg/L) was commonly observed in irrigation streams in California. Retention of the conditioned response was affected to approximately the same extent as learning when the fish were exposed to parathion after training and prior to retention testing. The mechanism(s) by which xenobiotics may affect learning are diverse. These include motivation, attention, sensitivity of sense organs and, of course, the memory process itself (Marcucella and Abramson, 1978).

Parathion is a strong anticholinesterase substance. Goldfish exposed to this insecticide at a concentration the same as used by Sun and Taylor for 24 h exhibited a 62% decrease in cholinesterase activity in the brain (Weiss, 1961). Anticholinesterase drugs in general are known to suppress both learning and retention in a variety of animals (Deutsch, 1971), so this is at least one of the probable mechanisms involved.

In concluding this section, it seems pertinent to note that since the early 1970s, there has been rather little interest in studying the action of pollutants on learning processes in fishes. It is not obvious just why this is so, but some tentative conjectures will be attempted here:

1. The conditioned response is the only sort of "learning" thus far used and it appears that short-term acute exposures have relatively little effect unless the concentrations of the test substance is at the upper end of the sublethal range. Thus, it is not an especially sensitive test for screening pollutants.
2. The conditioned response test is primarily a tool of the comparative psychologists so most aquatic biologists and physiologists are not oriented toward its use.
3. It is difficult to draw ecological relevance from results of conditioned response studies.

Of course, these "reasons" are not mutually exclusive and should not mean that such investigations may not prove fruitful. Because foraging behavior is rather sensitive to the presence of contaminants, and in natural environments learning is an important component of foraging (Hart, 1993), combining learning and foraging behavior studies might prove useful.

VIII. OPTOMOTOR RESPONSE

The optomotor response refers to the maintaining of an unchanging position relative to a moving visual stimulus (Scherer and Harrison, 1979). Fish use this in schooling and in maintaining position relative to the shore in a fast moving stream. In the laboratory it is usually assessed by moving a background past fish around a small circular chamber and determining how well the fish tracks the background. The optomotor response thus involves visual function, integration of information by the central nervous system, coordinated motor activity, and swimming capacity, all of which might be sensitive to toxicants in the environment. The effects of pollutants on swimming capacity, which is usually measured in swimming tunnels was discussed in Chapter 8. Here, the orientation is more on the behavioral aspects of this response, but as we will see, the variables of swimming capacity and optomotor response interact with each other.

Dodson and Mayfield (1979) determined that exposure of rainbow trout for 24 h to the herbicides Diquat and Simazine (fenitrothion) at concentrations commonly seen in the field caused loss of optomotor response such that it would probably cause downstream displacement of the fish if they had been in a normally fast flowing trout stream. Their procedure involved having the fish swim around the circle of a chamber as they attempted to keep up with a rotating outer drum. Those exposed did not swim as well and some tended to go in the opposite direction. However, using the same agricultural chemicals, Peyster and Long (1993) observed an actual enhancement of optomotor response in fathead minnows. They used a different procedure in which the drum was rotated slowly in alternating directions during six 1-min periods to test response latency. This variable decreased with chemical exposure so the fish were responding more quickly to the stimulus. They also tested swimming capacity in a water tunnel and found the pesticides caused a considerable decrease in that parameter. They speculate that the increased response to the stimulus could be an example of hormesis, or it could reflect an increase in sensitivity to stimuli, such as seen with substances that inhibit acetylcholinesterase (McKim et al., 1987). Fenitrothion is a strong inhibitor of that enzyme; it is not known if Diquat does the same. In the Dodson and Mayfield (1979) study, it seems to me that they were actually measuring swim capacity more than optomotor reflex. At the risk of sounding pedantic, one cannot help but note that it is a good example of how experimental design can certainly affect results.

Using a different approach, but still with a rotating striped drum around the central fish tank, Richmonds and Dutta (1992) tested bluegill and malathion, an organophosphorus insecticide. Their procedure was to rotate the drum and count the frequency of turns either in the direction of the drum or in the opposite direction (reversals). Exposures to the insecticide were for 24 h and there appeared to be a sort of threshold concentration of 0.016 ppm. At that concentration a distinct increase in orientation toward the drum rotation direction occurred. Reversals were relatively rare but these also increased at that concentration. At a concentration of 0.032 ppm orientation was similar to control levels and at 0.048 ppm the fish became quite lethargic. So at a concentration below that producing distinct toxicity, a state of hypersensitivity seems to take place. In the same laboratory (Dutta et al., 1992) the pesticide diazinon (another cholinesterase inhibitor) was tested and it was found to decrease the optomotor response in a dose-dependent fashion. They also observed a rather good correlation between inhibition of cholinesterase in the brain and inhibition of optomotor response. However, surprisingly, they do not comment on their seemingly contradictory findings in the two studies.

When trout fry emerge from the gravel in a stream they normally orient into the current. Ostrander et al. (1990) exposed the eggs during late embryonic development to benzo[a]pyrene, a component of petroleum and other things containing coal tars, at a concentration of 25 ppm. The emerging fish appeared normal except they did not orient into the current normally.

IX. ACETYLCHOLINESTERASE

This enzyme (AChE), frequently referred to as cholinesterase, plays an important role in the regulation of nerve impulse transmission at cholinergic synapses. It hydrolyzes acetylcholine, a common neurotransmitter, and thereby prevents it from accumulating in and around a synapse. Blood plasma contains a second form of cholinesterase called pseudocholinesterase, which has no known function, although *in vitro*, it hydrolyzes acetylcholine. The measurement of this pseudocholinesterase is a commonly used diagnostic tool in human and veterinary clinical laboratories for detecting individuals who have been exposed to organophosphate insecticides.

Inhibition of brain cholinesterase in fishes exposed to organophosphate or carbamate insecticides in food or water has been reported by many workers (e.g., Weiss, 1961; Holland et al., 1967; Post, 1969; Coppage, 1977; Malyarevskaya, 1979; Gantverg and Perevoznikov, 1984; Rao and Rao, 1989; Heath et al., 1993). A review of the use of AChE inhibition as a diagnostic tool for organophosphate or carbamate poisoning of fish in field studies has been published by Zinkl et al. (1991).

Gibson et al., (1969) have pointed out that care must be taken in interpreting results of AChE measurements in fish brain as different parts of this structure have markedly different normal concentrations. It is remarkable how much of an inhibition of AChE that a fish can tolerate before death. In general, it appears that around a 70–80% loss of activity must take place before death of the fish (Coppage and Mathews, 1974; Gantverg and Perevoznikov, 1984). Of course, less extensive inhibition is associated with a wide range of behavioral effects such as hyperactivity and loss of equilibrium (discussed in Zinkl et al., 1991), some of which may prove lethal for reasons other than direct poisoning by the pesticide. There appears to be a considerable species difference in the degree of AChE inhibition experienced with a given dose. When perch (*Perca fluviatilis*) and carp (*Cyprinus carpio*) were exposed to the same concentration (0.75 mg/L) of Carbophos, the brain AChE activity in the perch rapidly went to zero within 5 h whereas carp exhibited only a mild transient reduction in AChE activity (Gantverg and Perevoznikov, 1984). Indeed, the carp were able to tolerate 5 mg/L for at least 40 h with a 40–70% reduction in AChE, but no obvious signs of toxicity. Gantverg and Perevoznikov (1984) propose no mechanism for such a large species difference although several seem possible. They include differences in rate of uptake of the pesticide into the body and brain, different rates of detoxification, and different rates of activation. Regarding the latter, they do mention that carbophos has no direct inhibitory effect on AChE but that oxidative enzymes in the animal body convert it into malaoxon which is an anticholinesterase. Thus, the perch could be doing this at a more rapid rate.

A variety of tissues besides brain also have AChE. The extent of methyl parathion inhibition of this enzyme in these tissues was compared in tilapia which were exposed for 48 h to a dose one third the LC50 (Rao and Rao, 1984). Following this treatment, the relative amount of AChE inhibition was found to be greatest in brain, followed by muscle, gill, and liver in that order. This is the same order of degree of innervation of the tissue, and the control levels of AChE show a good correlation with that pattern, as well. Thus, the higher the AChE level in a tissue, the more susceptible it is to inhibition.

Small fish may show greater inhibition than larger ones to the same dose of poison (Rath and Misra, 1981). Small fish may also recover more rapidly when placed in clean water (Rath and Misra, 1981), however, Heath et al. (1993) found little recovery of AChE activity in larval striped bass 10 days after being exposed to methyl parathion for 4 days

Table 2 Molarity of Test Chemical Causing
50% Reduction of Muscle Acetylcholinesterase
Activity *In Vitro*

Chemical	I 50	Chemical	I 50
Carbaryl	1.0×10^{-5}	Dieldrin	2.2×10^{-3}
Malaoxon	1.8×10^{-5}	Methyl Hg	5.0×10^{-3}
Cu^{2+}	1.6×10^{-4}	Diazinon	5.0×10^{-3}
Pb^+	5.0×10^{-4}	Atropine	5.7×10^{-3}
Cd^{2+}	5.7×10^{-4}	Malathion	5.7×10^{-3}
Tubocurarine	1.0×10^{-3}	Cr^{2+}	7.1×10^{-3}
Hg^{2+}	1.6×10^{-3}	Zn^{2+}	1.0×10^{-2}

Data from Olson, D. L. and Christensen, G. M., *Environ. Res.,*
21, 327, 1980.

at a low sublethal level. As a rule, recovery is much slower in fish than in mammals (Wallace and Herzberg, 1988). Depending on the dose and exposure duration, it can take days to weeks for recovery to take place (Weiss, 1961; van der Well and Welling, 1989; Morgan et al., 1990; Heath et al., 1993).

Organophosphorus and carbamate insecticides clearly are potent AChE inhibitors, but they may not be the only pollutants that cause this effect in fishes. *In vitro* studies have shown that a variety of other substances exhibit a strong inhibitory action on this enzyme. From the data in Table 2, it is interesting to note that carbaryl and malaoxon were the strongest inhibitors, as might be expected, but several metals are also quite effective. Indeed, they are even more potent than malathion, a widely used organophosphate insecticide. Extrapolating from *in vitro* data such as these, to the intact animal is, of course, fraught with uncertainties. As was pointed out in Chapter 9, the *in vivo* effect can often be considerably different than that seen when the test chemical is merely added to a reaction mixture *in vitro*. However, these findings do suggest that we should not ignore the possibility that other chemicals, besides the well-studied pesticides, may inhibit AChE. For example, Shaw and Panigraphi (1990) found as much as a 26% inhibition of AChE in fish from a mercury-contaminated estuary. They also analyzed for mercury in the brain of the fish and found a close correlation between the level of mercury and AChE inhibition.

Even factors that are not necessarily anthropogenic may reduce AChE activity. Malyarevskaya (1979) reported that the toxins of bluegreen algae or even environmental hypoxia caused a severe drop in brain and liver AChE of perch. This finding of an inhibitory effect of hypoxia (the level of hypoxia was not mentioned in the paper) is interesting but I was unable to confirm it in my laboratory using trout and bluegill as experimental animals. Nevertheless, it appears that considerable caution is needed in concluding pesticide poisoning from inhibition of AChE alone.

Recently it has been revealed that injection of adrenaline into perch can produce an increase (doubling) in brain AChE (Pavlov et al., 1994). Increased protein synthesis was also observed and the authors note that this same thing occurs in mammals and is the result of induction of adenylate and guanylate cyclases by adrenaline. When fish are under stress, they typically exhibit elevations in adrenaline (discussed in Chapter 8). Thus, it is tempting to conjecture that this mechanism may actually reduce the extent of AChE inhibition in fish exposed to acute levels of pesticides that inhibit AChE.

X. CONCLUDING COMMENTS

In concluding this chapter, a few general points seem appropriate to emphasize:

1. Some fish behaviors (e.g., locomotor activity and avoidance) are extremely sensitive to pollutant chemicals, whereas others (e.g., aggression) seem to be rather refractory.
2. Interpretations of a change in behavior from a mechanistic standpoint must consider the fact that the test chemical will often simultaneously act on a variety of points in the nervous system. These include the sensory receptors, specific neurons in the brain, rates of neurotransmitter secretion, and rates of neurotransmitter inactivation. Thus, care must be taken to avoid jumping to conclusions as to "mode of action" when other additional and conceivably more important modes were not even examined.
3. Some behavioral alterations may be modulated by pollutant-induced changes in other physiological functions such as osmoregulation, respiration, and metabolism of hormones. Because behaviors involve such a large degree of integration in the organism, understanding the mechanisms of their alterations in response to pollution will be an especially difficult challenge.
4. Finally, there is a great need in future studies to try to close the gap in our knowledge as to what extent, if any, a change in some behavior affects populations and community structure. In some cases, for example social behavior, alterations in behavior may be an artifact of the lab setup and have little or no effect on the ecology of the organisms. Other variables such as predator-prey relationships may have profound effects on the community structure but this needs quantifying.

REFERENCES

Anadu, D., Chapman, G., Curtis, L. and Tubb, L., Effect of zinc exposure on subsequent acute tolerance to heavy metals in rainbow trout, *Bull. Environ. Contam. Toxicol.,* 43, 329, 1989.

Anderson, J. M., Laboratory studies on the effects of oil on marine organisms: an overview, *Am. Petrol. Inst. Publ.,* 4249, 70, 1975.

Anderson, J. M., Assessment of the effects of pollutants on the physiology and behavior. II. Sublethal effects and changes in ecosystems, *Proc. R. Soc. Lond. B.,* 177, 307, 1971.

Anderson, J. M., Effect of sublethal DDT on the lateral line of brook trout, *Salvelinus fontinalis, J. Fish. Res. Bd. Can.,* 25, 2677, 1968.

Anderson, J. M. and Prins, H. B., Effects of sublethal DDT on a simple reflex in brook trout, *J. Fish. Res. Bd. Can.,* 27, 331, 1970.

Anderson, J. M. and Peterson, M. R., DDT: sublethal effects on nervous system function in brook trout *(Salvelinus fontinalis), Science,* 164, 440, 1969.

Atchison, G., Henry, M. and Sandheinrich, M., Effects of metals on fish behavior: a review, *Environ. Biol. Fish.,* 18, 11, 1987.

Baatrup, E., Doving, K. B. and Winberg, S., Differential effects of mercurial compounds on the electroolfactogram (EOG) of salmon *(Salmo salar), Ecotoxicol. Environ. Safety,* 20, 269, 1990.

Babcock, M. M., Morphology of olfactory epithelium of pink salmon, *Oncorhynchus gorbuscha,* and changes following exposure to benzene: a scanning electron microscopy study, in, *Marine Biology of Polar Regions and Effects of Stress on Marine Organisms,* Gray, J. and Christensen, M., Eds., Proceedings of 18th EMBS, Oslo, 1985, 259.

Bahr, T. G. and Ball, R. C., Action of DDT on evoked and spontaneous activity from the rainbow trout lateral line nerve, *Comp. Biochem. Physiol.,* 38A, 379, 1971.

Bansal, S. K., Verma, S. R., Gupta, A. K., Rani, S. and Dalbla, R. C., Pesticide-induced alterations in the oxygen uptake rate of the freshwater major carp *Labeo rohita, Ecotoxicol. Environ. Safety,* 3, 374, 1979.

Bardach, J. E., Fujiya, M. and Holl, A., Detergents: effects on the chemical senses of the fish *Ictalurus natilis* (le Sueur), *Science,* 148, 1605, 1965.

Bass, M. and Heath, A., Cardiovascular and respiratory changes in rainbow trout, *Salmo gairdneri,* exposed intermittently to chlorine, *Water Res.,* 11, 497, 1977.

Beamish, F. W. H. and Mookherjii, P. S., Respiration of fishes with special emphasis on standard oxygen consumption. I. Influence of weight and temperature on respiration of goldfish, *Carassius auratus, Can. J. Zool.,* 42, 161, 1964.

Beitinger, T., Behavioral reactions for the assessment of stress in fishes, *J. Great Lakes Res.,* 16, 495, 1990.

Birge, W., Shoyt, R., Black, J., Kercher, M. and Robison, W., Effects of chemical stresses on behavior of larval and juvenile fishes and amphibians, *Am. Fish. Soc. Symp. 14, Water Quality and the Early Life Stages of Fishes,* Fuiman, L., Ed., American Fisheries Society, Bethesda, MD, 1993, 55.

Bjerselius, R., Winberg, S., Winberg, Y. and Zeipel, K., Ca^{2+} protects olfactory receptor function against acute Cu(II) toxicity in Atlantic salmon, *Aquat. Toxicol.,* 25, 125, 1993.

Black, J. A. and Birge, W. J., An avoidance response bioassay for aquatic pollutants, Research Report No. 123, University of Kentucky, Water Resource Research Institute, Lexington, KY, 1980.

Blaxter, J. and Hallers-Tjabbes, C. T., The effect of pollutants on sensory systems and behaviour of aquatic animals, Netherlands, *J. Aquat. Ecol.,* 26, 43, 1992.

Bodammer, J., Corneal damage in larvae of striped bass, *Morone saxatilis* exposed to copper, *Trans. Am. Fish. Soc.,* 114, 577, 1985.

Brown, V. M., The calculation of the acute toxicity of mixtures of poisons to rainbow trout, *Water Res.,* 2, 723, 1968.

Brown, J. A., Johansen, P., Colgan, P. and Mathers, R., Changes in the predator-avoidance behavior of juvenile guppies *(Poecilia reticulata)* exposed to pentachlorophenol, *Can. J. Zool.,* 63, 2001, 1985.

Cairns, J., Heath, A. G. and Parker, B. C., The effects of temperature upon the toxicity of chemicals to aquatic organisms, *Hydrobiology,* 47, 135, 1975.

Calmari, D., Marchetti, R. and Vailati, G., Influence of water hardness on cadmium toxicity to *Salmo gairdneri, Water Res.,* 14, 1421, 1980.

Chapman, G. A., Toxicities of cadmium, copper, and zinc to four juvenile life stages of chinook salmon and steelhead, *Trans. Am. Fish Soc.,* 107, 841, 1978.

Cherry, D. S. and Cairns, J., Biological monitoring part V — preference and avoidance studies, *Water Res.,* 16, 263, 1982.

Cherry, D. S., Larrick, S. R., Giattina, J. D., Dickson, K. L. and Cairns, J., Avoidance and toxicity response of fish to intermittent chlorination, *Environ. Intern.,* 2, 1, 1979.

Cherry, D. S., Larrick, S. R., Giattina, J. D., Cairns, J. and van Hassel, J., Influence of temperature selection upon chlorine avoidance of cold and warmwater fish, *Can. J. Fish. Aquat. Sci.,* 39, 162, 1982.

Colgan, P., The motivational basis of fish behavior, in, *Behavior of Teleost Fishes,* 2nd ed., Pitcher, T., Ed., Chapman and Hall, London, 1993, chap. 2.

Cope, O. B., Sport fishery investigations. Laboratory studies and toxicology, *U.S. Bur. Sport Fisheries Wildl. Ser. Cir.,* 226, 51, 1965.

Coppage, D. L., Anticholinesterase action of pesticidal carbamates in the central nervous system of poisoned fishes, in, *Physiological Responses of Marine Biota to Pollutants,* Vernberg, W. B., Calabrese, A., and Thurberg, F. P., Eds., Academic Press, New York, 1977, 93.

Coppage, D. L. and Mathews, E., Short-term effects of organophosphate pesticides on cholinesterases of estuarine fishes and pink shrimp, *Bull. Environ. Contam. Toxicol.,* 11, 483, 1974.

Davis, R. E. and Northcutt, G. R., Eds., *Fish Neurobiology, Vol. 2, Higher Brain Areas and Functions,* University of Michigan Press, Ann Arbor, 1983.

Deutsch, J. A., The cholinergic synapse and the site of memory, *Science,* 174, 788, 1971.

DiMichelle, L. and Taylor, M. H., Histopathological and physiological responses of *Fundulus heteroclitus* naphthalene exposure, *J. Fish. Res. Bd. Can.,* 35, 1060, 1978.

Dodson, J. and Mayfield, C., Modification of the rheotropic response of rainbow trout *(Salmo gairdneri)* by sublethal doses of the aquatic herbicides diquat and simazine, *Environ. Pollut.,* 18, 147, 1979.

Drummond, R. A., Spoor, W. A. and Olson, G. F., Some short-term indicators of sublethal effects of copper on brook trout, *Salvelinus fontinalis, J. Fish. Res. Bd. Can.,* 698, 1973.

Dutta, H., Marcelino, J. and Richmonds, C., Brain acetylcholinesterase activity and optomotor behavior in bluegills, *Lepomis macrochirus*, exposed to different concentrations of diazinon, *Arch. Intern. Phys. Biochim. Biophys.*, 100, 331, 1992.

Ellgaard, E. G., Ochsner, J. C. and Cox, J. K., Locomotor hyperactivity induced in the bluegill sunfish, *Lepomis macrochirus*, by sublethal concentrations of DDT, *Can. J. Zool.*, 55, 1077, 1977.

Ellgaard, E. G., Tusa, J. E. and Malizia, A. A., Locomotor activity of the bluegill *Lepomis macrochirus*: hyperactivity induced by sublethal concentrations of cadmium, chromium and zinc, *J. Fish Biol.*, 1, 19, 1978.

Ellgaard, E. G., Breaux, J. G. and Quina, D. B., Effects of south Louisiana crude oil on the bluegill sunfish, *Lepomis macrochirus*, *Proc. La. Acad. Sci.*, 42, 49, 1979.

Ellgaard, E. G. and Guillot, J. L., Kinetic analysis of the swimming behaviour of bluegill sunfish, *Lepomis macrochirus*, exposed to copper: hypoactivity induced by sublethal concentrations, *J. Fish Biol.*, 33, 601, 1988.

Evans, D. H., Ed., *The Physiology of Fishes*, CRC Press, Boca Raton, FL, 1993.

Evans, H. L., Laties, V. G. and Weiss, B., Behavioral effects of mercury and methylmercury, *Fed. Proc.*, 1858, 1975.

Farr, J. A., The effect of methyl parathion on the predator choice of two estuarine prey species, *Trans. Am. Fish. Soc.*, 107, 87, 1978.

Ferguson, D., The ecological consequences of pesticide resistance in fishes, Trans. 32nd *N. Am. Wildl. Natl. Resour. Conf.*, March 13, 1967.

Fingerman, S. W. and Russel, L. C., Effects of the polychlorinated biphenyl Aroclor 1242 on locomotor activity and on the neurotransmitters dopamine and norepinephrine in the brain of the gulf killifish, *Fundulus grandis*, *Bull. Environ. Contam. Toxicol.*, 25, 682, 1980.

Fisher, J. W., Dilego, R. A., Putnam, M. E., Livingston, J. M. and Geiger, G. L., Biological monitoring of bluegill activity, *Water Resour. Bull.*, 19, 211, 1983.

Fisher, J. W., Harrah, C. B. and Berry, W. O., Hydrazine: acute toxicity to bluegills and sublethal effects on dorsal light response and aggression, *Trans. Am. Fish. Soc.*, 304, 1980.

Fletcher, D. J., The physiological control of appetite in fish, *Comp. Biochem. Physiol.*, 78A, 617, 1984.

Foster, N. R., Scheier, A. and Cairns, J., Effects of ABS on feeding behavior of flagfish, *Jordanella floridae*, *Trans. Am. Fish. Soc.*, 95, 109, 1966.

Fox, D. A. and Stillman, A. J., Heavy metals affect rod, but not cone photoreceptors, *Science*, 206, 78, 1979.

Gantverg, A. N. and Perevoznikov, M. A., Inhibition of cholinesterase in the brain of perch, *Perca fluviatilis* (Percidea), and common carp, *Cyprinus carpio* (Cyprinidae), under the action of carbophos, *J. Ichthyol.*, 23, 174, 1984.

Gardner, G. R. and LaRoche, G., Copper induced lesions in estuarine teleosts, *J. Fish. Res. Bd. Can.*, 30, 363, 1973.

Gardner, G. R., Chemically induced lesions in extuarine or marine teleosts, in, *The Pathology of Fishes*, Migaki, G., Eds., University of Wisconsin Press, Madison, 1975, 657.

Gardner, G. R., A review of histopathological effects of selected contaminants on some marine organisms, *Mar. Fish. Rev.*, 40, 51, 1978.

Giatinna, J. D., Garton, R. R. and Stevens, D. G., The avoidance of copper and nickel by rainbow trout as monitored by a computer-based data acquisition system, *Trans. Am. Fish. Soc.*, 111, 491, 1982.

Giatinna, J. D. and Garton, R. R., A review of the preference-avoidance responses of fishes to aquatic contaminants, *Residue Rev.*, 87, 43, 1983.

Gibson, J. R., Ludke, J. L. and Ferguson, D. E., Sources of error in the use of fish brain acetylcholinesterase activity as a monitor for pollution, *Bull. Environ. Contam. Toxicol.*, 4, 17, 1969.

Giddings, J. M., Laboratory tests for chemical effects on aquatic population interactions and ecosystem properties, in, *Methods for Ecological Toxicology*, Ann Arbor Science, Ann Arbor, MI, 1981, 25.

Goodyear, C. P., A simple technique for detecting effects of toxicants or other stresses on a predator-prey interaction, *Trans. Am. Fish. Soc.*, 101, 367, 1972.

Guthrie, D. M., Integration and control by the central nervous system, in, *Control Processes in Fish Physiology*, Rankin, J. and Pitcher, T., Eds., John Wiley and Sons, New York, 1983, chap. 8.

Haider, G., Histopathological effects of sublethal poisoning by heavy metals upon the lateral line system in rainbow trout *(Salmo gairdneri)* *Zool. Anz.*, 203, 378, 1979.

Hall, L. W., Burton, D. T., Graves, W. C. and Margrey, S. L., Behavioral modification of estuarine fish exposed to sulfur dioxide, *J. Toxicol. Environ. Health*, 13, 969, 1984.

Hall, L. W., Margret, S. L., Burton, D. T. and Graves, W. C., Avoidance behavior of juvenile striped bass, *Marone saxatilis* temperature conditions, *Arch. Environ. Contam. Toxicol.*, 12, 715, 1983.

Hara, T. J., Chemoreception, in, *The Physiology of Fishes*, Evans, D., Ed., CRC Press, Boca Raton, FL, 1993, chap. 7.

Hara, T. J. and Thompson, B. E., The reaction of whitefish, *Coregonus clupeaformis,* to the anionic detergent sodium lauryl sulphate and its effects on their olfactory responses, *Water Res.,* 12, 893, 1978.

Hara, T. J., Law, Y. M. C. and Macdonald, S., Effects of mercury and copper on the olfactory response in rainbow trout, *Salmo gairdneri, J. Fish. Res. Bd. Can.,* 33, 1568, 1976.

Hart, P. B., Teleost foraging: facts and theories, in, *Behaviour of Teleost Fishes*, Pitcher, T. J., Ed., Chapman and Hall, London, 1993, chap. 8.

Hartwell, S. I., Cherry, D. S. and Cairns, J., Avoidance response of schooling fathead minnows (*Pimephales promelas*) to a blend of metals during a 9-month exposure, *Environ. Toxicol. Chem.,* 6, 177, 1987.

Hartwell, S. I., Jin, J., Cherry, D. and Cairns, J., Toxicity versus avoidance response of golden shiner, *Notemigonus crysoleucas,* to five metals, *J. Fish Biol.,* 35, 447, 1989.

Hasler, A. D. and Wisby, W. J., Use of fish for the olfactory assay of pollutants (phenols) in water, *Trans. Am. Fish. Soc.,* 79, 64, 1950.

Hasler, A. D., Orientation and fish migration, in, *Fish Physiology,* Hoar, W. E. and Randall, D. J., Eds., Academic Press, New York, 1971, chap. 7.

Hatfield, C. T. and Anderson, J. M., Effects of two insecticides on the vulnerability of Atlantic salmon (*Salmo salar*) predation, *J. Fish. Res. Bd. Can.,* 29, 27, 1972.

Hatfield, C. T. and Johansen, P. H., Effects of four insecticides on the ability of Atlantic salmon parr (*Salmo salar*) to learn and retain a simple conditioned response, *J. Fish. Res. Bd. Can.,* 29, 315, 1972.

Hawkes, J. W., The effects of xenobiotics in fish tissues: morphological studies, *Fed. Proc.,* 39, 3230, 1980.

Hawryshyn, C. W., Mackay, W. C. and Nilsson, T. H., Methyl mercury induced visual deficits in rainbow trout, *Can. J. Zool.,* 60, 3127, 1982.

Heath, A. G., A critical comparison of methods for measuring fish respiratory movements, *Water Res.,* 6, 1, 1972.

Heath, A. G., Toxicity of intermittent chlorination to freshwater fish: influence of temperature and chlorine form, *Hydrobiologia,* 56, 39, 1977.

Heath, A. G., Cech, J. J., Zinkl, J., Finlayson, B. and Fujimura, R., Sublethal effects of methyl parathion, carbofuran and molinate on larval striped bass, *Am. Fish. Soc. Symp.,* 14, 17, 1993.

Henry, M. G. and Atchison, G. J., Behavioral changes in bluegill (*Lepomis macrochirus*) as indicators of sublethal effects of metals, *Environ. Biol. Fish.,* 4, 37, 1979a.

Henry, M. G. and Atchison, G. J., Influence of social rank on the behavior of bluegill, *Lepomis macrochirus* Rafinesque, exposed to sublethal concentrations of cadmium and zinc, *J. Fish. Biol.,* 15, 309, 1979b.

Henry, M. G. and Atchison, G. J., Behavioral effects of methyl parathion on social groups of bluegill (*Lepomis macrochirus*), *Environ. Toxicol. Chem.,* 3, 399, 1984.

Henry, M. G. and Atchison, G. J., Behavioral changes in social groups of bluegills exposed to copper, *Trans. Am. Fish. Soc.,* 115, 590, 1986.

Henry, M. G. and Atchison, G. J., Metal effects on fish behavior — advances in determining the ecological significance of responses, in, *Ecotoxicology of Metals: Current Concepts and Applications,* Lewis Publishers, Boca Raton, FL, 1991, 131.

Hikada, I., The effects of transition metals on the palatal chemoreceptors of the carp. *Jpn. J. Physiol.,* 1970.

Hille, B., Pharmacological modifications of the sodium channels of frog nerve, *J. Gen. Physiol.,* 51, 199, 1968.

Hogan, J. W. and Knowles, C. O., Some enzymatic properties of brain acetylcholinesterase from bluegill and channel catfish, *J. Fish. Res. Bd. Can.,* 25, 615, 1968.

Holcombe, G. W., Giandt, J. and Phipps, G., Effects of pH increases and sodium chloride additions on the acute toxicity of 2,4-dichlorophenol to the fathead minnow, *Water Res.,* 14, 1073, 1980.

294

Holland, T. H., Coppage, D. L. and Butler, P. A., Use of fish brain acetylcholinesterase to monitor pollution by organophosphorus pesticides, *Bull. Environ. Contam. Toxicol.*, 2, 156, 1967.

Jackson, D. A., Anderson, J. M. and Gardner, D. R., Further investigations of the effect of DDT on learning in fish, *Can. J. Zool.*, 48, 577, 1970.

Johnson, C. R., The effects of sublethal concentrations of five organophosphorus insecticides on temperature tolerance, reflexes, and orientation in *Gambusia affinis affinis* (Pices: Poecilidae), *Zool. J. Linn. Soc.*, 1978.

Jones, K. A. and Hara, T. J., Behavioral alterations in Arctic char (*Salvelinus alpinus*) briefly exposed to sublethal chlorine levels, *Can. J. Fish. Aquat. Sci.*, 45, 749, 1988.

Jones, K. A., Hara, T. J. and Scherer, E., Behavioral modifications in Arctic char (*Salvelinus alpinus*) chronically exposed to sublethal pH, *Physiol. Zool.*, 400, 1985.

Kamchen, R. and Hara, T., Behavioral reactions of whitefish, (*Coregonus clupeaformis*) to food extract: an application to sublethal toxicity bioassay, *Can. Tech. Rep. Fish. Aquat. Sci.*, 975, 182, 1980.

Kania, H. J. and O'Hara, J., Behavioral alterations in a simple predator-prey system due to sublethal exposure to mercury, *Trans. Am. Fish. Soc.*, 103, 134, 1974.

Kleerekoper, H., Effects of sublethal concentrations of pollutants on the behavior of fish, *J. Fish. Res. Bd. Can.*, 33, 2036, 1976.

Kleerekoper, H., Some monitoring and analytical techniques for the study of locomotor responses of fish to environmental variables, in, *Biological Monitoring of Water and Effluent Quality*, ASTM STP 607, Cairns, J. and Westlake, G. F., Eds., American Society for Testing and Materials, Philadelphia, 1977, 110.

Kleerekoper, H., *Olfaction in Fishes*, Indiana University Press, Bloomington, 1969.

Kleerekoper, H., Waxman, J. B. and Matis, J., Interactions of temperature and copper ions as orienting stimuli in the locomotor behavior of the goldfish (*Carassius auratus*) *J. Fish Res. Bd. Can.*, 30, 725, 1973.

Kleerekoper, H., Westlake, G., Matis, J. and Gensler, P., Orientation of goldfish (*Carassius auratus*) in a shallow gradient of a sublethal concentration of copper in an open field, *J. Fish. Res. Bd. Can.*, 29, 45, 1972.

Koltes, K., Effects of sublethal copper concentrations on the structure and activity of Atlantic silverside schools, *Trans. Am. Fish. Soc.*, 114, 413, 1985.

Kynard, B., Avoidance behavior of insecticide susceptible and resistant populations of mosquitofish to four insecticides, *Trans. Am. Fish. Soc.*, 103, 557, 1974.

LaBarbera, M. and Vogel, S., An inexpensive thermistor flowmeter for aquatic biology, *Limnol. Oceanog.*, 21, 750, 1976.

Larrick, S. R., Dickson, K. L., Cherry, D. S. and Cairns, J., Determining fish avoidance of polluted water, *Hydrobiologia*, 61, 257, 1978.

Little, E. E., Flerov, B. and Ruzhinskaya, N., Behavioral approaches in aquatic toxicity; a review, in, *Toxic Substances in the Aquatic Environment: an International Aspect*, American Fisheries Society Water Quality Section, Bethesda, MD, 1985, 72.

Little, E. E., Archeski, R. D., Flerov, B. A. and Kozlovskaya, V. I., Behavioral indicators of sublethal toxicity in rainbow trout, *Arch. Environ. Contam. Toxicol.*, 19, 380, 1990.

Little, E. E., Fairchild, J. and Delonay, A., Behavioral methods for assessing impacts of contaminants on early life stage fishes, *Am. Fish. Soc. Symp.*, 14, 67, 1993.

Lubinski, K. S., Dickson, K. L. and Cairns, J., Effects of abrupt sublethal gradients of ammonium chloride on the activity level, turning, and preference-avoidance behavior of bluegills, in, *Aquatic Toxicology*, Parrish, P. R. and Hendricks, A. C., Eds., American Society for Testing and Materials, Philadelphia, 1980, 328.

Malyarevskaya, A. Y., Specific and nonspecific changes induced in fish by various toxic agents, *Hydrobiological J.*, 15, 52, 1979.

Marcucella, H. and Abramson, C. I., Behavioral toxicology and teleost fish, in, *The Behavior of Fish and Other Aquatic Animals*, Mostofsky, D. I., Ed., Academic Press, New York, 1978, 2.

Matis, J. H., Kleerekoper, H. and Gerald, K., Long-term cycles in the orientation of the goldfish (*Carassius auratus*) in an open field, *J. Interdisp. Cycl. Res.*, 1977.

McKim, J., Bradbury, S. and Niemi, G., Fish acute toxicity syndromes and their use in the QSAR approach to hazard assessment, *Environ. Health Perspect.*, 71, 171, 1987.

McNicol, R. and Scherer, E., Behavioral responses of lake whitefish (*Coregonus clupeaformis*) to cadmium during preference-avoidance testing, *Environ. Toxicol. Chem.*, 10, 225, 1991.

Mehrie, P., Buckler, D., Little, E., Smith, L., Petty, J., Peterman, P., Stalling, D., DeGraeve, G., Coyle, J. and Adams, W., Toxicity and bioconcentration of 2,3,7,8-tetrachlorodibenzodioxin and tetrachlorodibenzofuran in rainbow trout, *Environ. Toxicol. Chem.,* 7, 47, 1988.

Moran, D., Rowley, J. and Aiken, G., Trout olfactory receptors degenerate in response to water-borne ions: a potential bioassay for environmental neurotoxicology, *Ann. N.Y. Acad. Sci.,* 510, 509, 1987.

Morgan, M. J., Fancey, L. and Kiceniuk, J., Response and recovery of brain acetylcholinesterase activity in Atlantic salmon (*Salmo salar*) exposed to fenitrothion, *Can. J. Fish. Aquat. Sci.,* 47, 1652, 1990.

Morgan, W. S. G., Fish locomotor behavior patterns as a monitoring tool, *J. Wat. Pollut. Contr. Fed.,* 1979.

Morgan, J. D., Vigers, G., Farrell, A., Janz, D. and Manville, J., Acute avoidance reactions and behavioral responses of juvenile rainbow trout (*Oncorhynchus mykiss*) to Garlon 4, Garlon 3A and Vision herbicides, *Environ. Toxicol. Chem.,* 10, 73, 1991.

Murphy, S. D., Pesticides, in, *Toxicology: The Science of Poisons,* Casarett, L. J. and Doull, J., Eds., Macmillan, New York, 1975, 408.

Narahashi, T. and Haas, H. G., Interaction of DDT with the components of lobster nerve membrane conductance, *J. Gen. Physiol.,* 51, 177, 1968.

Northcutt, G. R. and Davis, R. E., Eds., *Fish Neurobiology, Vol. 1: Brain Stem and Sense Organs,* University of Michigan Press, Ann Arbor, 1983.

Ogilvie, D. M. and Anderson, J. J., Effect of DDT on temperature selection by young Atlantic salmon, *Salmo salar, J. Fish Res. Bd. Can.,* 22, 503, 1965.

Olson, D. L. and Christensen, G. M., Effects of water pollutants and other chemicals on fish acetylcholinesterase (*in vitro*), *Environ. Res.,* 21, 327, 1980.

Ostrander, G., Anderson, J., Fisher, J., Landholt, M. and Kocan, R., Decreased performance of rainbow trout, *Oncorhynchus mykiss,* emergence behaviors following embryonic exposure to benzo[a]pyrene, *Fish Bull.,* 88, 551, 1990.

Panigrahi, A. K. and Misra, B. N., Toxicological effects of mercury on a freshwater fish, *Anabas scandens* Val. and their ecological implications, *Environ. Pollut.,* 16, 31, 1978.

Pavlov, D., Chuiko, G., Gerassimov, Y. and Tonkopiy, V., Feeding behavior and brain acetylcholinesterase activity in bream (*Abramis brama*) as affected by DDVP, an organophosphorus insecticide, *Comp. Biochem. Physiol.,* 103C, 563, 1993.

Pavlov, D., Chuiko, G. and Shabrova, A., Adrenalin induced changes of acetylcholinesterase activity in the brain of perch (*Perca fluviatilis*), *Comp. Biochem. Physiol.,* 108C, 113, 1994.

Pedder, S. J. and Maley, E. J., The effect of lethal copper solutions on the behavior of rainbow trout, *Salmo gairdneri, Arch. Environ. Contam. Toxicol.,* 14, 501, 1985.

Peters, C. F. and Weber, D. D., Effect on the lateral line nerve of steelhead trout, in, *Physiological Responses of Marine Biota to Pollutants,* Thurberg, F. P. and Vernberg, W. B., Eds., Academic Press, New York, 1977, 75.

Peyster, A. and Long, W., Fathead minnow optomotor response as a behavioral endpoint in aquatic toxicity testing, *Bull. Environ. Contam. Toxicol.,* 51, 88, 1993.

Pickering, Q. H. and Gast, M. H., Acute and chronic toxicity of cadmium to the fathead minnow (*Promephales promelas*), *J. Fish Res. Bd. Can.,* 29, 1019, 1972.

Pitcher, T., Ed., *Behaviour of Teleost Fishes,* Second Edition, Chapman and Hall, London, 1993.

Popper, A. N. and Platt, C., Inner ear and lateral line, in, *The Physiology of Fishes,* Evans, D., Ed., CRC Press, Boca Raton, FL, 1993, chap. 4.

Post, G., Acetylcholinesterase inhibition and stamina in salmonids, in, *Progress Report in Sport Fisheries Research, Bureau of Sport Fisheries and Wildlife, Division of Fisheries Research Resource Publications,* 88, 1969.

Rao, K. S. and Rao, K. V., Impact of methyl parathion toxicity and eserine inhibition on acetylcholinesterase activity in tissue of the teleost (*Tilapia mossambica*) — a correlative study, *Toxicol. Lett.,* 22, 351, 1984.

Rath, S. and Misra, B. N., Toxicological effects of dichlorvos (DDVP) on brain and liver acetylcholinesterase (AChE) activity of *Tilapia mossambica,* Peters, *Toxicology,* 19, 239, 1981.

Richmonds, C. and Dutta, H., Effect of malathion on the optomotor behavior of bluegill sunfish, *Lepomis macrochirus, Comp. Biochem. Physiol.,* 102C, 523, 1992.

Sandheinrich, M. and Atchison, G., Sublethal copper effects on bluegill, *Lepomis macrochirus,* foraging behavior, *Can. J. Fish. Aquat. Sci.,* 46, 1977, 1989.

Sandheinrich, M. and Atchison, G., Sublethal toxicant effects on fish foraging behavior: empiral vs. mechanistic approaches, *Environ. Toxicol. Chem.,* 9, 107, 1990.

Saunders, R. and Sprague, J., Effects of copper-zinc mining pollution on a spawning migration of Atlantic salmon, *Water Res.,* 1, 419, 1967.

Scarfe, A. D., Jones, K. A., Steele, C. W., Kleerekoper, H. and Corbett, M., Locomotor behavior of four marine teleosts in response to sublethal copper exposure, *Aquat. Toxicol.,* 2, 335, 1982.

Scarfe, A. D., Steele, C. W. and Rieke, G. K., Quantitative chemobehavior of fish: an improved methodology, *Environ. Biol. Fishes,* 13, 183, 1985.

Scherer, E., Behavioural responses as indicators of environmental alterations: approaches, results, developments, *J. Appl. Ichthyol.,* 8, 122, 1992.

Scherer, E. and Harrison, S. E., The optomotor response test, in, *Toxicity Tests for Freshwater Fish,* Scherer, E., Ed., Canadian Special Publication Fisheries Aquatic Science, 44, 179, 1979.

Shaw, B. P. and Panigraphi, A. K., Brain AChE activity studies in some fish species collected from a mercury contaminated estuary, *Water, Air, Soil Pollut.,* 53, 327, 1990.

Shaw, C. M., Mottet, N. K., Body, R. L. and Luschec, E. S., Variability of neuropathologic lesions in experimental methylmercurial encephalopathy in primates, *Am. J. Pathol.,* 80, 451, 1975.

Smith, J. R., Fish neurotoxicology, in, *Aquatic Toxicology,* Weber, L. J., Ed., Raven Press, New York, 1984, 107.

Smith, L. S., *Introduction to Fish Physiology,* TFH Publications, Neptune, NJ, 1982.

Smyth, H. F., Sufficient challenge, *Food Cosmet. Toxicol.,* 5, 51, 1967.

Sparks, R. E., Waller, W. T. and Cairns, J., Effect of shelters on the resistance of dominant and submissive bluegills (*Lepomis macrochirus*) to a lethal concentration of zinc, *J. Fish. Res. Bd. Can.,* 29, 1356, 1972.

Spoor, W. A., A quantitative study of the relationship between the activity and oxygen consumption of the goldfish, and its application to the measurement of respiratory metabolism in fishes, *Biol. Bull.,* 1946.

Spoor, W. A., Neiheisel, T. W. and Drummond, R. A., An electrode chamber for recording respiratory and other movements of free-swimming animals, *Trans. Am. Fish. Soc.,* 100, 22, 1971.

Sprague, J., Factors that modify toxicity, in, *Fundamentals of Aquatic Toxicity,* Rand, G. and Petrocelli, S., Eds., Hemisphere Publishing, Washington, D.C., 1985, chap. 6.

Stabell, O. B., Olfactory control of homing behaviour in salmonids, in, *Fish Chemoreception,* Hara, T. J., Ed., Chapman and Hall, London, 1992, 249.

Steele, C. W., Effects of exposure of sublethal copper on the locomotor behavior of the sea catfish, *Arius felis, Aquat. Toxicol.,* 4, 83, 1983.

Steele, C. W., Attraction of zebrafish, *Brachydanio rerio,* to alanine and its suppression by copper, *J. Fish. Biol.,* 36, 341, 1990.

Sullivan, J. F., Atchison, G. J., Kolar, D. J. and McIntosh, A. W., Changes in the predator-prey behavior of fathead minnows (*Pimephales promelas*) and largemouth bass (*Micropterus salmoides*), *J. Fish. Res. Bd. Can.,* 35, 446, 1978.

Sun, T. and Taylor, D. H., The effects of parathion on acquisition and retention of shuttlebox avoidance-conditioning in the goldfish, *Carassius auratus, Environ. Pollut. (Series A),* 31, 119, 1983.

Sutterlin, A. M. and Gray, R., Chemical basis for homing of Atlantic salmon (*Salmo salar*) to a hatchery, *J. Fish. Res. Bd. Can.,* 30, 985, 1973.

Sutterlin, A. M. and Sutterlin, N., Taste responses in Atlantic salmon (*Salmo salar*) parr, *J. Fish. Res. Bd. Can.,* 27, 1927, 1970.

Symons, P. E. K., Behavior of young Atlantic salmon (*Salmo salar*) exposed to or force-fed fenitrothion, an organophosphate insecticide, *J. Fish. Res. Bd. Can.,* 651, 1973.

Taylor, R. J., Value of clumping to prey and the evolutionary response to ambush predators, *Am. Nat.,* 110, 13, 1976.

U.S.E.P.A., *Quality Criteria for Water,* Office of Water Regulations and Standards, EPA 440/5-86-001, 1986.

van der Well, H. and Welling, W., Inhibition of acetylcholinesterase in guppies (*Poecilia reticulata*) by chlorpyrifos at sublethal concentrations: methodological aspects, *Ecotoxicol. Environ. Safety,* 17, 205, 1989.

Vinyard, G. L. and O'Brian, W. J., Dorsal light response as an index of prey preference in bluegill (*Lepomis macrochirus*), *J. Fish Res. Bd. Can.,* 32, 1860, 1975.

Waiwood, K. G. and Beamish, F. W. H., The effect of copper, hardness and pH on the growth of rainbow trout, *Salmo gairdneri, J. Fish Biol.,* 13, 591, 1978.

Waiwood, K. G. and Johansen, P. H., Oxygen consumption and activity of the white sucker (*Catostomus commersoni*) lethal and nonlethal levels of the organochlorine insecticide, Methoxychlor, *Water Res.,* 8, 401, 1974.

Wallace, K. and Herzberg, U., Reactivation and aging of phosphorylated brain acetylcholinesterase from fish and rodents, *Toxicol. Appl. Pharmacol.,* 92, 307, 1988.

Weiss, C. M., Physiological effect of organic phosphorus insecticides on several species of fish, *Trans. Am. Fish. Soc.,* 90, 143, 1961.

Weir, P. A. and Hine, C. H., Effects of various metals on behavior of conditioned goldfish, *Arch. Environ. Health,* 20, 45, 1970.

Weis, J. S. and Khan, A. A., Reduction in prey capture ability and condition of mummichogs from a polluted habitat, *Trans. Am. Fish. Soc.,* 120, 127, 1991.

Weis, P. and Weis, J. S., DDT causes changes in activity and schooling behavior in goldfish, *Environ. Res.,* 7, 68, 1974.

Welsh, M. J. and Hanselka, C. W., Toxicity and sublethal effects of methyl parathion on behavior of Siamese fighting fish (*Beta splendens*), *Tex J. Sci.,* 23, 519, 1972.

Westlake, G., Kleerekoper, H. and Matis, J., The locomotor response of goldfish to a steep gradient of copper ions, *Water Resour. Res.,* 10, 103, 1974.

Winberg, S., Bjerselius, R., Baatrup, E. and Doving, K., The effect of Cu (II) on the electro-olfactogram (EOG) of the Atlantic salmon (*Salmo salar*) in artificial freshwater of varying inorganic carbon concentrations, *Ecotoxicol. Environ. Safety,* 24, 167, 1992.

Woodward, D., Little, E. and Smith, L., Toxicity of five shale oils to fish and aquatic invertebrates, *Arch. Environ. Contam. Toxicol.,* 16, 239, 1987.

Woltering, D. M., Hedtke, J. L. and Weber, L. J., Predator-prey interactions of fishes under the influence of ammonia, *Trans. Am. Fish. Soc.,* 107, 500, 1978.

Zinkl, J., Lockhart, W., Kenny, S. and Ward, F., Effects of cholinesterase inhibiting insecticides on fish, in, *Cholinesterase-Inhibiting Insecticides — Impact on Wildlife and the Environment,* Mineau, P., Ed., Elsevier Science Publishers, Amsterdam, 1991, 233.

Chapter 13

Reproduction

I. INTRODUCTION

The maintenance of a population of fishes is ultimately determined by the ability of its members to reproduce. Thus, from a practical fisheries standpoint, this aspect of physiology has profound ecological significance. In one sense, the other physiological functions (the topic of most of this book) serve mainly in order to make it possible for the organism to reproduce. When various homeostatic mechanisms are compromised, as from pollutant exposure, the organism may not even survive to carry out the ultimate function of reproduction. This chapter, however, deals with the action of chronic levels of pollutants that do not necessarily affect survival but do impact the process of reproduction. Fish have also been found to be useful models in biomedical research on reproductive toxicology in a broader taxonomic sense (Motte and Landolt, 1987).

II. BRIEF OVERVIEW OF FISH REPRODUCTIVE PHYSIOLOGY

The physiology of reproduction in fish has been the topic of several reviews and symposia (e.g., Sundararaj, 1981; Hoar et al., 1983; Redding and Patino, 1993). The tremendous variety of aquatic habitats has made possible the evolution of an almost infinite variety of reproductive modes among the some 30,000 species of fishes, so generalizations are difficult. Here the discussion will be kept to a very broad level; the interested reader can then pursue it further, if desired, with the aid of the references mentioned, for this is a topic of considerable depth and breadth.

Reproduction involves several more or less distinct processes. These are (1) the formation of gametes; (2) laying or release of eggs (which in some species are retained internally); (3) fertilization, often associated with some complex spawning behavior; (4) embryonic development in the egg; and (5) hatching and larval development. All of these are often cued to particular seasons of the year and are under hormonal controls which are modulated by environmental factors.

A. REPRODUCTIVE STRATEGIES

Most teleost fish species are seasonal breeders, however, a few may breed at any time the sexes can get together. In the Northern hemisphere, temperate zone warm-water fishes tend to spawn in spring and early summer; the salmonids and other cold water species are often late summer and autumn spawners. An individual species, depending on local conditions, can have very different spawning times. For example, the common carp (*Cyprinus carpio*) spawns in April in Israel, summer in France, and both spring and autumn in India (Sundararaj, 1981).

Physiological changes in preparation for spawning may require months and invoke a considerable energy expenditure, especially in females. Photoperiod and water temperature are the most important environmental cues for reproductive functions, with temperature usually being the dominant one. In order to increase productivity of commercial breeding programs, these cues are artificially altered to induce reproduction at times other than the usual. The effects of environmental cues on reproduction are extensively reviewed in Munro et al. (1990).

The reproductive strategies of fishes determine the behavior after eggs are laid. Balon (1975) has proposed a classification of these patterns that has been widely adopted. In this system each species falls into one of three broad groups: those that do not guard their eggs after they have been laid or shed into the water column; those that do guard them; and those that are bearers who carry them in brood chambers or that bear young "alive". Within each group are subclassifications depending on the substrate used for the eggs, how well they are hidden, whether a nest is constructed, etc.

In the context of this book, it should be mentioned here that the reproductive strategy will influence the extent to which the eggs and larvae come in contact with certain pollutants. For example, chemicals such as aromatic hydrocarbons will affect pelagic eggs and larvae more easily than benthic ones, but crude oil that settles onto the bottom may smother eggs laid there. Toxic chemicals in the sediments will affect those on the bottom, as well. Finally, the chemical composition of the estuarine areas (which are nurseries for many oceanic fish species) will be considerably different than the open ocean.

B. HORMONAL CONTROL OF GAMETOGENESIS

Our understanding of hormonal control of gametogenesis in fish is based largely from work on salmonids. The details may differ considerably in other species.

The formation of eggs and sperm, the first step in reproduction, is initiated by the environmental cues mentioned (Figure 1). These work through the hypothalamus which acts on the pituitary by way of gonadotropin-releasing hormones (GnRH). The pituitary gland is then stimulated to release gonadotropins I and II (GtH I and II) which stimulate gametogenesis in the sex organs. GtH I is comparable to follicle-stimulating hormone in mammals and is the dominant GtH in fish undergoing active growth and gametogenesis. GtH II is stimular to luteinizing hormone in mammals and predominates during final maturation of the gonads and at spawning (Redding and Patino, 1993).

Each ovum develops in an ovarian follicle which is composed of the central ovum surrounded by follicular (granulosa) and thecal layers of cells that are instrumental in forming the yolk. These cells also secrete the ovarian sex steroid hormones.

Yolk formation, termed vitellogenesis, involves both endogenous and exogenous material. Vitellogenin, a lipophosphoprotein, is synthesized in the liver and transported by the blood to the ovary where it is taken up into the oocytes. Estrogenic hormones from the ovary stimulate vitellogenin production and its uptake into the ovum and are

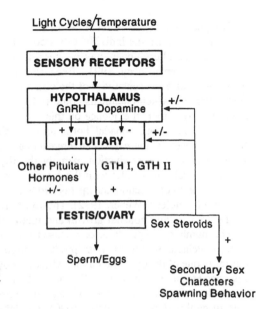

Light Cycles/Temperature

SENSORY RECEPTORS

HYPOTHALAMUS   +/-
GnRH   Dopamine

+   -   +/-
PITUITARY

Other Pituitary | GTH I, GTH II
Hormones
+/-   +

TESTIS/OVARY   Sex Steroids

+

Sperm/Eggs

Secondary Sex
Characters
Spawning Behavior

Figure 1   Diagram of the major endocrine controls of reproduction in fish. See text for discussion.

stimulated by GtH I (Tytler, 1991). Progestins from the ovary may be important in final oocyte maturation (Redding and Patino, 1993).

The sex steroids affect gametogenesis directly and indirectly via both positive and negative feedback on the hypothalamus and pituitary. In seasonally immature fish, the sex steroids generally promote GnRH and GtH release through positive feedback, whereas negative feedback predominates in sexually mature fish (Redding and Patino, 1993).

In the males, the germ cells in the testes may be in synchronous or variable stages of development, depending on the species. These spermatogonia are surrounded by Sertoli and Leydig cells which provide support and regulation of spermatogenesis (Callard, 1991). The Leydig cells produce the androgens, primarily testosterone, ketotestosterone, and androstenodione. In male brook trout, testosterone and 11-ketotestosterone levels rise at spermiation and then fall rapidly thereafter (Sangalang and Freeman, 1974). "Female" sex steroids are also present in the males, as is testosterone in the female. Sperm production is a synchronous event in some teleosts whereas in others it is cyclic or continuous.

C. SPAWNING

The sex hormones are all steroids and since their molecular structures and physiological activities overlap, understanding the details of their functions has been a considerable ongoing challenge. Estrogens such as estrone and estradiol control secondary sex characteristics in female fish and cause serum calcium to rise.

The steroid hormones in the male and female are important determinants of sexual behavior, thus they have effects on brain function. Prespawning female rainbow trout show higher levels of testosterone than males, probably because this hormone is a precursor for estradiol. In this species both testosterone and estradiol exhibit peaks some months before spawning (Koivusaari et al., 1984; Dickhoff et al., 1989). In males, but not females, the gonadotropin(s) from the pituitary may directly stimulate sexual behavior. (See Stacey [1983] for review of primary literature on hormonal control of behavior.)

Sex pheromones are secreted externally and affect conspecifics. Some actually serve as hormones within the individual thus coordinating both gametogenesis and spawning behavior. The pheromones may be steroids or prostaglandins and are detected by the olfactory system (Stacey, 1991). Because a number of pollutant chemicals have been shown to reduce the sensitivity of the olfactory system (see Chapter 12), this might be a mechanism by which chemicals could reduce spawning success, although there is currently no experimental evidence on this.

The pituitary and gonadal hormone controls of reproduction are obviously quite complex. Recently, there has been interest in an additional endocrine gland which appears to be involved in reproduction, namely the thyroid (Leatherland, 1987). The thyroid hormones cycle in parallel with the female sex steroids in trout (Cyr et al., 1988) and oocytes selectively accumulate the thyroid hormone T3 although its function in the oocyte is currently unclear (Weber, 1992). Thyroid hormones may act as permissive hormones during somatic and gonadal growth and maturation (Leatherland, 1993).

The number of eggs laid, or released into the water, depends on the reproductive pattern. Pelagic spawners must produce huge numbers to compensate for the high mortality rate of the eggs and larvae, whereas those at the other extreme that lay protected eggs need to lay relatively few. Of course, there is an almost infinite variety in between. Fecundity refers to the number of eggs in a female fish, and there are several ways of measuring this (Crim and Glebe, 1990). The larger members of a species produce more and larger eggs than would be predicted merely by the size difference, but the effect of this relationship on reproduction is not as straightforward as it appears because of interactions of fecundity with other factors such as fertility (Bagenal, 1978).

Fertility refers to the actual number of young produced by a female, rather than the number of eggs. Except in laboratory situations, this is extremely difficult to measure because young disperse so rapidly. Yet it is really a better indicator of reproductive success since some eggs may not be fertilized and eggs can be resorbed in the ovary and thereby not be released at all. This latter condition (oocyte atresia) has been suggested as a biomarker for xenobiotic effects on reproduction as it increases in fish subjected to a variety of chemical stressors (Hinton et al., 1992). In general, there is also a direct relationship between fecundity/fertility and food abundance.

Gamete viability may vary from female to female of the same species. It is also influenced by non-specific stressors. For example, when rainbow trout were subjected to the stress of approximately weekly 3-min lifts out of the water, egg sizes and survival of progeny were reduced (Campbell et al., 1992).

The pelagic eggs of many marine teleosts have extremely high water content (+90%), which is responsible for their buoyancy because they are more dilute than seawater. The water is taken into the maturing egg across the vitelline membrane into the yolk itself shortly before release by the female. The amount of water taken in raises the volume of the egg by a factor of 3–5 (Carik and Havey, 1984). This water uptake is a separate process from the subsequent one of water hardening.

Most teleost eggs are enclosed by a tough membrane called the chorion. A sperm enters through a single opening, the micropyle, which then closes after fertilization. The chorion then swells by imbibition of water (water hardening) but remains permeable to dissolved substances. The water that is imbibed forms the perivitelline fluid which is retained between the chorion and the vitelline membrane surrounding the egg.

D. DEVELOPMENT

Many aspects of the developmental biology of fishes have been reviewed in Whitt and Wourms (1981) and in Volume 11 of Hoar and Randall (1988). The rate of embryonic development varies with the type of egg (i.e., its size and how much yolk is present) and such environmental factors as temperature and salinity. As a general rule, the time between fertilization and hatching is much shorter in pelagic forms than those which provide protection for the eggs. For example, the anchovy (*Engraulis*) and striped bass (*Morone saxatilis*) hatch out in 2–4 days after fertilization while some salmon may require as much as 5 months of embryonic development during which time they remain buried in the gravel of a streambed. These marked differences in incubation period could influence the impact of a xenobiotic on embryonic development as those with the longer periods would have a greater period of time for chemical uptake and toxicological effect.

Because the chorion is quite tough, the larvae develop "hatching glands" around the head just before hatching. These secrete an enzyme (protease?) which softens the chorion to permit escape from the egg.

When a larva emerges from the egg, it is not necessarily a free-swimming form. Depending on the species and local environmental conditions, it may continue for a time as a free embryo receiving its nutrition from the yolk sac. In salmon, these sac fry, or alevins as they are frequently called when referring to salmonids, may stay buried in the gravel of the streambed for a month or two.

III. ACTION OF POLLUTANTS ON REPRODUCTIVE PROCESSES

Originally, most of the studies on pollutant effects on reproduction were for the purpose of estimating a maximum acceptable toxicant concentration (MATC), which could then be used in establishing water quality criteria. The concept of the MATC is based on the idea that certain stages in the life cycle will be more sensitive than others. Thus, breeding members of a population of fish in the laboratory were exposed to a graded series of concentrations of the toxicant and this was maintained through two generations. The MATC is defined as the threshold between the concentration where a measured effect is observed and the next lower one where no significant effect is observed.

McKim (1977) reviewed the results of 56 life-cycle tests done on 34 organic and inorganic chemicals and concluded that in most cases, the embryo and larval stages are as sensitive, or more so, than any of the other stages in the reproductive cycle. Criteria used were mostly survival, however, growth, egg hatchability, and developmental abnormalities were also utilized. Woltering (1984) reviewed 173 partial or complete life-cycle tests involving a wide assortment of metals and organic compounds and some 15 species of fish. Survival of the fry was the most sensitive stage in roughly two thirds of the tests. However, reproduction (fecundity) and hatchability and to a lesser extent growth of the adult fish, were the most sensitive for some chemical/species combinations. Further discussion of the MATC and methodologies for determining it can be found in Rand and Petrocelli (1985).

Our interest in this chapter is primarily physiological. For that purpose, we will follow the same sequence of events discussed in the section on general reproductive physiology and the coverage will be largely limited to those studies where physiological processes (other than survival) were investigated. For each physiological process, the action of metal and other inorganic contaminants will be discussed first, followed by consideration of organic pollutants. Some of the effects of environmental stressors in general, including a few pollutants, on reproductive processes in fish have been reviewed by Donaldson (1990).

A. HORMONAL CONTROLS

Prolonged cadmium exposure of male brook trout caused these hormones to continue to rise after spermiation at which time they normally decline (Sangalang and Freeman, 1974). In spite of this high level of steroids in the blood, the testes exhibited histopathological damage from cadmium. Thus, these abnormal hormonal elevations could be due to an inhibition of the hepatic microsomal cytochrome P450 system which is responsible for the normal metabolic breakdown of steroids (among other things). In support of this hypothesis, cadmium has been shown to inhibit cytochrome P450 in mammals and it could possibly do so in fish.

Steroid elevation from cadmium exposure may also occur by the metal acting on the gonad directly. Kime (1984), using an *in vitro* preparation, observed stimulation by cadmium of testosterone and llB-hydroxytestosterone synthesis in testes of rainbow trout. This does not rule out the possible suppression of what would normally be a rapid hepatic metabolism of these hormones, but does indicate that cadmium can have a direct stimulatory effect on testosterone production as well as inhibiting the normal elimination of the hormone by the liver. What effect this enhanced testosterone level has, if any, on sperm production, behavior, etc., is not known. Because normal production of gonadotropin hormones from the pituitary is probably under modulatory feedback control from the sex steroids, a notable effect on other aspects of reproduction could occur.

Other metals may not act the same as cadmium. For example, Allen-Gil et al. (1993) found a negative correlation between body burden of heavy metals (lead, mercury, and nickel) in feral Arctic grayling (*Thymallus articus*) and plasma testosterone, so for these metals in this species, inhibition rather than stimulation was seen. Estradiol in the females was, on the other hand, unaffected by the heavy metals.

Copper (Schreck and Lorz, 1978) and possibly most other metals except cadmium causes a large elevation in the hormone cortisol. Cortisol stimulates hepatic metabolism of androgens (Hansson, 1981) which could reduce these steroids in the blood. The cortisol response to cadmium stress is, on the other hand, essentially nil, even in fish dying of the metal exposure (Schreck and Lorz, 1978; Thomas and Neff, 1985). This fits in with the observed stimulation of androgen levels from this metal but suppression by others.

Mercury, both inorganic and organic forms, caused histological lesions in the Leydig cells of the freshwater catfish *Clarias* (Kirubagaran and Joy, 1992). Exposures were for up to 180 days during the period preparatory to spawning. Recall that the Leydig cells are responsible for androgen synthesis. Decreases in plasma cholesterol (a steroid hormone precursor) were also noted in this study. The decreased cholesterol could be due to a slowing of the synthesis of this molecule or to a stimulation of the hepatic metabolism of steroid molecules by the cytochrome P450 system. In the same study the gonadosomatic index was observed to be depressed. While no measurements of testosterone were made, it seems likely that this would be lower as would the rate of spermatogenesis in fish exposed to mercury.

Cyanide and lead at quite low concentrations (10 μg/L) cause marked suppression of spermatogenesis in trout. The mechanism of this has been investigated and found to involve a suppression of the transformation of spermatogonia to spermatocytes (Ruby et al., 1993a). The effect is apparently not due to a direct effect on the testes, however, instead it appears to involve a reduction in gonadotropin from the pituitary as selective gonadotropin-secreting cells there are lost following 12-day exposures to these pollutants. Ruby et al. (1993a) go on to note that others have shown elevations in the levels of brain dopamine in response to cyanide. Because dopamine is a controlling factor for gonadotropin, this could provide a mechanism whereby the cyanide (and probably lead) work their effect on spermatogenesis via the hypothalamic pituitary gonad axis.

Cyanide at 10 μg/L concentration for 12 days also caused several effects on female trout. These included suppressions in plasma estradiol (E2), the thyroid hormone T3, vitellogenin, and oocyte diameters (Ruby et al., 1993b). The latter effect was probably due to the reduced hormones and vitellogenin. Estradiol was severely reduced (>50%) and the mechanism may have been the same as seen with the effect on spermatogenesis, namely suppression of pituitary release of gonadotropin via effects of the cyanide on the dopamine levels in the hypothalamus. Vitellogenin synthesis depends on estradiol so the whole process of reproductive suppression by cyanide apparently begins in the hypothalamus.

As noted above, lead seems to inhibit spermatogenesis and androgen production by suppressing gonadrotropin release in males. The same may be true for females. Thomas (1988) found that chronic oral administration of lead to female Atlantic croaker caused a reduction in gonadal growth and estradiol levels in the plasma. Using an *in vitro* preparation, he was able to show that the capacity of the steroidogenic ovarian tissue was unaffected by the lead. Thus, the decline in estradiol must be due to there being less tissue available presumably due to reduced gonadotropin stimulation, although that was not measured.

The plasma concentrations of pituitary gonadotropins and sex steroids of coho salmon from various Great Lakes locations vary considerably depending on the location from which they were taken. Salmon from some sites have low levels of all of these hormones (Flett et al., 1992) and this is related to degrees of pollution in those areas. There are probable multiple causes for these suppressions which could involve direct effects on the hypothalamus (such as seen with cyanide and lead as discussed above) and/or there may be increases in metabolism of hormones by the mixed function oxygenase system. The hormone reductions are apparently not related to cortisol as this stress hormone showed normal levels.

Petroleum hydrocarbons seem to cause a variety of effects on reproductive parameters. Chronic exposure of marine fishes to petroleum resulted in decreased testicular growth (Payne et al., 1978). However, at approximately the same time Whipple et al. (1978) found accelerated gonadal maturation in starry flounder when exposed for up to 21 days to water-soluble components of crude oil.

An acute 7- to 14-day exposure of mature landlocked salmon (*Salmo salar sebago*) to crude oil caused depression in the plasma concentration of free + conjugated androgens (Truscott et al., 1983). The 11-ketotestosterone was the more sensitive component of the androgen pool. Truscott et al. (1983) present some evidence to indicate that the steroid reductions may be due to enhanced metabolism of the steroids by the liver, for the oil induced the hepatic aryl hydrocarbon hydroxylases as a detoxification process. However, they are careful to point out that the definitive confirmation of this would be to find changes in the amount of metabolites of the androgens in the bile following oil exposure, and this was not done. They make the interesting point that depression of androgen levels at spawning time has effects that are difficult to predict, as these steroids are elevated for a considerable period of time before spawning and may be normally far higher than required. Temporary inactivation by glucuronic acid (in the plasma) may also be a complicating factor. Indeed, they showed that a major fraction of the total androgen in the blood was conjugated to this acid.

Prolonged exposures of male winter flounder to very low levels of crude oil in sediments caused no effect on spermatogenesis, testes weights, or on the plasma levels of free testosterone and 11-ketotestosterone. However, there was a severe depression in the levels of plasma conjugates testosterone and 11-ketotestosterone glucuronids (Truscott et al., 1992). It is not clear whether this exposure regime had any reproductive effects on the flounder which often come in physical contact with the sediment, however, the authors cite work by others which suggests these conjugates may act as pheromones during spawning. Thus, their suppression could affect spawning behavior, a hypothesis that would probably be challenging to test.

Numerous studies have shown effects of pesticides on gonadal function in fish. The effects may occur at several levels from the hypothalamus through the pituitary to the gonadal cells directly. For example, the organochlorine insecticides have been shown to inhibit gonadotropin synthesis by the pituitary (Singh and Singh, 1982). Some, such as the organochlorine hexachlorocyclohexane (Singh et al., 1994), seem to have the potential to inhibit gonadotropin secretion and at the same time suppress the ability of the gonadal cells to respond to the gonadotropin.

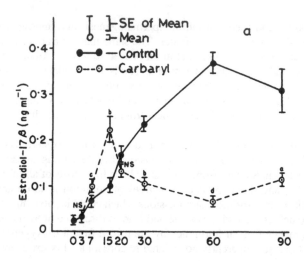

Figure 2 Changes in plasma estradiol in *Anabas testudineus* exposed to carbaryl (1.66 ppm) for 90 days. Letters above points indicate statistical comparison with controls: a = p <0.05, b = p <0.01, c = p <0.001, NS = not significant, and N = 5 per point. (From Choudhury, C. et al., *Environ. Biol. Fishes*, 36, 319, 1993. With permission.)

Bagchi et al. (1990) demonstrated reduced steroidogenesis in catfish exposed to organophosphorus insecticides and the effect was at least in part due to a suppression of the gonadal enzymes responsible for the hormone synthesis. However, while the gonad may be acted on directly by organophosphorus insecticides, Singh and Singh (1980) observe reduced gonadotropin "potency" of the pituitary in fish exposed to malathion or the organochloride endrin.

One of the axioms of toxicology is that the toxic effects are dependent on both dose (concentration in aquatic systems) and duration of exposure. The latter point is emphasized in the study by Choudhury et al. (1993) where *Anabas*, the climbing perch, was exposed to methyl parathion and carbaryl for 90 days but also sampled at several intervals during this time. Figure 2 shows how plasma estradiol levels rose until day 15 after which the exposed animals exhibited a marked decline while the controls continued to increase. The concentrations used were 2 and 10% of the LC50 for methyl parathion carbaryl, respectively.

Direct histological damage of the pituitary and gonads has been documented in several studies involving pesticides (see Shukla and Pandey, 1986; Singh et al., 1993, 1994 for citations). Other than seeing direct damage such as atresia of oocytes, it is often difficult to tell to what extent the damage affects such functions as hormonal secretion or response to a gonadotropin. When done, however, along with hormonal studies, histological examination can help pin-point the mechanism(s) involved in hormonal suppression.

Bleached kraft mill effluent has a wide variety of effects on fish including actions on reproductive processes. Extensive field investigations have shown that the effluent may suppress white sucker plasma sex steroids in both sexes and cause reduced gonad size, and these changes are associated with elevated mixed function oxygenase activity (McMaster et al., 1991; Munkittrick et al., 1991). Similar findings were obtained with lake whitefish (*Coregonus*) (Munkittrick et al., 1992). In addition to the elevated metabolism of plasma steroids by the liver in exposed fish, the kraft mill effluent also suppressed gonadotropin secretion by the pituitary and reduced ovarian steroid biosynthetic capacity (Van der Kraak et al., 1992). The overall effect of the pulpmill effluent is a reduction in gonad size, delayed maturation, and reduced expression of secondary sexual characteristics.

A further mechanism of reduced steroid production in pulpmill effluent exposed fish was revealed by work on carp. Exposures to phenol or sulfide, two components of pulpmill effluent, caused an inhibition in the conversion of cholesterol by the gonad to steroidal products (Mukherjee et al., 1991). This finding does not preclude additional actions of these contaminants at other points in the hypothalamic pituitary gonadal axis.

Of course, not all pulpmills are alike. Gagnon et al. (1994) working with white sucker found that local conditions can complicate the situation so that reductions in only some hormones may occur and there may or may not be any effect of the effluent on gonad development and fecundity.

One final note on pulpmill effluent: there appears to be one or more components in the effluent that have the ability to alter the secondary sexual characters of certain fishes. Kraft pulpmill effluent has been found to induce male secondary sex characters in female mosquitofish, a livebearer (Bartone and Davis, 1994). This was noted in both field and laboratory studies. Furthermore, when masculinized females from the field were placed in non-contaminated water the male traits decreased. At present it is unclear if the effect is due to androgenic or antiestrogenic stimuli, or both.

The PCB Clophen A50 has been found to produce a dose-dependent delay in spawning of minnows (*Phoxinus phoxinus*). It also suppresses gonadal development in flounder (Bengtsson, 1980). It thus is not suprising to note that PCBs have been shown to cause a suppression of plasma testosterone, estrogen, and cortisol in both rainbow trout and carp (Sivarajah et al., 1978). All three hormones decreased approximately the same amount (30–60%) following 4 weeks of PCB injections. Cytochrome P450 and several microsomal enzymes were induced in the liver, presumably as a result of the PCB and these enhanced catalytic activities were correlated with the lowered steroids, so they are a possible cause of this phenomenon.

Koivusaari et al. (1984) have shown there are multiple forms of cytochrome P450 in trout liver and some of these change in concentration depending on the state of spawning. They found that the levels of most of the cytochrome P450 and monooxygenase activities were lowest just prior to spawning, which may reflect a form of downregulation to avoid metabolism of the sex steroids. This does, however, suggest the fish would be least able to handle many foreign organic compounds.

The relationship of general stress to reproductive hormone levels is not at all clear. For example, handling of trout can cause depressions in plasma testosterone and estradiol (Magri et al., 1982; Barton and Iwama, 1991). Of course, the classic stress response is an elevation in plasma cortisol from the interrenal tissue (Donaldson, 1981; Barton and Iwama, 1991) so it is noteworthy that Foo and Lam (1993) showed that cortisol implants, which raise the blood levels of cortisol, also lower testosterone levels in *Tilapia*. Plasma cortisol also generally rises at spawning time (Billard et al., 1981) and catecholamines, which are involved in the primary stress response (Mazeaud and Mazeaud, 1981) may actually stimulate gonadotropin secretion by the pituitary (Chang et al., 1983).

The yolk precursor, vitellogenin, is in part under the control of estradiol. Injections of estradiol induce this material within 6 days (Chen et al., 1984). Fish fed a diet for 6 months contaminated by PCB or the insecticide Mirex exhibited a greatly reduced induction of vitellogenin by estradiol injection (Chen et al., 1984). A possible mechanism has been revealed using an *in vitro* preparation. Anderson et al. (1993) found that when cytochrome P450 (CYP 1A) was added to a hepatocyte cell culture, it depressed estradiol induction of vitellogenin production. Elevated levels of CYP 1A in liver cells are commonly encountered in response to a wide variety of organic contaminants (see Chapter 5).

Cadmium can also suppress vitellogenin production. (Poulsen et al., 1990). The mechanism is apparently different, however. Cadmium is noted for altering calcium regulation in fish causing a hypocalcemia in freshwater species (see Chapter 7). Calcium binds to vitellogenin and is then transported to the developing oocytes. Thus it is interesting that cadmium, in addition to the hypocalcemia, also causes a direct decrease in vitellogenin binding to calcium (Haux et al., 1988). Ghosh and Thomas (1993) further showed that cadmium binds to vitellogenin and displaces zinc and calcium from the protein. The ultimate effect is suppressed development of larvae from cadmium-exposed females (Haux et al., 1988).

Exposures of trout to cyanide for 12 days caused a dramatic decrease in vitellogenin synthesis after 6 days (Ruby et al., 1986). As was shown later in the same laboratory (Ruby et al., 1993b), estrogen levels drop as a result of sublethal cyanide exposure and this hormone is what stimulates vitellogenin synthesis. Thus, we see that vitellogenin levels can be suppressed either by the toxic chemical acting directly on the liver or by reducing estrogen concentration in the blood.

As discussed above, a wide variety of water pollutants cause suppression of vitellogenin production. However, recent work has revealed that some pollutants can induce vitellogenin production in male fish, where it is not normally found. When caged male trout were held in the outflows from sewage treatment plants for two weeks, they produced vitellogenin in their blood. This was also seen in fish even several kilometers downstream of the outfalls (Anon., 1993). Several components in domestic sewage are weakly estrogenic. These include breakdown products of several detergents and the urine of women on birth control pills. Jobling and Sumpter (1993) have shown that these compounds bind to the estradiol receptor in liver cells of trout.

One final endocrine gland that apparently is affected by pollution is the thyroid (Leatherland, 1993). Field studies of salmon in the Great Lakes have revealed extensive thyroid hyperplasia which is not due to iodide deficiency. Tentative candidates are waterborne goitrogens of microbial origin that correlate with the degree of eutrophication in the lakes. Whether this is a contributing factor to the decline in reproduction of the Great Lakes salmon remains to be confirmed. Indeed, the thyroid hormone levels are actually quite high in these fish (Leatherland, 1993).

From the above discussion, it is evident that pollutants can suppress the levels of sex steroids and vitellogenin by a variety of mechanisms, both involving production and clearance. The points of action associated with steroid production can be at the hypothalamic control of the pituitary, the formation of one or both of the gonadotropins by the pituitary, the sensitivity of the gonad to gonadotropin stimulation, the capacity of the gonad to produce the steroid sex hormones, and the ability of the liver to produce vitellogenin. In many studies only one or two of these points have been investigated so we should not conclude that just because a given toxic compound suppresses a given process (e.g., pituitary release of gonadotropin) that it is the only mechanism of action.

B. EGG PRODUCTION

There is not much information on the effects of pollutants on fecundity, in part because it is difficult to induce some species to reproduce in captivity. Laboratory studies in which brook trout were exposed through two or more generations to low levels of copper, zinc, or lead suggested that this function is relatively insensitive to these metals (McKim and Benoit, 1971; Holcombe et al., 1976, 1979). However, Mount (1968) using copper and Brungs (1969) using zinc found that egg production in fathead minnows was severely reduced in concentrations that had little or no effect on eggs, fry, or adults. Kumar and Pant (1984) found that 2- to 4-month exposures of an Indian teleost (*Puntius*) to copper, zinc, or lead caused histopathological changes in both testes and ovaries. Zinc and lead were more harmful to testes than was copper, even though the latter had a greater toxicity to the whole animal. All three metals caused disappearance of oocytes in the ovaries. The authors suggest that this effect is due to a direct action of the metals on the gonads, although altered gonadrotropin levels are not excluded.

Field studies of white sucker from metal-contaminated sites compared with reference sites revealed increased fecundity in one case and decreased fecundity in another (reviewed in Munkittrick and Dixon, 1989a). In the case where increased fecundity was seen, there was also a reduced spawning and those eggs released had abnormally soft shells, so the fecundity measurement alone was misleading. It seems obvious that the effect of metals on fecundity varies with the species and perhaps some other variables as well. These conflicting results also point out the danger of extrapolating from one species of fish, especially considering the extreme diversity of reproductive patterns among the teleosts.

From the standpoint of fecundity, probably more work has been done on acid stress than any other form of pollution. This work has been reviewed in Chapter 10. A common observation is the appearance of large numbers of atretic follicles (McCormick et al., 1989). Cyanide appears to act in a manner rather similar to low pH in that yolk deposition in the developing oocytes is inhibited when female trout are exposed to sublethal levels (Lesnaik and Ruby, 1982), and this is associated with a depressed level of serum calcium (Costa and Ruby, 1984).

Ovum development is an energetically expensive process so food availability would seem to be an important variable that could have a considerable effect on fecundity. Thus, field conditions where food might be limited by pollution or other variables could explain some cases of decreased fecundity. Where that is not a factor, it seems safe to generalize that fecundity will usually be depressed if estradiol is low because this causes a low vitellogenin and thus yolk deposition is reduced. As was mentioned above, some toxic chemicals also act directly on the liver to slow vitellogenin production.

C. TRANSFER OF CONTAMINANTS INTO EGGS FROM ADULTS

It was noted some time ago (Allison et al., 1963) that brook trout exposed to sublethal levels of DDT exhibited no alteration in egg production at doses that did, however, cause subsequent high mortality among the sac fry growing in non-contaminated water. Since then considerable evidence has been accumulated that a variety of organic chemicals can be transmitted from contaminated parent fish through the yolk lipids, and these then cause direct mortality of the fry or increased sensitivity of the fry to stress (Landner et al., 1985; Ankley et al., 1991; Hall and Oris, 1991).

The transfer of contaminants to eggs by the female fish has been further investigated by analyzing the parent tissues and eggs in gravid fish taken from Lake Ontario and Lake Erie (Niimi, 1983). Some 11 organic contaminants and mercury were found in the 5 species tested. They include the insecticides DDT, chlordane, Mirex, endrin, and dieldrin, among others. In general, there were no large differences in concentration of the organics between the whole-body of the parent and the eggs. However, little mercury appeared in the eggs compared to the rest of the body. Assuming all the eggs would have been laid, the percent of the total body burden for the organic contaminants which would have been deposited in the eggs ranged from 5.5% for rainbow trout to 25.5% for yellow perch. The percent lipid content in the fish and that in the eggs had a marked effect on this transfer capacity. Miller (1993) has shown that the amount transferred into the eggs can be predicted from body burden data on parents. Niimi (1983) reviewed several studies that suggest that the levels of organics transferred into these eggs would have had detrimental effects on subsequent development. Ankley et al. (1991) also provide evidence that PCBs transferred into the egg from the mother reduces hatching success in chinook salmon. Of course, this could be due to teratogenic effects on the developing embryo.

While mercury may not transfer readily from the mother into the eggs, some other metals such as selenium and copper apparently do. Bluegills given dietary selenium at 4 doses for 60 days exhibited no effects of the metal on any reproductive index except survival of the fry, which was severely reduced in those from parents receiving the highest amounts of selenium (Coyle et al., 1993).

Munkittrick and Dixon (1989b) have found that metal transfer may actually make white sucker larvae more resistant to subsequent copper exposure but this difference is lost after yolk absorption is complete at 21 days posthatch. There was also copper uptake into the eggs at egg activation time (water hardening), but the authors maintain the metabolic impacts are different. They further point out that the increased tolerance associated with the maternal metal transfer is not associated with metallothionein synthesis so the basis of the increased metal resistance is not clear.

D. EFFECTS ON SPERM AND FERTILIZATION

Sperm would appear to have little protection from the insults of foreign chemicals so there has been some interest in determining its sensitivity to toxicants. The assessment of effects on sperm is generally done by observing motility under a microscope following exposures to test solutions, or by determining subsequent fertilization success.

Anderson et al. (1991) reported that sperm was more sensitive to copper than was either the embryo or larva of topsmelt (*Atherinops affinis*). Their tests were done in seawater which may be important as work on freshwater species (Billard and Roubaud, 1985) suggests that this element has little effect on sperm, even at high concentrations. Because the sperm of freshwater forms has a period of motility less than 2 min (Duplinsky, 1982), the opportunity for chemical exposure in natural field freshwater situations must be extremely short.

Mercury presents some interesting problems for fertilization in *Fundulus*, an estuarine species. Judith and Peddrick Weis and their students (reviewed in Weis and Weis, 1989) have investigated two populations of this species that exhibit markedly different sensitivities to mercury. One population, designated Piles Creek, has experienced mercury exposure for many generations and produces eggs and sperm that are much more tolerant of methylmercury than a similar population from a clean area near Long Island. This enhanced tolerance was not found, however, in larvae or juveniles so it is limited to the gametes.

An increased tolerance to methylmercury was not without a cost. Somewhat surprisingly, the Piles Creek fish produced gametes that were less tolerant to inorganic mercury than the reference population (Khan and Weis, 1987). This phenomenon was further investigated by Khan and Weis (1993) who found by scanning electron microscopy that the micropyle is affected differently by the two mercury forms. Methylmercury caused a reduced fertilization by inducing a premature or artificial activation of the egg due to rupture of the cortical vesicles and this then caused blockage of the micropyle. Inorganic mercury, on the other hand, caused a swelling of the micropylar lip and a decrease in micropylar diameter. Just why enhanced tolerance to methylmercury causes a reduced tolerance to inorganic mercury remains a mystery.

E. EGG ACTIVATION

The newly shed fish egg immediately undergoes activation, which is often called water hardening and requires about 1–2 h in salmon. This process of water uptake and formation of the perivitelline fluid presumably depends on long-chain macromolecules, such as gelatins and collagens, which form moderately crosslinked polymers containing ionized groups which attract water. Eddy and Talbot (1983) observed that strongly hydrated ions such as metallic cations inhibit this uptake of water. Exposure of Atlantic salmon eggs to 1 mM aluminum (valence 3) caused a complete cessation of water hardening while cations of lesser charge, such as zinc, also caused some reduction in water uptake when tested at the same concentration but did not block the process entirely.

Often there is metal pollution in association with acidification so the observations of Peterson and Martin-Robichaud (1982) are noteworthy. They found that a pH of 4.5 or less inhibited water hardening, so the metals and acidity could act in concert on this process. While metals are quite inhibitory of water uptake by the egg, non-electrolytes, even those that are hydrophilic (e.g., alcohol), have little if any affect on this process (Eddy and Talbot, 1983).

Rosenthal and Alderdice (1976) reported reduced volume of Pacific herring eggs when exposed to 0.1 mM cadmium prior to fertilization. The effect was also shown to be dose dependent. The authors propose that cadmium may compete with calcium for binding sites on the mucopolysaccharide coat surrounding the egg capsule. This could in turn alter the permeability characteristics of this membrane. Whether other metallic cations, which do not interact with calcium in quite the manner of cadmium, would cause similar effects on water hardening of marine eggs is unclear. Eddy and Talbot (1983) predict that compared to freshwater species, it would require very high concentrations of most polyvalent metal ions to inhibit water hardening in pelagic eggs of marine fish.

An inhibition of water uptake might be particularly important for pelagic eggs which adjust their volume in order to maintain buoyancy in the water column (Rosenthal and Alderdice, 1976). In general, a pollutant that reduces water uptake would cause a reduction in buoyancy and could then cause the eggs to sink into waters of lower oxygen and/or salinity. Finally, it should be noted that egg activation involves the uptake of water which will also bring into the egg whatever is dissolved in the water (Munkittrick and Dixon, 1989b).

F. OOCYTE AND EMBRYO ENERGY METABOLISM

The mature oocyte before fertilization utilizes an energy metabolism based primarily on glycolysis (Boulekbache, 1981), thus the oxygen demand is low at that stage. Akburak and Earnshaw (1984) report that the addition of copper (0.1–1.0 mM) to unfertilized eggs of Eurasian perch (*Perca fluviatilis*) causes an immediate dose-dependent stimulation in oxygen consumption. The percent increase is as great as 2600% so the effect is quite dramatic. In an effort to understand the mechanism of this response, they exposed eggs to uncouplers of oxidative phosphorylation. These produced only a 50% increase in oxygen consumption so the authors propose the interesting hypothesis that the egg's respiration is limited by a low oxygen permeability of the chorion, and exposure to copper caused a removal of this limitation. In support of this they found that temperature had little effect on oxygen consumption rate of the eggs in the absence of copper, but when the metal was present, there was a nearly twofold increase for each 10°C increase (i.e., $Q10 = 1.86$). That is all well and good, but if oxygen permeability is so severely limiting for the unfertilized egg, there must be a tremendous increase in permeability following fertilization. Otherwise, how can the embryo show a fourfold increase in oxygen consumption during normal development (Boulekbache, 1981)? Another fascinating anomaly is that zinc caused no effect on the oxygen consumption of perch eggs (Akburak and Earnshaw, 1984).

When ATP production is inhibited by metabolic uncouplers such as dinitrophenol (DNP), fish embryos frequently exhibit arrested differentiation and even dedifferentiation (Rosenthal and Alderdice, 1976). This results in a variety of developmental abnormalities which are similar to the "spontaneous" malformations which have been described in the literature. Rosenthal and Alderdice (1976) review several studies on a variety of substances including metals, oil dispersants, DDT, etc., which showed the same sort of effects as those produced by DNP. Even physical changes such as temperature shock or severe environmental hypoxia cause the same thing. In other words, these developmental abnormalities are non-specific and presumably occur in part as a result of a reduction in ATP production in the tissues of the developing embryo.

While metabolic poisons can produce a variety of teratogenic effects, the first stage of development before the formation of the blastula is quite insensitive to cyanide, a strong metabolic inhibitor (Leduc, 1978). This is undoubtedly because the early embryo is supported primarily by anaerobic glycolysis and would thus be insensitive to metabolic poisons, unless they also inhibited glycolysis.

The heart rate of embryos decreases in response to a variety of environmental chemical stressors (Rosenthal and Alderdice, 1976). While under some circumstances, this response may be more sensitive than other measures such as survival to hatching, Sharp et al. (1979) report that relatively high concentrations of petroleum hydrocarbons are required to produce a slowing of heartbeat in *Fundulus*. They also found that oxygen consumption of 10-day-old embryos is also changed very little in response to hydrocarbons.

G. CHANGES IN POLLUTANT SENSITIVITY DURING EMBRYONIC DEVELOPMENT

The various stages of embryonic development show different sensitivities and even different qualitative effects of a pollutant insult. In general, early embryonic development before gastrulation has been completed is the most sensitive stage for various components of petroleum and for mercury (reviewed in von Westernhagen, 1988). This differential sensitivity appears to be due in part to a progressive reduction in permeability of the egg membrane to hydrocarbons and perhaps to metals as the embryo develops.

The early cleavage stage is also the most sensitive one to conditions of low pH (Peterson et al., 1982; Lee and Germing, 1980). Death occurs in acid waters due to a corrosion of the ectodermal cells, and the thin tissue layers at this stage allow rapid movement of electrolytes out of the egg. The older embryos, however, appear to be the most tolerant of acidic pH of any of the life stages in fish, even more so than the adults. Siddens et al. (1984) attribute this to the relative impermeability of the chorion to hydrogen ions. Because there is a precipitous decline in hydrogen ion tolerance at hatching when the organism is no longer protected by that membrane, the hypothesis has apparent merit.

Part of the reason for the high sensitivity of the early embryo to chemical harm may rest with the process of water hardening. As was mentioned above, this uptake of water will also tend to bring into the egg some of the dissolved contaminants. However, lest one conclude that the earliest stage is always the most susceptible to harmful chemicals, we should recall the lack of sensitivity of this stage to cyanide and other metabolic inhibitors.

As the embryo develops, it probably becomes more tolerant of various toxicants, at least in part due to a reduction in permeabilty of the capsule to them. For example, Skidmore (1966) removed the outer egg membrane from zebrafish (*Branchydanio rerio*) embryos and found their tolerance to zinc decreased markedly. Eddy and Talbot (1985) provide evidence that the outer membrane and the perivitelline fluid together give some protection from aluminum, zinc, or acidic water. Thus, the egg capsule appears to have the ability of excluding some metallic ions while at the same time increasing its permeability to oxygen as the embryo develops. Apparently mercury is not one of them as M. Greeley (personal communication) has found that medaka embryos will readily bioconcentrate this metal.

Lake trout eggs are rather resistant to high doses of selenium, up to 10 mg/L, but then an interesting metal interaction occurs at concentrations ten times higher than this; selenium produces a protective effect from mercury toxicity (Klaverkamp et al., 1983). Whether this has any relevance to "real" situations is open to question.

Continuous exposure to a toxicant may produce some changes in developmental rate of different embryonic stages. Figure 3 illustrates this phenomenon in herring eggs exposed to zinc. Note that cleavage was prolonged but the process of segmentation and gastrulation was shortened as a result of the zinc. Also at the higher concentrations, hatching time was shortened primarily due to a shortening of the final period of development.

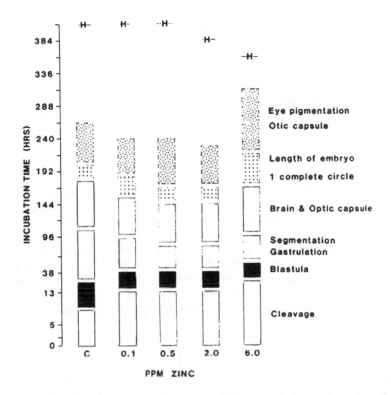

Figure 3 Histogram showing the development stages of herring (*Clupea harengus*) eggs incubated in different concentrations of zinc at a salinity of 21‰ and temperature 8°C. C = control and H = 50% hatching. (From Somasundaram, B. et al., *Aquat. Toxicol.*, 5, 167, 1984. With permission.)

H. HATCHING TIME AND VIABLE HATCH

Hatching time could also be called incubation time and refers to the period between fertilization and 50% hatch of a group of eggs. Assuming concentrations of a toxicant are not so high as to block hatching altogether, the general trend is a reduction in hatching time. This has been found for cadmium, (Rosenthal and Sperling, 1974), zinc, (Somasundaram et al., 1984), silver in freshwater (Davies et al., 1978) and seawater (Klein-MacPhee et al., 1984), and the PCB, Clophen A50 (Bengtsson, 1980). In general the effect is dose dependent up to where there is a significant reduction in hatching success. If that happens, it may take longer to reach 50% hatch because many of the eggs fail to hatch. A shortening of the hatching time often produces larvae that are smaller than normal so this does not necessarily represent a more rapid development, although that possibility is not entirely excluded (Trojnar, 1977).

Cyanide again appears to be an exception to many other toxic substances in that it caused a lengthening of the incubation period at all concentrations tested (Leduc, 1978). A lowering of the environmental pH also has generally been reported to prolong hatching time (Nelson, 1982). The hatching enzyme, which is used to soften the egg capsule just before hatching, has a pH optimum well above 7 (Hagenmeier, 1974). Because the perivitelline fluid where this enzyme does its work has a pH similar to the ambient water (Peterson et al., 1980), there is a good probability that the delay in hatching in eggs in acid water involves a low activity of this enzyme. Abnormalities of the developed embryo, particularly of the skeletal structure, could also be a contributing factor (Peterson et al.,

1982). Nelson (1982) found that raising the external calcium level partially reversed the effect of low pH on hatching in rainbow trout.

Viable hatch refers to the number of larvae actually emerging. The extensive data on this parameter has been reviewed by von Westernhagen (1988). He notes that where exposures of eggs have occurred, cadmium or zinc has considerable effect at low concentrations in freshwater but rather high (and unrealistic) concentrations of these metals are required to cause reductions of viable hatch in seawater. Copper, on the other hand, causes considerable effects on viable hatch at concentrations as low as 30–90 µg/L in both freshwater and seawater.

I. LARVAE

Larvae that hatch from eggs that have been continuously exposed to some pollutant are frequently smaller, so development must have been inhibited, or hatching occurred prematurely. Several examples of reduced larval length in larvae from eggs exposed to metals or petroleum and chlorinated hydrocarbons are cited by von Westernhagen (1988). He makes the interesting speculation, however, that since smaller length is frequently correlated with larger yolk sac, development may proceed normally (if there are no lasting effects). Further, it is not entirely clear whether smaller body size per se at hatching predicts reduced recruitment into the population. M. Greeley (personal communication) has observed that smaller larvae at hatch often catch up with their larger siblings at yolk-sac absorption.

Zinc caused some interesting dose-dependent effects on larval size at hatching in herring. Somasundaram et al. (1984) found that at concentrations up to 2.0 mg/L the length of the larvae increased whereas those at concentrations above 6.0 mg/L showed decreasing lengths. These findings raise the question of to what extent a similar relationship between concentration and larval size at hatching might take place with other pollutants.

When eggs are exposed to various inorganic or organic contaminants, the resulting larvae may suffer from a variety of teratogenic deformities. According to Rosenthal and Alderdice, (1976, p. 2057) "...in general, malformations observable after hatching that originated from tissue injury during earlier stages include (1) yolk-sac deformities, including patches of necrotic tissue and incomplete yolk circulation; (2) various types of eye deformation, generally involving reduction in size of one or both eyes, and as well, disorganization of retinal tissue; (3) otic capsule defects, ranging from missing otoliths to absence of otic capsules; (4) jaw anomalies, including stages from absence of the lower jaw to its deformation or delaid formation; (5) fin defects; and (6) malformations of the vertebral column." Some of these problems of development are probably the result of abnormal chromosome division of the fish embryos (Longwell et al., 1992). Many of these same deformities are also produced by natural stressors, such as low oxygen, or abnormal salinity and so do not necessarily implicate pollutants. However, studies in which several low concentrations of a contaminant chemical are tested can be useful in determination of a non-observed effect concentration.

Abnormally formed larvae will usually exhibit aberrant behavior, such as uncoordinated and reduced swimming activity. Larvae that swim into areas of petroleum hydrocabons tend to become narcotized and this can occur at very low concentrations (von Westernhagen, 1988). Reduced swimming activity would be expected to cause a reduction in food consumption due to lower foraging rate and therefore result in a lower growth rate. It has also been shown that narcotized larvae are more susceptible to predation (Rosenthal and Alderdice, 1976).

Rainbow trout and coho salmon fry emerge from the gravel beds 2–4 months after hatching but before exogenous feeding. Ostrander et al. (1990) exposed the eggs of rainbow trout to a 24-h pulse of benzo[a]pyrene (25 mg/L) 1 week before hatching. When fry from exposed eggs emerged from the gravel they showed decreased ability to swim upstream against the current either due to a reduced stamina or swimming speed, it is not clear which. This study shows how a lipophilic substance can be absorbed, probably into the yolk, over a short period but then affect the young fish 6–8 weeks later.

The ontogeny of various control systems in fish has received relatively little attention, thus the study by Fu and Lock (1990) is interesting from both a toxicological and basic physiology standpoint. They exposed tilapia eggs to cadmium and then used immunocytochemical methods to detect activity of the pituitary cells responsible for prolactin synthesis, a hormone involved in calcium regulation. Larvae from exposed eggs had reduced calcium content but when put into non-contaminated water, this recovered in 288 h. The recovery in calcium content was concomitant with increased activity of the prolactin cells, a phenomenon also observed in adult fish exposed to cadmium (Fu et al., 1989).

It is well known that aluminum toxicity is a major cause of poor recruitment of many fishes in soft, acid waters. Other metals may also become mobilized in acid waters and therefore be a contributing factor for fishery declines (Spry et al., 1981). Thus it is interesting that studies involving combinations at realistic concentrations of various trace metals (aluminum, cadmium, copper, iron, lead, nickel, and zinc) indicate that some combinations are far more toxic than others. Combinations involving either aluminum or copper were the most harmful, with copper being worse than aluminum (Sayer et al., 1991). These metal combinations also caused the fry to have reduced whole-body calcium, sodium, and potassium content. When lead and/or zinc were combined with aluminum or copper, they slightly reduced the harmful effect of the latter two metals.

Exposures of embryos to metals at a sublethal level can (but not necessarily always) induce a greater tolerance to that metal in the subsequent larvae from those eggs. This was found for cadmium in rainbow trout (Beatie and Pascoe, 1978) but not for chromium in the same species (Stevens and Chapman, 1984). Flagfish (*Jordanella floridae*) embryos acclimated to zinc produced larvae that could tolerate this metal better than controls (Spehar et al., 1978). Weis and Weis (1983) found no enhancement of mercury tolerance in *Fundulus* larvae by acclimating their eggs to this element. Instead, acclimation to methylmercury produced a slight but significant decrease in the time to death of larvae in a lethal concentration of this compound.

An improvement in tolerance to a metal in adult fish has been attributed to the synthesis of the protein metallothionein, which serves to sequester the metal (see Chapter 5). Weis (1984) has found that some clutches of *Fundulus* embryos taken from unpolluted sites are more tolerant of methylmercury or inorganic mercury than others. At the time of hatching, these tolerant embryos had twice as much of a protein that co-migrates on a gel with *Fundulus* adult metallothionein, but there was essentially no metallothionein in the eggs at the time of deposition, which presumably is the most critical time for subsequent teratogenic effects. Thus, the high levels of this protein in the resistant embryo clutches may be a mere coincidence. Also, acclimation of embryos with low doses of mercury did not induce this protein. Weis (1984) concludes that metallothionein has a questionable role in embryonic and larval tolerance to mercury, which may also explain why acclimation to mercury does not enhance tolerance to it. It was also noted that tolerance to other toxicants such as PCBs or lead did not correlate with tolerance to mercury. So the mechanism of mercury tolerance probably does not relate to chorionic permeability either (Weis and Weis, 1989).

While acclimation to a pollutant may occasionally make the organism less sensitive to that chemical, tolerance to other environmental factors such as temperature may be adversely affected by the exposure. Thus, Middaugh et al. (1975) observed a reduction in thermal maximum in larval spot following sublethal cadmium exposure. Heath et al. (1994) exposed fathead minnow larvae to a synthetic pyrethroid insecticide for 24 h and then tested their upper and lower lethal temperatures (critical thermal maximum and minimum). Concentrations of the insecticide equivalent to 20 and 70% of the 96-h LC50 were used. The overall effect of the exposures was to reduce the critical thermal maximum and increase the critical thermal minimum, thus narrowing the thermal tolerance limits of the larvae.

The size and shape of the yolk sac is sensitive to a variety of stressors including salinity and temperature (Rosenthal and Alderdice, 1976). When exposed to a chemical pollutant during embryonic development, the yolk sac of the emerging larvae may be smaller or larger than normal and interactions with salinity are complex (see Rosenthal and Alderdice, 1976). The mechanisms of these alterations probably involve two processes which are widely separated in time. On the one hand, changes in yolk volume can be produced by alterations in yolk water content due to changes in the osmoregulatory processes of the egg shortly before hatching (Alderdice and Forrester, 1974). The other major mechanism functionally goes back to when vitellogenesis occurs since that is when the yolk is laid. As was discussed earlier in this chapter, this process is under hormonal control so when changes in yolk are observed, we may, in fact, be seeing the result of a pollutant acting on the sex hormone picture of the adult fish some weeks or months before the ultimate effect becomes evident in the larvae.

Posthatching encounters with a variety of pollutants have been shown to cause erratic swimming and most particularly, a slowing of muscular activity in the larvae (Rosenthal and Sperling, 1974; von Westernhagen et al., 1974; Dethlefsen et al., 1975). Activity of larvae is difficult to quantify so the data are largely subjective. The effect of any change in this on sac fry is somewhat unpredictable as their sole source of energy is the yolk. A reduction in energy expenditure for muscular activity could actually leave more available for growth, and this indeed seems to happen with trout exposed to low levels of cyanide. Sac fry exposed to this chemical grow as fast or faster than controls (Leduc, 1978). However, after the fry have absorbed their yolk sac, the effect of slower swimming would be to reduce feeding ability as they would be unable to explore as much water volume for prey.

Swimming capacity has rarely been examined in larval fish after exposure to pollutants. Heath et al. (1993) used a simple system involving a petri dish with a smaller one glued in the center of it, thus forming a circular "race track". Six radial lines were placed on the bottom of the track and the dish filled with water. Single larvae were then placed in the outer track and chased with a glass rod and the number of lines crossed per minute was determined. Newly hatched striped bass larvae that had been exposed for four days to the insecticide methyl parathion or molinate (an herbicide) at a concentration less than one half the LC50 exhibited a decreased swimming performance. This impaired performance persisted even after the fish had been allowed to recover in non-contaminated water for 10 days.

As the larva of salmon (and probably many other species) develops but before exogenous feeding begins, the larva must absorb from the water approximately 65% of the sodium, 45% of the potassium, and 75% of the calcium required for normal function. The rest comes from the yolk. Rombough and Garside (1984) found that cadmium causes a dose-dependent (0.47–300 µg/L) decrease in the uptake of potassium and calcium. The uptake of sodium is, however, little affected by cadmium in the medium. Calcification of skeletal elements is inhibited, probably because of the reduced body calcium. This results in thin fin rays and an abnormal spinal column. The interference of calcium metabolism by cadmium can also affect subsequent equilibrium of the fish because the otoliths and otic capsules do not develop normally (von Westernhagen et al., 1974). Meteyer et al., (1988) have further shown that the effect of cadmium on calcium uptake can persist even after larvae are returned to control water and they no longer have accumulated cadmium in their bodies.

Finally, the heart rate of fish larvae can often be measured rather easily as they are nearly transparent, so the heart can be viewed under a microscope. Tortorelli et al. (1990) measured this in newly hatched catfish (*Plecostomus commersoni*) fry exposed to six different sublethal concentrations of paraquat, an herbicide. Exposure times were as long as 60 h and a dose- and time-dependent inhibition of heart activity was recorded. Using these data they were able to define a maximal acceptable toxicant concentration that was within the recommended single application rate for this herbicide.

J. CONCLUDING COMMENT

Reproduction has so many fairly discrete steps that there are numerous points where toxic contaminants could and do have impact. In carrying out studies on the effects of pollutants on reproduction, care must always be taken to avoid attributing mechanisms of toxic action to particular processes (e.g., teratogenic changes in embryo) as being the primary cause of poor recruitment into a population, unless other processes have also been examined (e.g., fecundity). Indeed, from a mechanistic standpoint, identifying the most sensitive stage(s) in the whole reproductive process for particular species and contaminants should be a challenging goal for the future. For those planning research in this area, von Westernhagen (1988) provides some thoughtful insights into fruitful approaches.

REFERENCES

Akburak, H. B. and Earnshaw, M. J., Copper-stimulated respiration in the unfertilized egg of the Eurasian perch, *Perca fluviatilis*, *Comp. Biochem. Physiol.*, 349, 1984.

Alderdice, D. F. and Forrester, C. R., Early development and distribution of the flathead sole (*Hippoglossoides elassodon*), *J. Fish. Res. Bd. Can.*, 31, 1899, 1974.

Allen-Gil, S. M., Curtis, L., Lasorra, B., Crecelius, E. and Landers, D., Plasma testosterone as a sensitive biomarker to heavy metal exposure in feral arctic fish, *Soc. Environ. Toxicol. Chem. Meet. Abstr.*, No. 116, 1993.

Allison, D., Kalliman, B. J. and Van Valin, C. C., Insecticides: effects on cutthroat trout of repeated exposure to DDT, *Science*, 142, 958, 1963.

Anderson, B., Middaugh, D., Hunt, J. and Turpen, S., Copper toxicity to sperm, embryos and larvae of topsmelt *Atherinops affinis*, with notes on induced spawning, *Mar. Environ. Res.*, 31, 17, 1991.

Anderson, M. J., Miller, M. R. and Hinton, D. E., Cytochrome P450 1A inducing compounds as potential inhibitors of vitellogenesis in fishes, *Soc. Environ. Toxicol. Chem. Meet. Abstr.*, 642, 1993.

Ankley, G., Tillitt, D., Giesy, J., Jones, P. and Verbrugge, D., Bioassay-derived 2,3,7,8-tetrachlorodibenzo-*p*-dioxin equivalents in PCB-containing extracts from the flesh and eggs of Lake Michigan chinook salmon, *Oncorhynchus tshawkytscha*, and possible implications for reproduction, *Can. J. Fish. Aquat. Sci.*, 48, 1685, 1991.

Anon., Twisted fish sex scare, *Science*, 259, 1119, 1993.

Bagchi, P., Chatterjee, S., Rkay, A. and Deb, C., Effect of quinalphos, organophosphorus insecticides, on testicular steroidogenesis in fish, *Clarias batrachus*, *Bull. Environ. Contam. Toxicol.*, 44, 871, 1990.

Bagenal, T. B., Aspects of fish fecundity, in, *Ecology of Freshwater Fish Production*, Gerking, S. D., Ed., Wiley, New York, 1978, 75.

Balon, E. K., Reproductive guilds in fishes: a proposal and definition, *J. Fish. Res. Bd. Can.*, 32, 821, 1975.

Barton, B. A. and Iwama, G. K., Physiological changes in fish from stress in aquaculture with emphasis on the response and effects of corticosteroids, *Annu. Rev. Fish Dis.*, 1, 3, 1991.

Bartone, S. A. and Davis, W. P., Fish intersexuality as indicator of environmental stress, *Bioscience*, 44, 165, 1994.

Beatie, J. H. and Pascoe, D., Cadmium uptake by rainbow trout, *Salmo gairdneri*, eggs and alevins, *J. Fish Biol.*, 13, 631, 1978.

Bengtsson, B. E., Long-term effects of PCB (Clophen A50) on growth, reproduction and swimming performance in the minnow, *Phoxinus phoxinus, Water Res.,* 14, 681, 1980.

Billard, R., Bry, C. and Gillet, C., Stress, environment and reproduction in teleost fish, in, *Stress and Fish,* Pickering, A. D., Ed., Academic Press, New York, 1981, 9.

Billard, R. and Roubaud, P., The effect of metals and cyanide on fertilization of rainbow trout (*Salmo gairdneri) Water Res.,* 19, 209, 1985.

Birge, W. J., Black, J. A., Hudson, J. E. and Bruser. D. M., Embryo-larval toxicity tests with organic compounds, in, *Aquatic Toxicology,* Kimerle, R. A., Ed., American Society for Testing and Materials, Washington, D.C., 1979, 131.

Blaxter, J. H. S. and Staines, M., Food searching potential in marine fish larvae, in, *Fourth European Marine Biology Symposium,* Crisp, D. J., Ed., Cambridge University Press, New York, 1971, 467.

Boulekbache, H., Energy metabolism in fish development, *Am. Zool.,* 21, 377, 1981.

Brungs, W. A., Chronic toxicity of zinc to the fathead minnow, *Pimephales promelas, Trans. Am. Fish. Soc.,* 98, 272, 1969.

Callard, G., Spermatogenesis, in, *Vertebrate Endocrinology: Fundamentals and Biomedical Implications,* Vol. 4A, Pang, P. and Schreibman, M., Eds., Academic Press, San Diego, 1991, chap. 6.

Campbell, P., Pottinger, T. and Sumpter, J., Stress reduces the quality of gametes, *Biol. Reprod.,* 47, 1140, 1992.

Carik, J. C. A. and Harvey, S. M., Biochemical changes occurring during final maturation of eggs of some marine and freshwater teleosts, *F. Fish Biol.,* 24, 599, 1984.

Chang, J. P., Cook, A. F. and Peter, R. E., Influence of catecholamines on gonadotropin secretion in goldfish, *Carassius auratus, Gen. Comp. Endocrinol.,* 49, 22, 1983.

Chen, T. T. and Sonstegard, R. A., Development of a rapid, sensitive and quantitative test for the assessment of the effects of xenobiotics on reproduction in fish, *Mar. Environ. Res.,* 14, 429, 1984.

Choudhury, C., Ray, A. K. and Bhattacharya, S., Non lethal concentration of pesticide impair ovarian function in the freshwater perch, *Anabas testudineus, Environ. Biol. Fishes,* 36, 319, 1993.

Costa, H. D. and Ruby, S. M., The effect of sublethal cyanide on vitellogenic parameters in rainbow trout *Salmo gairdneri, Arch. Environ. Contam. Toxicol.,* 13, 101, 1984.

Coyle, J., Buckler, D., Ingersoll, C., Fairchild, J. and May, T., Effect of dietary selenium on the reproductive success of bluegills (*Lepomis macrochirus), Environ. Toxicol. Chem.,* 12, 551, 1993.

Crim, L. W. and Glebe, B. D., Reproduction, in, *Methods in Fish Biology,* Schreck, C. B. and Moyle, P. B., Eds., American Fisheries Society, Bethesda, MD, 1990, chap. 16.

Cyr, D., Bromage, N., Duston, J. and Eales, J., Seasonal patterns in serum levels of thyroid hormones and sex steroids in relation to photoperiod-induced changes in spawning time in rainbow trout, *Salmo gairdneri, Gen. Comp. Endocrinol.,* 69, 217, 1988.

Davies, P. H., Goette, J. P. and Sinley, J. R., Toxicity of silver to the rainbow trout (*Salmo gairdneri), Water Res.,* 12, 113, 1978.

Dethlefsen, V., von Westernhagen, H., and Rosenthal, H., Cadmium uptake by marine fish larvae, *Helgol. Wiss. Meersunters.,* 27, 396, 1975.

Dickhoff, W. W., Yan, L., Plisetskaya, E. M., Sullivan, C. V., Swanson, P., Hara, A. and Bernard, M. G., Relationship between metabolic and reproductive hormones in salmonid fish, *Fish Physiol. Biochem.,* 7, 147, 1989.

Donaldson, E. M., Reproductive indices as measures of the effects of environmental stressors in fish, *Am. Fish. Soc. Symp.,* 8, 109, 1990.

Donaldson, E. M., The pituitary-interrenal axis as an indicator of stress in fish, in, *Stress in Fish,* Pickering, A. D., Ed., Academic Press, New York, 1981, 2.

Duplinsky, P., Sperm motility of northern pike and chain pickerel at various pH values, *Trans. Am. Fish. Soc.,* 11, 768, 1982.

Eddy, F. B. and Talbot, C., Formation of the perivitelline fluid in Atlantic salmon eggs (*Salmo salar*) in fresh water and in solutions of metal ions, *Comp. Biochem. Physiol.,* 75C, 1, 1983.

Eddy, F. B. and Talbot, C., Sodium balance in eggs and dechorionated embryos of the Atlantic salmon *Salmo salar* exposed to zinc, aluminum and acid waters, *Comp. Biochem. Physiol.,* 81C, 259, 1985.

Flett, P., Munkittrick, K., Van Der Kraak, G. and Leatherland, J., Reproductive problems in Lake Erie Coho salmon, in *Reproductive Physiology of Fish,* Scott, A., Sumpter, J., Kime, D. and Rolfe, M., Eds., Fish Symposium 91, Sheffield, 1992, 151.

Foo, J. and Lam, T., Serum cortisol response to handling stress and the effect of cortisol implantation on testosterone level in the tilapia, *Oreochromis mossambicus, Aquaculture,* 115, 145, 1993.

Fostier, A., Jalabert, B., Billard, R. and Zohar, Y., The gonadal steroids, in, *Fish Physiology,* Vol. IX, Part A, Hoar, W. S., Randall, D. J. and Donaldson, E. M., Eds., Academic Press, New York, 1983, 7.

Fu, H., Lock, A. and Wendelaar Bonga, S., Effect of cadmium on prolactin cell activity and plasma electrolytes in the freshwater teleost *Oreochromis mossambicus, Aquat. Toxicol.,* 14, 295, 1989.

Fu, H. and Lock, R., Pituitary response to cadmium during the early development of tilapia (*Oreochromis mossambicus*), *Aquat. Toxicol.,* 16, 9, 1990.

Gagnon, M., Dodson, J., Hodson, P., Van Der Kraak, G. and Carey, J., Seasonal effects of bleached kraft mill effluent on reproductive parameters of white sucker (*Catostomus commersoni*) populations of the St. Maurice River, Quebec, Canada, *Can. J. Fish. Aquat. Sci.,* 51, 337, 1994.

Ghosh, P. and Thomas, P., Binding of cadmium to vitellogenin and incorporation into oocytes of scianid fishes, *Soc. Environ. Toxicol. Chem. Meet. Abstr.* P678, 1993.

Hagenmeier, H. E., The hatching process in fish embryos. IV. The enzymological properties of a highly purified enzyme (chorionase) from the hatching fluid of the rainbow trout (*Salmo gairdneri*), *Comp. Biochem. Physiol.,* 49B, 313, 1974.

Hall, A. T. and Oris, J. T., Anthracene reduces reproductive potential and is maternally transferred during long-term exposure in fathead minnows, *Aquat. Toxicol.,* 19, 249, 1991.

Hansson, T., Effects of treated municipal waste water on the hepatic metabolism of 4-androstene-3,17-dione in rainbow trout, *Salmo gairdneri,* in, *Stress and Fish,* Pickering, A. D., Ed., Academic Press, New York, 1981, 339.

Haux, C., Bjornsson, B. and Forlin, L., Influence of cadmium exposure on plasma calcium, vitellogenin and calcitonin in vitellogenic rainbow trout, *Mar. Environ. Res.,* 24, 199, 1988.

Heath, A., Cech, J., Zinkl, J., Finlayson, B. and Fugimura, R., Sublethal effects of methyl parathion, carbofuran, and molinate on larval striped bass, *Am. Fish. Soc. Symp.,* 14, 17, 1993.

Heath, S., Bennett, W., Kennedy, J. and Beitinger, T., Heat and cold tolerance of the fathead minnow, *Pimephales promelas,* exposed to the synthetic pyrethroid cyfluthrin, *Can. J. Fish. Aquat. Sci.,* 51, 437, 1994.

Hinton, D. E., Baumann, P. C., Gardner, G. R., Hawkins, W., Hendricks, J., Murchelano, R. and Okihiro, M., Histopathologic Biomarkers, in, *Biomarkers: Biochemical, Physiological, and Histological Markers of Anthropogenic Stress,* Huggett, R., Kimerle, R., Mehrle, P. and Bergman, H., Eds., Lewis Publishers, Boca Raton, FL, 1992, chap. 4.

Hoar, W. S., Randall, D. J. and Donaldson, E. M., Eds., *Fish Physiology,* Vol. 9 (Parts A, B), Academic Press, New York, 1983.

Hoar, W. S. and Randall, D. J., Eds., *Fish Physiology, Vol. 11, The Physiology of Developing Fish, Part A, Eggs and Larvae,* Academic Press, New York, 1988.

Holcombe, G. W., Benoit, D. A., Leonard, E. N. and McKim, J. M., Long-term effects of lead exposure on three generations of brook trout (Salvelinus fontinalis), *J. Fish. Res. Bd. Can.,* 33, 1731, 1976.

Holcombe, G. W., Benoit, D. A. and Leonard, E. N., Long-term effects of zinc exposures on brook trout (*Salvelinus fontinalis*), *Trans. Am. Fish. Soc.,* 108, 76, 1979.

Holland, H. T., Coppage, D. L. and Butler, P. A., Increased sensitivity to pesticides in sheepshead minnows, *Trans. Am. Fish. Soc.,* 95, 110, 1966.

Jobling, S. and Sumpter, J., Detergent components in sewage effluent are weakly oestrogenic to fish: an *in vitro* study using rainbow trout (*Oncorhynchus mykiss*) hepatocytes, *Aquat. Toxicol.,* 27, 361, 1993.

Khan, A. and Weis, J., Toxic effects of mercuric chloride on sperm and egg viability of two populations of mummichog, *Fundulus heteroclitus, Environ. Pollut.,* 48, 263, 1987.

Khan, A. and Weis, J., Differential effects of organic and inorganic mercury on the micropyle of the eggs of *Fundulus heteroclitus, Environ. Biol. Fishes,* 37, 323, 1993.

Kime, D. E., The effect of cadmium on steroidogenesis by testes of the rainbow trout, *Salmo gairdneri, Toxicol. Lett.,* 22, 83, 1984.

Kirubagaran, R. and Joy, K., Toxic effects of mercury on testicular activity in the freshwater teleost, *Clarias batrachus, J. Fish Biol.,* 41, 305, 1992.

Klaverkamp, J. F., MacDonald, W. A., Lilie, W. R. and Lutz, A., Joint toxicity of mercury and selenium in salmonid eggs, *Arch. Environ. Contam. Toxicol.,* 12, 415, 1983.

Klein-MacPhee, G., Cardin, J. A. and Berry, W. J., Effects of silver on eggs and larvae of the winter flounder, *Trans. Am. Fish. Soc.,* 113, 247, 1984.

Koivusaari, U., Pesoneu, M. and Hanninen, O., Polysubstrate monooxygenase activity and sex hormones in pre and postspawning rainbow trout, *Salmo gairdneri, Aquat. Toxicol.*, 5, 67, 1984.

Kumar, S. and Pant, S. C., Comparative effects of the sublethal poisoning of zinc, copper and lead on the gonads of the teleost *Puntius conchonius, Toxicol. Lett.*, 23, 189, 1984.

Landner, L., Neilson, A. H., Sorensen, L., Tarnholm, A. and Viktor, T., Short-term test for predicting the potential of xenobiotics to impair reproductive success in fish, *Ecotoxicol. Environ. Safety*, 9, 282, 1985.

Leatherland, J., Thyroid hormones and reproduction, in, *Reproductive Endocrinology of Fishes, Amphibians and Reptiles,* Norris, D. and Jones, R., Eds., Plenum Press, New York, 1987, 411.

Leatherland, J., Field observations on reproductive and developmental dysfunction in introduced and native salmonids from the Great Lakes, *J. Great Lakes Res.*, 19, 737, 1993.

Leduc, G., Deleterious effects of cyanide on early life stages of Atlantic salmon (*Salmo salar), J. Fish. Res. Bd. Can.,* 35, 166, 1978.

Lee, R. M. and Germing, S. D., Survival and reproductive performance of the desert pupfish, *Cyprinidon nevadensisá* (Eigenmann and Eigenmann) in acid waters, *J. Fish Biol.*, 17, 507, 1980.

Lesnaik, J. A. and Ruby, S. M., Histological and quantitative effects of sublethal cyanide exposure on oocyte development in rainbow trout, *Arch. Environ. Contam. Toxicol.*, 11, 343, 1982.

Longwell, A. C., Chang, S., Hebert, A., Hughes, J. and Perry, D., Pollution and developmental abnormalities of Atlantic fishes, *Environ. Biol. Fishes,* 35, 1, 1992.

Magri, M. H., Billard, R., Reinaud, P. and Fostier, A., Induction of gametogenesis in the juvenile rainbow trout, *Gen. Comp. Endocrinol.*, 46, 294, 1982.

Mazeaud, M. M. and Mazeaud, F., Adrenergic responses to stress in fish, in, *Stress and Fish,* Pickering, A. D., Ed., Academic Press, New York, 1981, 3.

McCormick, J., Stokes, G. and Hermanatz, R., Oocyte atresia and reproductive success in fathead minnows (*Pimephales promelas*) exposed to acidified hardwater environments, *Arch. Environ. Contam. Toxicol.,* 18, 207, 1989.

McKim, J. M., Evaluation of tests with early life stages of fish for predicting long-term toxicity, *J. Fish. Res. Bd. Can.,* 34, 1148, 1977.

McKim, J. M. and Benoit, D. A., Effects of long-term exposures to copper on survival, growth, and reproduction of brook trout (*Salvelinus fontinalis), J. Fish. Res. Bd. Can.,* 28, 655, 1971.

McMaster, M., Van Der Kraak, G., Portt, C., Munkittrick, K., Sibley, P., Smith, I. and Dixon, D., Changes in hepatic mixed-function oxygenase (MFO) activity, plasma steroid levels and age at maturity of a white sucker (*Catostomus commersoni)* population exposed to bleached kraft pulp mill effluent, *Aquat. Toxicol.,* 21, 199, 1991.

Meteyer, M., Wright, D. and Martin, F., Effect of cadmium on early developmental stages of the sheepshead minnow (*Cypronodon variegatus), Environ. Toxicol. Chem.,* 7, 321, 1988.

Middaugh, D., Davis, W. and Yoakum, R., The response of larval fish, *Leiostomus xanthurus,* to environmental stress following sublethal cadmium exposure, *Contrib. Mar. Sci.,* 19, 13, 1975.

Miller, M., Maternal transfer of organochlorine compounds in salmonines to their eggs, *Can. J. Fish. Aquat. Sci.,* 50, 1405, 1993.

Motte, N. and Landolt, M., Advantages of using aquatic animals for biomedical research on reproductive toxicology, *Environ. Health Perspect.,* 71, 69, 1987.

Mount, D. I., Chronic toxicity of copper to fathead minnows, *Water Res.,* 2, 215, 1968.

Mukherjee, D., Guha, D., Kumar, V. and Chakrabarty, S., Impairment of steroidogenesis and reproduction in sexually mature *Cyprinus carpio* by phenol and sulfide under laboratory conditions, *Aquat. Toxicol.,* 21, 29, 1991.

Munkittrick, K. and Dixon, D., Use of the white sucker (*Catostomus commersoni*) populations to assess the health of aquatic ecosystems exposed to low-level contaminant stress, *Can. J. Fish. Aquat. Sci.,* 46, 1455, 1989a.

Munkittrick, K. and Dixon, D., Effects of natural exposure to copper and zinc on egg size and larval copper tolerance in white sucker (*Catostomus commersoni), Ecotoxicol. Environ. Safety,* 18, 15, 1989b.

Munkittrick, K., Portt, C., Van Der Kraak, G. J., Smith, I. and Rokosh, D., Impact of bleached kraft mill effluent on population characteristics, liver MFO activity and serum steroid levels of a Lake Superior white sucker (*Catastomus commersoni)* population, *Can. J. Fish. Aquat. Sci.,* 48, 1371, 1991.

Munkittrick, K., McMaster, M. E., Portt, C. B., Van Der Kraak, G. J., Smith, I. R. and Dixon, D. G., Changes in maturity, plasma sex steroid levels, hepatic mixed function oxygenase activity, and the presence of external lesions in lake whitefish (*Coregonus clupeaformis*) exposed to bleached Kraft mill effluent, *Can. J. Fish. Aquat. Sci.*, 49, 1560, 1992.

Munro, A. D., Scott, A. P. and Lam, T. J., Eds., *Reproductive Seasonality in Teleosts: Environmental Influences*, CRC Press, Boca Raton, FL, 1990.

Nelson, J. A., Physiological observations on developing rainbow trout, *Salmo gairdneri* (Richardson), exposed to low pH and varied calcium ion concentrations, *J. Fish. Biol.*, 20, 359, 1982.

Niimi, A. J., Biological and toxicological effects of environmental contaminants in fish and their eggs, *Can. J. Fish. Aquat. Sci.*, 40, 303, 1983.

Ostrander, G., Anderson, J., Fisher, J., Landolt, M. and Kocan, R., Decreased performance of rainbow trout *Oncorhynchus mykiss* emergence behaviors following embryonic exposure to benzo[a]pyrene, *Fish. Bull.*, 88, 551, 1990.

Payne, J. F., Kiceniuk, J. W., Squires, W. R. and Fletcher, G. L., Pathological changes in a marine fish after a 6-month exposure to petroleum, *J. Fish. Res. Bd. Can.*, 35, 665, 1978.

Peterson, R. H., Daye, P. G., Lacroix, G. L. and Garside, E. T., Reproduction in fish experiencing acid and metal stress, in, *Acid Rain/Fisheries*, American Fisheries Society, Bethesda, MD, 1982, 177.

Peterson, R. H. and Martin-Robichaud, D. J., Water uptake by Atlantic salmon ova as affected by low pH, *Trans. Am. Fish. Soc.*, 111, 772, 1982.

Peterson, R. H., Daye, P. G. and Metcalf, J. L., Inhibition of Atlantic salmon hatching at low pH, *Can. J. Fish. Aquat. Sci.*, 37, 770, 1980.

Poulsen, A., Korsgaard, B. and Bjerregaard, P., The effect of cadmium on vitellogenin metabolism in estradiol-induced flounder (*Platichthys flesus*) males and females, *Aquat. Toxicol.*, 17, 253, 1990.

Rand, G. M. and Petrocelli, S. R., Eds., *Fundamentals of Aquatic Toxicology*, Hemisphere Publishing, Washington, D.C., 1985.

Redding, J. M. and Patino, R., Reproductive physiology, in, *The Physiology of Fishes*, Evans, D. H., Ed., CRC Press, Boca Raton, FL, 1993, chap. 16.

Rombough, P. J. and Garside, E. T., Disturbed ion balance in alevins of Atlantic salmon *Salmo salar* exposed to sublethal concentrations of cadmium, *Can. J. Zool.*, 62, 1443, 1984.

Rosenthal, H. and Alderdice, D. F., Sublethal effects of environmental stressors, natural and pollutional, on marine fish eggs and larvae, *J. Fish. Res. Bd. Can.*, 33, 2047, 1976.

Rosenthal, H. and Sperling, K. R., Effects of cadmium on development and survival of herring eggs, in, *The Early Life History of Fish*, Blaxter, J. S., Ed., Springer-Verlag, Berlin, 1974, 383.

Ruby, S. M., Idler, D. and So, Y., The effect of sublethal cyanide exposure on plasma vitellogenin levels in rainbow trout (*Salmo gairdneri*) during early vitellogenesis, *Arch. Environ. Contam. Toxicol.*, 15, 603, 1986.

Ruby, S., Jaroslawski, P. and Hull, R., Lead and cyanide toxicity in sexually maturing rainbow trout, *Oncorhynchus mykiss* during spermatogenesis, *Aquat. Toxicol.*, 26, 225, 1993a.

Ruby, S., Idler, D. and So, Y., Plasma vitellogenin, 17B estradiol, T3 and T4 levels in sexually maturing rainbow trout *Oncorhynchus mykiss* following sublethal HCN exposure, *Aquat. Toxicol.*, 26, 91, 1993b.

Runn, P., Johansson, N. and Milbrink, G., Some effects of low pH on the hatchability of eggs of perch, *Perca fluviatilis*, *Zoology*, 5, 115, 1977.

Sangalang, G. B. and Freeman, H. C., Effects of sublethal cadmium on maturation and testosterone and 11-ketotestosterone production *in vivo* in brook trout, *Biol. Reprod.*, 11, 429, 1974.

Sayer, M., Reader, J. and Morris, R., Embryonic and larval development of brown trout, *Salmo trutta*: exposure to trace metal mixtures in soft water, *J. Fish Biol.*, 38, 773, 1991.

Schreck, C. B. and Lorz, H. W., Stress response of coho salmon (*Oncorhynchus kisutch*) elicited by cadmium and copper and potential use of cortisol as an indicator of stress, *J. Fish. Res. Bd. Can.*, 35, 1124, 1978.

Scott, A. P., Bye, V. J., Baynes, S. M. and Springate, J. R. C., Seasonal variations in plasma concentrations of 11-ketotestosterone and testosterone in male rainbow trout, *Salmo gairdneri*, *J. Fish Biol.*, 17, 495, 1980.

Scott, A. P., Bye, V. J. and Bayne, S. M., Seasonal variations in sex steroids of female rainbow trout, *Salmo gairdneri*, *J. Fish Biol.*, 17, 587, 1980.

Sharp, J. R., Fucik, K. W. and Neff, J. M., Physiological basis of differential sensitivity of fish embryonic stages to oil pollution, in, *Marine Pollution: Functional Responses,* Vernberg, F. J., Ed., Academic Press, New York, 1979, 85.

Shukla, L. and Pandey, A., Effects of endosulphan on the hypothalamo-hypophyseal complex and fish reproductive physiology, *Bull. Environ. Contam. Toxicol.,* 36, 122, 1986.

Siddens, L. K., Seim, W. K., Curtis, L. R. and Chapman, G. A., Toxicity of environmental acidity on various life stages of brook trout, *Proc. West. Pharmacol. Soc.,* 27, 265, 1984.

Singh, H. and Singh, T. P., Short-term effect of two pesticides on the survival, ovarian P-32 uptake and gonadotrophic potency in a freshwater catfish, *Heteropneustes fossilis, J. Endocrinol.,* 85, 193, 1980.

Singh, H. and Singh, T. P., Effect of some pesticides on hypothalamo-hypohyseal-ovarian axis in the freshwater catfish, *Heteropneustes fossilis, Environ. Pollut. (Ser. A),* 27, 283, 1992.

Singh, P. B., Kime, D. E. and Singh, T. P., Modulatory actions of *Mystus* gonadotropin on γ-BHC induced histological changes, cholesterol and sex steroid levels in *Heteropneustes fossilis, Ecotoxicol. Environ. Safety,* 25, 141, 1993.

Singh, P. B., Kime, D., Epler, P. and Ckhyb, J., Impact of γ-hexachlorocyclohexane exposure on plasma gonadotropin levels and *in vitro* stimulation of gonadal steroid production by carp hypophyseal homogenate in *Carassius auratus, J. Fish Biol.,* 44, 195, 1994.

Sivarajah, K., Franklin, C. S. and Williams, W. P., The effects of polychlorinated biphenyls on plasma steroid levels and hepatic microsomal enzymes in fish, *J. Fish Biol.,* 13, 401, 1978.

Skidmore, J. F., Resistance of zinc sulphate of zebrafish (Branchio rerio) outer egg membrane, *J. Fish. Res. Bd. Can.,* 23, 1037, 1966.

Somasundaram, B., King, P. E. and Shackley, S., The effects of zinc on postfertilization development in eggs of *Clupea harengus* L, *Aquat. Toxicol.,* 5, 167, 1984.

Spehar, R., Leonard, E. and Defoe, D., Chronic effects of cadmium and zinc mixtures on flagfish (*Jordanella floridae), Trans. Am. Fish. Soc.,* 107, 354, 1978.

Spry, D., Wood, C. and Hodson, P., The effects of environmental acid on freshwater fish with particular reference to softwater lakes in Ontario and the modifying effects of heavy metals. A literature review. *Can. Tech. Rep. Fish. Aquat. Sci.,* No. 999, 1981.

Stacey, N., Hormones and pheromones in fish sexual behavior, *Bioscience,* 33, 552, 1983.

Stacey, N., Hormonal pheromones in fish: status and prospects, in *Proc. Fourth Int. Symp. Reprod. Physiol. Fish,* Scott, A. P., Sumpter, J. P., Kime, D. E. and Rolfe, M. S., Eds., Fish Symp. 91, Sheffield, 1991, 177.

Stegeman, J. J. and Woodin, B. R., Differential regulation of hepatic xenobiotic and steroid metabolism in marine teleost species, *Mar. Environ. Res.,* 14, 422, 1984.

Stevens, D. G. and Chapman, G. A., Toxicity of trivalent chromium to early life stages of steelhead trout, *Environ. Toxicol. Chem.,* 3, 125, 1984.

Sundararaj, B. I., *Reproductive Physiology of Teleost Fishes,* Food and Agriculture Organization of United Nations, Rome, 1981.

Thomas, P. and Neff, J. M., Plasma corticosteroid and glucose responses to pollutants in striped mullet; different effects of naphthalene, benzo(a)pyrene and cadmium exposure, in, *Marine Pollution and Physiology: Recent Advances,* Vernberg, F. J., Thurberg, F. P., Calabrese, A. and Vernberg, W., University of North Carolina Press, Columbia, 1985, 633.

Thomas, P., Reproductive endocrine function in female Atlantic croaker exposed to pollutants, *Mar. Environ. Res.,* 24, 179, 1988.

Tortorelli, C., Hernandez, D., Vazquez, G. and Salibian, A., Effects of paraquat on mortality and cardiorespiratory function of catfish fry, *Plecostomus commersoni, Arch. Environ. Contam. Toxicol.,* 19, 523, 1990.

Trojnar, J. R., Egg hatchability and tolerance of brook trout (*Salvelinus fontinalis*) fry at low pH, *J. Fish. Res. Bd. Can.,* 34, 574, 1977.

Truscott, B., Walsh, J. M., Burton, M. P., Payne, J. F. and Idler, D. R., Effect of acute exposure to crude petroleum on some reproductive hormones in salmon and flounder, *Comp. Biochem. Physiol.,* 75C, 121, 1983.

Truscott, B., Idler, D. and Fletcher, G., Alteration of reproductive steroids of male winter flounder (*Pleuronectes americanus*) chronically exposed to low levels of crude oil in sediments, *Can. J. Fish. Aquat. Sci.,* 49, 2190, 1992.

Tytler, C., Vitellogenesis in salmonids, in *Proc. Fourth Int. Symp. Reprod. Physiol. Fish,* Scott, A. P., Sumpter, J. P., Kime, D. E. and Rolfe, M. S., Eds., Fish Symposium 91, Sheffield, 1991, 224.

Van Der Kraak, G. J., Munkittrick, K., McMaster, M., Portt, C. and Chang, J., Exposure to bleached kraft pulp mill effluent disrupts the pituitary gonadal axis of white sucker at multiple sites, *Ecotoxicol. Appl. Pharmacol.,* 115, 224, 1992.

von Westernhagen, H., Effects of pollutants on fish eggs and larvae, in, *Fish Physiology,* XI, Part A, Hoar, W. and Randall, D., Eds., Academic Press, 1988, chap. 4.

von Westernhagen, H., Rosenthal, H. and Sperling, K. R., Combined effects of cadmium and salinity on development and survival of herring eggs, *Helgol. Wiss. Meeresunters.,* 26, 416, 1974.

Weber, G., Okimoto, D., Richman, N. and Grau, E., Patterns of thyroxine and triiodothyronine in serum and follicle-bound oocytes of the tilapia, *Oreochromis mossambicus,* during oogenesis, *Gen. Comp. Endocrinol.,* 85, 392, 1992.

Weis, P. and Weis, J. S., Effects of embryonic pre-exposure to melthylmercury and Hg^{2+} on larval tolerance in *Fundulus heteroclitus, Bull. Environ. Contam. Toxicol.,* 31, 530, 1983.

Weis, P., Metallothionein and mercury tolerance in the killifish, *Fundulus heteroclitus, Mar. Environ. Res.,* 153, 1984.

Weis, J. and Weis, P., Tolerance and stress in a pollluted environment, *Bioscience,* 39, 89, 1989.

Whipple, J. A., Yocom, T. G., Smart, D. R. and Cohen, M. H., Effects of chronic concentrations of petroleum hydrocarbons on gonadal maturation in starry flounder (*Platichthys stellatus*), *Conf. Assessment Ecolog. Impacts Oil Spills,* American Institute Biological Science, Arlington, 1978, 757.

Whitt, G. S. and Wourms, J. P., Developmental biology of fishes, *Am. Zool.,* 21, 1981, 323.

Woltering, D. M., The growth response in fish chronic and early life stage toxicity tests: a critical review, *Aquat. Toxicol.,* 5, 1, 1984.

Chapter 14

Use of Physiological and Biochemical Measures in Pollution Biology

I. INTRODUCTION

Throughout this book, the emphasis has been on the study of physiological changes as a means for understanding the responses of individual fish to the presence of various types of pollution. Practical use of this information has been ignored, for the most part. Indeed, there is no intrinsic reason why such investigations must have a practical use. The spirit of curiosity has long been the driving force behind so-called "pure science", and will undoubtedly continue to have a primary impact in the future. Because humankind is constantly changing the physical and biological conditions of ecosystems, including the addition of pollutants, there will be a continued curiosity to understand the results of those perturbations at several levels of biological organization so that broader theoretical constructs can be formulated. Also, the specific biological levels that are investigated by different workers complement and often depend on each other.

Having said that, it is obvious that the changes in fish that occur as a result of pollution can also have real practical use. An organism makes a powerful integrator of the effect of any number of changes in chemical and physical (e.g., temperature) factors in the aquatic environment. The usual approach of performing chemical analysis and physical measurements of natural waters may or may not detect changes that are affecting the fish therein. In a real sense, "they" are the judge of that. Thus, water pollution control authorities increasingly are looking for ways to use aquatic organisms to determine the biological consequences of pollution.

Essentially, physiological measurements in organisms such as fish can be used in three ways in pollution control: (1) as a supplemental aid in the formulation of water quality criteria, from which standards can be established; (2) monitoring the health of populations of fish in the field so predictions of harm can be made before serious changes in population size occurs; and (3) as a biological sensor for detecting sudden changes in water quality as might be produced by an accidental spill or intentional poisoning of a water supply.

II. WATER QUALITY CRITERIA

Numerous pieces of information are used in formulating water quality criteria for aquatic organisms. These include traditional bioassays, where death is the "physiological" measure, all the way to field surveys (Alabaster and Lloyd, 1980; APHA, 1992; Lloyd, 1992). Most pollution biologists would probably agree that life cycle tests in which fish are exposed to several concentrations of a presumed toxicant for a year or more (allowing time for reproduction) yield the most relevant information for the development of water quality criteria. The working assumption is that certain stages of the life cycle are more sensitive than others and therefore, a pollutant in concentrations above some threshold would reduce a population by acting on that particular stage. Still, these tests are severely limited in that not all important species will reproduce in the laboratory; they are extremely expensive to carry out; the tests are conducted at a constant toxicant concentration in the lab whereas in nature, this rarely occurs; and finally, only a small number of toxicants could conceivably be evaluated. Consequently, complete life cycle tests have not been used in recent years.

Because of the time and expense involved, there has been considerable effort expended to develop modified forms of fish and invertebrate life cycle tests that will yield essentially the same information as one which is complete. In essence, that has meant identifying the most sensitive stages and concentrating on those. McKim (1977) evaluated 56 studies in which complete or partial life cycle tests were carried out and concluded that the embryo-larval and early juvenile life stages were the most sensitive in most cases. The biological measures of sensitivity were decreases in hatchability, survival and growth, and deformities in the juveniles. More recently, Landner et al. (1985) have shown that exposure of the adults before spawning to a mixed effluent can cause severe effects on the subsequent embryos and larvae, which again emphasizes the sensitivity of the young.

Woltering (1984) critically reviewed 173 tests involving a wide variety of chemicals and concluded that the inclusion of measures of fry and/or juvenile growth rate adds little significant information for the amount of work involved. Some (e.g., Birge et al., 1979) have been successful in carrying out greatly abbreviated embryo-larval lethality tests of only 4–8 days posthatch which avoids problems of feeding, not to mention all the "Murphy's Law" type of difficulties associated with long-term studies. Such abbreviated life cycle tests can even be carried out in the field using natural receiving waters. The procedures for carrying out acute and chronic toxicity tests with aquatic organisms are now well standardized (APHA, 1992; Weber et al., 1989).

Aside from complete or partial life cycle tests, other physiological data may also be utilized as supplemental information in development of criteria. These can range all the way from measures of changes in behavior in the presence of the test chemical to assessment of cellular enzyme activity. In order to have practical use for this purpose, such tests must involve exposures of the test organisms to several concentrations of the toxicant, preferably including one that causes no measurable effect on the parameter being measured. In this manner, some indication of a threshold concentration is obtained. This of course does not necessarily represent a "safe" concentration as other physiological measures may have lower thresholds, but were not measured. Physiological changes that are clearly associated with acute stress, such as elevated blood glucose, depressed liver glycogen concentration, high cortisol levels, etc., are not of much aid for determinations of water quality criteria, but along with other blood and tissue measures, they may prove useful in biomonitoring (see below). A variety of biochemical or physiological measurements applied to the younger stages might prove useful in reducing the usual time involved in carrying out such studies. These could include measures of such things as stress protein induction, changes in cellular enzyme activity, alterations in ventilation and heartbeat frequency, aberrant behavior, and examination for histological lesions. Care must be taken to not rely on only one or two measures, because they will not be equally sensitive to the presence of all chemicals.

III. BIOMONITORING OF FISH IN THE FIELD AND MESOCOSMS

Biomonitoring with aquatic organisms has experienced a tremendous explosion of information since the early 1980s resulting in several book-length reviews (Hocutt and Stauffer, 1980; Vernberg et al., 1981; J. Cairns et al., 1984; V. Cairns et al., 1984; Bayne, 1985; McCarthy and Shugart, 1990; Adams, 1990; Huggett et al., 1992). The major reason for the interest in this field is that traditional measures of water quality based on chemical analysis of effluents or receiving waters may not detect the most important components, and interactions among chemicals and factors such as temperature in the waterway may enhance or reduce water quality and alter the availability of chemicals to fish and other aquatic organisms. Thus, the reasoning goes, the organisms themselves should serve as sensors to supplement the chemical data. This means that other than death, appropriate measurements must be devised to indicate whether the recepient organisms are, in fact, being harmed by the quality of the water. Before discussing a few of the physiological and biochemical approaches, which are becoming known as "biomarkers", some broader generalities will be considered.

Biomonitoring techniques are, more often than not, methods to measure stress in organisms or ecosystems. In an individual organism, which is the topic here, it is sometimes said that there are two levels of stress, the generalized and the specific (Bayne, 1985). The generalized responses are stereotypic thus they are the same irregardless of the environmental stressor, while the specific ones may vary with the cause. This apparent dichotomy often breaks down when examining the effect of chemical pollutants on fish. For example, the typical "generalized" stress response involves an elevation in blood sugar (hyperglycemia) mediated by a rise in the hormone adrenaline, whereas, as was shown in Chapter 8, there are a number of cases where a toxicant causes hypoglycemia, even in fish under considerable "stress" from the chemical. As for specific responses, a major theme running through this entire book is that a given pollutant at sublethal concentrations will generally cause a wide spectrum of effects at all levels in the animal, from the cellular to the level of behavior. The case of a pollutant at sublethal concentrations acting on a specific target function in the animal is a relative rarity, although certain functions are generally more sensitive than others to a particular chemical or physical "insult", and this phenomenon can sometimes be useful for diagnostic purposes.

Most people dealing with stress in fish are reluctant to define it, perhaps wisely so. In an American Fisheries Society Symposium on the subject (Adams, 1990), researchers working on various levels of biological organization ranging from the molecular to the ecosystem presented a variety of approaches for measuring stress, but none of them presented a quotable definition. Most assumed they were seeing stress when one or more variables that were being measured differed significantly from controls (Heath, 1990). The problem with this assumption is that it does not incorporate biological significance. A statistically signicant change in some physiological variable does not necessarily mean the fish is unhealthy. For the purposes of biomonitoring in fish, it may be better to say the organism is under stress when one or more physiological variables are altered to the point where long-term survival may be impaired (but not necessarily due to failure of the variable being measured, as it may be only a secondary indicator). In a sense, evaluating long-term survival is an almost unattainable ideal because it is so difficult to make such predictions, but it remains a useful goal. And long-term survival should also incorporate into it the survival of not only the individual, but also the population.

Ecologists have long sought for a single all purpose method to assess enviromental health or condition. As Cairns and van der Schalie (1980, p. 1180) so appropriately put it: "This is the contemporary version of the search for the Holy Grail and almost certainly will be no more successful." In other words, no single method will ever serve for all situations. It seems safe to say that the same conclusion can be applied to the application of physiological and biochemical measures to biomonitoring the health of fish in the field. There will always be a need for a number of methods in order to be responsive to all types of potential forms of degradation of water quality.

In choosing measures to use in a biomonitoring program, there are at least three criteria that ought to be met. To some investigators these may seem self-evident but need emphasis, nonetheless:

1. The method should have sufficient sensitivity that an alteration will be predictive of death or reproductive impairment, should conditions continue unchanged. It does little good, if the animal is almost dead before the factor being measured changes significantly, for by that time, the population could be lost.
2. In order for a method to be widely used, it should be easy to perform on fish that have been captured in the field. If the method requires expensive equipment and highly trained scientists to perform it, financial constraints will probably prohibit its use unless it has especially high sensitivity and relatively low inherent variability. Then, the small number of samples required might make even an expensive technique applicable.
3. The measured factor must be relatively insensitive to the stress of capture. The usual methods for capture of fish, such as electroshocking and netting, are highly stressful to the organisms and cause immediate changes in a variety of physiological activities and biochemical measurements (Wedemeyer, 1972, 1976; Schreck et al., 1976; Pickering et al., 1982).

Even when control fish are captured in the same way, the handling associated with this undoubtedly contributes to a great deal of variability which tends to mask more subtle changes. The utilization of caged fish placed in areas to be monitored offers a partial solution to this problem (Mitz and Giesy, 1985; Oikari et al., 1985; Bidwell and Heath, 1993). Some creative developments in devising methods of capture that minimized this stress would be extremely valuable in biomonitoring as they would open up the possibilities of using a much wider assortment of physiological/biochemical techniques.

Three additional points deserve mention:

1. Seasonal changes in virtually any biochemical/physiological measurement can often be as large or larger than those produced by the presence of chemical stressors (e.g., Larsson et al., 1985; Bidwell and Heath, 1993), and climatic variations from year to year are superimposed on the seasonal effects. Thus, any attempt to determine a baseline at one time and then to make measurements of a sample population at some other time, even at the same season but different year, must deal with this problem. A better approach may be to sample control fish from a nearby similar area known to be free of contaminant stress at approximately the same time as those exposed are sampled.
2. Physiological compensation, often called acclimation, is a widespread phenomenon in many kinds of organisms. Initially, upon exposure to an altered environmental factor (e.g., temperature, chemical pollutant), a fish will exhibit some physiological response. Upon continued exposure, if the degree of stress is mild, the physiological factor may exhibit compensation (acclimation) in that it tends to return toward that of unexposed controls (see Chapter 1, Section V). Thus, this potential for compensation must be realized in water quality monitoring programs (Sastry and Miller, 1981).

3. Biomarkers can measure exposure in the sense they assess the biological availability of a toxicant to the organisms. The induction of proteins for detoxification (e.g., mixed function oxidate [MFO] enzymes for organic contaminants and metallothionein [MT] for metals) is clearly this type of biomarker and does not necessarily mean the fish are under stress. Biomarkers also measure effects of environmental toxicants on feral fish. These measurements may indicate failure of homeostasis (e.g., loss of plasma electrolytes) or a response by the fish to the stressor (e.g., elevation in plasma glucose).

The following are some biochemical/physiological measurements that may be suitable for biomonitoring of fish in the field. While these have been suggested by various investigators at different times, in most cases it remains to be ascertained to what extent they meet all the criteria listed above. The theoretical bases behind them are to be found in appropriate chapters of this book and in the references cited herein. The sequence followed here is from the cellular to the whole-body but does not reflect any priority, it is more a matter of convenience.

A. CELLULAR AND TISSUE CHEMISTRY

There are numerous enzymes and specialized proteins that could be assayed in tissues, however, only a few are currently proposed for biomonitoring. If the fish are large, tissue biopsies can be taken and the fish returned to the water (Harvey, 1990). For most variables, rapid freezing or other preparation is necessary to prevent intolerable postmortem changes in tissue constituents.

We start with the measurement of stress protein induction, a new but potentially promising technique (Sanders, 1993). It requires very small amounts of tissue and is unaffected much by the stress of capture, providing the fish are processed rapidly. Gill and muscle may be especially appropriate tissues, liver much less so (B. Bradley, personal communication). Stress protein analysis is not simple unless the laboratory has considerable expertise in running and interpreting sodium dodecyl sulfate (SDS) gels.

Glutathione is a cellular peptide involved in a wide variety of cellular activites. The overall concentration of glutathione generally rises in the presence of metals and organic contaminants (Thomas and Wofford, 1984; Thomas and Juedes, 1992) so it can be interpreted as a good indicator of exposure, not necessarily stress. The analysis is moderately easy although some would argue that the ratio of reduced glutathione (GSH) to the oxidized form (GSSH) should be determined (Stegeman et al., 1992). This complicates the analysis somewhat.

MT is a cellular protein that is useful in indicating metal contamination. MT is induced in several tissues, especially liver, by the presence of elevated metals (see Chapter 5) (Roesijadi, 1992). MT induction has been successfully used to detect metal pollution in natural waters (Roch et al., 1982). Because MT induction is also influenced by other environmental factors, such as season, the use of appropriate reference animals is critical (Endel and Brower, 1989).

The cytochrome P450 system (MFO) is induced by a number of organic xenobiotics. The literature on the cytochrome P450 system is considerable (see Chapter 5); recent reviews of the use of this method for field biomonitoring include Payne et al., 1987; Melancon et al., 1988; and Stegeman and Hahn, 1994. A variety of methods to measure various aspects of the cytochrome P450 system are being used including P450 mRNA induction with the use of cDNA probes and monoclonal and polyclonal antibodies. However, the more commonly used technique is to measure the catalytic activity of ethoxyresorufin-*O*-deethylase (EROD) or of aryl hydrocarbon hydroxylase (AHH). The catalytic assays are generally the easiest to carry out and, while usually done with liver, can also be done with other tissues as well (Stegeman, 1989). Recently, a spectrophotometric method has been proposed for rapid screening of samples and routine monitoring (Lindstrom-Seppa et al., 1993).

Cytochrome P450 activity (often called MFO) is extremely sensitive to organic pollutants and is induced fairly rapidly (few days) in exposed fish. It is insensitive to the stress of capture so lends itself to use with feral organisms. MFO activity is also induced in eggs and larvae so age is not an important variable (Binder and Lech, 1984).

Up to now, the factors mentioned have been, for the most part, adaptations to some sort of stressor. We now consider some variables that reflect damage or loss of homeostasis.

Vitamin C (ascorbate) is rapidly depleted in fish exposed to sublethal levels of several examples of both inorganic and organic substances, although increases have been seen with chlorinated phenolics (see Chapter 6). The determination of ascorbate in liver and other tissues is relatively easy (Carr et al., 1983) and "natural" stressors such as a change in temperature or salinity, or even handling, do not seem to have much effect on it (Thomas and Neff, 1984). Still, care must be taken in selecting appropriate species of fish as the ability of different species to synthesize this vitamin differs greatly, and, of course, diet may have a considerable effect on ascorbate levels.

A tissue factor that is is especially easy to measure is the water content. All that is required is a reasonably sensitive balance and an oven to dry the tissues. Sublethal levels of waterborne copper (Heath, 1984), cadmium (McCarty and Houston, 1976), and a mixture of metals simulating the effluent from a sulfide ore smeltery (Larsson et al., 1984) have been shown to cause a small but consistent rise in water content of muscle and/or liver of freshwater fish. In all likelihood, any chemical that affects osmoregulation in marine or freshwater species (Chapter 7) will alter the percentage of water in a tissue, except that estuarine fishes living in water nearly isotonic to the blood would probably not exhibit any effect.

Using caged fish, Grippo and Dunson (1991) showed that whole-body sodium loss can serve as a good indicator of acid and metal toxicity. This does require the use of an atomic absorption spectrophotometer which not all laboratories have, but the procedures are fairly straightforward, otherwise.

Acetylcholinesterase (AChE) is found in nearly all tissues. Because this enzyme is associated with neuronal synapses, the concentration is proportional to the extent of innervation of the tissue (Rao and Rao, 1984). AChE is inhibited by organophosphorus and carbamate insecticides, and the extent of this inhibition has been used to diagnose fish that are suffering from this type of poisoning (Zinkl et al., 1991). While inhibition of this enzyme is considered a classical example of a biomarker specific for a particular group of toxicants, in vitro studies implicate several metals as possible inhibitors as well (Olson and Christensen, 1980). Shaw and Panigraphi (1990) found inhibition of AChE in fish from a mercury-contaminated estuary, and the degree of inhibition was proportional to the mercury body burden. These findings suggest caution in interpreting AChE declines as being exclusively due to pesticide contamination.

Lysosomes are membrane-bound cellular organelles that contain acid hydrolases for intracellular digestion and breakdown of necrotic tissue. The membrane of lysosomes becomes less stable in response to a wide variety of environmental stressors. The stability of lysosomes can be quantified and has been used to reveal sublethal effects of cadmium in the laboratory (Versteeg and Giesy, 1985) and to detect a pollution gradient in the field (Kohler, 1991).

One of the most biologically relevant outcomes of a deterioration in water quality is a change in growth rate of the resident fish. Growth and energetics are intimately related (see Chapter 8). When the energy content of the animal is compromised, either due to reduced feeding or increased energy demand for tissue repair, growth may be reduced. Protein and lipid concentrations are easily determined in body tissues (Busacker et al.,

1990) and are not influenced much by the stress of capture. Carbohydrate stores such as plasma glucose and liver glycogen are much more variable and are extremely sensitive to capture stress so are not as useful for field studies.

Growth per se is extremely difficult to quantify in routine surveys where body length and weight are recorded, as it is often not clear as to whether the same population is being sampled each time. Thus, the measurement of one or more biochemical variables that reflect growth *rate* rather than absolute body size is needed. The measurement of the concentration of tissue RNA and DNA (Chapter 8), so the RNA/DNA ratio can be computed, has been used for assessment of starvation in larval fishes. This has not been applied much for pollution investigations in the field, but has considerable promise. The measurement of these nucleotides has relatively little intrinsic variability, are sensitive to alterations in water quality, and are probably not affected by the stress of capture (Passino, 1984; Busacker et al., 1990; Houlihan et al., 1993). Because of their much greater growth rate, young fish are more likely to show measurable effects of pollutant stress than are older ones on nucleotide concentration. Sensitive fluorometric enzyme assays now make it possible to measure RNA and DNA in single fish larvae of quite small size (Heath et al., 1993).

Houlihan et al. (1993) have pointed out that there is a close relationship between protein synthesis, growth, and oxidative metabolism. Good correlations are obtained when protein synthesis is related to the rate of oxygen consumption, and the latter correlates well to the maximal rate of activity of the cellular enzymes citrate synthase, cytochrome oxidase, and lactate dehydrogenase. Thus, measurements of the activity of these enzymes from tissues (or whole-body extracts) from feral fish should yield indications of recent relative growth rate, although some sort of laboratory calibration probably is needed before application in the field.

B. HISTOPATHOLOGY

The microscopic examination of tissues has long been used in pathology laboratories for human and veterinary medicine both for disease diagnosis and toxicology investigation. While strictly speaking it is not a biochemical/physiological technique, it is certainly related to the biochemical changes that are observed, as they are frequently the result of lesions that can be seen histologically. In a keynote address at a symposium on pollution physiology, Sindermann (1985) made a strong plea for the inclusion of histopathology in all future physiological/pollution investigations. Some might say that is overdoing things, but it is clear that there is an unrealized potential for problems in aquatic toxicology to be approached with these techniques. This is especially true for field studies as tissue structure is quite sensitive to chemical insults and relatively unaffected by extraneous factors such as season or the stress of capture. Hinton et al. (1993, p.159) claim that, "No other category of biomarker enables the researcher to examine so many potential sites of injury so rapidly."

Perhaps one of the greatest impediments to the greater use of histopathology in biomonitoring is the fact that it is so labor intensive, and there are few technicians with the requisite training for preparing and especially interpreting the microscope slides. Hinton (1990) has prepared an excellent guide to the basic techniques involved in fish histology and the use of this as a biomarker is discussed in considerable depth in Hinton et al. (1993). Atlases on fish histology include Hibiya (1982) and Groman (1982).

C. HEMATOLOGY, IMMUNOLOGY, AND BLOOD CHEMISTRY

Under the topic of hematology here is included the measurement of hematocrit (packed cell volume), hemoglobin concentration, erythrocyte count, and leukocyte count (or volume). Hematological methods have been used for many years by biologists to assess the general health of fish in hatcheries and research laboratories. The procedures are well standardized (Blaxhall and Daisley, 1973; Wedemeyer and Yasatuke, 1977; Houston, 1990) and are easy to carry out, even in the field. Hematological measures are not immune from the stress of capture but are influenced far less than some other measurements, such as blood glucose (Larsson et al., 1985). In Chapter 4, studies were reviewed which showed that some chemical pollutants induce anemia whereas others tend to cause an excessively high hematocrit and/or hemoglobin concentration. Moreover, hematological values are also influenced by the osmoregulatory status of the animal, which in turn is frequently upset by pollution (Chapter 7).

More recently, Houston et al. (1993) developed a way of using hematology which considerably increases its sensitivity. It is well known that a wide variety of environmental stressors cause changes in rates of erythrocyte synthesis. This results in more immature cells, division of circulating juvenile cells, and karyorrhexis (fragmentation of chromatin and breakdown of nuclei). These changes can have a considerable effect on respiratory gas transport even though the hematocrit is normal. The procedure involves making blood smears, which can be done in the field, and then making differential counts under a microscope.

The leukocyte count is sensitive to metals (Larsson et al., 1985) and possibly other pollutants as well. Since it is rather tedious to perform with the usual counting chamber, McLeay and Gordon (1977) proposed the utilization of a procedure termed the "leukocrit" which permits an estimate of the percent volume of leukocytes in the circulating blood. Blood is collected in microhematocrit tubes and after centrifugation, the thickness of the layer of leukocytes (on top of the erythrocytes) is measured using a low-power microscope with an ocular micrometer. Thus, this measurement can be obtained from the same tubes used for hematocrits with very little additional time expenditure. It remains to be determined how sensitive the procedure is to subtle environmental changes. In general, chemical and physical stressors cause a decrease in the leukocrit whereas infections produce the opposite response, however, it is also possible to get elevations in granulocytes concomitant with a decrease in lymphocytes, thereby yielding an unchanged leukocrit. Thus, the method has its limitations for the detection of chronic stress (Wedemeyer et al., 1983).

It is becoming more and more evident that the immune system of fish is sensitive to many environmental insults (see Chapter 11). The general pattern is a suppression of immune function at several points which then makes the fish more susceptible to disease. Disease surveys are a useful technique that can be performed (O'Connor et al., 1987; Goede and Barton, 1990). For those more interested in the actual immune system function, several tests are now available which range from simple to do to those that require considerable expertise and technology (Weeks et al., 1992). The potential for future application of these immune tests for biomonitoring is probably quite good.

Blood chemistry measurements are usually performed on plasma or serum, although the assay of erythrocytic delta-levulinic acid dehydratase (ALA-D) is used as a test for environmental exposure to lead (Hodson et al., 1984) (see below). The various procedures in blood chemistry range from the determination of blood clotting time, which is relatively easy to perform, to the analysis of hormones such as cortisol or catecholamines, which require elaborate laboratory facilities. The proliferation of automatic clinical analyzers for the determination of a considerable list of factors in blood plasma has generated some interest in their application to fish blood (e.g., Miller et al., 1983). However, the high cost of purchase and maintenance will probably limit their use for biomonitoring. There are a number of test kits available designed for the clinical laboratory which do the same analyses as the clinical analyzers at a cost that is far more reasonable when relatively small numbers of samples are to be run. Because the volume of blood available in a sample from fish is generally quite small, the usual procedures applicable to a clinical laboratory frequently have to be scaled down to accommodate plasma volumes in the range of 10–100 µL. Then it becomes possible to use the plasma when the hematocrit is prepared in microhematocrit tubes.

Some blood chemistry measurements that seem to offer promise in biomonitoring include: chloride, osmolality, total protein, glucose, clotting time, and plasma enzyme assays. The first two factors give essentially the same information so there is little reason to measure both when doing routine biomonitoring, especially when sample volumes are severely limited. If a vapor pressure osmomoter is available, osmolality is quite easy to measure and requires only a small drop of blood. The trend for fish in freshwater that are under stress will be a decrease in chloride or osmolality and the converse will occur in seawater (Chapter 7).

Total protein concentration in the plasma gives an indication whether hemoconcentration or dilution has occurred and it also shows the presence of nutritional stress. However, it may be relatively insensitive to chemical pollutants in the water, unless there is a sizable osmoregulatory alteration, which would be indicated by changes in osmolality or chloride. Plasma protein is extremely easy to measure with a refractometer, although colorimetric procedures are probably more precise.

Blood glucose is one of the more commonly used factors for measuring acute stress. It is technically easy to determine in small volumes (5–10 µL) of plasma and there are several test kits available on the market. The classic stress response is an elevation of blood sugar in response to the hormones adrenaline and possibly cortisol (see Chapter 8), however, hypoglycemia may occur with some pesticides (Lockhart and Metner, 1984; Pant and Singh, 1983). Unfortunately, this factor is particularly sensitive to the stress of capture and handling (Pickering et al., 1982) which limits its applicability. Furthermore, the changes in response to pollution, while often quite large, also tend to be rather transient so this may not be a good indicator of chronic stress.

There has been some interest in using plasma enzyme assays, especially for the enzymes that are released from damaged liver (see Chapter 6) and heart tissues. The primary enzymes of interest are serum glutamic oxalacetic transaminase, alkaline phosphatase, lactate dehydrogenase, and sorbital dehydrogenase. Lockhart and Metner (1984) showed how some of these can exhibit rather large changes during chronic exposure to a synthetic triaryl phosphate oil. Dixon et al. (1987) suggest that sorbital dehydrogenase may be especially good for assessing liver damage from a wide variety of substances, both metallic and organic. They found elevations in this enzyme in the plasma before histological lesions were evident in the liver. One potential problem with plasma enzymes in fish is that there can be a large degree of variability (Miller et al., 1983) and this may be in part due to the manner in which the analysis is often performed. Most of the assays are carried out at temperatures above 30°C, yet trout and other coldwater species possess enzymes that are adapted to function at temperatures well below that level. Thus, the enzyme activity that is measured may not be truly representative of the fish.

The enzyme leucine amino naphthylamidase is released from lysosomes in damaged tissues. Its appearance in the blood was, at one time, thought to be a good indicator of tissue damage from toxic chemicals (Bouck, 1984), but Dixon et al. (1985) found that the level of this enzyme in the blood is very sensitive to reproductive activity, salinity, stress of capture, and nutritional level. This caused such large variability that its usefulness as a biomarker for pollution was low.

Most blood enzymes that are assayed for clinical diagnosis are those that are released from damaged tissues. An exception is ALA-D. It is found in the erythrocytes and erythropoietic tissues and is easily measured. ALA-D is apparently inhibited by only lead, other metals have little or no effect on this enzyme (Hodson et al., 1984), thus it seemed to have promise as an assay for this element and is used diagnostically for suspected lead poisoning in human medicine. However, Hodson et al. (1984) reviewed its use in fish biomonitoring and noted that organic lead has little effect on it. Also, contamination by inorganic lead is rarely severe enough to cause significant inhibition. Thus, while it may have some utilization, it will probably have to be combined with analysis of levels of lead in the blood. Then, combined with the enzyme assay, one can determine the extent of lead contamination of the fish, and the type of lead compound. For example, a high blood lead but little inhibition of ALA-D would indicate organic lead, whereas inhibition of the enzyme would indicate the inorganic form. An extensive bibliography of fish blood chemistry has been recently published by Folmar (1993).

D. CHALLENGE TESTS

The idea behind a challenge test is that a fish that has been exposed to an altered environmental quality for a period of time (anywhere from days to years) may show a reduced (or perhaps enhanced) capacity to tolerate a different stress (Wedemeyer et al., 1984). The ones that have been suggested include tolerance to an elevated temperature, hypoxia, reference toxicants, diseases, and crowding.

The easiest and quickest temperature tolerance test is the determination of the critical thermal maximum (CTM). This is done by raising the temperature at a constant rate (usually 1°C 3 min) until the test fish lose equilibrium (Bonin, 1981). The endpoint is reasonably sharp, however, differences between pollutant-exposed fish and controls may be small, depending on the pollutant (Paladino et al., 1980; McLeay and Howard, 1977). Heath et al. (1994) found that sublethal exposure of fathead minnow larvae for 24 h to the pyrethroid pesticide cyfluthrin caused a reduction in the upper lethal temperature and an increase in the lower lethal temperature. Thus, the thermal limits of the fish were effectively constricted.

Tolerance to hypoxia is customarily measured by merely sealing fish in a jar and allowing them to deplete the oxygen until death occurs, then the residual dissolved oxygen is measured. Generally, the residual levels are proportional to dose of toxicant to which the fish has been exposed (Giles and Klaprat, 1979). Those chemicals that affect respiratory gas exchange will probably cause the greatest effect on the resistance to hypoxia (see Chapter 3).

There are a number of reference toxicants that have been suggested. One of the original purposes was to compare stocks of fish from different sources in bioassay investigations, but these chemicals could also be used as a challenge test. Some chemicals that have been proposed are sodium pentachlorophenate, sodium chloride, phenol, and sodium azide (Wedemyer et al., 1984). The time to death in a lethal concentration is the variable most commonly measured, however, other factors such as blood osmolality could also be used.

Theoretically, exposure to some pollutant should make the fish more sensitive to some other different chemical stressor, but this does not always occur. Heath (1987) exposed bluegill to copper or zinc for 7 days and then tested their resistance to a hypertonic NaCl challenge. Surprisingly, the fish that had been exposed to the metals survived far longer than did controls. Subsequent blood analyses revealed that the exposed fish were regulating blood electrolytes better in the hypertonic solution than controls, apparently due to a reduction in permeability of the gills to NaCl.

Challenge tests might be increased in sensitivity by using a somewhat less severe challenge (i.e., sublethal) and measuring one or more of the physiological factors discussed above while or after the fish is challenged. Larsson et al. (1984) evaluated this approach by exposing fish to a simulated sulfide ore smeltery effluent (containing several metals) for 27 days, and then challenged them by removing them from the water for exactly 3 min. They were then put back in water and sampled 2 or 4 days later. The stress of this asphyxiation induced a condition of hyperglycemia, muscle glycogen depletion, elevated muscle water, and lowered electrolyte levels in the blood as well as some minor hematological alterations. More importantly, the fish that had been previously exposed to the effluent exhibited significantly greater physiological responses to the challenge stress than did non-exposed controls.

Heath (1991) exposed bluegill to a sublethal concentration of copper which was then followed by a hypoxic stress similar to what might be experienced in a pond going hypoxic at night. Several physiological parameters were measured during the stress and subsequent recovery. In general the copper caused a more severe response to the hypoxia and delayed recovery.

A form of challenge test that has not been used much yet in pollution work is swimming capacity (see Chapter 8). This generally requires an appropriate swimming tunnel which can be costly to build. However, Heath et al. (1993) tested swimming capacity of striped bass larvae utilizing a petri dish with a smaller one glued in the middle. This creates a sort of circular "race track". Radial lines are marked on the bottom of the dish. Larvae that have been exposed to test waters for a period of time can then be tested by placing them one at a time in the race track and then chasing with a glass rod. The number of lines crossed in a minute is then counted.

E. BEHAVIOR TESTS

Most behavioral tests determine whether a stimulus (e.g., food) elicits an abnormal behavioral response outside the normal range of variability. Behavior integrates changes that may have occurred at lower levels of biological complexity, such as biochemical alterations in the nervous system but is only beginning to be used in toxicological investigations (Little, 1990). The best behavioral tests will indicate changes that might have an impact on survival, growth, or reproduction (Beitinger, 1990).

Some of the more potentially useful behavior assays are feeding behavior, phototaxis, predator avoidance, and spontaneous swimming activity (Beitinger, 1990; Birge et al., 1993; Little et al., 1993; Henry and Atchison, 1991). None of these require especially elaborate or expensive equipment nor is there a high level of expertise needed. Tests of avoidance and attractance to contaminants or temperatures and spawning behavior would be much more difficult to incorporate into field biomonitoring programs, although they would certainly have high ecological relevance. A potential advantage of behavioral tests is that they are relatively non-invasive so the organisms could be removed from their resident waters, tested, and then returned.

F. CONCLUSIONS

There will probably always be the issue of how relevant a given variable is to the ecological health of a body of water. This is a connection that is extremely difficult to make and is a valid criticism of these methodologies. Thus, it should be emphasized here that in any biomonitoring program using physiological measurements, reliance should never be placed on only one or two variables. Rather, several should be used spanning more than one level of biological organization. Then, if changes are seen in several variables and especially if a pollution gradient is detected, a fairly high degree of confidence in the findings can then be held. Examples of this approach are Adams et al. (1988, 1992) and Hodson (1990). These papers and the one by Aldrich (1989) provide useful discussions of theoretical and statistical aspects of the problems of monitoring organisms in the field utilizing physiological methodologies.

The various techniques discussed above provide a sort of snapshot of physiological condition in the test organisms. We now move to the matter of obtaining physiological data in real time.

IV. EARLY WARNING SYSTEMS

As the name implies, an early warning system is a means to detect the presence of toxic materials in water before the water gets to a place where greater harm may occur. Such systems are also often referred to as automated biomonitors (Gruber et al., 1991). Applications include the detection of a suddenly increased concentration of one or more toxicants in effluents, waste spills, and as a monitor of waters to be used for drinking.

An important limitation of monitoring using chemical analyses alone for assessing water quality is that an unexpected chemical may go undetected until considerable harm has taken place. An early warning system uses fish or other organisms as a sensor. The test fish are held in a continuous flow system receiving the waters to be monitored while one or more physiological variables are measured more or less continuously. In general, only short-term changes are detected with such systems as opposed to chronic or cumulative effects.

Cairns and van der Schalie (1980) and Gruber et al. (1991) have discussed in some detail the criteria that must be met for suitable early warning systems. The major ones are

1. The variable to be measured must be quantifiable and have the potential for automatic recording with computer or other electronic equipment.
2. The physiological variable must give reliable detection with a short latent period. In other words, such a system must operate on a real-time basis.
3. The system must be easy to operate by relatively untrained personnel.
4. The parameter to be measured must be sensitive to a wide variety of toxicants.

Obviously, these criteria eliminate a large number of physiological variables in fish from consideration in early warning systems. For example, measures of tissue chemistry and blood composition are unsuited for such a purpose. Even though cannulated fish can be maintained for weeks, automation of the blood analyses is a nearly insurmountable problem. This limits the potentially useful variables essentially to four: body movement, oxygen consumption rate, ventilation/coughing frequencies, and neurological activity.

Various schemes for detecting body movement have been devised using photocells or movement transducers (discussed in Chapter 12), but there has been only a limited attempt to automate this kind of data gathering (Smith and Bailey, 1988). A different type of bodily activity is the ability to maintain position in a slowly moving current of water (sometimes referred to as rheotaxis). Automated systems have been developed utilizing photocells at the rear of the swimming chamber to detect when the fish loses rheotaxis in response to a toxicant entering the water (Poels, 1977). A 2-year test of a rheotaxis system in The Netherlands successfully detected six alarms produced by increased pollutants in the river and one false alarm (Balk et al., 1994).

The oxygen consumption of a fish can be monitored on a continuous basis but it is sensitive to a variety of factors other than pollutants (see Chapter 8). One well-known fish physiologist commented to the effect that one has only to think evil of a fish to double its rate of oxygen consumption!

The application of changes in ventilatory activity has received the most development and application in automated biomonitoring. It is obviously related to oxygen consumption rate but changes in ventilation are generally more sensitive to the presence of toxicants than is oxygen consumption alone. The delicate gill tissue is the first tissue to be affected by a waterborne chemical and this can cause immediate and very marked changes in ventilation and/or coughing.

The methods for measuring ventilation by fish have been reviewed by Heath (1972). Even though that review is old, there have not been any significantly new methods developed since then. The simplest technique, which is non-invasive and involves no attached wires or tubes to the fish, is the use of electrodes at each end of a small test chamber. With a suitable amplifier, the potential changes (in the millivolt range) between the electrodes produced by the breathing of the fish can be detected and the sine wave displayed on a chart recorder (Spoor et al., 1971). Aberrant breaths such as coughs can also be detected. This ventilatory recording technique has been interfaced with a computer and is now available commercially for detecting the presence of toxic chemicals coming into domestic drinking water supplies (Gruber et al., 1991). Baldwin et al., (1994a,b) evaluated this technique over 1-year trial periods. They found a low rate of false alarms and a sensitivity using trout of between 10 and 250% of the LC50 (depending on pollutant) with a response within 40 min. As expected, the system detects acutely toxic concentrations much better than chronic changes.

The fish ventilation response has also been used to monitor water quality in trout streams in Tennessee (Morgan et al., 1988). Data collection platforms equiped with ventilation chambers, a water sampler, and pH and temperature sensors are placed beside streams in remote locations. Then, with the use of hardwire or satellite transmission, real-time data are transmitted to a central data processing facility. In this way, several streams can be monitored simultaneously.

Finally, the weakly electric fish *Gnathonemus petersi* has been found to change the character of its electric discharge in the presence of toxicants. These can be detected with electrodes in the water and Geller (1984) interfaced such a system with a microcomputer to automate it. Further evaluation of the sensitivity of this fish to detect pollutants was done by Lewis et al. (1993). It would be interesting to compare the fish ventilatory system with this one as to sensitivity, false alarms, etc. Then the next big hurdle will be to get wider acceptance by municipalities, regulatory agencies, and industry.

REFERENCES

Adams, S. M., Ed., Biological indicators of stress in fish, *Am. Fish. Soc. Symp. 8,* Bethesda, MD, 1990.

Adams, S. M., Beauchamp, J. J. and Burtis, C. A., A multivariate approach for evaluating responses of fish to chronic pollutant stress, *Mar. Environ. Res.,* 24, 223, 1988.

Adams, S. M., Crumby, W. D., Greeley, M., Ryon, M. G. and Schilling, E. M., Relationship between physiological and fish population responses in a contaminated stream, *Environ. Toxicol. Chem.,* 11, 1549, 1992.

Addison, R. F., Hepatic mixed function oxidase (MFO) induction in fish as a possible biological monitoring system, in, *Contaminant Effects on Fisheries,* Cairns, V. W., Hodson, P. V. and Nriagu, J. O., Eds., Wiley and Sons, New York, 1984, chap. 5.

Alabaster, J. S. and Lloyd, R., *Water Quality Criteria for Freshwater Fish,* Butterworths, London, 1980.

Aldrich, J. C., Diagnosis or elucidation — two differing uses of physiology, *Mar. Behav. Physiol.,* 15, 217, 1989.

American Society For Testing and Materials, *Standard Guide for Conducting Acute Toxicity Tests with Fishes, Macroinvertebrates and Amphibians,* ASTM E 729-88a, American Society for Testing and Materials, Philadelphia, 1988.

APHA, *Standard Methods for the Examination of Water and Wastewater,* 18th ed., American Public Health Association, Washington, D.C., 1992.

Baldwin, I. G., Harman, M. M. and Neville, D. A., Performance characteristics of a fish monitor for detection of toxic substances. I. Laboratory trials, *Water Res.,* 28, 2191, 1994a.

Baldwin, I. G., Harman, M. M., Neville, D. A. and George, S. G., Performance characteristics of a fish monitor for detection of toxic substances. II. Field trials, *Water Res.,* 28, 2201, 1994b.

Balk, F., Okkerman, P. C., van Helmond, C. A. M., Noppert, F. and van der Putte, I., Biological early warning systems for surface water and industrial effluents, *Water Sci. Technol.,* 29, 211, 1994.

Bayne, B. L., Ed., *The Effects of Stress and Pollution on Marine Animals,* Praeger Scientific, New York, 1985.

Beitinger, T. L., Behavioral reactions for the assessment of stress in fishes, *J. Great Lakes Res.,* 16, 495, 1990.

338

Bidwell, J. and Heath, A., An *in situ* study of rock bass (*Ambloplites rupestris*) physiology: effect of season and mercury contamination, *Hydrobiologia*, 264, 137, 1993.

Binder, R. L. and Lech, J. J., Xenobiotics in gametes of Lake Michigan lake trout *Salvelinus namacycush* induced hepatic monooxygenase activity in their offspring, *Fund. Appl. Toxicol.*, 4, 1042, 1984.

Birge, W. J., Black, J. A., Hudson, J. E. and Bruser, D. M., Embryo-larval toxicity tests with organic compounds, in, *Aquatic Toxicology*, ASTM STP 667, Marking, L. L. and Kimerle, R. A., Eds., American Society for Testing and Materials, Philadelphia, 1979, 131.

Birge, W. J., Hoyt, R. D., Black, J. A., Kercher, M. D. and Robison, W. A., Effects of chemical stresses on behavior of larval and juvenile fishes and amphibians, *Am. Fish. Soc. Symp.*, 14, 55, 1993.

Blaxhall, P. C. and Daisley, K. W., Routine haematological methods for use with fish blood, *J. Fish Biol.*, 5, 771, 1973.

Bonin, J. D., Measuring thermal limits of fish, *Trans. Am. Fish. Soc.*, 110, 662, 1981.

Bouck, G. R., Physiological responses of fish: problems and progress toward use in environmental monitoring, in, *Contaminant Effects on Fisheries*, Cairns, V. W., Hodson, P. V. and Nriagu, J. O., Eds., Wiley and Sons, New York, 1984, chap. 6.

Busacker, G. P., Adelman, I. R. and Goolish, E. M., Growth, in, *Methods for Fish Biology*, Schreck, C. B. and Moyle, P. B., Eds., American Fisheries Society Press, Bethesda, MD, 1990, 363.

Cairns, J., Jr., Buikema, A., Cherry, D., van der Schalie, W., Matthews, R. and Rodgers, J., *Biological Monitoring*, Pergamon Press, London, 1984.

Cairns, V. W., Hodson, P. V. and Nriagu, J. O., Eds., *Contaminant Effect on Fisheries*, Wiley & Sons, New York, 1984.

Cairns, J., Jr. and van der Schalie, W. H., Biological monitoring. Part I. Early warning systems, *Water Res.*, 14, 1179, 1980.

Cairns, J., Dickson, K. and Westlake, G., Continuous biological monitoring to establish parameters for water pollution control, *Prog. Wat. Technol.*, 7, 829, 1975.

Carr, B., Bally, M., Thomas, P. and Neff, J., Comparison of methods for determination of ascorbic acid in animal tissues, *Anal. Chem.*, 55, 1229, 1983.

Casillas, D. and Smith, L., Effect of stress on blood coagulation and haematology in rainbow trout (*Salmo gairdneri*), *J. Fish Biol.*, 10, 481, 1977.

Christensen, G., Hunt, E. and Fiandt, J., The effect of methylmercuric chloride, cadmium chloride, and lead nitrate on six biochemical factors of the brook trout (*Salvelinus fontinalis*), *Toxicol. Appl. Pharmacol.*, 42, 523, 1977.

Dixon, D. G., Hill, C. E., Hodson, P. V., Kempe, E. J. and Kaiser, K. L. E., Plasma leucine aminonaphthylamidase as an indicator of acute sublethal toxicant stress in rainbow trout, *Environ. Toxicol. Chem.*, 4, 789, 1985.

Dixon, D. G., Hodson, P. V. and Kaiser, K. L., Serum sorbitol dehydrogenase activity as an indicator of chemically induced liver damage in rainbow trout, *Environ. Toxicol. Chem.*, 6, 685, 1987.

Engel, D. and Brouwer, M., Metallothionein and metallothionein-like proteins: physiological importance, *Adv. Comp. Environ. Physiol.*, 4, 53, 1989.

Folmar, L. C., Effects of chemical contaminants on blood chemistry of teleost fish: a bibliography and synopsis of selected effects, *Environ. Toxicol. Chem.*, 12, 337, 1993.

Geller, W., A toxicity warning monitor using the weakly electric fish, *Gnathoneumus petersi*, *Water Res.*, 18, 1285, 1984.

Giles, M. A. and Klaprat, D., The residual oxygen test: a rapid method for estimating the acute lethal toxicity of aquatic contaminants, in, *Toxicity Tests for Freshwater Organisms*, Sherer, E., Ed., Canadian Special Publication Fisheries Aquatic Science, No. 44, 37, 1979.

Goede, R. W. and Barton, B. A., Organismic indices and an autopsy-based assessment as indicators of health and condition of fish, *Am. Fish. Soc. Symp.*, 8, 93, 1990.

Grippo, R. S. and Dunson, W., Use of whole body sodium loss from the fathead minnow (*Pimephales promelas*) as an indicator of acid and metal toxicity, *Arch. Environ. Contam. Toxicol.*, 21, 289, 1991.

Groman, D. B., Histology of the Striped Bass, Monograph Number 3, American Fisheries Society, Bethesda, MD, 1982.

Gruber, D. J., Diamond, J. M. and Parsons, M. J., Automated biomonitoring, *Environ. Auditor*, 2, 229, 1991.

Haasch, M. L., Prince, R., Wejksnora, P. J., Cooper, K. R. and Lech, J. J., Caged and wild fish: induction of hepatic cytochrome P-450, CYP1A1, as an environmental biomonitor, *Environ. Toxicol. Chem.,* 12, 885, 1993.

Harvey, W. D., Noble, R., Neill, W. and Marks, J., A liver biopsy technique for electrophoretic evaluation of largemouth bass, in, *Electrophoretic and Isoelectric Focusing Techniques in Fisheries Management,* Whitmore, D., Ed., CRC Press, Boca Raton, FL, 1990, 87.

Heath, A. G., Changes in tissue adenylates and water content of bluegill, *Lepomis machrochirus,* exposed to copper, *J. Fish Biol.,* 24, 299, 1984.

Heath, A. G., A critical comparison of methods for measuring fish respiratory movements, *Water Res.,* 6, 1, 1972.

Heath, A. G., Effects of waterborne copper or zinc on the osmoregulatory response of bluegill to a hypertonic NaCl challenge, *Comp. Biochem. Physiol.,* 88C, 307, 1987.

Heath, A. G., Summary and perspectives, *Am. Fish. Soc. Symp.,* 8, 183, 1990.

Heath, A. G., Effect of water-borne copper on physiological responses of bluegill (*Lepomis macrochirus*) to acute hypoxic stress and subsequent recovery, *Comp. Biochem. Physiol.,* 100C, 559, 1991.

Heath, A. G., Cech, J. J., Zinkl, J. G., Finlayson, B. and Fujimura, R., Sublethal effects of methyl parathion, carbofuran, and molinate on larval striped bass, *Am. Fish. Soc. Symp.,* 14, 17, 1993.

Heath, S., Bennett, W. A., Kennedy, J. and Beitinger, T., Heat and cold tolerance of the fathead minnow, *Pimephales promelas,* exposed to the synthetic pyrethroid cyfluthrin, *Can. J. Fish. Aquat. Sci.,* 51, 437, 1994.

Henry, M. G. and Atchison, G. J., Metal effects on fish behavior. Advances in determining the ecological significance of responses, in, *Ecotoxicology of Metals: Current Concepts and Applications,* Newman, M. C. and McIntosh, A. W., Eds., Lewis Publishers, Boca Raton, FL, 1991, 131.

Hibiya, T., Ed., *An Atlas of Fish Histology, Normal and Pathological Features,* Gustav Fisher Verlag, New York, 1982.

Hille, S., A literature review of the blood chemistry of rainbow trout, *Salmo gairdneri* Rich., *J. Fish Biol.,* 20, 535, 1982.

Hinton, D., Histological Techniques, in, *Methods for Fish Biology,* Schreck, C. B. and Moyle, P. B., Eds., American Fisheries Society, Bethesda, MD, 1990, chap. 7.

Hinton, D., Baumann, P., Gardner, G., Hawkins, W., Hendricks, J., Murchelano, R. and Okihairo, M., Histopathologic biomarkers, in, *Biomarkers: Biochemical, Physiological, and Histological Markers of Anthropogenic Stress,* Huggett, R., Kimerle, R., Mehrle, P. and Bergman, H., Lewis Publishers, Boca Raton, FL, 1993, chap. 4.

Hocutt, C. H. and Stauffer, J. R., Eds., *Biological Monitoring of Fish,* Lexington Books, Lexington, KY, 1980.

Hodson, P. V., Blunt, B. R. and Whittle, D. M., Monitoring lead exposure of fish, in, *Contaminant Effects on Fisheries,* Cairns, V. W., Hodson, P. V. and Nriagu, J. O., Eds., Wiley and Sons, New York, 1984, chap. 8.

Hodson, P. V., Indicators of ecosystem health at the species level and the example of selenium effects on fish, *Environ. Monitoring Assess.,* 15, 241, 1990.

Houlihan, D. F., Mathers, E. M. and Foster, A., Biochemical correlates of growth rate in fish, in, *Fish Ecophysiology,* Rankin, J. C. and Jensen, F. B., Eds., Chapman and Hall, London, 1993, chap. 2.

Houston, A. H., Blood and circulation, in, *Methods for Fish Biology,* Schreck, C. and Moyle, P., Eds., American Fisheries Society, Bethesda, MD, 1990, chap. 9.

Houston, A. H., Changes in erythron organization during prolonged cadmium exposure: an indicator of heavy metal stress?, *Can. J. Fish. Aquat. Sci.,* 50, 217, 1993.

Huggett, R., Kimerle, R., Mehrle, P. and Bergman, H., Eds., *Biomarkers, Biochemical, Physiological, and Histological Markers of Anthropogenic Stress,* Lewis Publishers, Boca Raton, FL, 1992.

Kohler, A., Lysosomal perturbations in fish liver as indicators for toxic effects of environmental pollution, *Comp. Biochem. Physiol.,* 100C, 123, 1991.

Landner, L., Neilson, A. H., Sorensen, L., Tarnholm, A. and Viktor, T., Short-term test for predicting the potential of xenobiotics to impair reproductive success in fish, *Ecotoxicol. Environ. Safety,* 9, 282, 1985.

Larsson, A., Haux, C. and Sjobeck, M., Fish physiology and metal pollution: results and experiences from laboratory and field studies, *Ecotoxicol. Environ. Safety,* 9, 250, 1985.

Larsson, A., Haux, C., Sjobeck, M. and Lithner, G., Physiological effects of an additional stressor on fish exposed to a simulated heavy-metal-containing effluent from a sulfide ore smeltery, *Ecotoxicol. Environ. Safety,* 8, 118, 1984.

Lewis, J. W., Kay, A. N. and Hanna, N. S., Responses of electric fish (family Mormyridae) to chemical changes in water quality. II. Pesticides, *Environ. Technol.,* 14, 1171, 1993.

Lindstrom-Seppa, P., Farmanfarmaian, L. and Stegeman, J., A visual test for hepatic EROD activity as a marker for exposure to aromatic and halogenated aromatic hydrocarbons, *Chemosphere,* 27, 2183, 1993.

Little, E. E., Behavioral toxicology: stimulating challenges for a growing discipline, *Environ. Toxicol. Chem.,* 9, 1, 1990.

Little, E. E., Fairchild, J. F. and DeLonay, A. J., Behavioral methods for assessing impacts of contaminants on early life stage fishes, *Am. Fish. Soc. Symp.,* 14, 67, 1993.

Lloyd, R., *Pollution and Freshwater Fish,* Blackwell Scientific Publishers, Cambridge, MA, 1992.

Lockhart, W. L. and Metner, D. A., Fish serum chemistry as a pathology tool, in, *Contaminant Effects on Fisheries,* Cairns, V. W., Hodson, P. V., Nriagu, J., McCarthy, J. and Shugart, L., Eds., *Biomarkers of Environmental Contamination,* Lewis Publishers, Boca Raton, FL, 1990.

McCarty, L. S. and Houston, A. H., Effects of exposure to sublethal levels of cadmium upon water-electrolyte status in the goldfish (*Carassius auratus*), *J. Fish Biol.,* 9, 11, 1976.

McKim, J. M., Evaluation of tests with early life stages of fish for predicting long-term toxicity, *J. Fish. Res. Bd. Can.,* 34, 1148, 1977.

McLeay, D. D. and Gordon, M. R., Leukocrit: a simple haematological technique for measuring acute stress in salmonid fish, including stressful concentrations of pulpmill effluent, *J. Fish. Res. Bd. Can.,* 34, 2164, 1977.

McLeay, D. J. and Howard, T. E., Comparison of rapid bioassay procedures for measuring toxic effects of bleached kraft mill effluent to fish, in, *Proc. 3rd Aquatic Toxicology Workshop,* Parker, W. R., Ed., Enviromental Protection Service Technical Report EPS-5-AR- 77-1, Halifax, N.S., 141, 1977.

Melancon, M., Binder, R. and Lech, J., Environmental induction of monooxygenase activity in fish, in, *Toxic Contaminants and Ecosystem Health, A Great Lakes Focus,* Evans, M. S., Ed., John Wiley & Sons, New York, 1988, 215.

Miller, W. R., Hendricks, A. C. and Cairns, J., Jr., Normal ranges for diagnostically important hematological and blood chemistry characteristics of rainbow trout (*Salmo gairdneri*), *Can. J. Fish. Aquat. Sci.,* 40, 420, 1983.

Mitz, S. V. and Giesy, J. P., Sewage effluent biomonitoring. I. Survival, growth, and histopathological effects in channel catfish, *Ecotoxicol. Environ. Safety,* 10, 22, 1985.

Morgan, E. L., Young, R. C. and Wright, J. R., Developing portable computer-automated biomonitoring for a regional water quality surveillance network, in, *Automated Biomonitoring,* Gruber, D. J. and Diamond, J. M., Eds., Halsted Press, London, 1988, chap. 9.

O'Connor, J. S., Ziskowski, J. and Murchelano, R. A., Index of pollutant-induced fish and shellfish disease, NOAA (National Oceanic and Atmospheric Administration) Special Report, Washington, D.C., 1987.

Oikari, A., Holmbom, B., Anas, E., Miilunpalo, M., Kruzynski, G. and Castren, M., Ecotoxicological aspects of pulp and paper mill effluents discharged to an inland water system: distribution in water, and toxicant residues and physiological effects in caged fish (*Salmo gairdneri*), *Aquat. Toxicol.,* 6, 219, 1985.

Olson, D. L. and Christensen, G. M., Effects of water pollutants and other chemicals on fish acetylcholinesterase (*in vitro*), *Environ. Res.,* 21, 327, 1980.

Paladino, F. V., Schubauer, J. P. and Kowalski, K. T., The critical thermal maximum: a technique used to elucidate physiological stress and adaptation in fishes, *Rev. Can. Biol.,* 39, 115, 1980.

Pant, J. and Singh, T., Inducement of metabolic dysfunction by carbamate and organophosphorus compounds in a fish, *Punctius conchonius, Pest. Biochem. Physiol.,* 20, 294, 1983.

Passino, D. R. M., Biochemical indicators of stress in fishes: an overview, in, *Contaminant Effects on Fisheries,* Cairns, V. W., Hodson, P. V. and Nriagu, J. O., Eds., Wiley & Sons, New York, 1984, chap. 4.

Payne, J. F., Fancey, L. L., Rahimtula, A. D. and Porter, E. L., Review and perspective on the use of mixed-function oxygenase enzymes in biological monitoring, *Comp. Biochem. Physiol.,* 86C, 233, 1987.

Pickering, A. D., Pottiger, T. G. and Christie, P., Recovery of brown trout, *Salmo trutta*, L., from acute handling stress: a time course study, *J. Fish Biol.*, 20, 229, 1982.

Poels, C. L. M., An automatic system for rapid detection of acute high concentrations of toxic substances in surface water using trout, in, *Biological Monitoring of Water and Effluent Quality,* STP 607, Cairns, J., Dickson, K. and Westlake, G., Eds., American Society for Testing and Materials, Philadelphia, PA, 85, 1977.

Rao, K. S. and Rao, K. V., Impact of methyl parathion toxicity and eserine inhibition on acetylcholines-terase activity in tissues of the teleost (*Tilapia mossambica*) — a correlative study, *Toxicol. Lett.*, 22, 351, 1984.

Roch, M., McCarter, J. A., Matheson, A. T., Clark, M. J. R. and Olafson, R. W., Hepatic metallothionein in rainbow trout (*Salmo gairdneri*) as an indicator of metal pollution in the Campbell river system, *Can. J. Fish. Aquat. Sci.*, 39, 1596, 1982.

Roesijadi, G., Metallothioneins in metal regulation and toxicity in aquatic animals, *Aquat. Toxicol.*, 22, 81, 1992.

Sanders, B. M., Stress proteins in aquatic organisms: an environmental perspective, *Critical Rev. Toxicol.*, 23, 49, 1993.

Sastry, A. N. and Miller, D. C., Application of biochemical and physiological responses to water quality monitoring, in, *Biological Monitoring of Marine Pollutants,* Vernberg, J., Calabrese, A., Thurberg, F. D. and Vernberg, W. B., Eds., Academic Press, New York, 1981, 265.

Schreck, C. B., Whaley, R. A., Bass, M. I., Maughan, O. E. and Solazzi, M., Physiological responses of rainbow trout (*Salmo gairdneri*) to electroshock, *J. Fish. Res. Bd. Can.*, 33, 76, 1976.

Schreck, C. B. and Lorz, H. W., Stress response of coho salmon (*Oncorhynchus kisutch*) elicited by cadmium and copper and potential use of cortisol as an indicator of stress, *J. Fish. Res. Bd. Can.*, 35, 1124, 1978.

Shaw, B. P. and Panigraphi, A. K., Brain AChE activity studies in some fish species collected from a mercury contaminated estuary, *Water, Air, Soil Pollut.*, 53, 327, 1990.

Sindermann, C. J., Keynote address: notes of a pollution watcher, in, *Marine Pollution and Physiology: Recent Advances,* Vernberg, F. J., Thurberg, F. P., Calabrese, A. and Vernberg, W., Eds., University of South Carolina Press, Columbia, 1985, 11.

Smith, E. H. and Bailey, H. C., Development of a system for continuous biomonitoring of a domestic water source for early warning of contaminants, in, *Automated Biomonitoring: Living Sensors as Environmental Monitors,* Gruber, D. S. and Diamond, J. M., Eds., Ellis Horwood, Chichester, 1988, 182.

Spoor, W. A., Neiheisel T. W. and Drummond, R. A., An electrode chamber for recording respiratory and other movements of free-swimming animals, *Trans. Am. Fish. Soc.*, 100, 22, 1971.

Stegeman, J. J., Monooxygenase systems in marine fish, in, *Pollutant Studies in Marine Animals,* Giam, C. S. and Ray, L., Eds., CRC Press, Boca Raton, FL, 1989, 65.

Stegeman J., Brouwer, M., DiGuilio, R., Forlin, L., Fowler, B., Sanders, B. and Van Veld, P., Molecular responses to environmental contamination: enzyme and protein systems as indicators of chemical exposure, in, *Biomarkers: Biochemical, Physiological and Histological Markers of Anthropogenic Stress,* Huggett, R., Kimerle, R., Mehrle, P. and Bergman, H., Eds., Lewis Publishers, Boca Raton, FL, 1992, chap. 6.

Stegeman, J. J. and Hahn, M. E., Biochemistry and molecular biology of monooxygenases: current perspectives on forms, functions and regulation of cytochrome P450 in aquatic species, in, *Aquatic Toxicology, Molecular, Biochemical and Cellular Perspectives,* Malins, D. C. and Ostrander, G. K., Eds., CRC Press, Boca Raton, 1994, chap. 3.

Thomas, P. and Neff, J. M., Effects of pollutant and other environmental variables on the ascorbic acid content of fish tissues, *Mar. Environ. Res.*, 14, 489, 1984.

Thomas, P. and Juedes, M., Influence of lead on the glutathione status of Atlantic croaker tissues, *Aquat. Toxicol.*, 23, 11, 1992.

Thomas, P. and Wofford, H., Effects of metal and organic compounds on hepatic glutathione, cysteine and acid-soluble thiol levels in mullet *Mugil cephalus, Toxicol. Appl. Pharmacol.*, 76, 172, 1984.

van der Schalie, W. H., Dickson, K. L., Westlake, G. F. and Cairns, J., Automatic monitoring of waste effluents using fish, *Environ. Mgt.*, 3, 217, 1979.

Vernberg, J., Calabrese, A., Thurberg, F. P. and Vernberg, W. B., Eds., *Biological Monitoring of Marine Pollutants,* Academic Press, New York, 1981.

Versteeg, D. and Giesy, J. P., Lysosomal enzyme release in the bluegill sunfish (*Lepomis macrochirus*) exposed to cadmium, *Arch. Environ. Contam. Toxicol.,* 14, 631, 1985.

Weber, C., Peltier, W., Norberg-King, T., Horning, W., Kessler, F., Menkedick, J., Neiheisel, T., Lazorchak, J., Wkymer, L. and Fryberg, R., Short-term methods for estimating chronic toxicity of effluents and receiving waters to freshwater organisms, 2nd ed. EPA-600/4-89-001, 1989.

Wedemeyer, G. A. and McLeay, D. J., Methods for determining the tolerance of fishes to environmental stressors, in, *Stress and Fish,* Pickering, A. D., Ed., Academic Press, New York, 1981, chap. 11.

Wedemeyer, G. A., Some physiological consequences of handling stress in the juvenile coho salmon (*Oncorhynchus kisutch*) and rainbow trout (*Salmo gairdneri*), *J. Fish. Res. Bd. Can.,* 29, 1780, 1972.

Wedemeyer, G. A. and Yasutake, W. T., Clinical methods for the assessment of the effects of environmental stress on fish health, Technical Paper 89, U.S. Fish and Wildlife Service, U.S. Department of the Interior, 1977.

Wedemeyer, G. A., Gould, R. W. and Yasutake, W. T., Some potentials and limits of the leucocrit test as a fish health assessment method, *J. Fish Biol.,* 23, 711, 1983.

Wedemeyer, G. A., LcLeay, D. J. and Goodyear, C. P., Assessing the tolerance of fish and fish populations to environmental stress: the problems and methods of monitoring, in, *Contaminant Effects on Fisheries,* Cairns, V. W., Hodson, P. V. and Nriagu, J. O., Eds., Wiley & Sons, New York, 1984, chap. 12.

Weeks, B. A., Anderson, D. P., DuFour, A. P., Fairbrother, A., Goven, A. J., Lahvis, G. and Peters, G., Immunological biomarkers to assess environmental stress, in, *Biomarkers: Biochemical, Physiological and Histological Markers of Anthropogenic Stress,* Hugett, R. J., Kimerle, R. A., Mehrle, P. M. and Bergman, H. L., Eds., Lewis Publishers, Boca Raton, FL, 1992, chap. 5.

Woltering, D. M., The growth response in fish chronic and early life stage toxicity tests: a critical review, *Aquat. Toxicol.,* 5, 1, 1984.

Wydoski, R. S., Wedemeyer, G. A. and Nelson, N. C., Physiological response to hooking stress in hatchery and wild rainbow trout (*Salmo gairdneri*), *Trans. Am. Fish. Soc.,* 105, 601, 1976.

Zinkl, J., Lockhart, W., Kenny, S. and Ward, F., Effects of cholinesterase-inhibiting insecticides on fish, in *Cholinesterase-Inhibiting Insecticides — Impact on Wildlife and The Environment,* Mineau, P., Ed., Elsevier Science Publishers, Amsterdam, 1991, 233.

Index

A

Acclimation, 234
  to acid pollution, 248
  biomonitoring, 328
  energetics, physiological, 186
  hypoxia, environmental, 36
  to metals, 277
Accumulation
  of metals in different organs,
    xenobiotics, 91
  of organics, xenobiotics, 101
Acetate, energetics, physiological, 194
Acetylcholine
  nervous system function, 282
  respiratory response, 58
Acetylcholinesterase
  biomonitoring, 330
  energetics, physiological, 199
  hypoxia, environmental, 17
  nervous system function, 282, 286–288
Acid
  acclimation to, acid pollution, 248
  effects of chronic exposures, acid
    pollution, 247
  effects on swimming performance, acid
    pollution, 251
  effects on ventilation and blood gases,
    acid pollution, 250
  hematology, 73
  hormonal responses, acid pollution,
    249
  respiratory response, 50, 57, 61
  water pollution, 3
Acid exposure, acid pollution, 244

Acid phosphatase, 224–226
  liver, 132
Acid pollution, 239–257
  comparative species sensitivities, acid
    pollution, 240
Acid rain. See Acid pollution
Acrolein, respiratory response, 54
Active metabolic rate, energetics,
  physiological, 174
Adenosine triphosphate, hypoxia,
  environmental, 29
Adenylate energy charge, 232
  liver, 126
Adenylates, 232, 233
  hypoxia, environmental, 30
Adrenalin. See also Epinephrine
  hypoxia, environmental, 35
  liver, 127
  osmoregulation, 143
  water pollution, 9
Adrenaline
  energetics, physiological, 178, 200–202,
    206, 207
  hypoxia, environmental, 24
  nervous system function, 289
  respiratory response, 58
Aerobic scope, energetics, physiological,
  175, 195, 198, 200
Aggression, nervous system function, 284
Air breathing, hypoxia, environmental, 33
Alanine, 279
  hypoxia, environmental, 30
  nervous system function, 279
Alanine aminotransferase, 222
  liver, 131

Alarm response, respiratory response, 55
Albumin, liver, 129, 131
Aldicarb, 229
  excretion, xenobiotics, 112
Aldolase, 228
Aldosterone, osmoregulation, 143
Aldrin
  hematology, 74
  water pollution, 3
Alkaline conditions, acid pollution, 249
Alkaline phosphatase, 219, 224, 226, 227
  liver, 131
Allethrin, water pollution, 4
Aluminum
  acclimation, osmoregulation, 156
  acid pollution, 239, 240, 243, 247, 248,
    251, 252
  effect on swimming ability, energetics,
    physiological, 196
  effects on blood calcium,
    osmoregulation, 156
  effects on osmoregulation,
    osmoregulation, 156
  energetics, physiological, 181, 196
  immunity, 263
  low assimilation by gut, xenobiotics, 86
  osmoregulation, 156
  reproduction, 310, 315
  respiratory response, 54
  xenobiotics, 82, 86
Amino acid
  energetics, physiological, 200
  oxidase, energetics, physiological, 205
Amitrol, water pollution, 4
Ammonia, 230
  effects on osmoregulation,
  excretion, acid pollution, 249
  liver, 127
  nervous system function, 284
  respiratory response, 50, 54, 60
  water pollution, 3
  xenobiotics, 109
Anaerobic energy expenditure, energetics,
    physiological, 174, 175
Anaerobic metabolism
  effect of temperature, hypoxia,
    environmental, 29
Androgen, acid pollution, 241. See also
    Testosterone
Anemia
  biological significance, hematology, 73
  hematology, 68, 70

Angiotensin, respiratory response, 61
Anoxia, hypoxia, environmental, 15, 16,
    30, 31
Anthracene
  excretion, xenobiotics, 112
  hypoxia, environmental, 18
  respiratory response, 54
  xenobiotics, 112
Antigen processing, immunity, 260
Antioxidant, 231
  defenses, hematology, 70
Appetite
  acid pollution, 252
  energetics, physiological, 175, 178, 179,
    193
  nervous system function, 282
Aquatic surface respiration, hypoxia,
    environmental, 33
Aquatic toxicology, water pollution, 5
Arginine vasotocin, acid pollution, 250
Aroclor, liver, 127
Aroclor 1242, 275
Aromatic hydrocarbons
  excretion, xenobiotics, 111
  immunity, 265
  reproduction, 300
Arsenic
  liver, 127
  nervous system function, 285
  respiratory response, 50
  water pollution, 2
  xenobiotics, 96
Arterial oxygen
  respiratory response, 47
  tension, 56
Ascorbate, 231
  biomonitoring, 330
Ascorbic acid
  liver, 134
  xenobiotics, 109
Aspartate aminotransferase, 131, 219, 222
Assimilation
  efficiency
    energetics, 173
    xenobiotics, 86, 91
  energetics, physiological, 186
  intestine, xenobiotics, 81
Atrazine, immunity, 264
Atrial natriuretic factor, osmoregulation,
    143
Automated biomonitors, biomonitoring,
    336

Avoidance
   acid pollution, 252
   hypoxia, environmental, 34
   waterborne chemicals, 275

B

B lymphocytes, immunity, 260, 261
Baseline, biomonitoring, 328
Behavior, 271–297. See also Specific
      behaviors
   acid pollution, 252
   biomonitoring, 327
   effects of hypoxia, environmental,
      33
   reproduction, 301
   tests, biomonitoring, 335
   water pollution, 6
Benzaldehyde, respiratory response,
      54
Benzene
   energetics, physiological, 189, 205
   nervous system function, 279
Benzo[a]pyrene
   energetics, physiological, 189
   excretion, xenobiotics, 112
   nervous system function, 287
   reproduction, 314
   xenobiotics, 110, 112, 113
Benzophenone, energetics, physiological,
      194
Beryllium, respiratory response, 50
Bile
   immunity, 265
   liver, 129, 133
   metals in, xenobiotics, 96
   xenobiotics, 95, 98, 102, 111
Bilirubin, liver, 133
Bimodal respiration, hypoxia,
      environmental, 33
Bioaccumulation, xenobiotics, 79
Bioassay, water pollution, 5
Bioavailability
   water pollution, 2
   xenobiotics, 81, 86, 101
Biochemical oxygen demand, water
      pollution, 1
Bioconcentration
   organic pollutants, xenobiotics, 101
   xenobiotics, 79, 94, 101
Bioenergetics, 171–216, 171
Biological organization, water pollution, 6

Biomagnification, xenobiotics, 79
Biomarker, 233
   acid pollution, 250
   biomonitoring, 327, 329, 330, 331, 332
   energetics, physiological, 182, 207
   hematology, 72
   immunity, 265, 266
   reproduction, 302
   water pollution, 5
   xenobiotics, 107
Biomonitoring of fish in the field and
      mesocosms, 327
Biotelemetry, energetics, physiological,
      190, 191
Biotransformation
   effects on accumulation, xenobiotics,
      104
   of organic contaminants, xenobiotics,
      106
Biotransfusion in intestine, xenobiotics,
      110
Blood, 67
   affinity for oxygen, hypoxia,
      environmental, 22, 25
   oxygen transport, hypoxia,
      environmental, 24
Blood acid-base
   acid pollution. See Arterial acid-base
   hypoxia, environmental, 35
Blood acidity
   hypoxia, environmental, 34
   respiratory response, 56
Blood cell count
   hematology, 68
   hypoxia, environmental, 35
Blood chemistries. See Individual
      constituents
Blood clotting, effect of stress, liver, 129
Blood electrolytes, effects of acid on, acid
      pollution, 244
Blood glucose, hypoxia, environmental,
      35
Blood liver, volume, 126
Blood oxygen
   hematology, 71
   osmoregulation, 156
   respiratory response, 59, 60
Blood pressure, acid pollution, 245
Blood proteins, binding of organic
      chemicals, xenobiotics, 88
Bluegill, hypoxia, environmental,
      20

Bluegreen algae, nervous system function, 289
BOD. See Biochemical oxygen demand
Body burden, xenobiotics, 79
Bohr effect
  acid pollution, 251
  hypoxia, environmental, 25
Bone
  marrow, immunity, 260
  xenobiotics, 95
Botanicals, water pollution, 3
Brain, 224, 225, 229, 232, 233, 271
  anatomy, 271
  energetics, physiological, 177, 185, 206
  hypoxia, environmental, 30, 31
  nervous system function, 271, 282, 288, 289
  xenobiotics, 93, 98
Burst speed, energetics, physiological, 195
Burst swimming, energetics, physiological, 199

C

Cadmium, 69, 219, 220, 222, 231, 274
  avoidance of, 277
  biomonitoring, 330
  effects on blood oxygen, respiratory response, 57
  effects on calcium metabolism, osmoregulation, 152
  effects on hematology, 69
  effects on immune function, immunity, 261
  effects on osmoregulation, 146
  effects on reproduction, 303
  energetics, physiological, 181–183, 191, 204
  immunity, 263
  liver, 127, 132, 135
  nervous system function, 274, 281, 283, 285, 289
  osmoregulation, 162
  reproduction, 303, 304, 307, 311, 313–316
  respiratory response, 50, 54, 57, 59, 61
  water pollution, 2
  xenobiotics, 79, 82, 86, 94, 97, 98, 108
Calcitonin, osmoregulation, 144
Calcium
  acid pollution, 240–248

importance of environmental levels in, acid pollution
  osmoregulation, 152, 156
  regulation of blood concentration, osmoregulation, 144
  reproduction, 301, 308, 309, 315, 316
  xenobiotics, 82
Calcium channels, respiratory response, 59
Capture, biomonitoring, 333
Carbamate, 275
  insecticides, 229
  nervous system function, 275, 288
  water pollution, 3
Carbaryl
  energetics, physiological, 185
  reproduction, 306
  respiratory response, 54
  xenobiotics, 91
Carbofuran, 199, 233
  energetics, physiological, 199
  liver, 135
  water pollution, 4
Carbohydrates, energetics, physiological, 200
Carbon dioxide effect on ventilation, hypoxia, environmental, 23
Carbon tetrachloride, liver, 130, 131
Cardiac output, energetics, physiological, 190
Cardiovascular response, 47–66
Carp, hypoxia, environmental, 20
Catalase, 219, 231, 232
Catecholamine, 275
  acid pollution, 245, 247, 250, 251
  energetics, physiological, 207
  hypoxia, environmental, 22, 24–26, 35
  immunity, 266
  nervous system function, 275
  reproduction, 307
  respiratory response, 58, 59, 61
Catfish, hypoxia, environmental, 18, 20
Caudal neurosecretory system, acid pollution, 250
Cell fragility, hematology, 70
Cell mediated immunity, immunity, 260
Cellular enzymes, 228
Cerebellum, 272
  nervous system function, 272
Cerebrum, 272
  nervous system function, 272
Ceruloplasmin, xenobiotics, 91

Challenge tests, biomonitoring, 334
Chattonella, respiratory response, 59
Chemicals, energetics, physiological, 206
Chemoreceptors, nervous system function, 279
Chemotactic, immunity, 265
Chlordane, 275
  energetics, physiological, 185, 199
  hematology, 74
  nervous system function, 275
  water pollution, 3
Chloride cell
  acid pollution, 242, 244, 248, 250
  osmoregulation, 141
  proliferation, osmoregulation, 162
  xenobiotics, 97
Chlorinated hydrocarbons
  hematology, 74
  liver, 128
  xenobiotics, 91
Chlorine, 275, 278
  effects on blood oxygen, respiratory response, 56
  effects on hematology, hematology, 71
  liver, 127
  nervous system function, 275, 278, 282
  osmoregulation, 161
  respiratory response, 50, 54, 59
  water pollution, 2
Chlormaine, hematology, 71
Chlorothalonil
  hematology, 72
  respiratory response, 50, 54
  xenobiotics, 109
Cholesterol
  energetics, physiological, 207
  liver, 129, 132, 133
  reproduction, 304, 307
Chorion
  acid pollution, 242
  reproduction, 302, 303
Chorionase, acid pollution, 242
Chromaffin cells, energetics, physiological, 200
Chromium, 225, 274
  effects on osmoregulation, osmoregulation, 155
  energetics, physiological, 181, 194
  hematology, 74
  immunity, 262
  nervous system function, 274, 289

reproduction, 315
respiratory response, 54
water pollution, 2
xenobiotics, 82, 94
Circulatory physiology, respiratory response, 58
Citrate synthase
  biomonitoring, 331
  energetics, physiological, 190, 191, 195
Clay, respiratory response, 54
Clotting time
  biomonitoring, 333
  hematology, 72
Coal dust, respiratory response, 54
Cobalt
  hematology, 72
  water pollution, 2
Collagen, liver, 134
Compensatory growth, energetics, physiological, 193
Complement, immunity, 260
Conditioned responses, nervous system function, 285
Conjugation reactions, xenobiotics, 109
Conversion efficiency, energetics, physiological, 186
Copper, 220, 221, 224, 225, 232, 233, 274, 279
  avoidance of, 277
  biomonitoring, 330, 334, 335
  effects on immune function, immunity, 262
  effects on liver histology, liver, 127
  effects on locomotor activity, 273
  effects on metabolic rate, 176–179, 176
  effects on osmotic and ionic regulation, osmoregulation, 145
  effects on plasma enzymes, liver, 131
  energetics, physiological, 176, 183, 191, 197, 203
  hematology, 74
  hypoxia, environmental, 18
  immunity, 262
  liver, 127, 131, 133, 134, 135
  nervous system function, 274, 279, 280, 282–285, 289
  osmoregulation, 149, 162
  reproduction, 304, 308–311, 314, 315
  respiratory response, 50, 54, 57, 61
  water pollution, 2
  xenobiotics, 81, 91, 95, 97

Copper uptake, xenobiotics, 85
Corpuscles of stannius, osmoregulation, 144
Corticosteroid, energetics, physiological, 178
Cortisol, 230
  acid pollution, 247, 249, 250
  energetics, physiological, 198, 200–205, 207
  hematology, 74
  hypoxia, environmental, 35
  immunity, 262, 266
  liver, 127, 133, 135
  nervous system function, 282
  osmoregulation, 143
  reproduction, 307
  water pollution, 9
Coughing
  biomonitoring, 336
  respiratory response, 54, 55
Creosote, immunity, 265
Cresol, xenobiotics, 111
Critical oxygen tension
  acid pollution, 251
  hypoxia, environmental, 27, 36
Critical swimming speed, energetics, physiological, 195, 196
Critical thermal maximum
  biomonitoring, 334
  hematology, 72
  reproduction, 316
Crotisol, osmoregulation, 163
Crucian carp, hypoxia, environmental, 16, 31
Crude oil, respiratory response, 54
Cutaneous respiration, hypoxia, environmental, 18
Cyanide
  energetics, physiological, 188, 194, 195, 198, 200
  liver, 131
  reproduction, 304, 308, 309, 311, 313
  respiratory response, 54, 59, 60
  water pollution, 3
Cypermethrin
  energetics, physiological, 207
  respiratory response, 54
Cyprinid
  hypoxia, environmental, 16, 19
  larvae, hypoxia, environmental, 32
Cytochrome oxidase
  biomonitoring, 331

energetics, physiological, 188, 190, 191, 195
Cytochrome P450, 234, 304, 307, 329, 330
  xenobiotics, 107
Cytoplasmic inclusions, liver, 128

D

DDT, 274, 275, 278
  effect on amino acid uptake, osmoregulation, 158
  effect on osmoregulation, osmoregulation, 158
  liver, 133
  nervous system function, 274, 275, 278, 281, 285
  reproduction, 309
  xenobiotics, 113
Dehydroabietic acid, hematology, 72
Delta-aminolevulinic acid dehydratase, 69
Delta-aminolevulinic acid dehydrase, 224
Depuration, xenobiotics, 96
Detergent, 278
  effects on osmoregulation, osmoregulation, 157
  energetics, physiological, 205
  excretion, xenobiotics, 112
  nervous system function, 278, 281, 282
  reproduction, 308
  respiratory response, 50, 59
  water pollution, 3
  xenobiotics, 85, 90
Diazinon
  liver, 127
  nervous system function, 287
  water pollution, 4
Dibenzofurans, xenobiotics, 101
Dichlorvos
  immunity, 264
  water pollution, 4
Dieldrin, 230
  liver, 127
  respiratory response, 59
  water pollution, 3
  xenobiotics, 91
Differential leukocyte counts, immunity, 261
Dimethoate, energetics, physiological, 206
Dioxins, nervous system function, 282
Diquat, nervous system function, 287

Disease
  biomonitoring, 332
  immunity, 265
Dispersant, respiratory response, oil and,
    54
Dissolved organic matter, xenobiotics, 90
Dissolved oxygen, energetics,
    physiological, 176
Dorsal light response, nervous system
    function, 284
Dose, water pollution, 7
Dragonet, hypoxia, environmental, 20
Dursban
  liver, 127
  water pollution, 4

E

Early warning systems, biomonitoring, 336
Eels, hypoxia, environmental, 18
Effect
  of acid on, osmoregulation, 244
  of acid on larvae, acid pollution, 243
  on blood and urine chemistry, hypoxia,
    environmental, 34
  on enzyme activity, hypoxia,
    environmental, 37
  on gill histopathology, hypoxia,
    environmental, 36
  on kidney function, hypoxia,
    environmental, 35
  on larvae, hypoxia, environmental, 27
  on muscle capillaries, hypoxia,
    environmental, 37
  on organic chemical uptake, xenobiotics,
    90
  on temperature selection, hypoxia,
    environmental, 34
Effect of copper on, energetics,
    physiological, 179
Egg
  acid pollution, 240, 241, 242
  activation, reproduction, 310
  capsule, reproduction, 312
  energetics, physiological, 189
  hypoxia, environmental, 33
  reproduction, 300, 302, 303, 309
Electric fish, biomonitoring, 337
Electrobranchiogram, respiratory response,
    53
Electromyography, acid pollution,
    252

Electroolfactogram, nervous system
    function, 279
Electroretinogram, nervous system
    function, 279
Embryo-larval lethality tests,
    biomonitoring, 326
Embryo
  biomonitoring, 326
  energetics, physiological, 189
  hypoxia, environmental, 17
Embryonic development, reproduction,
    302, 303, 312
Endosulfan, 229
  energetics, physiological, 207
  respiratory response, 54
Endrin
  energetics, physiological, 206
  heptachlor, water pollution, 4
  immunity, 264
  liver, 127, 132, 133
  osmoregulation, 158
  xenobiotics, 88
Energetics, physiological, 171–216
Energy
  budget, energetics, physiological, 174,
    187
  charge, hypoxia, environmental, 30
  expenditure methods of measuring,
    energetics, physiological, 174
  metabolism, energetics, physiological,
    173
Enterohepatic circulation, xenobiotics, 97
Enzyme activity
  effects of metals on, 219–225
  use for estimating metabolic rate, 190
Enzyme methodology, 217
Enzymes, 229
Epinephrine, 230. See also Adrenalin
Erythrocyte, 224
  count, biomonitoring, 331
  hematology, 67
  nucleoside triphosphate, hypoxia,
    environmental, 36
  synthesis, biomonitoring, 332
Erythron, hematology, 69
Erythropoiesis, hematology, 74
Estradiol
  acid pollution, 241
  reproduction, 304
Estrogen, reproduction, 301, 307, 308
Estrogenic effluents, reproduction, 308
Estrogens testosterone, reproduction, 306

Ethanol
  hypoxia, environmental, 34
  metabolic pathway, hypoxia,
    environmental, 31
Ethological units, 271
  nervous system function, 271
Ethyl, energetics, physiological, 194
Excretion
  of metals, xenobiotics, 96
  of organic contaminants, xenobiotics,
    110
Exercise, acid pollution, 248
Exocrine pancreas, liver, 129
Extraction coefficient, xenobiotics, 82
Extraction efficiency, hypoxia,
  environmental, 25
Eye, nervous system function, 279, 282

F

Fat
  energetics, physiological, 203
  liver, 126
  xenobiotics, 106
Fecundity, reproduction, 302, 303, 308,
  309
Feeding
  effects on metabolic rate, energetics,
    physiological, 175
  energetics, physiological, 188
  nervous system function, 282
Fenitrothion, 199
  energetics, physiological, 199
  excretion PCB, excretion, xenobiotics,
    111
  nervous system function, 287
  respiratory response, 54
  xenobiotics, 111
Fenvalerate, 233
  effects on osmoregulation,
    osmoregulation, 158
  energetics, physiological, 185
  respiratory response, 54
Fertility, reproduction, 302
Fertilization
  acid pollution, 241
  reproduction, 310
Fibrinogens, liver, 129
Field measurements of energy expenditure,
  energetics, physiological, 189
Field studies, effects on immune function,
  immunity, 265

Fight or flight, water pollution, 9
Fish acute toxicity syndromes, water
  pollution, 10
Fly ash, 277
  nervous system function, 277
Follicle-stimulating hormone,
  reproduction, 300
Foraging theory, nervous system function,
  284
Formalin, respiratory response, 50
Free radical, hypoxia, environmental,
  17
Fry
  acid pollution. See Larvae
  effects of cadmium on osmoregulation
    in, 153
    osmoregulation, 153
  energetics, physiological, 196
  environmental factors, energetics,
    physiological, 176
  hypoxia, environmental, 17
  osmoregulation, 160

G

Gallbladder, xenobiotics, 97, 111
Gametogenesis, hormonal control,
  reproduction, 300
Garlon, 278
Gill, 145, 157, 162, 222, 224
  acid pollution, 246
  control of blood circulation, hypoxia,
    environmental, 24
  energetics, physiological, 177, 185
  function recovery, respiratory response,
    56
  histopathology, 50, 197
    respiratory response, 49
  hypoxia adjustments, hypoxia,
    environmental, 24
  microenvironment, xenobiotics, 85
  nervous system function, 288
  osmoregulation, 145, 162, 163
  respiratory response, 61
  structure, respiratory response, 47
  surface microenvironment, xenobiotics,
    82
  tissue metabolism: effects of metals,
    energetics, physiological, 182
  uptake, limiting factors, xenobiotics,
    88
  vasculature, respiratory response, 59

xenobiotics, 93, 97, 99, 110
Gills, 153, 163
Glucagon, 143
  energetics, physiological, 207
  osmoregulation, 143
Glucose, 223
  acid pollution, 247, 250
  biomonitoring, 333
  energetics, physiological, 173, 179, 200,
    201, 202, 204, 205, 207
Glucose-6-phosphatase, 223, 230
  energetics, physiological, 205
Glucose-6-phosphate dehydrogenase, 219,
  222
Glutamate, synthetase, 227
Glutamate dehydrogenase, 223, 227, 230
  energetics, physiological, 295
Glutamic oxaloacetic transaminase, liver,
  130
Glutamic pyruvic transaminase, liver,
  130
Glutathine, 231
Glutathione, 231, 233
  biomonitoring, 329
  xenobiotics, 98, 107
Glyceraldehyde dehydrogenase, 225
Glycogen, 228
  acid pollution, 252
  energetics, physiological, 173, 188, 200,
    202, 203, 204, 206, 295
  hypoxia, environmental, 29
  liver, 127
Glycolysis, 223, 225
Goldfish
  effects of anoxia, hypoxia,
    environmental, 31
  hypoxia, environmental, 16, 31
Gonadotropin-releasing hormones,
  reproduction, 300
Growth
  acid pollution, 241, 247
  biomonitoring, 326, 330, 331
  energetics, physiological, 177, 182, 186,
    188, 189, 191, 194, 197, 207
  nervous system function, 284
  reproduction, 303
  water pollution, 6
Growth hormone, 143
  energetics, physiological, 194, 207
Growth rate, biomonitoring, 326
Gut, 162
  osmoregulation, 162

H

Handling time, nervous system function,
  284
Hardness, xenobiotics, 84
Hatchability
  biomonitoring, 326
  reproduction, 303
Hatching time, reproduction, 312, 313
Heart, 222
  hypoxia, environmental, 31
Heart pacemaker, respiratory response, 58
Heart rate
  biomonitoring, 327
  effect of hypoxia, hypoxia,
    environmental, 26
  energetics, physiological, 190
  reproduction, 312, 317
  respiratory response, 59, 61
Hematocrit
  acid pollution, 245
  biomonitoring, 331
  hematology, 68, 73, 74
  hypoxia, environmental, 35
Hematological methods, hematology, 68
Hematology, 67–78, 331, 332
  acid, 73
  aldrin, 74
  chlormaine, 71
  chlorothalonil, 72
  chromium, 74
  hematopoietic tissue, 67
  organochlorine pesticides, 72
  organophosphate insecticides, 74
  tolerance, 72
  zinc, 74
Hemocytoblast, hematology, 67
Hemoglobin
  biomonitoring, 331
  hematology, 67
  hypoxia, environmental, 25
Hepatocytes
  energetics, physiological, 201
  liver, 133
Hepatosomatic index, liver, 126
Herbicides
  water pollution, 3
  xenobiotics, 101
Hexachlorocyclohexane, reproduction, 305
Hexokinase, 223, 226
Histopathology, biomonitoring, 331
Homeostasis, water pollution, 8, 10

Hormesis
  energetics, physiological, 193, 199
  nervous system function, 287
Hormonal controls, 143
Humoral immune expression, immunity,
  264
Hydrazine, nervous system function, 283,
  284
Hydrocarbons, xenobiotics, 111
Hydrogen peroxide, 231
Hyperoxia, hypoxia, environmental,
  15
Hypoxemia, 156
  osmoregulation, 156
  respiratory response, 56, 61
Hypoxia
  acclimation, 36
  avoidance, 34
  blood, affinity for oxygen, 22, 25
  crucian carp, 16, 31
  effects on gill histopathology, 36
  environmental, 17
  noradrenaline, 24
  prostaglandins, 24
  swimming, speed, effect of hypoxia,
    31
  trichlorobenzene, 17
  trout, 20
  UV light, 18
  ventilation, 20
Hypoxic stress, biomonitoring, 335

I

Immune function, effects of pollutants on
  260–265
Immune system, 259–269
  biomonitoring, 332
Immunoglobulins, immunity, 260
Immunology, biomonitoring, 331
Immunotoxicology, immunity, 260
Incubation time, reproduction, 313
Individual chemicals, energetics,
  physiological, 206
Insecticides
  respiratory response, 59
  water pollution, 3
  xenobiotics, 101
Insulin, 143
  energetics, physiological, 200, 203, 204,
    206, 207
  osmoregulation, 143

Insulin diabetes, energetics, physiological,
  204
Insulin-like growth factor, 143
  osmoregulation, 143
Interferon, immunity, 260
Intestine
  uptake of organic chemicals,
    xenobiotics, 90
  xenobiotics, 100
Iodine, water pollution, 2
Ionic fluxes, species differences, 142
Ionizing radiation, nervous system
    function, 283
Ionoregulation, acid pollution, 242
Iron, 232
  reproduction, 315
  respiratory response, 50
  water pollution, 2

K

Kidney, 152–154, 162, 224
  acid pollution, 247
  effects of cadmium, 152
    osmoregulation, 152
  energetics, physiological, 204, 205
  excretion of pesticides, xenobiotics, 113
  hematology, 71
  immunity, 260
  osmoregulation, 152, 154
  xenobiotics, 92, 95, 97, 99, 113
Kraft pulpmill effluents. See Pulpmill
  effluents
Kuppfer cells, liver, 127

L

Lactate
  acid pollution, 251
  energetics, physiological, 173, 183
Lactate dehydrogenase, 224, 225, 229
  biomonitoring, 331
  energetics, physiological, 195
Lactic acid
  energetics, physiological, 203, 205, 206
  hypoxia, environmental, 29, 30, 34, 35
  respiratory response, 56
Lamellae, hypoxia, environmental, 20
Larvae, 314–317
  acid pollution, 241
  biomonitoring, 326, 335
  effects of acid on, acid pollution, 242

energetics, physiological, 189, 191, 194, 196
hypoxia, environmental, 17
nervous system function, 288
reproduction, 300, 308–310, 314
swimming performance, 199
energetics, physiological, 199
Lateral line receptors, nervous system function, 281
Lead, 219, 220, 221, 224
biomonitoring, 333
effects on osmoregulation, 155
hematology, 70
nervous system function, 281, 285, 289
xenobiotics, 79, 85, 93, 96, 98, 100
Learning, nervous system function, 285
Leucine amino naphthylamidase, biomonitoring, 333
Leukocrit, biomonitoring, 332
Leydig cells, reproduction, 304
Life cycle tests, biomonitoring, 326
Lindane
immunity, 264
water pollution, 4
Lipid
biomonitoring, 330
energetics, physiological, 173, 198, 200, 202, 207
liver, 132
Lipid peroxidation
hematology, 70
liver, 136
xenobiotics, 99
Liver, 125–139, 125–140, 126, 132, 222, 224, 232, 233
acid phosphatase, 132
arsenic, 127
ascorbic acid, 134
blood liver, volume, 126
brian, 134
cholesterol, 129, 132, 133
collagen, 134
endrin, 127, 132, 133
energetics, physiological, 177, 200
exocrine pancreas, 129
histology, 125
interrenal, 135
mercury, 127, 131, 132
petroleum hydrocarbons, 126, 127, 135
plasma proteins, 129
sorbitol dehyrogenase, 130

zinc, 128
Locomotor activity, 272
methods for measuring, 272
nervous system function, 272
reproduction, 316
Luteinizing hormone, reproduction, 300
Lymph nodes, immunity, 260
Lysosomes, 225, 229
biomonitoring, 330, 333
liver, 132

M

Magnesium, 152
osmoregulation, 152
Malaoxon, nervous system function, 288
Malate dehydrogenase, 223, 224, 229
Malathion, 199, 226
energetics, physiological, 199, 207
immunity, 264
liver, 135
nervous system function, 287, 289
respiratory response, 50
water pollution, 4
Manganese
effects on osmoregulation, 157
osmoregulation, 157
immunity, 262
water pollution, 2
Maximal acceptable toxicant concentration, reproduction, 317
Mercury, 220–225
attractance to, 277
nervous system function, 277
effect on mucus production, 161
effects on osmoregulation, 153
Metabolic rate
effects of metals on, 176–186
effects of pesticides 184–189
energetics, physiological, 185
Metabolic scope, energetics, physiological, 175, 177
Metallothione, biomonitoring, 329
Metallothionein, 231, 232, 234
reproduction, 315
Metaloids, water pollution, 2
Metals, 239
acclimation to, xenobiotics, 100
accumulation in different organs, xenobiotics, 91
bioavailability, xenobiotics, 81
effect on MFO activity, xenobiotics, 108

mechanisms of uptake, xenobiotics,
81
uptake in food, xenobiotics, 86
water pollution, 2
xenobiotics, 81
Metasystox, 226, 228
Methemoglobin, 160
hematology, 71
osmoregulation, 160
Methlmercury, reproduction, 310
Methoxychlor
energetics, physiological, 185
water pollution, 4
Methyl parathion, 199
energetics, physiological, 185, 199
nervous system function, 283, 285,
288
reproduction, 316
respiratory response, 54
water pollution, 4
Methylmercury, 154, 224
hematology, 71
immunity, 263
nervous system function, 280
osmoregulation, 154
reproduction, 315
xenobiotics, 83, 85, 91, 97, 100
Mg ATPase, 157, 158
Micropyle, reproduction, 302, 310
Mill effluent, water pollution, 4
Mirex, immunity, 264
Mitochondria, 229
Mitochondrial calcium, xenobiotics, 99
Mode of action, water pollution, 10
Molinate, 199
energetics, physiological, 193, 199
reproduction, 316
water pollution, 4
Monocytes, immunity, 259, 260
Mucus, 161
acid pollution, 250, 251
immunity, 259
importance in osmoregulation, 161
osmoregulation, 161
osmoregulation, 161
respiratory response, 50
xenobiotics, 82, 85, 95
importance for metal uptake by gills,
82
Muscle, 232
energetics, physiological, 185, 193,
200

hypoxia, environmental, 31
nervous system function, 288
xenobiotics, 92

N

Na,K ATPase, 145, 150, 151, 154, 155,
156, 157, 158, 162, 163, 225
acid pollution, 250
energetics, physiological, 173
osmoregulation, 145, 150, 151,
154–158, 162, 163
Naphthalene, 159
nervous system function, 281
osmoregulation, 159
respiratory response, 54
xenobiotics, 90
Natural killer cells, immunity, 260
Nervous system function, 278
acetylcholine, 282
diazinon, 287
Dieldrin, 289
feeding, 282
fenitrothion, 287
hydrazine, 283, 284
hypoxia, 289
lead, 281, 285, 289
liver, 288
naphthalene, 281
olfactory, 281
petroleum hydrocarbon, 279
predator-prey behavior, 283
silver, 281
telencephalon, 284
temperature, 281
zinc, 281, 285, 289
Neutrophils, immunity, 260, 261, 262
Nickel
hematology, 74
immunity, 262
reproduction, 304, 315
respiratory response, 50
Nitrate
hypoxia, environmental, 18
water pollution, 3
Nitrite, 161
effects on hematology, hematology, 72
effects on osmoregulation, 160
osmoregulation, 160
hematology, 72
osmoregulation, 161
water pollution, 3

O

Octanol/water partition coefficients,
    xenobiotics, 87
Oil
  dispersant, 157
    osmoregulation, 157
  spill dispersant, respiratory response, 59
Olfactory bulbs, 272
  nervous system function, 272
Olfactory receptors, nervous system
    function, 279
Olfactory system, reproduction, 302
Oocyte, reproduction, 302
Operant conditioning, acid pollution, 252
Optic lobes, 272
  nervous system function, 272
Optomotor response, nervous system
    function, 287
Organic chemical, mechanisms of uptake,
    xenobiotics, 87
Organic contaminant excretion,
    xenobiotics, 110
Organochlorine insecticides, 294
  effects on cellular
  energetics, physiological, 185. See also
    Individual insecticides
  nervous system function, 282, 294
  reproduction, 305
Organophosphorus
  nervous system function, 282
  xenobiotics, 104
Organotin, xenobiotics, 95

P

Paraquat, 230
  reproduction, 317
  respiratory response, 50, 59
Parathion, 226, 228, 278
  nervous system function, 278, 286
  water pollution, 4
  xenobiotics, 91
Parquat, water pollution, 4
Parr-Smolt transformation, 143
  effect of copper on, 150
  osmoregulation, 143
Partition coefficient, xenobiotics, 101
Pasteur effect, hypoxia, environmental, 30
PC. See Critical oxygen tension
PCBs, 231, 278
  energetics, physiological, 192, 198

nervous system function, 278
reproduction, 307, 309, 315
P-cresol, energetics, physiological, 194
Pentachlorophenol, 226, 228, 275
  energetics, physiological, 187, 198
  hematology, 74
  liver, 135
  nervous system function, 275, 283
  water pollution, 4
  xenobiotics, 102
Percentage utilization, hypoxia,
    environmental, 25
Permethrin, respiratory response, 50
Pesticides
  effects on immune function, immunity,
    263
  effects on locomotor activity, 275
  effects on metabolic rate, 184–189
  energetics, physiological. See also
    Individual chemicals
  hematology, 74
  water pollution, 3
Petroleum
  reproduction, 312
  respiratory response, 50
  water pollution, 4
Petroleum hydrocarbons, 225, 274
  effect of salinity on
    osmoregulation
  effect on liver, 126
  effect on metabolic rate, 225
    energetics, physiological, 189
  effect on osmoregulation, 159
    osmoregulation, 159
  energetics, physiological, 193, 198, 295
  liver, 126, 127, 135
  nervous system function, 274
  reproduction, 305, 314
Petroleum hydrocarbons accumulation,
    xenobiotics, 104
Phagocytes, immunity, 263
Phase I reactions, xenobiotics, 107
Phase II reactions, xenobiotics, 109
Phenol, 278
  effects on osmoregulation,
    osmoregulation, 158
  excretion, xenobiotics, 111
  hypoxia, environmental, 18
  liver, 127, 131, 133
  nervous system function, 278
  respiratory response, 50
  xenobiotics, 109

Phenoxyacetic acid herbicides, xenobiotics, 113

Pheromones, reproduction, 302

Phosphamidon, 226, 229

Phosphatase, 229

Phosphoenol-pyruvate carboxykinase, 222

Phosphorylase, 228, 230

Plaice, hypoxia, environmental, 18

Plasma calcium, 153

Plasma clearance by the liver of specific dyes, liver, 130

Plasma copper, xenobiotics, 95

Plasma enzyme assays, biomonitoring, 333

Plasma enzymes of liver origin, liver, 130

Plasma proteins, liver, 129

Plasma skimming, hematology, 69

Pollutants 200–207

Pollution control, biomonitoring, 325

Polyaromatic hydrocarbons, xenobiotics, 91

Polychlorinated dibenzo-p-dioxins, xenobiotics, 101

Polychlorineated biphenyls, water pollution, 4

Polycyclic aromatic hydrocarbons, xenobiotics, 81

Polyhalogenated biphenyls, xenobiotics, 101

Postprandial effect, energetics, physiological, 175

Potassium, 152, 155
  osmoregulation, 152

Predator-prey behavior, nervous system function, 283

Prey, nervous system function, 284

Progestins, reproduction, 301

Prolactin, 143, 144, 153, 163
  acid pollution, 250
  osmoregulation, 143, 153
  reproduction, 315

Prolonged swimming speed, energetics, physiological, 195

Propionate, hypoxia, environmental, 30

Prostaglandins
  hypoxia, environmental, 24
  reproduction, 302

Protein
  acid pollution, 247
  biomonitoring, 330, 331, 333
  energetics, physiological, 173, 194, 200, 202–205, 207

liver, 132

Psychological stress, water pollution, 8

Pulpmill effluent, 278
  effect on energy stores, energetics, physiological, 202
  effect on growth, energetics, physiological, 193
  effect on hematology, hematology, 72
  energetics, physiological, 188, 193, 196, 202
  hematology, 72
  liver, 133
  nervous system function, 278
  reproduction, 306, 307
  respiratory response, 54

Pyrethroid, 158
  biomonitoring, 334
  energetics, physiological, 185
  reproduction, 315

Pyrethrum, water pollution, 4

Q

Quinalphos, 206, 226

R

Ram gill ventilation, hypoxia, environmental, 23

Regulation of metal concentration, xenobiotics, 96

Renin, 160
  osmoregulation, 160
  respiratory response, 60

Reninangiotensin system, 143

Reproduction, 299–323
  strategies, 300
  water pollution, 6

Reproductive hormones, effects of pollutants on, reproduction, 303

Residual dissolved oxygen, biomonitoring, 334

Residue analysis, xenobiotics, 110

Resin acids, water pollution, 4

Respiration, energetics, physiological, 174

Respiratory conformity, hypoxia, environmental, 28

Respiratory poisons, energetics, physiological, 198

Respiratory regulation and conformity, hypoxia, environmental, 27

Respiratory response, 47–66
  aluminum, 54
  arterial oxygen tension, 56
  cadmium, effects on blood oxygen, 57
  chlorine, 50, 54, 59
  coal dust, 54
  cypermethrin, 54
  formalin, 50
  heart rate, 59, 61
  maximum acceptable toxicant
      concentration, 55
  oil spill dispersant, 59
  surfactants, 54
  zinc, 50, 54, 56, 57, 59
    effects on blood gases, 56
Reticular system, 272, 274
  nervous system function, 272, 274
Reticuloendothelial cells, xenobiotics,
  106
Rheotaxis, biomonitoring, 336
RNAase, 219
RNA/DNA ratio
  biomonitoring, 331
  energetics, physiological, 193, 194
Root effect, hypoxia, environmental, 25
Rotenone
  energetics, physiological, 187
  water pollution, 4
Routine metabolic rate, energetics,
    physiological, 175

S

Salinity
  effect on PCP accumulation,
      xenobiotics, 103
  energetics, physiological, 176
Schooling, nervous system function, 283
Seasonal effects, biomonitoring, 328
Secondary blood system, xenobiotics,
  90
Selenium, 155, 277
  effect on mercury toxicity, 155
    osmoregulation, 155
  nervous system function, 277, 285
  osmoregulation, 155
  reproduction, 309, 312
  water pollution, 2
  xenobiotics, 95
Sensory receptors, nervous system
    function, 279
Serum. See also Plasma

liver, 130
  triglycerides, energetics, physiological,
      207
Sevin, 228
  water pollution, 4
Sewage sludge, immunity, 265
Silt, water pollution, 2
Silver, 220, 224
  nervous system function, 281
  reproduction, 313
  xenobiotics, 81
Silvex, water pollution, 4
Skin
  excretion of organics, xenobiotics,
      113
  hypoxia, environmental, 18
  uptake of organic chemicals,
      xenobiotics, 90
  xenobiotics, 95, 96, 97, 111, 113
Smoltification, acid pollution, 247
Social rank of a fish, 274
  nervous system function, 274
Somatostatin, 143
  osmoregulation, 143
Sorbital dehydrogenase, biomonitoring,
  333
Spawning
  effects of acid 240–241
  reproduction, 300, 301
Specific immune response, immunity,
  260
Sperm, reproduction, 310
Spermatogenesis, reproduction, 301, 304,
  305
Spinal deformities, liver, 134
Spleen
  hematology, 74
  immunity, 260, 262, 266
Spontaneous activity, 199
  effects of metals on, energetics,
      physiological, 181
  energetics, physiological, 181, 182, 185,
      187, 199
Squalane, xenobiotics, 106
Standard metabolic rate, energetics,
    physiological, 174
Starvation, 219
Stress
  biomonitoring, 327, 333
  energetics, physiological, 200, 201,
      202
  immunity, 264

reproduction, 307
  water pollution, 9
Sturgeon, hypoxia, environmental, 20
Succinic dehydrogenase, 223, 224, 228, 229
Sugar, energetics, physiological, 202
Sumithion, 228
Superoxide, 231
Surfactants
  energetics, physiological, 205
  respiratory response, 54
Suspended solids, water pollution, 2
Sustained swimming capacity, energetics, physiological, 200
Swimming
  acid pollution, 248
  reproduction, 316
  speed, effect of hypoxia, hypoxia, environmental, 31
Swimming capacity
  biomonitoring, 335
  reproduction, 316
Swimming performance
  acid pollution, 251
  energetics, physiological, 195

T

Target organ, water pollution, 10
Taste receptors, nervous system function, 230
Telencephalon, 272
  nervous system function, 272, 284
Temperature, 222, 275
  nervous system function, 275
Testosterone, reproduction, 301, 304, 305, 307
Thermal pollution, water pollution, 5
Thigmotaxis, acid pollution, 252
Thiotox, liver, 135
Threshold, water pollution, 7
Thrombocytes
  hematology, 67
  immunity, 260, 262
Thymus, immunity, 260
Thyroid, 143, 229
  acid pollution, 250
  energetics, physiological, 188, 193
  osmoregulation, 143
  reproduction, 302, 304, 308

Tin, 157
  osmoregulation, 157
  water pollution, 2
  xenobiotics, 95
Tissue oxygen tension, hypoxia, environmental, 25
T-lymphocyte, immunity, 261
Toadfish, hypoxia, environmental, 16
Tolerance, hematology, 72
Toluene, 159
  excretion, xenobiotics, 111
  liver, 132
  osmoregulation, 159
  respiratory response, 54
  xenobiotics, 111
Toxaphene
  liver, 134
  water pollution, 4
Toxic action, water pollution, 10
Toxicokinetic models, xenobiotics, 106
Toxicology, water pollution, 5
Trace metals, water pollution, 2
Transfer of contaminants into eggs from adults, reproduction, 309
Transport by blood of pollutants, xenobiotics, 91
Tributyltin, 157
  effect on MFO activity, xenobiotics, 108
  immunity, 263
  osmoregulation, 157
  water pollution, 3
  xenobiotics, 95
Trichlorobenzene, hypoxia, environmental, 17
Trichlorofon, immunity, 264
Triclopyr, energetics, physiological, 185
Trout, hypoxia, environmental, 20

U

Ucrit, energetics, physiological, 195, 196, 197, 198
Ultimobranchial bodies, 144
Uptake efficiencies, xenobiotics, 87
Urea nitrogen, liver, 131
Uric acid, liver, 133
Urinary excretion, xenobiotics, 97
Urine
  acid pollution, 247

xenobiotics, 97
Urotinsins, 143
  osmoregulation, 143
Utilization efficiency
  respiratory response, 48
  xenobiotics, 82

V

Ventilation
  acid pollution, 250
  biomonitoring, 326, 336, 337
  changes, mechanisms, respiratory
    response, 55
Ventrolateral peduncular neuropil,
    272
  nervous system function, 272, 282
Viable hatch, reproduction, 313, 314
Vitamin E, xenobiotics, 109
Vitellogenesis
  acid pollution, 240
  reproduction, 300, 316
Vitellogenin
  acid pollution, 241
  reproduction, 304, 307, 308

W

Water content, biomonitoring, 330
Water hardening
  acid pollution, 242
  reproduction, 302, 310, 312
Water hardness
  acid pollution, 244
  effect on copper toxicity,
    osmoregulation, 145
Water pollution, 1–13
  aquatic toxicology, 5
  biochemical oxygen demand, 1
  carbofuran, 4
  cortisol, 9
  diquat, 4
  fish acute toxicity syndromes, 10
  insecticides, 3
  manganese, 2
  mode of action, 10
  organophosphate, 3
  physiological mode of death, 8
  resin acids, 4
  stress, 9
  tin, 2

water quality criteria, 5
zinc, 2, 11
Water quality criteria, water pollution,
    5
Wood fibers
  energetics, physiological, 196
  water pollution, 2
Wood pulp, respiratory response, 54

X

Xanthine oxidase, 219, 223
  energetics, physiological, 205
Xenobiotics, 79–123
  aluminum, 82, 85, 86
  arsenic, 96, 97
  bioconcentration, 79
  body burden, 79
  chromium, 82, 93, 94, 96
  dibenzofurans, 101
  fenitrothion, 111
  gill uptake, limiting factors, 88
  hardness, 84
  lead, 79, 82, 85, 86, 93, 94, 96–100
  metals, acclimation to, 100
  mucus, 82, 84, 85, 93, 95, 97
  organophosphorus, 104
  petroleum hydrocarbons, 104
  reticuloendothelial cells, 106
  urine, 97, 111
  zinc, 82, 84, 85, 91, 95–97, 100

Y

Yolk
  proteins, acid pollution, 240
  sac, reproduction, 316

Z

Zinc, 150, 162, 221, 224, 274, 279
  avoidance of, 277
  effects on blood calcium, 151
  effects on blood gases, respiratory
    response, 56
  effects on osmoregulation, 150
  nervous system function, 274, 279, 281,
    285, 289
  osmoregulation, 150
  water pollution, 2, 11
  xenobiotics, 82, 85, 95, 97